MACMILLAN ENCYCLOPEDIA OF
ENERGY

JOHN ZUMERCHIK

Editor in Chief

VOLUME 2

Macmillan Reference USA

an imprint of the Gale Group

New York • Detroit • San Francisco • London • Boston • Woodbridge, CT

EDITORIAL BOARD

MACMILLAN ENCYCLOPEDIA OF

ENERGY

JOHN ZUMERCHIK

Editor in Chief

VOLUME 2

Macmillan Reference USA

an imprint of the Gale Group

New York • Detroit • San Francisco • London • Boston • Woodbridge, CT

Macmillan Encyclopedia of Energy

Macmillan Reference USA
1633 Broadway
New York, NY 10019

Macmillan Reference USA
27500 Drake Rd.
Farmington Hills, MI 48331-3535

Library of Congress Catalog Card Number: 00-062498

Printed in the United States of America
Printing number
1 2 3 4 5 6 7 8 9 10

Library of Congress Cataloging-in-Publication Data
Macmillan encyclopedia of energy / John Zumerchik, editor in chief.
     p. cm.
    Includes bibliographical references and index.
    ISBN 0-02-865021-2 (set : hc). — ISBN 0-02-865018-2 (vol. 1). — ISBN 0-02-865019-0 (vol. 2). — ISBN 0-02-865020-4 (vol. 3).
    1. Power resources—Encyclopedias. I. Title: Encyclopedia of Energy. II. Zumerchik, John
TJ163.25 .M33 2000
                                              00-062498

621.042'03 dc21

EMISSION CONTROL, POWER PLANT

Power plant emissions result from the combustion of fossil fuels such as coal, gas, and oil. These emissions include sulfur dioxide (SO_2), nitrogen oxides (NO_x), particulate matter, and hazardous air pollutants, all of which are subject to environmental regulations. Another emission is carbon dioxide (CO_2), suspected of being responsible for global warming.

Historically, under both federal and state regulations, the demand for gas to heat homes and to meet needs of business and industry took priority over utility use to generate electricity. These restrictions have been eased by amendments to the Fuel Use Act in 1987, and, as a result, new gas-fired generation units are being constructed. However, coal-fired units continue to provide over 50 percent of the total utility generation of electricity.

Until the late 1960s, a typical electric utility scenario was one of steadily growing electricity demand, lower costs of new power plants through technological advances, and declining electricity prices. Utility companies appeared before Public Utility Commissions (PUCs) to request approval for rate reductions for customers. In the 1970s, multiple factors caused costs to increase dramatically: fuel costs escalated, primarily because of oil embargoes; legislators passed more stringent environmental laws requiring huge investments in emissions control technology; and these costs escalated as inflation and interest rates soared. When utility companies requested rate increases to cover these higher costs, PUC hearings were no longer sedate and routine. Cost recovery pressures continued throughout the 1980s as environmentalists succeeded in lowering emissions limits.

FORMATION OF POLLUTANTS IN UTILITY BOILERS

Sulfur is found in coal in organic forms as well as inorganic forms such as pyrites, sulfate, and elemental. Organic sulfur is the most difficult to remove, and reliable analytical methods are required to support coal-cleaning technologies designed to remove the sulfur prior to burning the coal. At the present time, most utilities burn "uncleaned" coal, and, upon combustion, most of the sulfur is converted to SO_2, with a small amount further oxidized to form sulfur trioxide (SO_3)

$$S(coal) + O_2 \rightarrow SO_2$$
$$SO_2 + \tfrac{1}{2} O_2 \rightarrow SO_3$$

The sulfur content of coals available to utilities ranges from about 4 percent in high-sulfur coals to less than 1 percent in some Western coals. Although transportation costs may be higher for Western coals, many Eastern utilities elect to burn Western coals to comply with increasingly stringent SO_2 regulations.

NO_x emissions are less dependent on the type of coal burned, and two oxidation mechanisms are associated with the release of NO_x into the atmosphere during the combustion process. Thermal NO_x results from the reaction of nitrogen in the combustion air with excess oxygen at elevated temperatures, and fuel NO_x is a product of the oxidation of nitrogen chemically bound in the coal.

Hazardous air pollutants (HAPs) are substances that may cause immediate or long-term adverse effects on human health. HAPs can be gases, particulates, trace metals such as mercury, and vapors such as benzene. For coal-fired power plants, the HAPs of most concern are metals such as mercury, arsenic, and vanadium.

All combustion processes produce particulate matter. Amounts and size distribution of the particulates emitted depend on a number of factors, including fuel burned, type of boiler, and effectiveness of collection devices.

A wide variety of control technologies have been installed by utilities throughout the United States to reduce the emissions of these pollutants. At the same time, research on new technologies is being conducted to ensure compliance with future environmental standards.

ENVIRONMENTAL ISSUES

The legislation most responsible for addressing power plant emissions is the Clean Air Act. Initially established in 1970 with major amendments in 1977 and 1990, it provides for federal authorities to control impacts on human health and the environment resulting from air emissions from industry, transportation, and space heating and cooling. In the original 1970 programs, National Ambient Air Quality Standards (NAAQS) were established for six "criteria" air pollutants—SO_2, NO_x, particulate matter, ozone, lead, and carbon monoxide—at a level to protect human health and welfare and the environment with a "margin of safety." New Source Performance Standards (NSPS) were set for major new facilities projected to emit any pollutant in significant amounts. To receive an operating permit, a new unit must meet or exceed control standards established by the Environmental Protection Agency (EPA). In the 1977 Amendments, permits required control levels for new plants that were not only as stringent as NSPS but also reflected the best available technologies.

The reduction of atmospheric concentrations of the sulfur and nitrogen oxides blamed for acid rain was a major issue in the debate that led to the 1990 Clean Air Act Amendments (CAAA). The final legislative action is one of the most complex and comprehensive pieces of environmental legislation ever written.

The 1990 CAAA contain the following sections:

Title I: Provisions for Attainment and
    Maintenance of National Ambient Air Quality
    Standards
Title II: Provisions Relating to Mobile Sources
Title III: Hazardous Air Pollutants
Title IV: Acid Deposition Control
Title V: Permits

Title VI: Stratospheric Ozone Protection
Title VII: Provisions Relating to Enforcement.

Titles I and IV are most relevant to SO_2 and NO_x control. Title I establishes a 24-hour average ambient air standard for SO_2 of 0.14 ppm. The NO_x provisions require existing major stationary sources to apply reasonably available control technologies and new or modified major stationary sources to offset their new emissions and install controls representing the lowest achievable emissions rate. Each state with an ozone nonattainment region must develop a State Implementation Plan (SIP) that includes stationary NO_x emissions reductions.

Title IV, the Acid Rain Program, addresses controls for specific types of boilers, including those found in coal-fired power plants. A two-phase control strategy was established. Phase I began in 1995 and originally affected 263 units at 110 coal-burning utility plants in twenty-one eastern and midwestern states. The total of affected units increased to 445 when substitution or compensating units were added. In Phase II, which began January 1, 2000, the EPA has established lower emissions limits and also has set restrictions on smaller plants fired by coal, oil, and gas. For example, the Phase I SO_2 emissions limit is 2.5 lb/million Btu of heat input to the boiler whereas the Phase II limit is 1.2 lb/million Btu. In both phases, affected sources will be required to install systems that continuously monitor emissions to trace progress and assure compliance.

One feature of the new law is an SO_2 trading allowance program that encourages the use of market-based principles to reduce pollution. Utilities may trade allowances within their system and/or buy or sell allowances to and from other affected sources. For example, plants that emit SO_2 at a rate below 1.2 lb/million Btu will be able to increase emissions by 20 percent between a baseline year and the year 2000. Also, bonus allowances will be distributed to accommodate growth by units in states with a statewide average below 0.8 lb/million Btu.

The Clean Air Act of 1970 and the Amendments of 1977 failed to adequately control emissions of hazardous air pollutants, that are typically carcinogens, mutagens, and reproductive toxins. Title III of the 1990 Amendments offers a comprehensive plan for achieving significant reductions in emissions of hazardous air pollutants from major sources by defining a new program to control 189 pollutants.

Although the petrochemical and metals industries were the primary focus of the toxic air pollutants legislation, approximately forty of these substances have been detected in fossil power plant flue gas. Mercury, which is found in trace amounts in fossil fuels such as coal and oil, is liberated during the combustion process and these emissions may be regulated in the future. EPA issued an Information Collection Request (ICR) that required all coal-fired plants to analyze their feed coal for mercury and chlorine. Since these data will be used in making a regulatory decision on mercury near the end of the year 2000, it is critical that the power industry provide the most accurate data possible.

In 1987, health- and welfare-based standards for particulate matter (measured as PM_{10}, particles 10 micrometers in diameter or smaller) were established. A 10 micrometer (micron) particle is quite small; about 100 PM_{10} particles will fit across the one millimeter diameter of a typical ballpoint pen. For PM_{10} particles, an annual standard was set at 50 micrograms per cubic meter (50 $\mu g/m^3$) and a 24-hour standard was set at 150 $\mu g/m^2$.

Since these PM_{10} standards were established, the EPA has reviewed peer-reviewed scientific studies that suggest that significant health effects occur at concentrations below the 1987 standards. In addition, some studies attributed adverse health effects to particles smaller than 10 microns. In July 1997, the EPA, under the National Ambient Air Quality Standards (NAAQS), added standards for particulate matter with a diameter of 2.5 microns or less ($PM_{2.5}$). The annual $PM_{2.5}$ standard was set at 15 $\mu g/m^3$ and the 24-hour $PM_{2.5}$ standard was set at 65 $\mu g/m^3$.

Through implementing new technologies and modifying unit operating conditions, the electric utility industry has significantly reduced the emissions of SO_2, NO_x, and particulates since passage of the 1970 Clean Air Act and its subsequent amendments. With full implementation of Title IV of the 1990 CAAA, the 1990 baseline level of more than 14.5 million tons of SO_2 will be reduced to 8.9 million tons per year. NO_x emissions during Phase I will be reduced by 400,000 tons per year, and Phase II will result in a further reduction of 1.2 million tons per year. Particulate control devices, installed on nearly all coal-fired units, have reduced particulate emissions from more than three million tons per year in 1970 to less than 430,000 tons per year in 1990.

Over the years, utilities have funded research to develop technologies that not only will meet existing standards but also will meet future emissions reductions based on continuing concerns about acid rain, ozone, fine particulates, and other environmental issues. To remain competitive in the global economy, utilities must seek technologies that balance the conflicting drivers of productivity demands, environmental concerns, and cost considerations. Engineering designs should minimize costs and environmental impact, and, at the same time, maximize factors such as reliability and performance.

CLEAN COAL TECHNOLOGY PROGRAM

The Clean Coal Technology (CCT) Program, a government and industry cofunded effort, began in 1986 with a joint investment of nearly $6.7 billion. The recommendation for this multibillion dollar program came from the United States and Canadian Special Envoys on Acid Rain. The overall goal of the CCT Program is to demonstrate the commercial readiness of new, innovative environmental technologies. The program is being conducted through a multiphased effort consisting of five separate solicitations administered by the U.S. Department of Energy. Industry proposes and manages the selected demonstration ventures. Many of the projects funded in the first stages of the program are generating data or have finished their testing program. Within the next few years, the United States will have in operation a number of prototype demonstration projects that promise to meet the most rigorous environmental standards.

The CCT projects, in general, are categorized as follows:

1. Advanced electric power generation
   a. Fluidized bed combustion
   b. Integrated gasification combined cycle
   c. Advanced combustion systems
2. Environmental control technologies
   a. Sulfur dioxide control technologies
   b. NO_x control technologies
   c. Combined SO_2/NO_x control technologies
3. Coal processing for clean fuels
   a. Coal preparation technologies

b. Mild gasification

4. Indirect liquefaction

5. Industrial applications.

The following sections describe the various compliance options for controlling emissions from utility power plants.

SULFUR DIOXIDE CONTROL TECHNOLOGIES

Three major compliance options for SO_2 emissions available to utilities using coal-fired boilers are to switch fuels, purchase/sell SO_2 allowances, or install flue gas desulfurization (FGD) technologies. Costs, availability, and impact on boiler operation must be considered when evaluating switching to low-sulfur coal or natural gas. As more utilities enter the free market to purchase SO_2 allowances, prices will rise. Therefore, to minimize costs and, at the same time, meet environmental standards, power producers should continuously monitor the tradeoffs among these three options.

Although FGD processes, originally referred to as scrubbing SO_2 from flue gas, have been available for many years, installations in the United States were quite limited until passage of the Clean Air Act of 1970. Even then, installations were usually limited to new facilities because existing plants were exempt under the law.

Projects in the CCT program demonstrated innovative applications for both wet and dry or semidry FGD systems. The wet FGD systems, which use limestone as an absorber, have met or exceeded the 90 percent SO_2 removal efficiency required to meet air quality standards when burning high-sulfur coal. The dry or semidry systems use lime and recycled fly ash as a sorbent to achieve the required removal.

In wet FGD systems, flue gas exiting from the particulate collector flows to an absorber. In the absorber, the flue gas comes into contact with the sorbent slurry. The innovative scrubbers in the CCT program featured a variety of technologies to maximize SO_2 absorption and to minimize the waste disposal problems (sludge).

A number of chemical reactions occur in the absorber beginning with the reaction of limestone ($CaCO_3$) with the SO_2 to form calcium sulfite ($CaSO_3$). The calcium sulfite oxidizes to calcium sul-

Lime being unloaded at a power plant. The material will be used to reduce sulfur dioxide emissions as part of its flue gas desulfurization system. (Corbis Corporation)

fate ($CaSO_4$) which crystallizes to gypsum ($CaSO_4 \cdot 2H_2O$). The gypsum crystals are either stored in on-site waste disposal landfills or shipped to a facility where the gypsum is used in manufacture of wallboard or cement. The scrubbed gas then passes through mist eliminators that remove entrained slurry droplets. Recovered process water is recycled to the absorption and reagent preparation systems.

In the dry or semidry FGD system, the sorbent,

usually lime, is injected into the flue gas stream before the particulate collection device. The lime, fed as a slurry, quickly dries in the hot gas stream and the particles react with the SO_2. The resulting particles are then removed, along with the fly ash, by the particulate collection device.

NO_x CONTROL TECHNOLOGIES

Combustion modifications and postcombustion processes are the two major compliance options for NO_x emissions available to utilities using coal-fired boilers. Combustion modifications include low-NO_x burners (LNBs), overfire air (OFA), reburning, flue gas recirculation (FGR), and operational modifications. Postcombustion processes include selective catalytic reduction (SCR) and selective noncatalytic reduction (SNCR). The CCT program has demonstrated innovative technologies in both of these major categories. Combustion modifications offer a less-expensive approach.

Because NO_x formation is a function of the temperature, fuel-air mixture, and fluid dynamics in the furnace, the goal of a combustion modification is to mix fuel and air more gradually to reduce the flame temperature (lower thermal NO_x production), and to stage combustion, initially using a richer fuel-air mixture, thus reducing oxidation of the nitrogen in the fuel. LNBs serve the role of staged combustion.

Overfire air (OFA) is often used in conjunction with LNBs. As the name implies, OFA is injected into the furnace above the normal combustion zone. It is added to ensure complete combustion when the burners are operated at an air-to-fuel ratio that is lower than normal.

Reburning is a process involving staged addition of fuel into two combustion zones. Coal is fired under normal conditions in the primary combustion zone and additional fuel, often gas, is added in a reburn zone, resulting in a fuel rich, oxygen deficient condition that converts the NO_x produced in the primary combustion zone to molecular nitrogen and water. In a burnout zone above the reburn zone, OFA is added to complete combustion.

By recirculating a part of the flue gas to the furnace, the combustion zone turbulence is increased, the temperature is lowered and the oxygen concentration is reduced. All of these factors lead to a reduction of NO_x formation.

Boilers can be operated over a wide range of conditions, and a number of operational changes have been implemented to reduce NO_x production. Two promising technologies for staged combustion are taking burners out-of-service (BOOS) and biased firing (BF). With BOOS, fuel flow is stopped but air flow is maintained in selected burners. BF involves injecting more fuel to some burners while reducing fuel to others. Another operational modification is low excess air (LEA) which involves operating at the lowest excess air while maintaining good combustion. Depending on the type of boiler and the orientation of the burners, operators have viable choices to reduce NO_x production. Advances in boiler control systems enable operators to minimize NO_x and maximize performance.

Postcombustion processes are designed to capture NO_x after it has been produced. In a selective catalytic reduction (SCR) system, ammonia is mixed with flue gas in the presence of a catalyst to transform the NO_x into molecular nitrogen and water. In a selective noncatalytic reduction (SNCR) system, a reducing agent, such as ammonia or urea, is injected into the furnace above the combustion zone where it reacts with the NO_x to form nitrogen gas and water vapor. Existing postcombustion processes are costly and each has drawbacks. SCR relies on expensive catalysts and experiences problems with ammonia adsorption on the fly ash. SNCR systems have not been proven for boilers larger than 300 MW.

COMBINED SO_2/NO_x/PARTICULATE CONTROL TECHNOLOGIES

The CCT program involves a number of projects that achieve reduction of SO_2, NO_x, and particulate emissions in a single processing unit. The technologies described are uniquely combined to achieve project goals and, at the same time, to provide commercial-scale validation of technologies for utilities to consider in order to meet environmental standards.

PARTICULATE CONTROL TECHNOLOGIES

The two major compliance options for particulate control are electrostatic precipitators and fabric filters (baghouses). Dust-laden flue gas enters a precipitator and high voltage electrodes impart a negative charge to the particles entrained in the gas. These negatively charged particles are then attracted to and collected on positively charged plates. The plates are rapped at a preset intensity and at preset intervals, causing the collected material to fall into hoppers. Electrostatic precipitators can remove over 99.9 percent of the particulate matter.

Fabric filters (baghouses) represent a second accepted method for separating particles from a flue gas stream. In a baghouse, the dusty gas flows into and through a number of filter bags, and the particles are retained on the fabric. Different types are available to collect various kinds of dust with high efficiency.

FUTURE INTEGRATED FACILITY

As the global economy expands and worldwide population increases, the demand for additional electric power will grow. The Utility Data Institute (UDI), a Washington, D.C.-based trade organization, estimates that new generating plants totalling 629,000 MW in capacity will be built worldwide by 2003. UDI projects that 317,000 MW, or more than half of this new capacity, will be installed in Asia. Estimates for other regions are: North America, 81,000 MW; European Union, 78,000 MW; Latin America, 55,000 MW; the Middle East, 34,000 MW; Russia and former Soviet Union, 34,000 MW; and other, 30,000 MW. UDI forecasts that fossil fuels will account for 57 percent of the new generating plants, with coal taking 31 percent, gas 19 percent, and oil 7 percent.

The opportunities and threats of the 1990s and for the foreseeable future are related to competition and deregulation. The challenge for utilities will be to produce electric power as cheaply as possible while still complying with environmental regulations.

The Electric Power Research Institute (EPRI), founded by the electric utility industry to manage technology programs, envisions the evolution of a fully integrated facility that produces numerous products in addition to electricity. The first step is to remove mineral impurities from coal. Some of the clean coal could be gasified to provide not only fuel for fuel cells but also products such as elemental sulfur and chemical feedstocks. The remainder of the clean coal can be used in conventional boilers or fluidized bed combustion systems to generate process steam and electricity. The ash resulting from the combustion process can be mined for valuable trace metals before it is used in applications such as road construction. By employing advances from the CCT program, the integrated facility will allow utilities to provide their customers with reliable electrical service and to meet present and future environmental standards.

Charles E. Hickman

See also: Air Pollution; Climatic Effects; Coal, Consumption of; Energy Management Control Systems.

BIBLIOGRAPHY

Air Pollutants. (2000). Vancouver: Greater Vancouver Regional District. <http://www.gvrd.bc.ca/services/air/pollution/pollution.html>.

Clean Coal Technology Compendium. (1999). Los Alamos, NM: Los Alamos National Laboratory. <http://www.laml.gov/projects/cctc>.

Clean Coal Technology Program. (2000). Washington, DC: U.S. Department of Energy, Office of Fossil Energy. <http://apollo.osti.gov/fe/cct.html>.

Electric Power Research Institute. (1987). "Electrostatic Precipitator Guidelines. Volume 1: Design Specifications; Volume 2: Operations and Maintenance; Volume 3: Troubleshooting." *EPRI Journal CS-5198.* Final Report (June).

Electric Power Research Institute. (1999). "Coal Technologies for a New Age." *EPRI Journal* (Summer):6.

Electric Power Research Institute. (1999). "Helping Coal Plants Respond to Mercury Directive." *EPRI Journal* (Summer):6.

Fabric Filters. (1998–2000). Pittsburgh: Wheelabrator Air Pollution Control, Inc. <http://www.wapc.com/ff.htm>.

"Monitoring Effort Initiated Under $PM_{2.5}$ Program." (1998). *Clean Coal Today* (30):8–10.

Nitrogen Oxides Reduction Program: Final Rule for Phase II (Group 1 and Group 2 Boilers). (1996). Washington, DC: U.S. Environmental Protection Agency. <http://www.epa.gov/acidrain/nox/noxfs3.html>.

Overview: The Clean Air Act Amendments of 1990. (1990). Washington, DC: U.S. Environmental Protection Agency Office of Air and Radiation. <http://www.epa.gov/oar/caa/overview.txt>.

U.S. Department of Energy, Clean Coal Technology Topical Reports. (1995). "SO_2 Removal Using Gas Suspension Absorption Technology." Report No. 4 (April). Washington, DC: U.S. Government Printing Office.

U.S. Department of Energy, Clean Coal Technology Topical Reports. (1999). "Advanced Technologies for the Control of Sulfur Dioxide Emissions from Coal-Fired Boilers." Report No. 12 (June). Washington, DC: U.S. Government Printing Office.

U.S. Department of Energy, Clean Coal Technology Topical Reports. (1999). "Reburning Technologies for the Control of Nitrogen Oxides Emissions from Coal-Fired Boilers." Report No. 14 (May). Washington, DC: U.S. Government Printing Office.

U.S. Department of Energy, Clean Coal Technology Topical Reports. (1999). "Technologies for the Combined Control of Sulfur Dioxide and Nitrogen Oxides Emissions from Coal-Fired Boilers." Report No. 13 (May). Washington, DC: U.S. Government Printing Office.

U.S. Environmental Protection Agency. (1997). "EPA's Revised Particulate Matter Standards." Fact Sheet (July). Washington, DC: U.S. Government Printing Office.

Utility Data Institute. (1995). "UDI Sees Electricity Demand Surging with Global Economy." *Wind Energy Weekly* No. 648 (May 29).

EMISSION CONTROL, VEHICLE

Researchers linked automobile use to air pollution in the early 1950s when A. Haagen-Smit of the California Institute of Technology and fellow researchers began to unravel the complex atmospheric chemistry that leads to the formation of photochemical smog (ozone). Ozone is a strong lung and throat irritant that decreases lung function, increases respiratory problems, and complicates heart disease. Moderate ozone concentrations also damage materials and crops, increasing the cost of living. Ozone forms in the atmosphere when oxides of nitrogen (NO_x) and hydrocarbons (HC) mix and react in the presence of sunlight. Because onroad automobiles, trucks, passenger vans, and sport utility vehicles are typically responsible for about 30 percent of a region's HC and NO_x emissions, transportation is a significant contributor to smog problems in urban areas.

Onroad transportation sources are also responsible for more than 60 percent of regional carbon monoxide emissions. Carbon monoxide is a colorless, odorless gas that interferes with oxygen transfer in the bloodstream. With a higher affinity than oxygen to bind with red blood cell hemoglobin, with continual exposure, CO gradually displaces oxygen in the bloodstream. Because CO disperses well, it tends to be a hotspot pollutant, with troubling concentrations occuring in areas of high vehicle activity and poor air circulation, such as urban street canyons. When transportation facilities are constructed, engineers model the microscale air quality impacts to ensure that the highway design will not result in unhealthy CO levels downwind from the facility.

Carbon monoxide (CO) and hydrocarbon (HC) emissions arise from incomplete combustion, when petroleum hydrocarbons (gasoline or diesel fuel) do not completely oxidize to carbon dioxide and water. Hydrocarbon emissions also result from gasoline evaporation (liquid leaks, daily heating of fuel and tank vapors, seepage from fuel lines and other components, and displacement of fuel vapors during refueling). A number of toxic air contaminants, such as benzene and 1,3-butadiene are also associated with unburned and partially burned fuel hydrocarbons. Oxides of nitrogen (NO_x) form in the high temperature and pressure environment of the engine cylinder when the elemental nitrogen in air reacts with oxygen. Higher levels of NO_x form at the higher engine temperatures, which unfortunately correspond to peak fuel efficiency.

Given the health impacts that arise from exposure to vehicle emissions and their byproducts, regulatory agencies have focused on motor vehicle emissions control in their efforts to clear the air. Five basic strategies are employed for reducing onroad vehicle emissions: (1) reducing the emissions from new vehicles that displace the older high-emitting vehicles that are scrapped, (2) accelerating vehicle fleet turnover to get new vehicles into the fleet more quickly, (3) reducing emissions from in-use vehicles, (4) reducing travel demand to reduce vehicle activity, and (5) improving traffic flow to reduce emission rates.

The primary focus of federal environmental policy over the past 30 years has been on reducing the emissions from new vehicles. Strategies to enhance vehicle fleet turnover have not prove cost effective over the long term, but such strategies can provide useful short-term emissions reductions. Strategies aimed at limiting in-use vehicle emissions began around 1983 and continue today. Strategies designed to reduce travel demand and improve traffic flow, generically classified together as transportation control measures, have achieved mixed results to date. While demand management measures do work for individual large employers, regional demand management programs have failed to garner public support and have not provided cost-effective emissions reductions. On the other hand, technology-based traffic flow improvement programs, such as traffic signal optimization, can still provide significant emissions reductions at the regional level.

Surveyors from the Environmental Protection Agency inspect the engines and exhaust systems of volunteers' cars for emission control tampering and fuel switching. (Corbis Corporation)

VEHICLE STANDARDS AND EMISSION CONTROLS TECHNOLOGY

In 1963, new cars emitted nearly eleven grams per mile of hydrocarbons (HC), four grams per mile of oxides of nitrogen (NO_x), and eighty-four grams per mile of carbon monoxide (CO). Public pressure to reduce vehicle emissions began to mount in the early 1960s. Manufacturers responded to the general public pressure (and a few specific emissions control regulations) by adding positive crankcase ventilation (PCV) systems to new vehicles. A PCV valve prevents the release of unburned fuel from the crankcase by sending these vapors back to the intake air for combustion. With the passage of the Clean Air Act of 1970, the U.S. Environmental Protection Agency (EPA) began implementing a series of comprehensive regulations limiting the gram-per-mile emissions from new motor vehicles. Manufacturers must

produce vehicles that comply with the EPA standards, as measured in the laboratory on EPA's federal test procedure, but manufacturers are free to develop and implement any combinations of control systems they choose. Assembly-line testing and in-use surveillance testing and recall ensure that manufacturers comply with the certification standards.

Emissions standards in 1970 and 1972 were designed to reduce HC and CO emissions by 60 percent and 70 percent, respectively. In response to evaporative emissions standards, manufacturers developed onboard charcoal canisters in the early 1970s to capture gasoline vapors driven from the gas tank (and from the carburetor fuel bowl) by the daily rise and fall of ambient temperature. Exhaust emissions control strategies generally focused on de-tuning the engine to increase exhaust manifold temperature and adding a smog pump that delivered fresh air into the exhaust manifold to oxidize unburned CO and HC.

Manufacturers added exhaust gas recirculation (EGR) systems to counter the increased in-cylinder NO_x formation associated with higher operating temperatures. The EGR recycles a portion of the exhaust stream back into the engine intake air. The relatively inert exhaust gas, containing carbon dioxide and water but little oxygen, serves as a combustion buffer, reducing peak combustion temperatures.

By 1975, new cars were required to meet a 1.5 gram-per-mile HC standard, a 15 gram-per-mile CO standard, and a 3.1 gram-per-mile NO_x emissions standard. In response to the regulatory requirements, the automotive industry added new innovative emissions control technologies. Vehicles came equipped with oxidation catalysts (platinum and palladium on an alumina honeycomb or pellet substrate) designed to convert the CO and partially burned HC in the exhaust stream to CO_2 and water. The catalytic converter allows oxidation to occur at temperatures as low as 300°C, so that oxidation of the exhaust stream can continue downstream of the exhaust manifold. In 1975, lead was eliminated from the gasoline supply for these new vehicles, not because of lead's known harmful health effects, but because lead would foul the new catalytic converters of the new vehicles. A single tank full of leaded gasoline is enough to significantly and permanently reduce the efficiency of the catalytic converter. To reduce contamination of catalysts on new vehicles, the size of the opening to the gasoline tank fill neck was narrowed so that only the nozzles from unleaded gasoline pumps could be inserted into the fill neck during refueling.

Reduction catalysts that convert NO_x back to nitrogen and oxygen under conditions of low oxygen concentration began to appear in the late 1970s. At this time, vehicles began to employ dual-bed catalyst systems. These dual-bed systems employed a reduction catalyst followed by an oxidation catalyst, with fresh air (and thus additional oxygen) injected between the two catalyst beds. Dual-bed systems were capable of controlling NO_x, CO, and HC in a sequential mode.

The emissions reductions provided by the catalytic converter also allowed engineers to re-tune their engine designs and add an improved proportional EGR system. By modifying the EGR that recycles exhaust in proportion to intake air (as a function of engine speed) rather than at a constant rate, emissions could be better controlled over a wider range of oper-

ating conditions. The new EGR systems provided the added bonus of improved vehicle performance, balancing some of the efficiency losses associated with the use of catalytic converters.

Probably the most significant control technology breakthrough came in 1977, when Volvo released a computer-controlled, fuel-injected vehicle equipped with a three-way catalyst. The new catalytic converters employed platinum, palladium, and rhodium to simultaneously reduce NO and oxidize CO and HC emissions under carefully controlled oxygen conditions. The new Bosch fuel injection system on the vehicle provided the precise air/fuel control necessary for the new catalyst to perform effectively. The combined fuel control and three-way catalyst system served as the foundation for emissions control on the next generation of vehicles.

By 1981, exhaust emissions standards had tightened to 0.41 grams-per-mile HC, 3.4 grams per mile CO, and 1.0 gram-per-mile NO_x. Manufacturers turned from carburetors, to single-point throttle-body injection, and then to multi point fuel injection systems. With each shift in technology, better control over the air and fuel mixture and combustion was achieved. Better control over the delivery and mixing of air and fuel provided significant emissions and performance benefits. New computer-controlled variable EGR significantly reduced NO_x formation. Smog pumps had also given way to lightweight, inexpensive, pulse air injection systems, significantly improving engine performance. Using the natural pressure variations in the exhaust manifold, fresh air flows in to the manifold through a one-way reed valve and helps to oxidize CO and HC. Finally, many of the new vehicles now came equipped with the improved three-way catalytic converters (TWC) that debuted in 1977.

The new three-way catalytic converters required precise control of fuel/air ratio, so onboard computers became necessary to monitor the combustion process and rapidly adjust the air/fuel mixture through closed loop control. The goal of the computer program is to keep combustion at stoichiometric proportions, where there is just enough air (and therefore oxygen) delivered to completely oxidize the fuel. The stoichiometric ratio for an average fuel is roughly 14.7 kilograms of air per kilogram of fuel. An oxygen sensor in the exhaust manifold monitors the oxygen concentration of the exhaust gases to determine if the combustion mixture contained sufficient oxygen. A reading of zero oxygen in the exhaust gas probably indicates that too much fuel was mixed with the intake air (consuming all of the oxygen before combustion was completed) while a high oxygen concentration in the exhaust gas indicates that too little fuel was mixed with the intake air. The computer processes and evaluates multiple readings each second making minute adjustments to the amount of fuel delivered to the intake air (closing the loop between computer action, sensor reading, and computer response). The computer never achieves a perfect stoichiometric mixture; the air and fuel mix instead alternates between slightly rich and slightly lean. However, the extremely rapid measurement and computer response minimizes emissions formation by responding rapidly to changes in engine operation.

The efficiency of the three-way catalytic converter is also a function of air/fuel ratio. At the stoichiometric air/fuel ratio of 14.7 kilograms of air per kilogram of fuel, the relative air/fuel ratio known as λ equals 1.0. Figure 1 illustrates catalytic converter efficiency for each pollutant as a function of relative air/fuel ratio λ (where a positive λ indicates a lean mixture and a negative λ indicates a rich mixture). The closer the mixture stays to stoichiometric, the more efficient the catalyst at reducing the combined emissions of the three pollutants.

By 1994, EPA had further tightened the standards to 0.25 grams-per-mile for HC and 0.4 grams per mile for NO_x. Hence, new vehicle HC emissions had now dropped nearly 98 percent and NO_x emissions had dropped 90 percent compared to the level of the 1960s. Manufacturers were also required to ensure that the emissions control systems would endure for at least 100,000 miles. To meet the stringent 1994 standards, manufacturers relied on improved technology and materials, and more advanced computer systems to monitor combustion and rapidly adjust a variety of operating parameters (fuel metering, spark timing, and EGR) to optimize vehicle performance and minimize emissions. Advanced exhaust treatment systems (such as electrically heated or close-coupled catalysts), that manufacturers originally believed in the 1980s would be necessary to comply with these standards, have not been needed.

In 1999, the EPA proposed stringent standards applicable to model year 2004 vehicles. Thus, the EPA continues to implement technology-forcing regulations, in which EPA tasks manufacturers with an emissions standard, and industry must develop

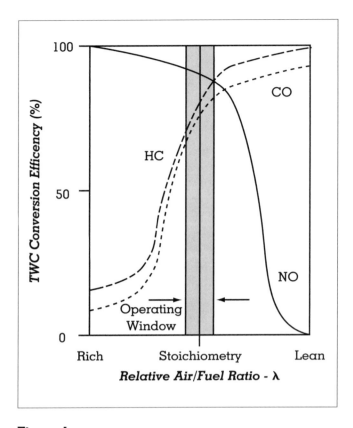

Figure 1.
SOURCE: Chowanietz, 1995.

technologies to enable the vehicles to comply. When these standards are in place, new vehicles will emit less than 1 percent of the HC and NO_x emissions of their 1960s counterparts (see Figure 2). Advanced computer controls, variable valve timing, and improved catalysts will continue to provide significant reductions. New control systems are also likely to focus on reducing emissions immediately following the engine start. Advances in diesel emissions control technologies may yield viable light-duty diesel vehicles.

Vehicles powered by alternative fuels have yet to make significant inroads into public ownership. Battery technology has not advanced sufficiently to deliver low-cost electric vehicles capable of providing comparable vehicle performance and more than 100 miles between recharging. However, new hybrid electric vehicles that perform on a par with current vehicles and never require recharging began entering the marketplace in 2000. These hybrid electric vehicles are achieving emissions levels as low as 10 percent of 1999 emissions levels, qualifying well below the

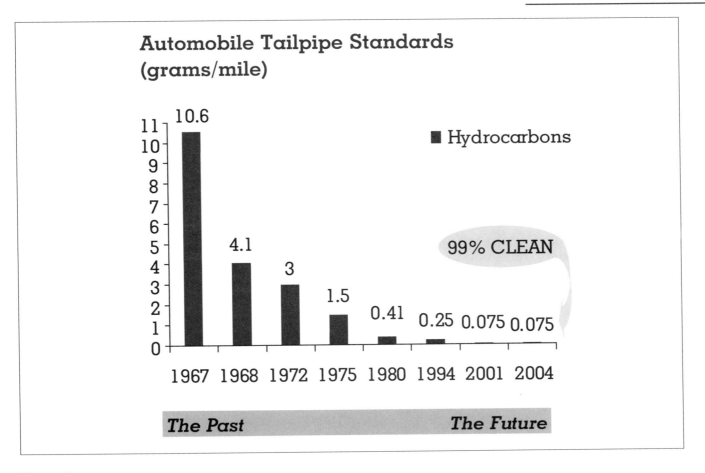

Figure 2.
SOURCE: Alliance of Automobile Manufacturers (2000)

2004 standards. Fuel cell vehicles, which convert the chemical energy of a fuel into electricity, are expected to provide near-zero emissions within the decade.

LIGHT-DUTY TRUCKS AND SPORT UTILITY VEHICLES

Sales of sport utility vehicles and light-duty trucks have topped 45 percent of the new vehicle sales market in the late 1990s, and shares are still climbing. In some months, light-truck and SUV sales exceed those of automobiles. Current emissions regulations are currently less stringent for the vast majority of these light-duty trucks (LDTs) and sport utility vehicles (SUVs) than they are for automobiles. These vehicles are heavier than automobiles, employ larger engines, and the drivetrain (engine, transmission, rear differential, and tire diameter) is designed to handle heavier loads. As such, the emissions from these vehicles are naturally higher. New and in-use LDTs and SUVs exhibit much higher emissions levels than do automobiles (roughly 30 percent more HC and CO, and 85 percent more NO_x per mile traveled). Light-duty trucks also have a significantly longer lifespan in the fleet than do automobiles, compounding the emissions impact over time.

The environmental community has argued, for some time, that the actual onroad duties of the vast majority of LDTs and SUVs are not significantly different from the duties performed by automobiles. That is, most of these vehicles are simply making commute, shopping, and recreation trips that do not require enhanced performance. This issue, combined with the availability of emissions control systems to significantly lower emissions from LDTs and SUVs, led the EPA to harmonize the two standards under the new Tier 2 program. Beginning in 2004, through a phase-in schedule, LDTs and SUVs are required to meet the same emissions standards as automobiles.

Although the lightest of current onroad trucks already meet the same standards as automobiles, the new certification requirement will bring the remaining 80 percent of light-duty truck sales into alignment with the more stringent emissions standards.

HEAVY-DUTY VEHICLES

Onroad and off-road heavy-duty vehicles (greater than 8,500 pounds gross vehicle weight rating) contribute significantly to emissions of NO_x, which in turn participate in ozone formation. As one would expect, heavy-duty engines are large, and the engine load for a vehicle carrying a 60,000-pound payload is extremely high. Most heavy-duty trucks operate on the diesel cycle, an engine cycle that produces much higher temperature and pressure conditions, leading to the formation of significantly greater NO_x levels per mile traveled. Given the stringent controls implemented for light-duty vehicles, it is not surprising that the heavy-duty vehicle contribution as a percentage of regional emissions has been increasing. Projections for the Los Angeles basin in 2010 indicate that without further controls, heavy-duty vehicles will contribute more than 55 percent of onroad NO_x emissions.

According to industry experts, the state of emissions control for heavy-duty engines in the 1990s was at the level of technical advancement that we were achieving for light-duty vehicles in the 1970s. In the last few years, new, highly effective diesel particulate trap and catalyst systems have been developed for heavy-duty diesel vehicles. Many of these new system designs are currently on the road undergoing performance and durability testing. All of the new technologies that are forthcoming were developed in response to new EPA heavy-duty vehicle certification standards that are affective as of 2004. Significant reductions in heavy-duty vehicle emissions are on the horizon.

ENGINE START EMISSIONS

Exhaust emissions are high during the first one to three minutes of engine operation, until combustion stabilizes and the catalytic converter reaches approximately 300°C (known as light-off temperature, when the catalyst begins controlling emissions). Peak catalyst efficiency occurs between 400°C and 800°C. A vehicle that sits more than an hour is usually considered to be starting in a cold-start mode, because the temperature of the catalytic converter has dropped significantly since the vehicle was last used. The aver-

age vehicle on the road in 2000 emitted 2 to 4 grams of HC, 1 to 3 grams of NO_x, and 30 to 50 grams of CO for each cold engine start. Vehicles starting in warm-start mode (less than one hour of parking time) produce significantly lower emissions than a cold start, but still contribute significantly to overall trip emissions. Hot starts after 10 minutes or parking time still produce nearly 0.5 gram of HC, 0.5 gram of NO_x, and 20 grams of CO per start for the average vehicle. New emissions models are forthcoming that estimate engine start emissions as a continuous function of park-time distributions.

Because gram/mile emissions from modern vehicles are so low, engine start emissions have become a large fraction of the emissions associated with a vehicle trip. For a typical twenty-mile commute trip in a 1994 vehicle, roughly 30 percent of the CO and 10 percent of the NO_x and HC can be attributed to the cold start. For a ten-mile trip, the overall emissions are about 35 percent lower, but the cold start contributions rise to approximately 50 percent of the CO and 20 percent of the NO_x and HC. On very short trips, total trip emissions are lower still, but the cold start contribution dominates the total. For a half-mile trip, most vehicles never achieve catalyst light-off, and more than 95 percent of the CO and 80 percent of the HC and NO_x trip emissions can be attributed to cold-start operation (as compared to the same trip made by a fully-warmed-up vehicle). It is important to note that a single trip of twenty miles will result in significantly lower emissions than ten trips of two miles each. Trip chaining, where the end of one trip serves as the beginning of the next trip after a short parking period, also results in significantly lower emissions than if each trip results in a cold engine start.

Given the importance of engine start emissions in urban areas, new emissions control systems are likely to focus on achieving instant catalyst light-off. Catalyst manufacturers will increase catalyst surface area and use materials and designs that are resistant to damage from high-temperature exhaust gas. Such designs will allow placement of catalysts closer to the exhaust manifold where higher temperatures will help the catalyst reach light-off much more quickly.

ENRICHMENT EMISSIONS

In recent years, research has demonstrated that real-world vehicle emissions under typical onroad operating conditions can differ significantly from the emissions observed in the laboratory under standard federal test procedures. The occurrence of enrichment, when the air/fuel mixture becomes rich for a few moments, results in orders of magnitude increases in CO and HC emissions rates for short periods. N. Kelly and P. Groblicki (1993) first reported indications that enrichment conditions were likely to be causing a significant portion of vehicle emissions not captured during standard laboratory certification tests. Numerous studies since then have identified enrichment as a widespread concern.

Carbon monoxide emissions rates (grams/second) under enrichment conditions for the very cleanest of vehicles can soar as high as 2,500 times the emissions rate noted for stoichiometric conditions. Although most vehicles spend less than 2 percent of their total driving time in severe enrichment, this can account for up to 40 percent of the total CO emissions (LeBlanc et al., 1995). Hydrocarbon emissions rates can rise by as much as a factor of a hundred under enrichment conditions. Enrichment activity is usually associated with high power demand and engine load conditions, such as high-speed activity, hard accelerations, or moderate accelerations under moderate to high speeds. However, enrichment also occurs during hard deceleration events. When the throttle plate snaps shut during a rapid deceleration event, the rapid decrease in intake manifold pressure vaporizes liquid fuel deposits, causing the fuel mixture to become rich.

All vehicles undergo some enrichment. Fuel enrichment sometimes results from malfunctions of vehicle sensors and control systems. When engine and exhaust gas sensors fail to provide appropriate data to the onboard computer under certain operating conditions, the computer sends inappropriate control commands to fuel injectors and spark advance units. Depending upon the type and extent of component failure, such malfunctions can result in a super-emitter, with significantly elevated emissions rates under all operating conditions. It is interesting to note that engine manufacturers have engineered occurrences of enrichment through the onboard vehicle computer software. Because peak engine torque develops when the air/fuel mixture is slightly rich, manufacturers sometimes use enrichment to improve vehicle performance. Enrichment can increase acceleration rates, improve engine performance while hill-climbing or running accessories such as air conditioning, and can be used to control cylinder detonation. In addition,

the cooling properties associated with vaporizing and partially combusting excess fuel lowers peak combustion temperatures, protecting cylinders, valves, and catalysts from high-temperature damage during high RPM activity.

When enrichment episodes occur in the real world, but not in the laboratory under federal certification tests, real-world emissions are significantly higher than predicted. Further complicating emissions prediction is that aggressive driver behavior and complex traffic flow characteristics play a large role in enrichment occurrence. Current vehicle activity simulation models can predict average speeds and traffic volumes very well, but poorly predict the hard-acceleration events that lead to enrichment.

The federal test procedure for new vehicle certification is limited to a maximum acceleration rate of 3.3 mph/second and a maximum speed of 57 mph (and even that speed is for a very short duration). Based upon extensive data collected in Baltimore, Spokane, and Atlanta, more than 8.5 percent of all speeds exceeded 57 mph, and more than 88 percent of trips contained acceleration activity exceeding 4 mph/second. In fact, more than one-third of the trips monitored included an acceleration rate at some point during the trip of more than 7 mph/second. Similarly, more than 15 percent of the deceleration activity exceeded -3.5 mph/second. Hence, enrichment events are significant in real-world emissions inventories.

To counter the elevated emissions associated with enrichment, the EPA has adopted supplemental federal test procedures. The new laboratory test procedures contain higher speeds, higher acceleration and deceleration rates, rapid speed changes, and a test that requires the air conditioning to be in operation. These tests increase the probability that vehicles will go into enrichment under laboratory test conditions. Hence, manufacturers have an incentive to reduce the frequency of enrichment occurrence in the real world. Future catalytic converters and emissions control systems will be resistant to the high-temperature conditions associated with engine load, and will be less likely to require enrichment for protection. Thus, enrichment contributions to emissions will continue to decline.

IN-USE VEHICLE EMISSIONS

New vehicle emissions standards have served as the primary means for reducing vehicle emissions over the last thirty years. However, urban areas must wait for years before the purchase of new vehicles significantly reduces onroad emissions. Meanwhile, daily motor vehicle emissions remain dominated by the small fraction of very high-emitting vehicles. The average car on the road emits three to four times more pollution than new standards allow, and minor control system malfunctions greatly increase emissions. Numerous research studies conclude that a small fraction of onroad vehicles contribute a large fraction of fleet emissions. Some researchers argue that as few as 5 percent of the vehicles are causing 40-50 percent of onroad emissions, but published estimates of super-emitter contribution estimates vary widely. These research studies rely upon laboratory data collected on certification tests, field data collected using portable testing systems, data from remote sensing devices that estimate pollutant concentrations in vehicle tailpipe exhaust plumes, or laboratory or roadside inspection and maintenance data. The controversy surrounding the wide range of super-emitter contribution estimates stems from significant differences in the vehicles sampled, data collected, and the analytical methods and assumptions employed in the various analyses. Although the contribution percentage is uncertain, it is clear that a small fraction of super-emitting onroad vehicles contribute disproportionately to emissions.

To reduce emissions from onroad vehicles, urban areas have turned to inspection and maintenance (I/M) programs. By 1983, sixty-four cities nationwide had established I/M programs, requiring passenger vehicles to undergo a visual inspection and a two-speed idle test to detect severely malfunctioning emissions control systems. Many areas are now adopting advanced I/M programs, which require vehicle testing on a garage treadmill to better identify problem vehicles. Enhanced I/M programs achieve greater emissions reductions than standard I/M programs. However, other states are beginning to restructure and sometimes eliminate statewide inspection and maintenance programs, because the annual fees and testing hassle are not popular with the public. Furthermore, some studies indicate that the emission reduction benefits of I/M programs, while still significant, may be achieving only half of their current modeled emissions reductions. When a state eliminates or scales back an I/M program, the state is responsible for identifying other sources of emissions reductions.

New onboard diagnostics (OBD) systems bridge the gap between new vehicle certification and the in-use compliance verification of I/M. Onboard diagnostics systems detect failures of the engine sensors and the control actuators used by the onboard computer to optimize combustion and minimize emissions. Federal and California OBD programs introduced in 1994 detect component failures (such as an oxygen sensor) by continuously monitoring and evaluating the network of sensor readings to detect erroneous or illogical sensor outputs. Such OBD systems employ detailed computer programs that can change the control logic, discard the inputs from bad sensors, and ensure that emissions remain low even when failures do occur. A malfunction indicator lamp (MIL), or Check Engine light, illuminates on the dashboard when the OBD system identifies problems. Engine computers facilitate repair by reporting trouble codes to mechanics through handheld diagnostic tools that interface with the engine computer. Under I/M programs, vehicles with OBD-reported malfunctions cannot be re-registered until the problem is diagnosed and repaired. The new OBD systems are designed to improve the effectiveness of I/M and minimize lifetime emissions from the vehicle.

Super-emitters behave differently than their normal-emitter counterparts. Whereas normal-emitting vehicles may exhibit high emissions under a hard acceleration or high speeds, vehicles classified as super-emitters tend to exhibit elevated emissions under almost every operating condition. New emissions models will likely track the activity of high-emitting vehicles separately, applying different emission rate algorithms to this activity. Similarly, these new emissions models will also model the effect of I/M programs as decreasing the fraction of onroad high-emitting vehicles.

CLEANER FUELS

Numerous fuel properties affect evaporative and exhaust emissions. Refiners can modify fuel vapor pressure, distillation properties, olefin content, oxygen content, sulfur content, and other factors to reduce emissions. In 1989, the EPA set fuel volatility limits aimed at reducing evaporative emissions. In 1992, manufacturers introduced oxygenated gasoline into cities with high wintertime CO levels. By 1995, the EPA's reformulated gasoline (RFG) program required the sale of special gasoline in nine metropolitan areas that do not meet national clean air standards for ozone. RFG yielded a 15 percent reduction in HC emissions without increasing NO_x emissions, at a cost of somewhere between four and seven cents per gallon. Fuels had to include 2 percent oxygenate by weight (ethanol or MTBE), but manufacturers could adjust a variety of other gasoline properties to achieve the mandated emissions reduction.

Proposed fuel regulations associated with EPA's 2004 vehicle standards program (Tier 2) will substantially reduce the allowable sulfur content of fuel, significantly enhancing the effectiveness of advanced catalytic converters. Sulfur in gasoline temporarily deactivates the catalyst surface, thereby reducing catalyst efficiency. The sulfur reductions are critical for enabling the vehicle emissions control technology to meet Tier 2 standards. By 2006, Tier 2 regulations will require an average fuel sulfur level of 30 ppm, with an 80 ppm cap. This is a substantial decrease from current average sulfur levels of 340 ppm. Vehicle manufacturers estimate that the 90 percent reduction in fuel sulfur will reduce NO_x emissions from the new, low-emitting vehicles by 50 percent, at a marginal cost of between two and four cents per gallon. Vehicle manufacturers argue that further reducing sulfur levels from 30 ppm to 5 ppm will provide additional emissions benefits at a cost somewhere between two and three additional cents per gallon.

FUEL ECONOMY IMPROVEMENTS AND EMISSIONS

In the 1970s, manufacturers requested and received some delays in the implementation of new vehicle certification standards. EPA granted these delays to help manufacturers balance emissions reduction efforts with their efforts to increase corporate average fuel economy. At that time, almost every control system (smog pumps, EGR, and catalytic converters) resulted in a fuel economy penalty. The direct relationship between increased emissions control and decreased fuel economy was broken in the late 1980s with the widespread adoption of advanced computer-controlled fuel injection and spark timing systems. Smog pumps were removed and other devices that reduced fuel economy were improved with computer control. Today, the same technologies that reduced motor vehicle emissions (electronic fuel injection, spark timing, and computer control) have improved fuel economy (or provided improved engine power

output in lieu of fuel economy improvements). In general, reduced fuel consumption results in emissions reductions from the vehicle, and reduced vehicle refueling minimizes evaporative emissions. Improving fuel economy is sometimes referred to as the forgotten emissions control strategy.

CONCLUSION

Despite the emissions rate reductions achieved during the last thirty years from new and in-use vehicles, rapid growth in vehicle use has offset a good portion of the total potential reductions. Population growth continues at a rate between 1 percent and 2 percent per year, but the number of trips per day and vehicle miles of travel are increasing at double or triple that rate in many areas. More people are making more trips and driving farther each day. As vehicle miles of travel continue to increase, so do congestion levels. The net effect is that more people are making more trips and driving farther under conditions that increase emission rates. Manufacturers sell nearly 16 million vehicles per year in the United States. More importantly however, the average vehicle lifespan of nearly fourteen years continues to increase. Given the tremendous growth in vehicle use and the emissions rate increases that come with congestion and an aging onroad fleet, reducing onroad vehicle emissions remains extremely important for air quality. Without the previous 30 years of transportation emissions controls, urban air quality would have continued to degrade. Instead, the most polluted areas of the United States have experienced significant air quality improvements.

Randall Guensler

See also: Traffic Flow Management.

BIBLIOGRAPHY

Bosch, R. (1996). *Automotive Handbook, 4th Edition.* Warrendale, PA: Society of Automotive Engineers.

Cadle, S. H.; Gorse, R. A.; and Lawson, D. R. (1993). "Real-World Vehicle Emissions: A Summary of the Third Annual CRC-APRAC On-Road Vehicle Emissions Workshop." *Journal of the Air and Waste Management Association* 43(8): 1084–1090.

Chatterjee, A.; Wholley, T.; Guensler, R.; Hartgen, D.; Margiotta, R.; Miller, T.; Philpot, J.; and Stopher, P. (1997). *Improving Transportation Data for Mobile Source Emissions Estimates* (NCHRP Report 394). Washington, DC: National Research Council, Transportation Research Board.

Chowanietz, E. (1995). *Automobile Electronics.* Warrendale, PA: Society of Automotive Engineers.

Chrysler Corporation. (1998). *Emissions and Fuel Economy Regulations.* Auburn Hills, MI: Chrysler Environmental and Energy Planning, Environmental Team.

Degobert, P. (1995). *Automobiles and Pollution.* Warrendale, PA: Society of Automotive Engineers.

Faiz, A.; Weaver, C.; Walsh, M. (1996). *Air Pollution from Motor Vehicles.* New York: The World Bank.

Guensler, R.; Bachman W.; and Washington, S. (1998). "An Overview of the MEASURE GIS-Based Modal Emissions Model." In *Transportation Planning and Air Quality III,* ed. T. Wholley. New York: American Society of Civil Engineers.

Heck, R. M., and Farrauto. R. J. (1995). *Catalytic Air Pollution Control.* New York: Van Nostrand Reinhold.

Heywood, J. (1988). *Internal Combustion Engine Fundamentals.* New York: McGraw-Hill.

Kelly, N., and Groblicki, P. (1993). "Real-World Emissions from a Modern Production Vehicle Driven in Los Angeles." *Journal of the Air and Waste Management Association* 43(10)1351–1357.

LeBlanc, D.; Saunders, F. M.; and Guensler; R. (1995). "Driving Pattern Variability and Potential Impacts on Vehicle CO Emissions." *Transportation Research Record* 1472:45–52.

National Research Council. (1991). *Rethinking the Ozone Problem in Urban and Regional Air Pollution.* Washington, DC: National Academy Press.

Seinfeld, J. H. (1986). *Atmospheric Chemistry and Physics of Air Pollution.* New York: John Wiley and Sons.

Society of Automotive Engineers. (1998). *Advanced Converter Concepts for Emission Control (SP-1352).* Warrendale, PA: Author.

South Coast Air Quality Management District. (1997). *Air Quality Management Plan, Chapter 2: Air Quality and Health Effects.* Diamond Bar, CA: South Coast AQMD.

St. Denis, M. J.; Cicero-Fernandez, P.; Winer, A. M.; Butler, J. W.; and Jesion, G. (1995). "Effects of In-Use Driving Conditions and Vehicle/Engine Operating Parameters on "Off-Cycle" Events: Comparison with FTP Conditions." *Journal of the Air and Waste Management Association.* 44(1)31–38.

U.S. Department of Transportation, Bureau of Transportation Statistics. (1996). *Transportation Statistics Annual Report 1996: Transportation and the Environment.* Washington DC: U.S. Department of Transportation.

U.S. Environmental Protection Agency. (1995). *Air Quality Trends (EPA-454/F-95-003).* Research Triangle Park, NC: EPA Office of Air Quality Planning and Standards.

U.S. Environmental Protection Agency. (1996). "Final Regulations for Revisions to the Federal Test Procedure for Emissions from Motor Vehicles." *Federal Register.* 61:54852–54906.

U.S. Environmental Protection Agency. (1999). "Fleet Characterization Data for MOBILE6 (EPA420-P-00-011; M6.FLT.007)." Ann Arbor, MI: Office of Mobile Sources.

U.S. Environmental Protection Agency. (2000). "Control of Air Pollution from New Motor Vehicles: Tier 2 Motor Vehicle Emissions Standards and Gasoline Sulfur Control Requirements." *Federal Register* 65:6698–6870.

Wark, K.; Warner, C. F.; and Davis, W. T. (1998). *Air Pollution: Its Origin and Control.* Menlo Park, CA: Addison-Wesley.

ENERGY CONSUMPTION

See: Supply and Demand and Energy Prices

ENERGY ECONOMICS

Energy economics is the application of economics to energy issues. Central concerns in energy economics include the supply and demand for each of the main fuels in widespread use, competition among those fuels, the role of public policy, and environmental impacts. Given its worldwide importance as a fuel and the upheavals in its markets, oil economics is a particularly critical element of energy economics. Other efforts have treated natural gas, coal, and uranium. Energy transforming and distributing industries, notably electric power, also receive great attention. Energy economics addresses, simultaneously as well as separately, both the underlying market forces and public policies affecting the markets.

Economic concerns differ sharply from those of natural scientists and engineers. The most critical difference is in the outlook towards supply development. Many economists argue that market forces allow smooth adjustments to whatever happens to the physical stock of resources. Potentially, these market forces can produce resources cheaper than other methods presently employed to cause adjustments. At worst, the cost rises will be gradual and manageable. In contrast, these economists stress the harmful effects of governments on energy.

A major influence on this optimism in market forces is observation that, historically, technical progress has promoted energy market development. However, the processes by which this progress emerges have not been conducive to formal economic analysis. Rather than wait for a satisfactory model, economists must treat new technology as an unexplained but vital element of energy markets. A related consequence is that established energy sources get almost all the attention. Economics can say little about products that have not emerged.

RESEARCH IN ENERGY ECONOMICS

Scholarly study of the issues in energy economics ranges from massive tomes to short articles. Government and international agencies, consulting firms, industrial corporations, and trade associations also produce a vast body of research in energy economics.

THE PRACTITIONERS

The practitioners of energy economics variously identify themselves as energy economists, mineral economists, natural resource economists, and industrial organization economists. Separate professional societies exist to represent each of three specialties: resource and environmental economics, mineral economics, and energy. These associations do not interact with one another In addition, academic programs exist in each of these areas.

SPECIALTIES AND SUBDISCIPLINES

In each case, quite different bases created the specialties. Energy economics as a separate field emerged with the energy turmoil of the 1970s. Many people were suddenly drawn into dealing with energy issues and felt a strong need for organizations exclusively dealing with their concerns. In contrast, mineral economics emerged in the 1930s from interactions among mineral engineers, geologists, and economists on how the insights of their individual fields could be combined to deal with the problems of all forms of mineral extraction. Mineral economists took a broad view of minerals that gave a prominent role to energy, but many energy economists viewed mineral economists as concerned only with rocks. A more critical problem with mineral economics was that its reach was and remains limited to specialized academic programs and long-established mineral agencies such as those in the U.S. Department of the Interior, and that it was dominantly North American. The

energy economists attained greater breadth in the identity and nationality of participants.

Resource economics has at least two bases. Many natural scientists raise widely accepted broad concerns over natural resource availability. A massive federal government study of the problem was instituted during the Truman administration. One result of the effort was the Ford Foundation endowment that established Resources for the Future, a Washington, D.C., research institute, devoted to the study of all natural resource problems. The institute has attracted leading figures in all areas of resource economics, including the environment. Academic programs in resource economics grew mainly from efforts to broaden the scope of agricultural economics, a discipline long practiced principally at the land-grant colleges of major U.S. agricultural states. The scope went on to encompass consideration of environmental problems.

WHY SUBFIELDS EXIST

As is standard in the development of subdisciplines, these fields arose from an intuitive perceived need. Nothing more than the ability to maintain a critical mass of participants adequately justifies the separations. Although the areas deal with different bodies of fact, this distinction is insufficient to justify a field. If it were, we might also have automotive economics or baseball economics. All involve application of economic principles to particular problems. Finding unifying analytic bases for separation is problematic.

Similar arguments, however, can be made about the formal justifications made for even the largest, longest-established branches of economics. The line between pure theory and industrial economics is fuzzy; international trade economics is largely a demonstration that nationality should be economically irrelevant. Given the vast expansion of modern economics, it is necessary to subdivide into convenient sectors. The energy and related realms reflect, on a considerably smaller scale, the value of such segmentation in uniting people with similar interests. Thus, the argument that special analytic issues arise does not justify the existence of special fields; a community of interests is what dominates.

MINERAL DEPLETION AND ITS POLICY IMPLICATIONS

A perennial subject of concern is the long-term availability of energy resources. Natural scientists and even many economists not specializing in natural resources stress the importance of the physical limits to the amount of available energy and to the thermodynamic forces through which initial use precludes economic reuse. Natural resource economists counter that this approach provides an incomplete and misleading vision of the "natural" resource sector. While the physical endowment is limited by definition, its usability is not. Existence does not guarantee economic value. First, a use for the "resources" must be found. Then techniques must be designed and profitably implemented for finding ("exploration"), converting the finds to producing properties ("development"), processing the output to a useable form, and moving the production to customers.

Resource economics stresses that, to date, experience is that the conversion of unutilized resources to profitable ones has moved ahead of demand growth and resource commodity prices have fallen over time. Moreover, M. A. Adelman's classic studies of petroleum supply development have shown that exploration is an ongoing effort to expand the potential for supply expansion. A backlog of developable prospects always exists. When profitable, the much more expensive development stage is undertaken.

Resource pessimists counter that this process cannot proceed forever because the eternal persistence of demand for any given commodity that is destroyed by use must inevitably lead to its depletion. However, the eternal persistence assumption is not necessarily correct. The life of a solar system apparently is long but finite. Energy sources such as nuclear fusion and solar energy in time could replace more limited resources such as oil and natural gas. Already, oil, gas, nuclear power, and coal from better sources have displaced traditional sources of coal in, for example, Britain, Germany, Japan, and France.

Alarms about depletion arose long before massive energy consumption emerged. Experience suggests that the fear is premature. Acting politically to save energy resources may prove more wasteful than allowing consumption. Indeed, the exhaustion problem already is extensively abused to justify undesirable policies. Every local energy producer that reaches the end of its economic life argues that it should be preserved as a hedge against exhaustion (or whatever other evil that can be thought up).

Markets themselves are structured to be very responsive to whatever problems arise with resource supply. Depletion is an economic problem, and markets are capable of reacting to it. Economic limits necessarily are greater than physical ones. That is, much that is physically present will never be economical to employ. Total depletion, if it occurs, will repeat on a large scale what has already occurred for specific suppliers. Costs will become prohibitive, output will start declining, and a steady, recognizable move to extinction will follow.

Elaborate economic theories of exhaustible resources shows that the cost and price pressures associated with such resource depletion provide incentives for slowing depletion. Prudent investors are rewarded doubly. The impending decline in supply pushes up prices and thus the gross income from sales. The restraint slows the depletion of low cost resources, lowers production costs, and thus the net payoff to delayed sales. The theory also indicates that the distribution of payoffs between the two sources can differ greatly. Thus, while we can expect to observe steady upward pressures on energy prices, many different patterns are possible. Price rises will be persistent but not necessarily at a constant rate. The absence of evidence of such price pressures is strong evidence that exhaustion is not an immediate threat. Whatever pattern proves optimal can emerge in the marketplace, and the skepticism among many economists about government foresight are amply proved by prior energy experience. Appraisals of prospects will change with improved knowledge and whatever else permanently alters the situation. Competition among specialized private investors is more adaptable than public policy.

To complicate matters, a government program to assist investment in minerals will not necessarily slow depletion. Investment aid, to be sure, encourages depletion-reducing investment in delaying production. However, the aid also stimulates depletion-increasing investments in production facilities. Which effect predominates depends on the circumstances. Depletion retardation is more likely for producers with presently large excesses of price over cost. Depletion stimulation is more likely when prices are close to costs.

ECONOMIC EFFICIENCY VERSUS SUSTAINABLE DEVELOPMENT

Since the 1980s, the persistence of concerns over both the maintenance of natural resource commodity supply and environmental quality has become restated as a search for sustainable development. The concern is that unregulated markets will produce a pattern of natural resource commodity production that unduly favors the present and near-term future over the longer-term, and generates too much pollution.

Some resource economists fervently support the concept of sustainability. Others argue that the principle is less coherent, comprehensible, and compelling than prior concepts, particularly the core economics principle of efficiency. For economists, the choice of terminology is secondary. The primary concern is resolving the underlying problems of possible market inefficiencies and the ability of governments to cure them.

Richard L. Gordon

BIBLIOGRAPHY

Adelman, M. A. (1995). *The Genie out of the Bottle: World Oil since 1970.* Cambridge, MA: MIT Press.

Barnett, H. J., and Morse, C. (1963). *Scarcity and Growth: The Economics of Natural Resource Availability.* Baltimore: Johns Hopkins Press for Resources for the Future.

Brennan, T. J.; Palmer, K. L.; Koop, R. J.; Krupnick, A. J.; Stagliano, V.; and Burtraw, D. (1996). *A Shock to the System: Restructuring America's Electricity Industry.* Washington, DC: Resources for the Future.

Gilbert, R. J., and Kahn, E. P., eds. (1996). *International Comparisons of Electricity Regulation.* Cambridge, Eng.: Cambridge University Press.

Simon, J. L. (1996). *The Ultimate Resource 2.* Princeton, NJ: Princeton University Press.

ENERGY INTENSITY TRENDS

Energy intensity is defined as the ratio of energy used to some measure of demand for energy services. There is no one measure of energy intensity. The measure depends on the universe of interest being measured. If the universe of interest is an entire country, then the measure needs to reflect the energy used and the demand for energy services for that country. A country's demand for energy services is usually measured as the dollar value of all goods or services produced during a given time period, and is referred to as Gross Domestic Product (GDP). If the universe of interest is households, energy intensity might be measured as energy used per household or size of the housing unit. If the universe of interest is the iron and steel industry, the measure for energy intensity could be energy used per ton of raw iron produced or energy per dollar value of the raw iron.

Energy-intensity measures are often used to measure energy efficiency and its change over time. However, energy-intensity measures are at best a rough substitute for energy efficiency. Energy intensity may mask structural and behavioral changes that do not represent "true" efficiency improvements. A shift away from producing products that use energy-intensive processes to products using less-intensive processes is one example of a structural change that might be masked in an energy-intensity measure. It is impossible to equate one energy-intensity measure to some "pure" energy efficiency. Therefore a set of energy-intensity measures should be developed—keeping in mind the caveats underlying the measures.

ENERGY-INTENSITY TRENDS IN THE UNITED STATES, 1972 TO 1986

Before the 1970s the United States experienced a time of falling energy prices and ample supplies of petroleum. In 1973, crude petroleum prices shot up by 400 percent. In the early 1980s, the growth of economic activity outpaced the demand for energy. Between 1972 and 1986, energy consumption per dollar of GDP declined at an average annual rate of 2.1 percent. Energy per dollar of GDP is a useful measure. It is important, however, to understand the factors that lie behind any changes in the measure.

At first, short-term behavioral changes, such as lowered thermostats and reduced driving, were common in reducing energy demand, but their overall effects on demand were small. While these transient changes were taking place, other, more fundamental changes were working their way into energy-using processes. Examples were the introduction of automobile fuel economy standards, appliance efficiency standards, and the movement away from energy-intensive processes in the manufacturing sector.

In 1975, Congress responded to the oil crisis of 1973 by passing the Energy Policy and Conservation Act. This legislation established Corporate Average Fuel Economy (CAFE) standards. By 1985, CAFE standards required that all new passenger cars had to have an

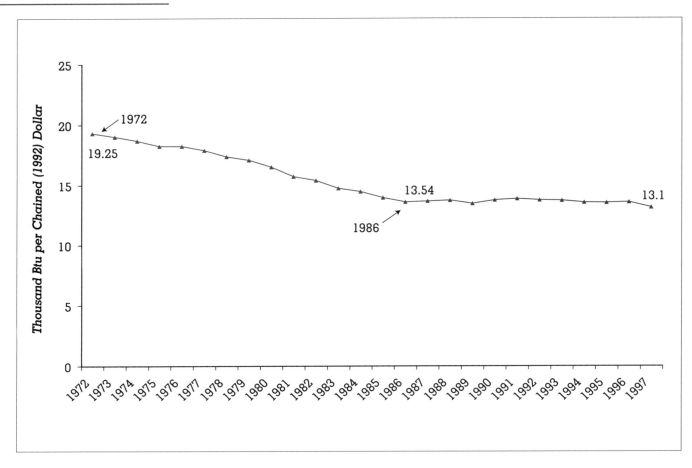

Figure 1.

U.S. Energy Consumption per Dollar of GDP, 1972–1997

source: Energy Information Administration, Annual Energy Review 1997.

average gas mileage of 27.5 miles per gallon (mpg) of gas, nearly double the mileage of typical cars made in the mid 1970s. Light trucks had to average 20.6 mpg.

Additionally, as a response to rising energy prices and uncertainty of supply, several states adopted appliance efficiency standards. At the federal level, the National Appliance Energy Conservation Act of 1987 established the first national standards for refrigerators and freezers, furnaces, air conditioners, and other appliances. The Energy Policy Act of 1992 added national standards for incandescent and fluorescent lights, small electric motors, office equipment, and plumbing products.

The movement away from energy-intensive processes in the manufacturing sector was an important force in the reductions in energy intensity during these years. One of the most noticeable shifts was in

the primary metal industry. In 1976, 128 million tons of raw iron were produced by 1986 production had fallen to 81.6 million tons. Domestic primary production of aluminum was 3.7 million metric tons in 1972, but fell to 3 million metric tons in 1986.

ENERGY-INTENSITY TRENDS IN THE UNITED STATES, 1986 TO 1997

After 1986, the CAFE and appliance standards in place resulted in stock turnovers to more efficient automobiles and appliances. However, the decline in energy consumption per dollar of GDP slowed appreciably and, between 1986 and 1997, the energy intensity trend remained rather flat. Other forces in the U.S. economy were pushing energy consumption higher, resulting in increases in the energy-intensity measure.

Between 1987 and 1997, twelve million households were added to the country's housing stock, representing a 13 percent increase. Since most of this increase took place in the West and the South, the demand for electricity for central air conditioning increased as well. In 1987, 52 percent of all households in the South had central air conditioning. In 1997, this percentage was 70 percent—a 35 percent increase. Additionally, new households have been getting larger, resulting in an increased demand for heating, air conditioning, lighting, and appliances. New energy-consuming devices such as VCRs, microwaves ovens, and home computers were also purchased on a wide scale.

The energy intensity measure, miles per gallon, for the stock of passenger cars increased from 17.4 in 1986 to 21.3 in 1996. However, average miles driven per car per year increased from 9,464 to 11,314. There has also been a change in the mix of new vehicles purchased. In 1980, 38 percent of all new vehicles were subcompacts, falling to 18 percent in 1997. The market share for small vans was less than 1 percent in 1980, but grew to 19 percent in 1997. Small utility vehicles had a market share of 3.4 percent in 1980, growing to 26 percent in 1997. Although the larger vehicles became more efficient, more were being purchased and were being driven more.

Although the manufacturing sector continued its decline in the production of energy-intensive products, between 1986 and 1997 the service sector continued to grow. Not only did the commercial building stock increase, but the use of office equipment—from computers to copy machines—has grown rapidly. In just three years, between 1992 and 1995, the number of personal computers and computer terminals in commercial buildings increased from 29.8 million to 43 million (45%).

Clearly the United States is doing more with less energy. Although total energy has grown as demand for goods and services has climbed, energy use per person has hardly changed, 348 million Btu per person in 1972 and 352 million Btu in 1997. During this same time period, energy per GDP has declined 32 percent. State and federal energy-efficiency standards, consumer behavior, and structural shifts, have fueled this decline.

Stephanie J. Battles

See also: Appliances.

BIBLIOGRAPHY

The Aluminum Association. (1997). *Aluminum Statistical Review for 1996.* Washington, DC: Department of Economics and Statistics.

American Iron and Steel Institute (1998). *Annual Statistical Report 1988.* Washington, DC: Office of Energy Markets and End Use.

Energy Information Administration. (1995*). Measuring Energy Efficiency in the United States' Economy: A Beginning*, DOE/EIA-0555 (95)/2. Washington, DC: Office of Energy Markets and End Use.

Energy Information Administration. (1998). *Annual Energy Review 1997*, DOE/EIA-0384 (97). Washington, DC: Office of Energy Markets and End Use.

Heal G., and Chichilnisky, G. (1991). *Oil and the International Economy.* Oxford, England: Clarendon Press.

International Energy Agency. (1997). *Indicators of Energy Use and Efficiency.* Paris, France: Organization for Economic Cooperation and Development.

U.S. Bureau of the Census. (1997). *Statistical Abstract of the United States,* 117th ed. Washington, DC: U.S. Government Printing Office.

U.S. Department of Energy. (1998). *Transportation Energy Data Book: Edition 18.* ORNL-6941. Oak Ridge, TN. Oak Ridge National Laboratory

ENERGY MANAGEMENT CONTROL SYSTEMS

EVOLUTION PERSPECTIVE OF EMCS

The primary purpose of energy management control systems (EMCS) is to provide healthy and safe operating conditions for building occupants, while minimizing the energy and operating costs of the given building. Aided by technological developments in the areas of electronics, digital computers, and advanced communications, EMCS have been developed to improve indoor quality while saving more energy.

Electromechanical Timers

The earlier devices used in the first half of the twentieth century to control building loads (such as lighting and space conditioning) were electromechanical timers, in which a small motor coupled to a gearbox was able to switch electrical contacts according to a predefined time schedule. Normally the output shaft of the gearbox causes one or more pairs of electrical contacts to open or close as it rotates. These electromechanical devices were simple and reliable, and they are still used to control lights and ventilation in some buildings. However, for lighting and space conditioning, the main drawbacks of this type of timer was inflexibility. Manual intervention is required to change settings, and the operation mode is essentially without feedback (open-loop controllers), since the schedule of operation is not easily influenced by the variables in the controlled process.

Electronic Analog Controllers

The development of electronic circuitry capable of processing sensor signals made possible the appearance of electronic controllers able to respond to variable conditions. For example, to control street lighting a light sensor coupled to a simple electronic amplifier and switch can turn off the lights during the day, avoiding energy waste. In some cases the analog electronic controllers were coupled with electromechanical timers, providing the possibility of controlling the output as a function of time and/or operation conditions.

Although electronic circuitry has been available since the beginning of the twentieth century, the invention of the transistor in 1948 was a key milestone, bringing a decrease in costs, an improvement in reliability, and a reduction in the size of control circuits. During the 1950s and 1960s, electronic analog controllers became widely available for controlling lighting, heating, ventilation, and air conditioning (HVAC). Although these controllers made possible improved control based on information from sensor feedback (closed-loop control), they still suffer from lack of flexibility, since they are able to implement only simple control strategies and require manual change of settings.

Digital Controllers

The invention of the microprocessor in 1970 signaled a radical change in the area of building controls, allowing the development of increasingly powerful EMCS. From the outset, microprocessor-based EMCS could be easily programmed for changeable and variable time schedules (e.g., workday versus weekend operation, public holiday operation). In addition to this flexibility, increasingly powerful microprocessors, along with new energy management hardware and software, allowed for the pro-

gressive implementation of sophisticated control algorithms, optimization of building operation, and integration of more functions into building control and supervision.

Most new hardware and software development in the EMCS market now aims at utilizing the full potential of EMCSs and at making the information obtained more usable and accessible. This accessibility has created a secondary benefit with a greater potential for customer/utility communication. In many ways the EMCS evolution and market penetration in the industrial, commercial, and to a certain extent, residential markets, has equaled that of the personal computer.

During the 1970s EMCS digital controllers were mostly electronic time clocks that could be programmed according to a variable schedule. However, increasingly powerful, low-cost microcomputers have rapidly improved computing power and programmability and hence increased the application of EMCS. This rapid improvement has in turn introduced benefits such as the following:

- flexible software control, allowing simple modification;
- built-in energy management algorithms, such as optimal start of space conditioning, limiting the electricity peak demand; also, control of HVAC and lighting loads in practically all EMCS applications;
- integrated process, security, and fire functions;
- monitoring capability allowing identification of faulty equipment and analysis of energy performance;
- communication with the operator and other EMCS that can be remotely located.

These possibilities make EMCS increasingly attractive for energy monitoring, control, and utility/customer interface. The integration of functions associated with safety and security (e.g., intrusion and fire) as well as those associated with building diagnostics and maintenance are also expanding the potential market of EMCS. The application of advanced EMCS with these features is leading to the appearance of "smart buildings," which can adjust to a wide range of environmental conditions and which offer comfort and security with minimal use of energy.

CURRENT EMCS CHARACTERISTICS

An EMCS is usually a network of microprocessor-based programmable controllers with varying levels of

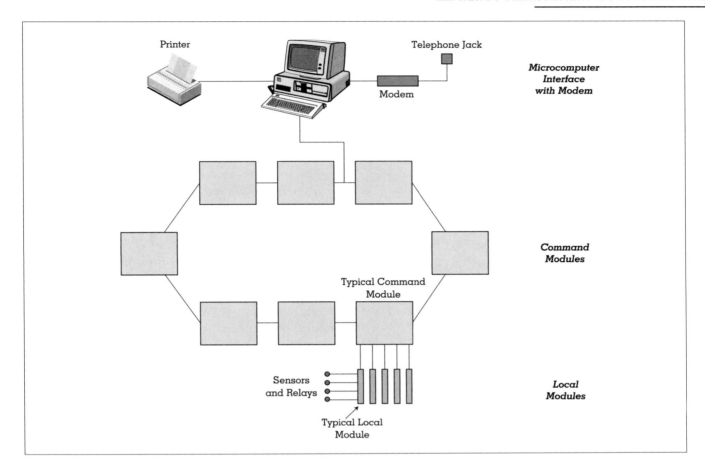

Figure 1.
Schematic of a typical EMCS with local control modules, command control modules, and personal computer interface.

intelligence and networking capabilities. This "typical" EMCS has three components (see Figure 1):

- Local control module: directly wired to sensors and actuators (e.g., equipment to control space conditioning, such as pumps and fans);
- Command control module: makes control decisions;
- Personal computer interface: simplifies operator control of the system through a user-friendly interface.

An important feature of the typical system is its modularity. The most powerful EMCS installations have all three components, but often only a single module is necessary for simple applications, such as controlling a single air-conditioning unit, or for most applications in the residential sector. Thus local and control modules are capable of stand-alone operation without higher-level components. The functions,

hardware, software, and communication characteristics of these three components are as follows.

Local Control Modules

A local control module performs the following basic functions in an EMCS:

- receives information from sensors;
- controls actuators, through relays, to switch equipment on and off or to change its variable output;
- converts analog sensor data to a digital form;
- performs direct digital control;
- communicates with the command module.

Modern EMCS use a variety of sensors, including temperature, humidity, occupancy, light, pressure, air flow, indoor air quality, and electric power (normally pulses from power meters). The actuators are

465

the units that can influence the state of the system, including chillers, pumps, fans, valves, and dampers.

Command Control Module

The command control module, or control module, is the real intelligence of the EMCS; all programming and control software resides here. Command control module software can create reports (e.g., for recording the historical status of variables) and perform various demand-limiting schemes. The common energy management strategies offered at this level include proportional-integral-derivative (PID) control loops, duty cycling, and optimal start/stop of HVAC units. Also available are economizer control (use of free cooling with outside air when the outside temperature is below the target temperature) and programmed start/stop with demand limiting of selected loads.

The command module can be programmed either through a keypad or by downloading programs from the PC host. The information presented in command module reports is easy to change, but it must follow a prearranged format. These reports usually show temperatures, peak demand, whole-building energy, equipment status, maintenance records, and alarm records. Automatic reporting of a system alarm, such as machine failure, by phone or by electronic mail is another typical software feature.

Command modules communicate with other modules through a local area network (LAN). Through this LAN, command modules receive information from the local control modules and store data. These data can be stored from a week to two years, depending on the recording interval and the number of points to be monitored. Unlike host-based systems, which use a central computer to interrogate each command module individually, the computer interface can tap into the network like any other command module.

A recent data communications protocol for Building Automation and Control Networks (BACnet), ASHRAE Standard 135-1995, is an important step to ensure that controllers made by different manufacturers can communicate with each other in a simple way, avoiding the expense of additional interface hardware and communication software.

Personal Computer Interface

The personal computer interface allows for easy operation of an EMCS, but all of the system's control functions can be performed in its absence. This user interface serves three purposes:

- storage of a backup of the command module's programs to be used in case of power or system failure;
- archiving of trend data for extended periods of time;
- simplification of programming and operation of the system through a user-friendly interface.

The software at the user-interface level usually involves block programming, whereby the operator can define control loops for different sensors as well as other specific control strategies in a BASIC-type, easily understood programming format. This block structure allows modification or enabling of different program blocks, avoiding the necessity of rewriting the main program. Programs at this level are menu-driven, with separate sections for trend reports, programming, graphics, etc. Visual displays, such as the schematic of HVAC equipment interconnections, are used to make information easier to comprehend. Building floor plans can be incorporated to display information such as room temperatures.

INTEGRATION OF FUNCTIONS

Because of increasing computation power, EMCSs can integrate other functions and also be coupled to other processes. By helping coordinate plant operation, the integration of functions also makes the acquisition of EMCSs more attractive, since a significant portion of the hardware can be shared by different applications. For tracking the performance of the building operation and carrying out the required maintenance, monitoring and data-logging functions are essential in an integrated system, since operators need to be aware of the operating status of each part of the plant and to have access to reports of previous performance. Also, alarm display and analysis is a very convenient function for carrying out diagnostics that can be performed by an integrated system. Safety and security functions can be easily and economically integrated, although some installations prefer separate systems due to potential litigation problems.

The use of peak electricity demand-reducing technologies, such as thermal storage and peak-shaving involving the use of a standby generator, also can benefit the control and monitoring capabilities of EMCSs. In thermal storage (ice or chilled water for cooling), the EMCS can schedule the charging of the system during off-peak hours to optimize savings in both peak demand and energy costs. Temperatures in the storage system are monitored to minimize energy and demand costs without adversely affecting plant operations or products. Thermal storage, already an attractive option for space cooling in commercial and industrial buildings and in several food industries (dairy, processed meat, fish) and for space cooling, has become an even more effective energy saver thanks to EMCSs. An EMCS also can determine when the use of an existing standby generator during periods of peak demand can reduce costs, taking into account the load profile, demand and energy costs, hour of the day, fuel costs, etc. The standby generator also can be put on-line on request from the utility in times of severe peak demand or loss of generating capacity reserve margins.

EMCS FOR ENERGY SAVINGS

Typical savings in energy and peak demand are in the range of 10 to 15 percent. These savings are achieved by reducing waste (e.g., switching off or reducing the lights and space conditioning in nonoccupied spaces), by optimizing the operation of the lighting (e.g., dimming the lights and integration with natural lighting), and space conditioning (e.g., control of thermal storage).

THE EMCS FOR REAL-TIME PRICING

State-of-the-art systems with flexible software allow for utility interface. These systems are all currently capable of data monitoring and of responding to real-time pricing, depending only on the software installed on the user-interface computer and the number of sensors the customer has installed for end-use monitoring.

To reduce use of electricity when electricity costs are highest, EMCS use a network of sensors to obtain real-time data on building-operating and environmental conditions. Some large electricity consumers are connected with the utility through a phone line for the communication of requests to reduce the peak electricity load and for present and forecasted demand. The building operator traditionally closes the link, instructing the EMCS to respond to the utility's signals. However, this current manual load shedding/shifting response to utility prices is too labor-intensive and operationally inefficient for large-scale implementation.

Energy management systems have been developed that can control loads automatically in response to real-time prices. Real-time prices are sent to the customer, whose EMCS can modulate some of the loads (e.g., air conditioning, ventilation, nonessential lighting). Thus peak demand can be reduced at times of high electricity price, maintaining all essential services. An EMCS could be used to modulate HVAC load by controlling temperature, humidity, volatile organic compounds (VOCs), and carbon dioxide levels within a window of acceptance, the limits of which may be adjusted as a function of the real-time prices. In theory this strategy can save energy and substantially decrease peak demand. The most attractive candidates for ventilation control performed by EMCSs are: large commercial buildings with long thermal time constants (or with thermal storage), buildings with low pollutant emission from building materials, buildings that house furnishings and consumer products, and buildings that require a large volume of ventilation (or circulation) air per occupant.

THE EMCS FOR DATA MONITORING

Parallel to the increase in performance of microcomputers in the late twentieth century been an exponential decrease in the price of semiconductors and memory. This means that data-logging functions can be added at a small extra cost, considerably increasing the usefulness of a given EMCS.

Because monitoring and data-logging facilities add to initial cost, some customers may be hesitant to choose between a simple configuration without those capabilities and a more complex model including them. The benefits associated with monitoring often outweigh the price premium, as they allow the user to:

- tune the performance of a system and check energy savings;
- record the load profile of the different operations or major pieces of equipment within the plant;
- find the potential for improvements;
- allocate energy charges to each operation or product line;
- submeter electricity end use, allowing more accurate charging of costs to specific processes or divisions;
- check worn or faulty equipment (e.g., declining fan pressure in an air-handling unit may mean a clogged filter);
- check utility meter readings (potentially a particularly useful feature in plants using power electronics con-

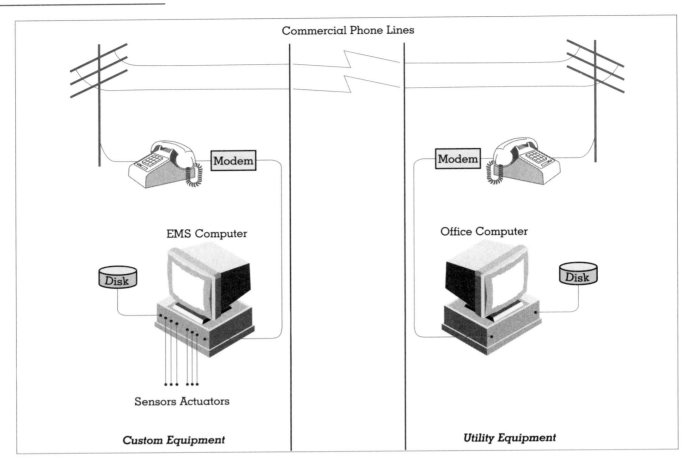

Figure 2.
EMCS-based monitoring of building and end uses. A utility computer can interface the customer EMCS and obtain building and energy-use information.

trol devices, which can generate errors in conventional power meters).

Figure 2 shows a diagram of the layout of EMCS-based end-use monitoring in which the data collected by the EMCS can be remotely monitored by the utility or by the maintenance staff in the company's central office.

The load data can be disaggregated into main uses by using suitable algorithms that take into consideration sensor information, plant equipment, and plant schedule. Figure 3 shows hourly electricity use of an office building, broken down by major end uses. This type of data analysis can be useful for monitoring building performance and for identifying opportunities to save energy at peak demand.

TRENDS IN THE EVOLUTION OF ENERGY MANAGEMENT AND CONTROL SYSTEMS

Future trends for the evolution of EMCS are likely to involve improvements in user interface, easier access, better controls, and advances in integration, namely including the following:

- better access to system information, with more remote diagnosis and maintenance capabilities;
- easier installation and programming, with advanced graphics user interface;
- smaller, more distributed controllers and unitary control for more advanced energy management optimization;
- easier integration of products from different manufacturers, and continuing effort toward communication standardization;

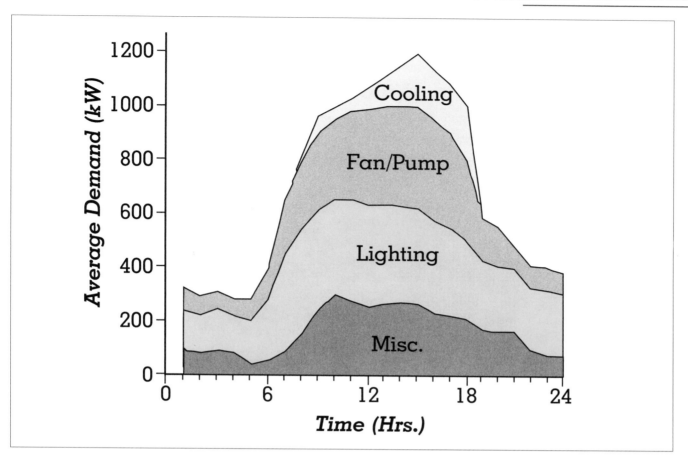

Figure 3.
An example of disaggregated hourly end-use data obtained from whole-building load.

• less integration of fire and security functions.

One of the most significant EMCS trends is toward better access to the information gathered by the EMCS. The EMCS industry is developing smaller, distributed units with increased programmability and control capabilities. These smarter local units offer increased system versatility and reliability and allow smaller units to perform functions that would previously have required larger, more expensive systems. Another trend in EMCS hardware development has been the integration of sensors and controllers from various manufacturers. The work performed toward the development of a standard communications protocol (BACnet) is of the greatest importance in ensuring communication compatibility between and among equipment made by various manufacturers.

One system feature of particular relevance is remote troubleshooting. For example, in case of an alarm signal in the building, the user can trace back through the system to find the cause of the alarm. In fact, manufacturer representatives can perform much of the routine troubleshooting over the phone. Whenever an operator has a software problem, the representatives can call up the system to correct programming problems or help develop new applications. Both preventive maintenance (carried out at programmed intervals of operation time) and predictive maintenance management (carried out when the plant sensors detect a deterioration in the equipment performance) can be incorporated in powerful EMCS, including databases containing details (even images) of the spare parts required for maintenance.

The creation of graphics can be menu-driven, often utilizing a building floor plan or system schematic to display the collected data. The floor plan is first drawn by the customer, and then variables, such as current room temperature, are superimposed.

Optical scanners also can allow easier graphics creation, and user-selected video frames can be incorporated into the software displays.

Voice communication capabilities, including voice recognition and speech synthesis, are also being increasingly used to provide a simpler user interface. Thus, for example, verbal instructions can be given for resetting set points (temperature, etc.), or to request other actions from the system. Fire and security monitoring is one EMCS feature, which may require its own dedicated system, although this leads to higher costs. This trend is conditioned by insurance and liability issues.

Anibal T. de Almeida
Hashem Akbari

See also: Efficiency of Energy Use; Electric Power, System Protection, Control, and Monitoring of; Energy Economics; Industry and Business, Productivity and Energy Efficiency in; Risk Assessment and Management.

BIBLIOGRAPHY

Akbari, H.; de Almeida, A. T.; Connell, D.; Harris, J.; and Warren, M. (1988). "Use of Energy Management Systems for Performance Monitoring of Industrial Load-Shaping Measures." *Energy, The International Journal* 13(3):253–263.

Akbari, H.; Heinemeier, K.; Le Coniac, P.; and Flora, D. (1988). "An Algorithm to Disaggregate Commercial Whole-Building Electric Hourly Load into End Uses." *Proceeding of the ACEEE 1988 Summer Study on Energy Efficiency in Buildings* 10:13–26.

American Society of Heating, Refrigeration, Air-Conditioning Engineers (ASHRAE). (1995). *ASHRAE Standard 135-1995: BACnet—Data Communication Protocol for Building Automation and Control Networks.* Atlanta: Author.

Electric Power Research Institute (EPRI). (1985). *Control Strategies for Load Management.* Palo Alto, CA: Author.

Electric Power Research Institute (EPRI). (1986). *Energy Management Systems for Commercial Buildings.* Palo Alto, CA: Author.

Fisk, B., and de Almeida, A. T. (1998). "Sensor-Based Demand Controlled Ventilation." *Energy and Buildings* 29(1):35–44.

Heinemeier, K., and Akbari, H. (1992). "Evaluation of the Use of Energy Management and Control Systems for Remote Building Monitoring Performance Monitoring." *Proceedings of the ASME International Solar Energy Conference.*

ENGINES

An engine is a machine that converts energy into force and motion. Possible sources of energy include heat, chemical energy in a fuel, nuclear energy, and solar radiation. The force and motion usually take the form of output torque delivered to a rotating shaft. (Torque is the twisting effort developed around a center of rotation. In an engine it is conceptually quantified by the product of the radius from the center of the output shaft to a second point at which a tangential force is applied, and the magnitude of that applied tangential force.) In contrast, the output of a jet or a rocket engine is simply the thrust force derived from its high-velocity exhaust jet.

Most often the input energy to an engine is derived through combustion of a fuel. The result is a combustion engine. Combustion engines can be classified according to the nature of their combustion and its initiation. Possible combinations are depicted in Figure 1. First, combustion may occur either continuously or intermittently. Second, that combustion may occur either external to the engine or internally, within the engine.

In the continuous external-combustion engine, a fuel is burned outside the confines of the machine responsible for the conversion of energy into useful work. The heat energy generated through combustion is then transferred into a working medium that undergoes a repetitive cycle of thermodynamic processes. The portion of this engine that converts the heat energy into work is appropriately termed a heat engine because the input heat energy need not necessarily come from combustion. The second law of thermodynamics states that not all of the heat energy transferred into the working medium of the heat-engine cycle can be converted into output work. The first law of thermodynamics states that the difference between the heat transferred into the engine cycle and the work produced as a result of that cycle must be rejected to the surroundings.

The steam engine used by utilities to generate electricity is an example of a continuous external-combustion engine. A fuel—usually coal, oil, or natural gas—is burned to generate heat. That heat is transferred into the working medium of the engine cycle—namely, water—through a heat exchanger known as a boiler or a steam generator. The engine cycle produces work in the form of torque on a rotating shaft, usually by means of a steam turbine. The difference between the heat added to the cycle in the steam generator and the net work produced by the turbine is rejected to the environment in a condenser that returns the steam to water at its original pressure and temperature for a repetition of the cycle. Other examples of the continuous external-combustion engine include the Stirling engine and the closed-cycle gas turbine. In these engines the working medium remains in the gaseous phase throughout the cycle and is often hydrogen, helium, or some gas other than air.

Because the steam engine described is a heat engine of the external-combustion type, the cycle experienced by the working medium can be executed without combustion. In some steam engines, for example, the required input heat is supplied by a nuclear reactor. Stirling engines have been operated on radiant energy supplied by the sun.

The dominant continuous internal-combustion engine is the gas turbine, which is used in both torque-producing and thrust-producing applications. It involves continuous compression in an aerodynamic compressor; continuous combustion in a burner that in principle resembles a household oil burner; and continuous work extraction in a turbine that both drives the compressor and, in torque-producing configurations of the engine, delivers engine output. In applications where fuel economy is of great importance, a heat exchanger may be added that transfers heat from the turbine exhaust gas to the burner inlet air to preheat the combustion air. The torque-producing gas turbine is used for electric power generation, ship propulsion, in military tanks, and as an aircraft turboprop (propjet) engine. In on-road automotive applications, the gas turbine has never progressed beyond the demonstration stage because it has not been commercially competitive with existing automotive piston engines.

In thrust-producing gas turbines, the turbine extracts only enough energy to drive the compressor and engine accessories. The remaining available energy is converted to a high-velocity exhaust jet that provides a forward thrust. The aircraft turbojet and its cousin the turbofan (fanjet) embody this concept. The turbofan differs from the turbojet in that the turbine also drives a low-pressure-rise compressor, whose airflow bypasses the burner and the turbine

and joins the turbine exhaust in the jetstream. This results in the production of thrust by a jet that has a higher flow rate but a lower velocity than the turbojet. As a result, the turbofan offers higher propulsion efficiency at a lower aircraft speed than the turbojet. The turbojet and turbofan have driven the large piston engine once used in passenger airliners and military aircraft into obscurity.

Despite the existence in Figure 1 of an intermittent external-combustion engine as a possibility, no such engine is known to have been placed in service. However, the intermittent internal-combustion engine is another matter. It is applied to everything from chain saws and lawn mowers to large trucks, locomotives, and huge oceangoing vessels. The intermittent internal-combustion engine can be further subdivided as to the method used for initiating combustion and the nature of the air-fuel charge that is prepared for that ignition.

The most common internal-combustion engine in use is the homogeneous-charge spark-ignition (HCSI) engine. The ubiquitous reciprocating-piston gasoline engine serves as an example. In it, fuel and air are premixed upstream of the cylinder or cylinders with the objective of supplying to the engine a homogeneous mixture of the two. Combustion is initiated at the appropriate time in the engine cylinder by discharging an electric spark in the entrapped air-fuel mixture. The hot products of combustion expand, generating useful work in the form of force on the moving piston. That force is transformed into torque on a rotating shaft by the engine mechanism. The spent combustion products are exhausted to the environment as heat energy and replaced by a fresh air-fuel charge for repetition of the engine cycle. The internal-combustion engine operates on a repetitive mechanical cycle that consecutively repeats the events comprising it. However, it does not follow a true thermodynamic cycle, as is done in the heat engine, because the working medium is never returned to its original state. Instead, the working medium is exhausted from the engine as products of combustion, to be replaced by a fresh charge.

Although this description of the reciprocating-piston engine fits the vast majority of gasoline engines in service, the same cycle of events can be executed in a variety of other kinematic arrangements. One such alternative approach is the rotary engine, which avoids the oscillatory force production of the recipro-

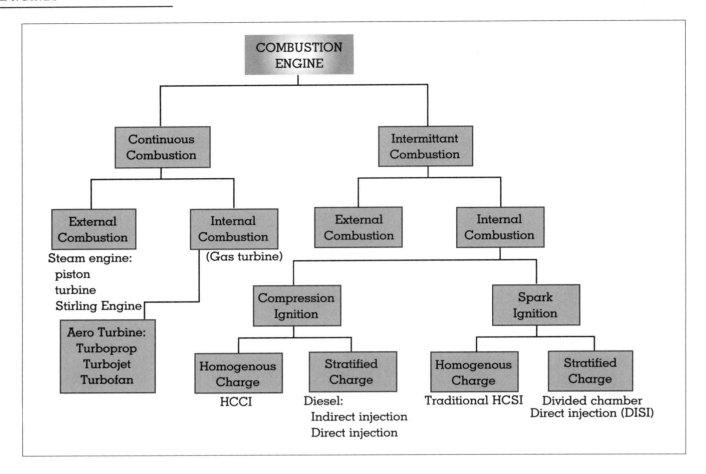

Figure 1.
Combustion engine tree.

cating-piston. The best known rotary engine is the Wankel engine. Just like the reciprocating-piston engine, it consecutively effects an intake of fresh mixture, combustion of that mixture, expansion of the resulting hot gasses to produce output work, and expulsion of the hot exhaust products to prepare for repetition of the cycle. The varying cylinder volume of the reciprocating-piston configuration is mimicked in three chambers, each enclosed by one flank of a three-sided rotor and the inner wall of a specially shaped stationary housing within which that rotor turns.

In the stratified-charge spark-ignition engine, homogeneous mixing of the air and fuel is avoided in favor of creating a mixture in the region surrounding the spark plug that is richer in fuel than the cylinder average. One approach to charge stratification once experiencing limited application in automobiles was the divided-chamber engine, in which a richer-than-average mixture was inducted into a prechamber connected to the main chamber above the piston by a restrictive passageway. A spark plug in the prechamber ignited the segregated rich mixture, which then sent a jet of burning gas into the main chamber to serve as a powerful ignition source for a leaner-than-average mixture inducted into that chamber.

The developing trend is to eliminate the prechamber and stratify the charge by injecting the fuel directly into the cylinder. This approach, known by several different names but adequately described as the DISI (Direct-Injection-Stratified-Charge) engine, involves careful control of both injection timing and in-cylinder air motion. Charge stratification of this nature facilitates using a cylinder-average air-fuel ratio that is higher (leaner) than could normally be expected to burn satisfactorily if the air and fuel were mixed homogeneously. Burning such an overall-lean mixture has the potential for better fuel economy than the traditional homogeneous-charge engine provides.

The compression-ignition stratified-charge engine is commonly known as the diesel engine. It has also been implemented both with and without a divided chamber. In the divided-chamber, or indirect-injection (IDI) version, fuel is injected only into the prechamber, where combustion begins. The direct-injection (DI) configuration has no prechamber, with fuel being injected directly into the volume above the piston. In either case, the fresh charge of air is compressed to a sufficiently high pressure and temperature that the injected fuel autoignites. Such autoignition sites are generally found on the periphery of the injector spray plume and eliminate the need for a spark plug.

On the century-old time scale of the internal-combustion engine, the homogeneous-charge compression-ignition (HCCI) engine is a comparative newcomer. It involves preparation of a homogeneous charge upstream of the engine cylinder or cylinders. That charge is then autoignited in the cylinder by compression, as in the diesel, with combustion beginning at many distributed ignition sites essentially concurrently. If the mixture is too fuel-rich, combustion occurs too quickly and generates excessive noise. If the mixture is too fuel-lean, combustion becomes incomplete and erratic. Typically, exhaust gas is recirculated into the inlet mixture to help control combustion. Compression ignition of the homogeneous charge cannot be used over the complete operating range of the engine. Currently a popular subject of research study, the homogeneous-charge compression-ignition concept has seen very limited commercial application to date.

Charles A. Amann

See also: Aircraft; Combustion; Diesel Cycle Engines; Gasoline Engines; Spacecraft Energy Systems; Steam Engines; Stirling Engines; Thermodynamics; Turbines, Gas; Turbines, Steam.

BIBLIOGRAPHY

Amann, C. A. (1980). "Why not a New Engine?" *SAE Transactions* 89:4561–4593.

Heywood, J. B. (1988). *Internal Combustion Engine Fundamentals*. New York: McGraw-Hill.

ENTHALPY

See: Thermodynamics

ENTROPY

See: Thermodynamics

ENVIRONMENTAL ECONOMICS

In dealing with environmental questions, economists emphasize efficiency, social welfare, and the need for cost accountability. A basic principle for efficiency is that all costs be borne by the entity who generates them in production or consumption. For example, production and consumption of diesel fuel will be socially inefficient if significant resulting costs are shifted to others who happen to be downwind or downstream from the refinery that makes the fuel or the truck that burns it. The benefits of making and using the fuel should exceed the cost—society at large—or else the process reduces total social welfare.

Information is the key to such internalizing "external costs." If the generator of pollution damage is known, along with the victim and the size of the damage, then the polluter can be held accountable. Historically, if wrongful damage is done, courts in the United Kingdom, the United States, Canada, and nations with similar legal systems have been willing to force compensation by polluters or, when the damage is great enough, to order cessation of the pollution. Small damage is ignored. But if no one knows whether the damage is serious, or who caused it, then regulation may be instituted to cope with it. However, since the regulator may not know more than courts could learn at trial, the results of regulation vary from increasing efficiency to reducing it. An analysis of the problems facing a regulator provides context for examining the key issues in environmental economics.

Cost accountability will be most prevalent when property rights to land and resources are clearly defined. Clear property rights make the owners of a resource responsible, in most countries, for the way that resource is used and for the harms it may cause others. An owner's right to the use of property does not include the right to use it in ways that impose a cost on others. In such cases, courts have historically held owners of a polluting plant or business responsible for harm they may cause other parties. Clear property rights make owners face the cost of inefficient use of a resource and thus encourage owners to ensure that their property or equipment is put to the most highly valued use. Property rights provide what economists consider the *incentives* to ensure that resources are used efficiently (maximizing net value) and in a way that constrains negative impacts on other individuals.

When ownership rights are less well defined, or not easily defended, the incentives for resource owners to efficiently use resources in a safe, non–polluting way, are decreased or removed. Individuals are less likely to take expensive, time-consuming action to protect a resource that they do not own, and by which they are not directly affected financially. For instance, most landowners would be quick to take action to prevent garbage generated by a local business from piling up in their own backyard. However, they would be less likely to take actions to prevent the same business from polluting a nearby lake or river. The reason is that any one individual has less direct, or at least less obvious, interest in the lake or river, than they do in their own property—and usually less ability to affect the outcome.

This incentive problem is increased as the number of polluters and the number of land owners increase so that it is difficult to pinpoint specific incidents of pollution and their effect on individuals. The case of air pollution from cars and multiple factories is a classic example of this information problem. In a large metropolitan area, there are millions of automobiles and many factories that could contribute to air pollution. There are also millions of individuals who could be harmed by that pollution. But it is generally difficult, if not impossible, for one individual to identify a specific problem they have experienced due to air pollution and then to pinpoint the source of that problem. Such situations, where property rights are not well defined, as is the case with air, and where it is difficult to identify a particular source of pollution, often lead to calls for government regulations to prevent a certain activity, or to reduce a certain activity such as exhaust from automobiles, in an effort to prevent harm to others.

Just as in the decision-making process of individual land owners, incentives are important in the government decision-making process. Economists have identified a characteristic of the government decision-making process that can allow the concerns of special interest groups can take precedence over the interest of the general public. The principle of rational voter ignorance states that since the cost of obtaining information about political issues is high, and any individual voter is likely to pay a small portion of the cost as well as reap a small portion of the benefit of any government action, individual voters are not likely to take the time to become well informed on specific issues. In contrast, politically organized special interest groups, such as firms in a polluting industry, will pay a heavy price for any new regulations that might be directed toward them. Therefore, they have a financial incentive not only to be well informed on the issues affecting them, but also to spend time and money trying to influence the government to ensure that they do not bear the cost of regulations.

With strong incentives for businesses and other special interest groups, and weak incentives for individual voters, it is not surprising that many environmental regulations have often been less successful at preventing harm than the more traditional property rights-based approaches. Thus, while privately owned lands and resources are generally healthy and well preserved, many resources that are not owned, such as air or many waterways, are polluted.

Many economists have therefore become disappointed in the effectiveness of traditional regulatory solutions to environmental problems. They look to market incentives such as those provided by private property rights, and market-like mechanism, where polluters must bid for or trade for the right to release potentially harmful emissions, as policy alternatives.

Richard L. Stroup

See also: Acid Rain; Air Pollution; Atmosphere.

BIBLIOGRAPHY

Brubaker, E. (1995). *Property Rights in Defence of Nature*. Toronto: Earthscan Publications.

Field, B. C. (1997). *Environmental Economics: An Introduction.* Boston: Irwin McGraw-Hill.

Wills, I. (1997). *Economics and the Environment: A Signalling and Incentives Approach.* St. Leonards, NSW, Aust.: Allen & Unwin.

ENVIRONMENTAL PROBLEMS AND ENERGY USE

For millions of years humans existed in harmony with nature. But the Industrial Revolution, and the exploration and development of energy to fuel that revolution, began a period of ever-growing fossil fuel combustion that resulted in greater water pollution, air pollution, deforestation, and growing atmospheric carbon dioxide concentrations. Nuclear energy, which was supposed to be the solution to these problems associated with fossil fuel production and combustion, turned out to present an equally great threat to the environment in terms of safety and waste disposal. Because almost no energy source is totally benign, and all the major energy sources have unwanted drawbacks, billions of dollars are being spent each year on scientific research to find ways to lessen the environmental impact of the sources in use, to make the more benign sources more cost-competitive, to improve the energy efficiency of technology, and to determine if the high energy-consuming habits of humans are putting the planet in peril by irreversibly and harmfully altering the atmosphere.

THE ENERGY-ENVIRONMENT LINK

Environmental protection and resource use have to be considered in a comprehensive framework, and all of the relevant economic and natural scientific aspects have to be taken into consideration. The concepts of "entropy" and "sustainability" are useful in this regard. The entropy concept says that every system will tend toward maximum disorder if left to itself. In other words, in the absence of sound environmental policy, Earth's energy sources will be converted to heat and pollutants that must be received by Earth. The concept of sustainability has to do with the ability of a population to engage in economic activity and energy development without creating future irreversible problems for the environment and therefore for the economy.

The energy issues associated with environmental protection are complex ones involving trade-offs. Energy is an integral aspect of civilization. People in the industrialized nations use energy for heating, cooling, lighting, cooking, entertainment, transportation, communication, and for a variety of other applications, including home security systems and fire protection systems. Energy powered the Industrial Revolution and is now necessary for the age of information technology.

The developing world is rapidly exploiting energy supplies so that they, too, can benefit from industrial growth and economic development to enjoy the kind of comforts and conveniences available in the highly industrialized nations. Half of the new electric power generation facilities to be installed in the first ten years of the twenty-first century will be in China, largely because it is one of the fastest-growing economies in the developing world, and more than one-fifth of the world's population resides within its borders.

Some environmentalists believe that the root cause of environmental problems associated with energy is population growth: The more people there are, the more the demand for energy, and the greater the adverse environmental impact. Unfortunately, there is no easy solution. Most people living in the developing world live in dire poverty. Developing nations trying to solve the poverty problem have historically pushed for economic growth, but over the last quarter of the twentieth century more nations made greater efforts to control population as well. It is widely believed that the sustainable solution to the energy and environment problem requires an effort on both fronts. In 1900 the world's population totaled about 1.6 billion people. The total went over 5 billion by 1990, reached nearly 6 billion in 2000, and the United Nations projects 9.35 billion by 2050. Virtually all of the projected increase will come from the developing world (from 4.75 billion to 8.20 billion).

As the world's population increases, economic affluence is also increasing in many regions. Affluence usually means more business providing more goods and services; more homes and business with more climate control; more use of appliances and electronic technology in homes and businesses; and more miles logged by more automobiles, trucks, trains, and air-

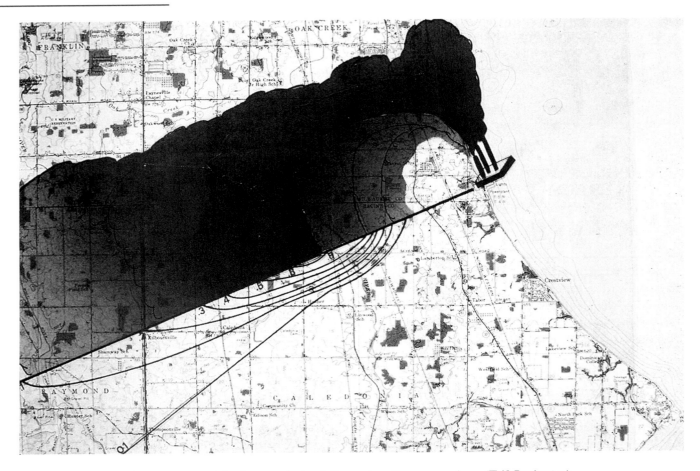

Computers can be used to generate models of power plant pollution such as the one seen here. (FMA Production)

planes. Despite impressive gains in energy efficiency, the demand for energy has continued to grow throughout the world. Total energy consumption, including energy use in all sectors and non-energy uses of fossil fuels, increased from 66 quadrillion Btus in 1970 to more than 95 quadrillion Btus by 2000 in the United States. Over the same period, total world energy consumption had increased to more than 400 quadrillion Btus, from 207 quadrillion Btus in 1970. In the United States, about 64 percent of the 95 quadrillion Btus was consumed directly in end uses, and about 36 percent of the total was consumed for electric power production.

The fear associated with the growing populations and affluence in the developing world is that there will be far more people demanding far more energy to power technology. If economic growth and affluence accelerate in the developing world and if its population assumes the high energy use behavior of the United States—where energy use per capita is ten to twenty times that of the developing world—environmental problems will significantly worsen. For example, China's economy grew more than tenfold from 1953 to 1989, and at the same time its energy consumption grew 18-fold. Because China has vast reserves of coal, which is by far the dirtiest-burning fossil fuel and the one that emits the most carbon dioxide, it is highly likely it will double the amount of coal burned between 1998 and 2015 just to maintain its present economic growth rate.

The greatest growth in energy consumption has come in the form of electrical energy. From 1970 to 1990, the amount of electricity consumed worldwide more than doubled. This increase in electrical energy demand required a corresponding increase in generating plants, substations, and transmission lines. However, because of improvements in the efficiency of equipment—for both new sites and upgrades—this has not resulted in a doubling of either coal and natural gas production, or of generating and distribution equipment.

The fuels being used today for electric power production are coal, oil, natural gas, uranium (for nuclear power), and various other materials, including refuse-derived fuels. Most of the electric power in the United States is produced using coal-fired generating plants. Coal met 52 percent of the total requirements for electricity generation in 1997. In the same year, nuclear power plants provided 18 percent, natural-gas-fired power plants 14 percent, and hydroelectric plants about 10 percent. Oil, or petroleum, makes a relatively small contribution, only about 3 percent. In the future non-hydro renewable-energy technologies, which accounted for less than 2 percent of the total in 1997, may make a larger contribution if policies are implemented to curtail carbon emissions to combat global warming.

With the unprecedented levels of energy consumption taking place in all sectors of the economy, it begs the question of sustainability and the capacity of Earth's environment to withstand this consumption. Most environmentalists feel that the heavy reliance on fossil fuels, and the emissions from combustion, are already beyond Earth's sustainable capacity. Others are more optimistic, believing that the greater energy demands of humans can be fulfilled with only a minimal negative impact on the environment, and that science and technology can solve all the environmental problems associated with energy production and consumption.

ENVIRONMENTAL ACTION

Public concern with controlling population growth, managing economic growth, and ensuring that growth does not adversely affect the environment has found a voice through the many environmental groups that have come into existence, such as the Sierra Club, the Environmental Defense Fund, Friends of the Earth, the Bullitt Foundation, the Wilderness Society, and Greenpeace. Environmental advocacy organizations collectively have an annual budget of more than $50 million, and push for greater spending for environmental mitigation and cleanup efforts, which already run into the billions of dollars annually. In 1997 the Department of Energy (DOE) spent more than $6 billion, about one-third of its budget, for nuclear waste management and clean up at federal facilities.

The first Earth Day, April 22, 1970, was supported, directly or indirectly, by more than 20 million Americans. It is often cited as the beginning of the

In 1997 the Department of Energy spent about one-third of its budget for nuclear waste management and clean up at federal facilities. (Field Mark Publications)

environmental awareness movement in the United States. However, prior to that day several individuals spoke to the need for environmental protection. Notable are the writings of Henry David Thoreau and Rachel Carson and the public policies of Presidents Theodore Roosevelt and Franklin D. Roosevelt. At Walden Pond, Thoreau saw an unspoiled and undeveloped forest as the means to preserve wild animals. Rachel Carson's book *Silent Spring* described the dangers associated with using DDT and other pesticides. Theodore Roosevelt and Franklin D. Roosevelt expanded the national park system to provide a habitat for wildlife and to prevent the destruction of natural lands through development and exploitation.

During the last third of the twentieth century, many federal legislative proposals have addressed environmental protection and resource conservation that has had a profound impact on the energy industry. At the federal level, environmental regulations have been managed by the U.S. Environmental Protection

Agency (EPA), which was established in 1970, the same year the first Clean Air Act was passed into law. In 1972 the Clean Water Act became law, and in 1973 the Endangered Species Act became law. Other important federal environmental legislation includes the Resource Conservation and Recovery Act, passed in 1976; the Response, Compensation, and Liability Act of 1980; the Nuclear Waste Policy Acts of 1982 and 1987; and the Low-Level Radioactive Waste Policy Acts of 1980 and 1985. From 1980 to 2000 these environmental regulations, and the enforcement efforts of the EPA, have had a much greater impact on decisions made in the energy industry than all the policy initiatives implemented by the DOE.

Some of the environmental issues associated with energy production and utilization that are being given the most attention by maker of public policy and by environmental groups include clean air, global climate change, nuclear power, electric and magnetic fields, oil spills, and energy efficiency.

Clean Air

Public concerns about air quality led to the passage of the Clean Air Act in 1970 to amendments to that act in 1977 and 1990. The 1990 amendments contained seven separate titles covering different regulatory programs and include requirements to install more advanced pollution control equipment and make other changes in industrial operations to reduce emissions of air pollutants. The 1990 amendments address sulfur dioxide emissions and acid rain deposition, nitrous oxide emissions, ground-level ozone, carbon monoxide emissions, particulate emissions, tail pipe emissions, evaporative emissions, reformulated gasoline, clean-fueled vehicles and fleets, hazardous air pollutants, solid waste incineration, and accidental chemical releases.

Energy use is responsible for about 95 percent of NO_x (precursor of smog—i.e., urban ozone) and 95 percent of SO_x (acid rain). The choice for the energy industry was to switch to "cleaner" energy sources, or to develop the technology that will make using the "dirtier" energy sources less harmful to the environment. For utilities making electricity generation decisions, this meant going with natural gas instead of coal for new generation, switching from high-sulfur eastern U.S. coal to low-sulfur western U.S. coal, and installing flue gas desulfurization equipment to comply with the sulfur dioxide provisions of the Clean Air Act amendments of 1990. In the transportation indus-

try, most of the dramatic reductions in vehicle emissions have come about from advances in engine technology, not from improvements in fuels.

Global Climate Change

There is growing concern that certain human activities will alter the Earth's climate. The global climate change issue is perhaps better termed "the greenhouse issue" because the concern is that certain "greenhouse gases," including carbon dioxide, CFCs, HFCs, PFCs, SF6, nitrous oxide, and methane will accumulate in Earth's upper atmosphere and cause a heating effect similar to that in a greenhouse.

The global climate change issue is of particular concern to those in the energy field because energy production and consumption involve combustion, which has been a major factor in increasing atmospheric carbon dioxide concentrations since the beginning of the Industrial Revolution. Energy use in the United States is responsible for about 98 percent of human-generated carbon dioxide emission.

Concerns about global climate change have led to extensive research and high-level international debates about the need for targets and timetables to reduce carbon dioxide emissions. Some policymakers believe that current uncertainties in how to approach the issue do not justify an all-out effort to reduce carbon dioxide emission, while others feel that this is a crisis needing immediate attention.

There are many uncertainties about global climate change, the workings of global greenhouse gas sources and sinks, and techniques to reduce or sequester emissions. It has been established that the CFCs used for airconditioning have higher global warming potential than carbon dioxide. CFC-reduction methods have been implemented. Research is under way to better understand how agriculture and forestry produce and absorb gases such as carbon dioxide, nitrous oxide, and methane. Methane is produced by rice cultivation, animal waste, and biomass burning. Nitrous oxide is produced from cultivation, fossil fuel/biomass burning, and fertilizer use.

The Intergovernmental Panel on Climate Change predicted in 1993 that a doubling of carbon dioxide concentrations by 2100 will occur under a "business as usual" scenario. However, technologies now exist where greenhouse gas growth rate can be reduced. The electric utility industry advocates an evolution toward highly efficient electrotechnologies, with more of the electricity produced by natural gas, as a way to mitigate carbon dioxide emissions.

Transportation

The transportation sector is a major polluter of the environment. In 1994 there were 156.8 million noncommercial vehicles on the road, averaging 19.8 miles per gallon, traveling an average of 11,400 miles, and burning an average of 578 gallons of fuel annually per vehicle. The trend has been one of more vehicles more miles driven, and more roads demanded by drivers.

Transportation accounts for about one-fourth of the primary energy consumption in the United States. And unlike other sectors of the economy that can easily switch to cleaner natural gas or electricity, automobiles, trucks, nonroad vehicles, and buses are powered by internal-combustion engines burning petroleum products that produce carbon dioxide, carbon monoxide, nitrogen oxides, and hydrocarbons. Efforts are under way to accelerate the introduction of electric, fuel-cell, and hybrid (electric and fuel) vehicles to replace some of these vehicles in both the retail marketplace and in commercial, government, public transit, and private fleets. These vehicles dramatically reduce harmful pollutants and reduce carbon dioxide emissions by as much as 50 percent or more compared to gasoline-powered vehicles.

Technology is making possible more fuel-efficient and cleaner-running automobiles, but it cannot do anything to reduce the number of automobiles or the ever-increasing number of highways. As long as the law requires that all the revenue from the federal fuel tax be used to build and maintain highways, the number of miles of highway in the United States will continue to increase at an explosive rate. The amount of road-building that has taken place in America since the early 1950s have produced an environment well suited for the automobile but not for plants, wildlife, or even people. In fact, physicist Albert Bartlett showed that the number of miles of highway will approach infinity under extended projections.

Besides all the gaseous and liquid wastes of transportation that result from energy use, and the loss of natural environment to roadways, there is also the solid-waste problem of disposal—vehicles and components such as tires and batteries. Responding to the growing disposal problem, many manufacturers are building automobiles that contain far more recyclable parts.

Water Pollution

The major energy-related sources of water pollution are from thermal pollution, surface water pollution from oil spills, polychlorinated biphenyls, and groundwater contamination.

Thermal Pollution. Thermal pollution occurs as a result of hot-water emissions from electric power plants. High-temperature steam, which causes the turbine blades to rotate, passes by the blades, cools and condenses to liquid water. This condensation liberates energy that must be removed by circulating water in pipes in contact with the condenser. Each second the water must remove more than a billion joules of heat. When this heated water is discharged into a body of water, it can alter aquatic life in two ways: Significantly warmer water retains less oxygen, making it more difficult for many species to survive, and warmer water favors different species than cooler water. The altered ecosystem may harm all species; attract less desirable organisms; or could even improve survivability of the most desirable organisms, especially during the winter months.

Oil Spills. Oil spills occur from oil pipeline leaks, oil tanker accidents, or submarine oil drilling operations. The two major ocean drilling accidents—oil wells blowing out—were the 1969 Santa Barbara Channel spill and the 1979 Yucatan Peninsula spill, in Mexico. The Yucatan spill spewed out more than three million barrels before being capped in 1980. Both caused damage to beaches and marine life, but the smaller Santa Barbara spill was far more devastating because of unfavorable winds following the accident.

The largest oil spill in U.S. history occurred in March 1989, when the tanker *Exxon Valdez* ran aground in the Prince William Sound inlet in the Gulf of Alaska, spilling 11 million gallons (42 million liters) of crude oil. The resulting slick covered more than 1,000 miles (1,600 kilometers) of the Alaska coastline and caused an estimated $3 billion to $15 billion in environmental damages. The spill killed hundreds of thousands of fish and seabirds and thousands of otters. The tanker's captain, Joseph J. Hazelwood, had reportedly been drinking before the accident and had turned control of the ship over to the third mate. The state of Alaska in March 1989 brought a criminal indictment against Hazelwood for his role in the disaster, but in March 1990 he was acquitted of the most serious criminal charges against him. Prosecutors were unable to convince jurors that

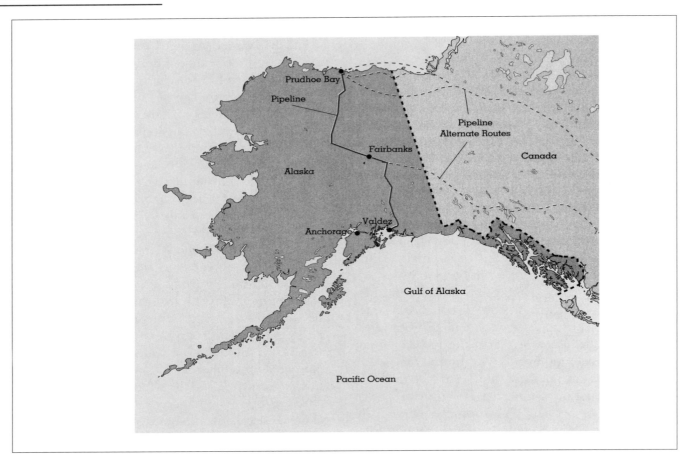

Figure 1.
Although the pipeline from the North Slope to the port of Valdez was the route chosen and completed in 1977, the scientific community preferred the trans-Alaska-Canada route; its use would have prevented the *Exxon Valdez* tanker accident of 1989.

Hazelwood was legally drunk at the time of the accident. In September 1994, a federal court jury ordered Exxon Corp. to pay $5 billion in punitive damages to Alaskan fishermen, local residents, and property owners. The fine was reported to be the highest punitive award ever levied against a corporation and also the largest ever in an environmental pollution case. Since the *Exxon Valdez* spill, a number of safeguards have been instituted to help prevent future oil spills, such as the mandating of double-walled tankers and better pipelines.

Some environmentalists anticipated a major accident like that of the *Exxon Valdez*. When planning for the Trans-Alaska Pipeline took place during the energy crisis of the early 1970s, it was controversial because of the conflicts of balancing the needs of oil resource development and the known and predicted environmental problems. The tremendous political pressure to quickly deliver new oil supplies for national security reasons hastened a decision. Many feel that the Prudhoe Bay to Valdez Trans-Alaska Pipeline was chosen because it meant the most rapid resource development. An alternative trans-Alaska-Canada route—which would have avoided the marine environment tanker risk and earthquake zone pipeline risk—was favored by environmentalists, but was rejected during the political process (see Figure 1). It would have taken longer to construct, yet this route would have made it possible to include an oil and gas pipeline in one corridor.

Polychlorinated Biphenyls. Polychlorinated biphenyls (PCBs) are carcinogenic and adversely affect the liver, nervous system, blood, and immune response system. Moreover, PCBs persist a long time in the natural environment and become concentrated in the higher parts of the food chain. A major source of PCBs occurred during the production of electrical capacitors and transformers. It is believed that the Hudson River has more than 295,000 kilograms of PCBs that were discharged into the river by General Electric from 1950 to 1977. The controversy is whether to spend millions removing and treating the river sediment, or to allow the river to clean itself by the natural process of sediment transport to the ocean. Removal and treatment, has been advocated by the EPA but has not yet been initiated.

Groundwater. The major threat to groundwater is from leaking underground fuel storage tanks. It is unknown how many underground storage tanks leak, how much gets into the groundwater, and what impact the leaks have on human health. The primary worry is not with the fuels themselves but with the "cleaner burning" additive methyl-t-butyl ether (MTBE). Ironically, MTBE, whose use the EPA mandated in 1990 as a way to improve air quality by lowering harmful emissions from vehicles, may turn out to be the most serious threat ever to drinking water. It is a suspected carcinogen, it migrates quickly into groundwater, and there is no known way to treat groundwater that is polluted with MTBE. The federal government was looking at banning MTBE as of 2000, but because of its prevalent use during the 1990s and the extent of underground fuel tank leakage, MTBE in groundwater will be an environmental problem for years to come. There is also a problem in finding a gasoline additive substitute for MTBE. The leading replacement contender, ethanol, has a lower octane rating (106 compared to 116 for MTBE) and is considerably more expensive.

Nuclear Power

Public opposition to commercial nuclear power plants began with the misperception that the plants could explode like nuclear weapons. The nuclear industry made progress in dispelling this misperception, but suffered major setbacks when an accident occurred at the Three-Mile Island nuclear power plant in Pennsylvania and at the Chernobyl nuclear power plant in the USSR.

March 28, 1979, an accident at Three-Mile Island resulted in a small release of radiation when a pressure relief valve became stuck open. According to a report by the U.S. Nuclear Regulatory Commission (NRC), the dose of radiation received by the people in the immediate area of Three-Mile Island was much less than the radiation dose that the average member of the U.S. population receives annually from naturally occurring radiation, medical use of radiation, and consumer products. In subsequent years law suits have been filed by plaintiffs contending that high radiation exposure levels at Three-Mile Island caused them to develop cancer, but the courts have ruled that the plaintiffs failed to present any evidence that they were exposed to enough radiation to cause their cancers.

A much more serious nuclear accident occurred at Chernobyl in the USSR on April 26, 1986, when one of the Chernobyl units experienced a full-core meltdown. The Chernobyl accident has been called the worse disaster of the industrial age. An area comprising more than 60,000 square miles in the Ukraine and Belarus was contaminated, and more than 160,000 people were evacuated. However, wind and water have spread the contamination, and many radiation-related illnesses, birth defects, and miscarriages have been attributed to the Chernobyl disaster.

The fear of accidents like Chernobyl, and the high cost of nuclear waste disposal, halted nuclear power plant construction in the United States in the 1980s, and in most of the rest of the world by the 1990s. Because nuclear fusion does not present the waste disposal problem of fission reactors, there is hope that fusion will be the primary energy source late in the twenty-first century as the supplies of natural gas and petroleum dwindle.

Electric and Magnetic Fields

Questions are being raised about the possible health effects of electric and magnetic fields from electric power transmission, distribution, and end-use devices. Electric and magnetic fields exist in homes, in workplaces and near power lines. Electric fields exist whenever equipment is plugged in, but magnetic fields exist only when equipment is turned on. Both electric and magnetic fields become weaker with distance from their source. Science has been unable to prove that electric and magnetic fields cause adverse health effects, but further investigation is under way. Research is focusing on the possible association between magnetic field exposure and certain types of childhood cancer. While laboratory experiments have shown that magnetic fields can cause changes in living cells, it is unclear whether these experiments suggest any risk to human health.

Energy Efficiency and Renewable Energy

Energy efficiency and renewable energy are being promoted as sustainable solutions to the world's energy needs. Through energy efficiency less energy is used, generally resulting in reduced demands on natural resources and less pollution. And if less energy is being consumed, a far greater fraction can come from renewable sources.

Some energy efficiency strategies include moving toward more fuel-efficient cars, electricity-stingy appliances, and "tighter" homes with better windows to reduce heating and air conditioning requirements. Because of dramatic improvements in the energy efficiency of appliances, particularly refrigerators, far fewer coal-fired power plants needed to be built in the 1980s and 1990s. However, the price of energy is relatively low in the United States, so businesses and citizens do not feel an urgency to conserve energy and purchase only energy-efficient products. In Japan and Europe, where citizens pay two to three times as much as Americans for gasoline and electricity, energy consumption per capita is half what it is in the United States. Many environmentalists feel the U.S. government needs to institute a carbon tax, mandate that manufacturers provide more energy-efficient appliances, and develop other incentives to encourage purchase of energy-efficient products.

During the 1970s and early 1980s, many in the environmental movement suggested a low-energy or soft-path-technology future, one that would involve greater conservation, increased efficiency, cogeneration, and more use of decentralized, renewable energy sources. Moreover, the claim was that this could be done without a reduction in the quality of life if future development included more energy-efficient settlement patterns that maximized accessibility of services and minimized transportation needs, if agricultural practices involved less energy in the production of food and emphasized locally grown and consumed foods, and if industry followed guidelines to promote conservation and minimize production of consumer waste. By 2000, many of these gains in energy efficiency were realized—energy is being used far more efficiently than

ever before—yet the ever-growing U.S. population and new uses of technologies necessitated more fossil fuel energy production, which made low-energy-future scenarios in the United States impossible (see Figure 2). Nevertheless, if not for energy efficiency improvements, many more electricity generating facilities would have been built.

THE ENERGY ENVIRONMENT FUTURE

Since the 1970s, technological advances have solved many environmental problems associated with energy production and consumption, and proven that more energy consumption does not necessarily mean more pollution. The fossil fuel industries are producing and distributing more energy less expensive-

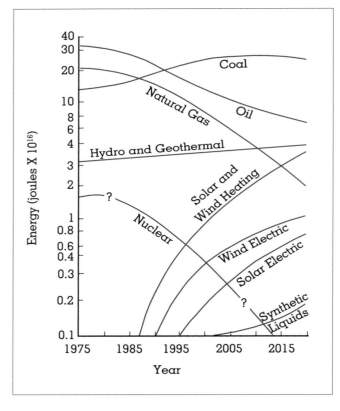

Figure 2.
A low-energy-future scenario envisioned in the late 1970s. By 2000, the energy being consumed from petroleum, natural gas, coal, and nuclear power were all on the rise, and renewable energy supplied a smaller fraction of total energy.
SOURCE: Steinhart et. al, 1978.

ly than ever before while creating less of an impact on the environment; the electricity-producing industry is generating more electricity with less fossil fuel and less harmful pollutants; and the automobile industry's best 2000 model year cars produce one-twentieth the harmful emissions of 1975 automobile models and at the same time offer far better performance and fuel economy. Certainly there are still major environmental problems. Nuclear waste is piling up, awaiting political decisions for its long-term storage. Underground fuel tanks, whose numbers are in the hundreds of thousands, are prone to leaks and migration into the drinking water supply. Deforestation continues in many parts of the world where people desire the wood for fuel and the land clear-cut for agriculture.

Hopefully, as the ongoing energy and environment debate continues, policymakers will choose a middle ground between the proposals of a few extreme environmentalists who would greatly cut back on energy development, and the proposals of a few of those in the energy industries who would unnecessarily exploit the environment to increase energy supplies. Reasonable strategies can ensure that there is sufficient energy for economic development to continue, but with minimal adverse environmental effects.

Fred I. Denny

See also: Acid Rain; Air Pollution; Atmosphere; Carson, Rachel; Climatic Effects; Disasters; Environmental Economics; Fossil Fuels; Gasoline and Additives; Gasoline Engines; Government and the Energy Marketplace; Nuclear Fission; Nuclear Fusion; Nuclear Waste.

BIBLIOGRAPHY

Bartlett, A. A. (1993). "The Arithmetic of Growth: Methods of Calculation." *Population and the Environment* (14):359–387.

Bartlett, A. A. (1969). "The Highway Explosion." *Civil Engineering* (December):71–72.

Buell, P., and Girard J. (1999). *Chemistry: An Environmental Perspective.* Englewood Cliffs, NJ: Prentice-Hall.

Botkin, D. B., and Keller, E. A. (1999). *Environmental Science: Earth as a Living Planet*, 3rd ed. New York: John Wiley & Sons.

Steinhart, J. S.; Hanson, M. E.; Gates, R. W.; Dewinkel, C. C.; Brody, K.; Thornsjo, M.; and Kambala, S. (1978). "A Low-Energy Scenario for the United States: 1975–2000." In *Perspectives on Energy*, 2nd ed., ed. L. C. Ruedisli and M. W. Firebaugh. New York: Oxford University Press.

ENVIRONMENTAL PROTECTION AGENCY

See: Government Agencies

ERICSSON, JOHN (1803–1889)

John Ericsson, born at Langbanshyttan, in the province of Varmland, Sweden, on July 31, 1803, was the youngest of three children of a mine owner. Ericsson acquired his skills as a mechanical draftsman at the age of thirteen, working on the construction of the Göta Canal. Against all advice Ericsson left the canal company to embark on a military career in 1820. At an early age Ericsson had experimented with mechanics and continued to do so when in the military, constructing what he termed a "flame engine" to challenge the steam engine. This engine worked by internal combustion, and the success of a small model prompted Ericsson to travel to England in 1826 to demonstrate and patent his invention.

Although the engine was not a success, Ericsson's trip to London allowed him to meet John Braithwaite, a machine manufacturer, who had the expertise to put Ericsson's ideas into practice. In 1828 Ericsson, with Braithwaite, patented the principle of artificial draft in steam boilers. The principle of forced draft was applied to a fire engine and a locomotive entered for the Rainhill locomotive trials of 1829.

Ericsson continued his search for a substitute for steam, and in 1833 patented his caloric engine. Fitted with what Ericsson termed a regenerator, this engine allowed heat to be re-used resulting in savings in fuel. These savings, Ericsson believed, made the engine suitable for marine use. The tubular heat exchanger was an invention of Robert Stirling, but Ericsson was not always inclined to admit priority or give credit to those whose ideas he put into practical form. Ericsson unsuccessfully opposed Stirling's second patent application in 1827. Ericsson demonstrated a five horsepower engine, generating much fierce

Undated illustration of John Ericsson's solar engine. (Corbis Corporation)

debate and being dismissed by many as an unworkable attempt at perpetual motion.

Ericsson lived an expensive lifestyle. The costs of his experiments plunged him into debt, confined him twice to a debtors' prison, and eventually led to his declaration of bankruptcy in 1835. In 1836 Ericsson patented a rotary propeller for ships. With the financial backing of the U.S. Consul in Liverpool, Ericsson built in 1837 a 45-foot boat that successfully demonstrated the superiority of screw over paddlewheel, and gained him orders for two iron ships from the United States. Although Ericsson built another caloric engine in 1838, he failed to convince the British establishment of the benefits of either his propeller or engine. He was persuaded to travel to New York the following year, never to return to England or Sweden.

Once in New York, Ericsson met with Harry DeLameter of the DeLameter Iron Works and immediately sought out backers for his inventions. Work on the caloric engine with a wire gauze regenerator continued. Surprisingly, Ericsson adopted an open cycle, although in his patent he had stated the benefits of using air under pressure. Such was Ericsson's faith in his engine that by 1852 he had raised sufficient capital to build a caloric-powered ship to challenge the dominance of the marine steam engine. The 250-foot ship, built of wood, was fitted with sails and paddlewheels, although Ericsson had successfully applied his propeller to the U.S. warship *Princeton*. The trial voyage of the caloric-ship in January 1853 was a stage-managed affair in which Ericsson avoided giving precise details of the ship's machinery.

A subsequent trip was made to Washington, a round trip of some five hundred miles. Unfortunately, the ship sank in April in freak weather conditions and, although successfully salvaged, her engines were replaced with steam engines. Ericsson's detractors claimed the ship was underpowered and the giant fourteen-foot diameter pistons impractical. By adopting an open cycle, Ericsson was unable to prove convincingly that his giant caloric engines fitted with regenerators could successfully overthrow the marine steam engine. Ericsson was to subsequently destroy all details and drawings of the ship's engines. The concept that "caloric" was a fluid that could be used over and over was not sustained in the eyes of Ericsson's critics.

The failure of the caloric ship and the financial losses incurred by his backers did not diminish Ericsson's enthusiasm for the air engine. The stationary steam engine had weaknesses: It required skilled operators and incurred heavy insurance costs. Ericsson again successfully raised funds to develop a small open-cycle stationary caloric engine with only a rudimentary regenerator. For Ericsson and his backers it was a financial success. Marketed as requiring no water, it could be operated by unskilled labor and, perhaps most important of all, would not explode.

During the American Civil War, Ericsson designed a steam-driven iron-clad ship, a concept initially rejected by the Navy Board as being too novel. The *Monitor,* built through the intervention of President Lincoln, but at Ericsson's expense, brought to a sudden end the era of wooden warships.

Ericsson devoted much time to the study of tidal action, although schemes for tidal power were abandoned, as he later commented as "not being able to compete with the vast energy stored in lumps of coal. But the time will come when such lumps will be as scarce as diamonds." A life-long interest in solar power proved more fruitful. Ericsson experimented with both condensing steam engines and his caloric engine, using silvered glass reflectors. Ericsson built in 1873 a closed-cycle solar motor along the lines of Stirling's engine, developed with a view of irrigating the parched lands of western states; however, little interest was shown in the sun-motor by the agricultural community. Ericsson patented the design in 1880 as a stationary engine that could be heated by liquid, gas, or solid fuels. It was sold in large numbers.

Shortly after joining the army, Ericsson met and later became betrothed to Carolina Lillieskold, the daughter of an army captain. From this relationship, Ericsson's only child, a son Hjalmer, was born in 1824; however, Captain Lillieskold's hostility prevented a marriage. Ericsson left Sweden two years later without knowing he had fathered a son. It was Ericsson's brother Nils who eventually learned of the birth and arranged for the child to be raised by their mother, although for nearly fifty years Ericsson had no contact with his son. In London, in 1836, Ericsson married Amelia Byam, fourteen years his junior. Amelia joined her husband in New York but found him absorbed in his work. The marriage was not a happy one, and Amelia returned to England. Ericsson continued to support her financially until her death in 1867, but they never met again.

For all Ericsson's years in America, he rarely travelled outside of New York. Because of failing health he withdrew from public life, but resolved to die in harness and continued inventing. John Ericsson passed away, in seclusion, on the morning of March 8, 1889.

Congress, which had had an uneasy relationship with the living Ericsson, and unable to decide how to honor the inventor of the *Monitor,* acquiesced to a suggestion from the Swedish government that Ericsson's body be returned to Sweden. In 1926 Ericsson was finally honored with a statue unveiled in Washington, D.C. by President Calvin Coolidge and Gustaf Adolf, Crown Prince of Sweden.

Robert Sier

See also: Stirling, Robert.

BIBLIOGRAPHY

Church, W. C. (1890). *The Life of John Ericsson.* New York: Scribner.

Ericsson, J. (1876). *Contributions to the Centennial Exhibition.* Philadelphia: John Ericsson.

Miller, E. M. (1978). *U.S.S. Monitor: The Ship That Launched a Modern Navy.* Annapolis: Leeward Publications.

Norton, W. A. (1853). "On Ericsson's Hot Air or Caloric Engine." *American Journal of Science and Arts,* 45.

Sier, R. (1987). *A History of Hot Air and Caloric Engines.* London: Argus Books.

ETHANOL

See: Biofuels

ETHICAL AND MORAL ASPECTS OF ENERGY USE

The production and use of energy gives rise to a wide range of ethical and moral issues. Worldwide there are four general energy options available, each of which can raise significant ethical questions. We can continue to rely primarily on fossil fuels, presently estimated to account for more than 80 percent of worldwide energy use. Second, we could shift to greater reliance on nuclear energy. Third, we could develop alternative energy sources such as wind, solar, and geothermal power. A fourth alternative would focus on conservation and energy efficiency and seek to decrease the overall demand for energy.

Continuing dependence on fossil fuels raises several major ethical issues. Ethical questions concerning our responsibilities to future generations are raised by the fact that fossil fuels are a nonrenewable energy source, so that every barrel of oil or ton of coal burned today is forever lost to future generations. Further, the by-products of fossil fuel combustion pose hazards to both present and future generations.

Nuclear energy also faces major ethical challenges. Nuclear power generates toxic wastes that remain hazardous for thousands of generations. Even assuming that the operation of nuclear power plants can be made safe, disposal of nuclear wastes can jeopardize the health and safety of countless future people. Further, the proliferation of nuclear technology that is necessary for generating nuclear energy also raises ethical concerns of international peace and security.

Energy sources such as solar, wind, and geothermal power are often proposed as renewable and non-polluting alternatives to fossil and nuclear fuels. But here, too, ethical challenges must be faced. Over the short term, alternative energy sources will likely be more expensive relative to fossil and nuclear fuels. Such price differentials mean that safer and cleaner energy sources will more likely be available to wealthy countries and individuals while the poor will continue relying on more dangerous and polluting energy sources. As a result, questions need to be raised concerning equality and fairness in the distribution of alternative energy sources. Further, development of these alternative technologies may require government subsidies and incentives, which can raise additional questions of freedom, fairness, and equality.

Finally, conservation and energy efficiency also raise ethical challenges. A significant decrease in energy consumption is possible in only two ways: we significantly decrease demand or we significantly decrease the population of people demanding energy. Either option raises major questions concerning values such as individual freedom of choice, property rights, fairness, and equal opportunity, as well as ethical issues regarding population policy, standards of living, and quality of life.

INDIVIDUAL CHOICE AND ENERGY POLICY

We can begin to focus on the specific value issues involved in energy by reflecting on the nature of ethics itself. At the start of a dialogue on ethics, Socrates once said that "we are dealing with no small thing but with how we ought to live." There is no more fundamental ethical question than this: How ought we to live? But as Socrates understood, this question can be interpreted in two ways. "We" can mean each one of us individually, or it may mean all of us collectively. Taken in the first sense, ethics is concerned with how I should live my life, with the choices I ought to make, with the type of person I should be. We can refer to this sense of ethics as morality. Taken in the second sense, this question refers to how we ought to live together in society. Ethics in this sense raises questions of public policy, law, institutions, social justice. To distinguish this sense of ethics from morality, we can refer to these as questions of social ethics.

Although important questions of individual morality can be involved with energy issues, the production and use of energy primarily raises questions of social ethics and public policy. This emphasis can be explained simply by the magnitude of energy issues. Such questions as resource conservation, global warming, nuclear waste disposal, and pollution will not be resolved through individual action alone. However, before turning to the public policy perspective, it will be helpful to consider some aspects of individual energy choices.

Some would argue that energy policy ought to be left to individual choice. In such a situation, some individuals may choose a frugal and conservative lifestyle that demands relatively little energy resources. One thinks here of the common environmental adage "live simply so that others may simply live." Other individuals may choose a more energy-intensive lifestyle. Either way, one could argue that the choice ought to be left to the moral values of individuals.

This option is favored by those who defend economic markets as the most ethically appropriate approach to public policy. This view argues that individual consumers, relying on their own preferences, should be free to choose from a variety of energy options. The working of a competitive market would then guarantee the optimal distribution of both the benefits and burdens of energy production. Consumers who value safe and clean energy sources would be free to choose wind or solar power and, presumably, would be willing to pay more for these benefits. Assuming a social system in which government subsidies were eliminated and external costs such as pollution were fully internalized, this economic arrangement would most efficiently satisfy the greatest number of individual desires while also respecting individual freedom of choice.

Defenders of this approach could point to the deregulation of the electric utility industry within the United States during the late 1990s as an example. As the industry becomes deregulated, a number of new firms stepped into the market to offer consumers a wider range of energy choices. In many areas, consumers who are willing to pay higher prices are able to purchase energy generated from environmentally friendly, "green" sources.

There are major problems with this individualistic approach to energy policy, however. The ideal market of economic theory exists nowhere in reality. Further, even market defenders acknowledge cases in which markets fail. Significantly, some paradigmatic examples of market failure, such as the externality of pollution and monopolistic control of production, are associated with the production of energy. More importantly, perhaps, crucial ethical questions can be missed if we only consider the perspective of individual values and choice.

Consider how a single individual might deliberate about the consumption of fossil fuels. Burning fossil fuels increases the amount of greenhouse gases released into the atmosphere and strong evidence suggests that this can lead to global warming. Given this scenario, does morality require that individuals refrain from driving automobiles?

To answer this question an individual might well weigh the benefits of driving a private car against the costs of increased greenhouse gas emissions. An average car might burn between two and three gallons of gasoline each hour. Given that an average driver might drive only a few hours each day, the amount of fuel burned in this activity will make little difference in the amount of atmospheric carbon dioxide. Weighed against the convenience and freedom of driving one's own car and the market-established price of gasoline, it may well be reasonable for an individual to decide that there is no significant moral issue involved in driving.

However, if we extend this line of reasoning across a large population, a decision that seems minor to an individual can turn out to have enormous social implications. If millions of people make the same seemingly reasonable decision and burn millions of gallons of gasoline each day, the atmospheric consequences are significant. We are here faced with ethical and policy questions that would never arise from an individual's point of view. For example, we might consider increasing taxes on gasoline, requiring automobile manufacturers to improve mileage efficiency, subsidizing mass transit, providing tax incentives for alternative fuel transportation, or even prohibiting private automobiles in urban areas. The crucial point is that none of these questions would ever arise from the perspective of an individual facing a single choice. Recognizing that these are important ethical questions that deserve consideration, we recognize the need for treating public policy questions as distinct from individual moral questions.

A second inadequacy of the individualistic approach is that it can underestimate the influence which social practices have upon individual choice. As individuals, we pursue goals based on our interests and desires. But a complete ethical analysis should include an examination of the source of those interests and desires.

If we take consumer demand as a given, then the major task for energy policy is to produce enough energy to satisfy that demand. Alternative policies will then be judged in terms of how well they accomplish that task. But when we recognize that the demand for an energy-intensive lifestyle is a product of social and cultural factors, and that these factors themselves can be influenced by public policy, then we see the need to ask questions that might ordinarily be ignored by individuals.

For example, should we pursue policies that would discourage energy use and encourage conservation? Are all uses of energy equally valid on ethical grounds? Should energy producers be discouraged from advertising, or should they be required to promote conservation? Should poor, less-developed countries receive subsidized energy resources from the developed, industrialized countries? Again, these are questions that are raised only from a public policy perspective. Clearly, an adequate ethics of energy must move beyond moral questions and focus on social and public policy perspectives.

SOCIAL ETHICS AND ENERGY POLICY: PRESENT GENERATIONS

Turning to social ethics, we can distinguish two general types of ethical questions that pertain to energy policy: questions of justice in the present, and questions concerning our responsibilities to future generations. Issues concerning social justice for present generations can be categorized in terms of debates between utilitarian (maximizing beneficial consequences) and deontological (acting in accord with moral principles and duties) approaches to ethics. Ethical questions concerning future generations involve both the content and the very existence of such duties.

Utilitarian ethics holds that energy policy ought to be set according to the general ethical rule of maximizing the overall good. For example, if oil exploration in an Arctic wilderness area would produce greater overall social happiness than would preservation of the wilderness area, utilitarian ethics would support exploration. Utilitarianism is a consequentialist ethics in which good and bad policy is a function of the consequences that follow from that policy. Policies that increase net social benefits are right, those that decrease net social benefits are wrong. Thus, utilitarianism employs what can be thought of as an ethical cost-benefit methodology, weighing the benefits and harms of various alternatives and promoting that option which proves most useful in maximizing benefits over harms. Because energy is something valued only for its usefulness, utilitarian ethics seems well suited for establishing energy policy.

Explained in such general terms as maximizing the good, utilitarianism is an intuitively plausible ethical theory. Disagreements occur when defenders attempt to specify the content and meaning of the good (or "happiness"). An important contemporary version of utilitarianism identifies happiness with the satisfaction of individual desires or, simply, getting what one wants. Sometimes identified as preference utilitarianism, this view equates the good with the satisfaction of individual preferences and is closely associated with the goal of microeconomic efficiency. This particular version of utilitarianism has had a profound impact on energy policy, especially energy policy as it is found in liberal democratic societies. From this perspective, the goal of energy policy is to optimally satisfy the demand for energy while minimizing any potential harms that might result.

Two trends within this general utilitarian approach dominate energy policy. One holds that there are experts who can predict policy outcomes, determine relative risks and benefits, and administer policies to attain the goal of maximum overall happiness. These experts, trained in the sciences, engineering, and the social sciences, are best situated to predict the likely consequences of alternative policies. Scientific understanding of how the world works enables these experts to determine which policies will increase the net aggregate happiness. This version of utilitarian thinking typically supports government regulation of energy policy and, as a result, is often criticized on ethical grounds as involving paternalistic interference with individual decision-making and property rights.

A second trend within the utilitarian tradition argues that efficient markets are the best means for attaining the goal of maximum overall happiness. This version would promote policies that deregulate energy industries, encourage competition, protect property rights, and allow for free exchanges. In theory, such policies would direct rationally self-interested individuals, as if led by an "invisible hand" in famed economist Adam Smith's terms, to the optimal realization of overall happiness. As with the approach that relies on energy experts, the market approach agrees that the goal of energy policy ought to be the optimal satisfaction of consumer demand.

Both approaches share two fundamental utilitarian assumptions. First, utilitarianism is a consequentialist ethics that determines right and wrong by looking to the results of various policies. Second, they hold that ethics ought to be concerned with the overall, or aggregate, welfare. Deontological ethics (the word is derived from the Greek word for duty) rejects both of these assumptions.

Committed to the ethical maxim that the ends don't justify the means, deontological ethics rejects the consequentialism of utilitarianism for an ethics based on principles or duties. There are many cases in which ethics demands that we do something *even if* doing otherwise would produce greater overall happiness. On this view, right and wrong policy is a matter of acting on principle and fulfilling one's duties. Respect for individual rights to life and liberty or acting on the demands of justice are common examples of ethical principles that ought not be sacrificed simply for a net increase in the overall happiness.

An especially troubling aspect of utilitarianism is the emphasis on collective or aggregate happiness. This seems to violate the ethical principle that individuals possess some central interests (to be distinguished from mere preferences) that ought to be protected from policies aimed simply at making others happier. Most of us would argue that individuals have rights that ought not to be sacrificed to obtain marginal increases in the aggregate overall happiness. Our duty to respect the rights of individuals, to treat individuals as ends in themselves, and not as mere means to the end of collective happiness, is the hallmark of deontological ethics.

Nowhere is this concern with individual rights more crucial than in questions concerning the justice of energy policy. Distributive justice demands that the benefits and burdens of energy policy be distributed in ways that respect the equal dignity and worth of every individual. A *prima facie* violation of justice occurs when social benefits and burdens are distributed unequally. Particularly troubling inequalities occur when the benefits of policy go to the powerful and wealthy, while the burdens are distributed primarily among the poor and less powerful.

Consider, as an example, the logic of a policy decision to build and locate an electric generating plant or oil refinery. Economic considerations such as the availability of ample and inexpensive land, and social considerations such as zoning regulations and political influence, would play a major role in such a decision. In practice, this makes it more likely that plants and refineries, as well as waste sites and other locally undesirable land uses, will be located in poorer communities whose population is often largely people of color.

Evidence suggests that this is exactly what has happened. Beginning in the 1970s, sociologist Robert Bullard studied the location of hazardous and polluting industries within the United States. He found that many such industries, including many electric generating plants and oil refineries, are disproportionately located in minority communities. This is not to claim that there has been an intentional social policy to unfairly burden minority communities. But it does suggest that the economic and political factors that give rise to such decisions have much the same practical effect. Upper income levels disproportionately benefit from inexpensive energy to fuel consumerist lifestyles, while lower income minority communities carry a disproportionate burden of energy production.

Much the same has been said on the international level. Debates concerning international energy justice occurred frequently during the Earth Summit in Rio de Janeiro in 1992 and the Kyoto conference on global warming in 1997. Representatives of the less developed countries argue that industrialized countries have long benefited from readily accessible and inexpensive energy resources, which have been primarily responsible for the proliferation of greenhouse gases and nuclear wastes. However, after having attained the benefits of this lifestyle, the industrialized countries now demand a decrease in greenhouse emissions, conservation of resources, and a reduction in the use of nuclear energy. These policies effectively guarantee that the non–industrialized world will remain at an economic and political disadvantage. Many argue that justice would demand industrialized countries carry a heavier burden for decreasing energy demands, reducing greenhouse emissions, storing nuclear wastes, and for conserving non–renewable resources. Influenced by such reasoning, the majority of industrialized countries accepted greater responsibility at the Kyoto conference for reducing greenhouse gases.

From a strictly utilitarian perspective, unequal distribution of the benefits and burdens of energy production and use might be justified. Utilitarians have no in-principle objection to unequal distribution. If an unequal distribution would create a net increase in the total aggregate amount of happiness, utilitarians would support inequality. Deontologists would argue that these practices treat vulnerable individuals as mere means to the end of collective happiness and, thus, are unjust and unfair. Such central interests as health and safety ought not be sacrificed for a net increase in overall happiness. Moral and legal rights

The Hanford Nuclear Reservation in eastern Washington stores class A low-level radioactive waste. (Corbis)

function to protect these interests from being sacrificed for the happiness of others.

Legal philosopher Ronald Dworkin suggests that individual rights can be thought of as "trumps" which override the non-central interests of others. Public policy issues that do not violate rights can be appropriately decided on utilitarian grounds and properly belong to legislative bodies. However, when policies threaten central interests, courts are called upon to protect individual rights. The judiciary functions to determine if rights are being violated and, if so, to enforce those rights by overruling, or "trumping," utilitarian policy.

Critics raise two major challenges to deontological approaches. Many charge that deontologists are unable to provide a meaningful distinction between central and non-central interests. Lacking this, public policy is unable to distinguish between rights and mere preferences. From this perspective, the language of rights

functions as a smoke-screen raised whenever an individual's desires are frustrated by public policy. The inability to distinguish central from non–central interests has given rise to a proliferation of rights claims that often obstructs effective public policy.

Critics might cite the NIMBY (not in my back yard) phenomenon as a case in point. A small minority is sometimes able to thwart the common good by claiming that their rights are being violated when, in fact, this minority simply does not want to bear its share of the burden. The cessation of nuclear power plant construction within the United States might provide an example of this. By claiming that nuclear plants threaten such rights as health and safety, opponents to nuclear power have been able to block further construction. If, however, there is little evidence of any actual harm caused by nuclear plants, these opponents may have succeeded in obstructing a beneficial public policy by disguising their preferences as rights. A similar claim could be made concerning debates about locating such locally undesirable but socially beneficial projects as oil refineries and electric generating plants.

A second challenge argues that deontologists are unable to provide a determinate procedure for deciding between conflicting rights claims. Even if we can distinguish rights from mere preferences, effective policy needs a procedure for resolving conflicts. Returning to the analogy of trump cards, deontologists are challenged to distinguish between the ace and deuce of trumps.

Consider, for example, one possible scenario that could follow from the Kyoto Protocol on carbon reduction. Developing countries claim, as a matter of right, that the United States should bear a greater responsibility to reduce carbon emissions. Failing to do so would violate their rights to equal opportunity and fairness. One means by which the United States could meet the Kyoto targets would involve significantly scaling back its energy-intensive agriculture and military sectors. However, because the United States is a major exporter of food products and because its military protects many democracies throughout the world, these options might well threaten the basic rights to food, health, and security for many people in the developing world. In turn, that could be avoided if the United States scaled back its industrial rather than agricultural or military sectors. But this would threaten the freedom, property rights, and economic security of many U.S. citizens. Deontologists are challenged to provide a decision procedure for resolving conflicts between such rights as equal opportunity, fairness, food, health, security, property, and freedom. According to critics, no plausible procedure is forthcoming from the deontological perspective.

SOCIAL ETHICS AND ENERGY POLICY: FUTURE GENERATIONS

Many energy polices also raise important ethical questions concerning justice across generations. What, if any, responsibilities does the present generation have to posterity? This question can be raised at many points as we consider alternative energy policies.

Fossil fuels are a nonrenewable resource. Whatever fossil fuel we use in the present will be forever lost to posterity. Is this fair? The harmful effects of global warming are unlikely to occur for many years. Should we care? Is it ethical to take risks

with the welfare of future generations? Nuclear wastes will remain deadly for thousands of generations. Does this require us to change our behavior now? Do we have a responsibility to invest in alternative energy sources now, even if the benefits of this investment go only to people not yet born? Given the energy demands made by increasing populations, what is an ethically responsible population policy?

In many ways, debates surrounding our ethical responsibility to future generations parallel the debates described previously. Utilitarians are concerned with the consequences that various policies might have for the distant future. Committed to the *overall* good, utilitarians must factor the well-being of future people into their calculations. Some argue that future people must count equally to present people. Others, borrowing from the economic practice of discounting present value of future payments, argue that the interests of present people count for more than the interests of future people. Counting future people as equals threatens to prevent any realistic calculation from being made since the further into the future one calculates, the less one knows about the consequences. Discounting the interests of future people, on the other hand, threatens to ignore these interests since, eventually, any discount rate will mean that people living in the near future count for nothing.

In contrast to this utilitarian approach, some argue that future people have rights that entail duties on our part. For example, in 1987 the UN-sponsored Brundtland Commission advocated a vision of sustainable development as development "which meets the needs of the present without sacrificing the ability of the future to meets its needs." This suggests that future people have an equal right to the energy necessary to meet their needs. However, if future generations have a right to energy resources in an equal amount to what is available to us, we would be prohibited from using any resources, because whatever we use today is denied forever to the future. On the other hand, if future generations have a right to use energy resources at some point in the future, why do we not have an equal right to use them today?

As can be seen from these examples, even talking about ethical responsibilities to future people can raise conundrums. Some critics claim that talk of ethical responsibilities to distant people is nonsense and that present energy policy should be governed solely by a concern for people living in the present and immediate future.

Two challenges are raised against claims that present generations have ethical responsibilities to future generations. The first is called the problem of "disappearing beneficiaries." Consider the claim that present generations ought to decrease our reliance on fossil fuels to ensure a future world protected from the harmful effects of global warming. The defense of this view argues that future people would be made better-off by this decision. But we need to ask "better-off" than what? Intuitively, we would say that they will be better-off than they would have been otherwise. However, this argument assumes that the people who benefit will be the same people as those who would exist if we adopted the alternative policy of continued reliance on fossil fuels. Yet alternative policy decisions as momentous as international energy policy, population controls, or significant conservation measures, would surely result in different future people being born. When we consider alternative policy decisions, we actually are considering the effects on two, not one, sets of future people. The future population that would be harmed by our decision to continue heavy use of fossil fuels is a different population than the one that would benefit from major conservation programs. Thus, it makes little sense to speak about one future generation being made better or worse-off by either decision. The potential beneficiaries of one policy disappear when we choose the alternative policy.

The second challenge is called the argument from ignorance. Any discussion of future people and their happiness, their needs and preferences, their rights and interests, forces us to make assumptions about who those people might be and what they will be like. But realistically, we know nothing about who they will be, what they will want or need, or even *if* they will exist at all. Since we are ignorant of future people, we have little basis to speak about our responsibilities to them.

The implication of these arguments is that energy policy ought to be set with due consideration given only to present generations. While we might have responsibilities which *regard* future people (present duties to our children affect the life prospects of future generations), we have no direct responsibilities *to* future people. We can have no responsibility to that which does not exist.

Plausible answers can be offered to these challenges. While we may not know who future people will be, if they will be, or what their specific interests and needs might be, we do know enough about future people to establish present ethical responsibilities to them. Just as in cases of legal negligence where we hold individuals liable for unintended but foreseeable and avoidable harms that occur in the future, it can be meaningful to talk about foreseeable but unspecific harms to unknown future people. Surely we have good reasons for thinking that there will be people living in the future, and that they will have needs and interests similar to ours. To the degree that we can reasonably foresee present actions causing predictable harms to future people, we can acknowledge the meaningfulness of ethical responsibilities to future generations. While the present may not have specific responsibilities to identifiable future people, it does make sense to say that we have responsibilities to future people, no matter who they turn out to be.

What might our responsibilities to the future be? How do we balance our responsibilities to future people against the interests, needs, and rights of the present? Perhaps the most reasonable answer begins with the recognition that we value energy not in itself but only as a means to other ends. Thus, the requirement of equal treatment demanded by social justice need not require that future people have an equal right to present energy resources but, rather, that they have a right to an equal opportunity to the ends attained by those resources. Our use of fossil fuels, for example, denies to them the opportunity to use that fuel. We cannot compensate them by returning that lost energy to them, but we can compensate them by providing the future with an equal opportunity to achieve the ends provided by those energy resources. Justice requires that we compensate them for the lost opportunities and increased risks that our present decisions create.

While there are practical difficulties in trying to specify the precise responsibilities entailed by this approach, we can suggest several general obligations of compensatory justice. First, our responsibility to the future should include a serious effort to develop alternative energy sources. Continued heavy reliance on fossil fuels and nuclear power places future people at risk. Justice demands that we minimize that risk, and investment in alternative energy sources would be a good faith step in that direction. Arguments of this sort could justify government expenditures on research into fusion and renewable energy sources. Second, we have a responsibility to conserve nonrenewable resources. Wasting resources that future people will need, especially when we presently have the technology to significantly increase energy efficiency, makes it more difficult for future people to obtain a lifestyle equal to ours. Finally, it would seem we have a responsibility to adopt population policies and modify consumption patterns to moderate worldwide energy demand over the long term.

Joseph R. DesJardins

See also: Conservation of Energy; Culture and Energy Usage; Government and the Energy Marketplace.

BIBLIOGRAPHY

Brundtland, G. (1987). *Our Common Future.* New York: Oxford University Press.

Bullard, R. (1993). *Confronting Environmental Racism.* Boston: South End Press.

Bullard, R. (1994). *Dumping in Dixie.* Boulder, CO: Westview Press.

Daly, H., and Cobb, J. (1989). *For the Common Good.* Boston: Beacon Press.

DesJardins, J. (1997). *Environmental Ethics: An Introduction to Environmental Philosophy.* Belmont, CA: Wadsworth Publishing

DesJardins, J. (1999). *Environmental Ethics: Concepts, Policy, Theory.* Mountain View, CA: Mayfield Publishing.

Dworkin, R. (1977) *Taking Rights Seriously.* Cambridge, MA: Harvard University Press.

MacLean, D., and Brown, P., eds. (1983). *Energy and the Future.* Lanham, MD: Rowman and Littlefield.

Partridge, E., ed. (1980). *Responsibilities to Future Generations.* Buffalo: Prometheus Press.

Rachels, J. (1999). *The Elements of Moral Philosophy.* New York: McGraw-Hill.

Schneider, S. (1989). *Global Warming: Are We Entering the Greenhouse Century?* San Francisco: Sierra Club Books.

Shrader-Frechette, K. S. (1989). *Nuclear Power and Public Policy.* Dordrecht, Holland: Reidel.

Shrader-Frechette, K. S. (1991). *Risk and Rationality.* Berkeley: University of California Press.

Simon, J. (1981). *The Ultimate Resource.* Princeton: Princeton University Press.

Sikora, R. I., and Barry, B., eds. (1978). *Obligations to Future Generations.* Philadelphia: Temple University Press.

Wenz, P. (1988). *Environmental Justice.* Albany: State University of New York Press.

Westra, L., and Wenz, P., eds. (1995). *Faces of Environmental Racism.* Lanham, MD: Rowman and Littlefield Publishers.

EXPLOSIVES AND PROPELLANTS

Explosions occur when gases in a confined space expand with a pressure and velocity that cause stresses greater than the confining structure can withstand. The gas expansion can be caused by rapid generation of gaseous molecules from a solid or liquid (e.g., an explosive) and/or rapid heating (as in a steam "explosion"). An explosion can be low-level, yet still dangerous, as in the deflagration (rapid burning) of a flour dust and air mixture in a grain elevator, or very intensive, as in the detonation of a vial of nitroglycerin.

Explosives and propellants are mixtures of fuel and oxidizer. The intensity of combustion is determined by the heat of combustion per pound of material, the material's density, the gas volume generated per volume of material, and the rate of deflagration or detonation. The latter, the most important variable, is determined by the speed at which fuel and oxidizer molecules combine.

The first explosive was black gunpowder, invented by the Chinese in the Middle Ages. In gunpowder, the fuel is powdered sulfur and charcoal, and the oxidizer is saltpeter (potassium nitrate). When heated, the oxidizer molecule decomposes to form potassium oxide (a solid), nitrogen and nitrous oxides (gases), and excess pure oxygen that burns the fuel to form more gases (carbon oxides, sulfur oxides). The gases generated by this rapidly burning mixture can explode (rupture its container), as in a firecracker, or propel a projectile, as in a rocket or a gun. Because it takes time for oxygen to diffuse to the fuel molecules the explosion of gunpowder is a rapid burning, or deflagration, not the high-rate detonation characteristic of a high explosive.

While propellants are formulated to burn rapidly and in a controlled manner, they can go from deflagration to detonation if mishandled. High explosives, on the other hand, are designed to detonate when activated. Here oxidizer and fuel are always situated in the same molecule, and in the right proportions, as determined by the desired end-products. Once initiated, gases are formed too fast to diffuse away in an orderly manner, and a shock wave is generated that passes through the explosive, detonating it almost instantaneously. This shock wave and the resultant

Explosive Substances

Name	Formula	Use
ammonium nitrate	NH_4NO_3	solid oxidizer
ammonium perchlorate	NH_4ClO_4	solid oxidizer
lead azide	$Pb(N_3)_2$	primary explosive
ammonium picrate		secondary high explosive
lead styphnate		primary explosive
2,4,6-trinitrotoluene		secondary high explosive
picric acid		secondary high explosive
nitrocellulose		secondary explosive used in propellants
nitroglycerin	H_2CONO_2 $\vert$ $HCONO_2$ $\vert$ H_2CONO_2	liquid secondary explosive ingredient in commercial explosives and propellents
nitromethane	CH_3NO_2	liquid secondary explosive
pentaerythritol tetranitrate		secondary high explosive used as booster

Table 1.
Explosive substances.

high-velocity expansion of gases can cause great damage, even if there is no confining container to rupture.

A propellant typically burns at the rate of about 1 cm/sec, some 10,000 times the rate coal burns in air. However high explosives "burn" at a rate some 100,000 to 1,000,000 times faster than propellants because the reaction rate is controlled by shock transfer rather than heat transfer. Thus, although there is more energy released in burning a pound of coal than a pound of dynamite, the power output, which is the rate of doing work, can be 100 times larger for a propellant, and 100 million times larger for a high explosive.

Black gunpowder, was the only explosive (actually a propellant) known until 1847, when the Italian chemist, Ascani Sobrero, discovered nitroglycerin. This rather unstable and shock-sensitive liquid was the first high explosive. It was too dangerous to use until Alfred Nobel developed dynamite in 1866, a 50 percent mixture of nitroglycerin stabilized by absorption in inert diatomaceous earth. The much safer trinitrotoluene (TNT) was synthesized soon after, and military needs in the two World Wars led to a number of new high explosives. Almost all of these are made by nitration of organic substrates, and they are formulated to be simultaneously safer and more energetic (see Table 1). For commercial uses, low cost is of paramount importance, and it has been found that a mixture of ammonium nitrate, a fertilizer ingredient, and fuel oil (ANFO) gives the most "bang for the buck."

By designing an explosive charge to focus its blast on a small area, a so-called shaped or armor-piercing charge, gas velocities as high as 20,000 mph and at pressures of 3 million psi can be achieved. Such a force pushes steel aside through plastic flow, much as a knife cuts through butter. Under the high pressure, the metallic steel behaves like a viscous liquid, the same way plastics flow when extruded through dies to make various shapes.

The thrust of recent explosives research continues to emphasize increased safety and control. These properties are achieved by the use of primary and secondary explosives. The secondary explosive is the main ingredient in a charge, and it is formulated to be stable in storage and difficult to initiate. Some secondary explosives just burn slowly without detonating if accidentally ignited. The initiator or primary explosive is a small quantity of a more sensitive material fashioned into a detonator or blasting cap. It is attached to the main charge just before use and is activated by a fuse, by percussion (as in a gun), or by an electrical current.

The key to safety in explosives manufacturing is to use isolated high-velocity nitric acid reactors that have only a very small hold up at any one time (that is, only a small amount of dangerous material is "held up" inside the reactor at any time). Units are widely spaced, so any accident involves only small amounts of explosive and does not propagate through the plant. Fire and electrical spark hazards are rigorously controlled, and manpower reduced to the absolute minimum through automation.

The recent rise in the use of expolosives in terrorist activity poses new challenges to industry and law enforcement. This challenge is being met by the use of sophisticated chemical detection devices to screen for bombs and more rigorous explosive inventory safeguards and controls. Plans have also been proposed to tag explosives with isotopes to make them easier to trace if misused.

Herman Bieber

See also: Nuclear Energy; Nuclear Fission; Nuclear Fusion.

BIBLIOGRAPHY

Harding, G., ed. (1993). *Substance Detection Systems, Vol. 2092.* Bellingham, WA: Proceedings of the International Society for Optical Engineering.

Squire, D. R. (1991). *Chemistry of Energetic Materials.* San Diego, CA: Academic Press.

Urbanski, T. (1985). *Chemistry and Technology of Explosives* New York: Pergamon Press.

U.S. Bureau of Alcohol, Tobacco and Firearms. (1986). *Firearms and Explosives Tracing.* Washington, DC: Author.

U.S. Senate, Committee on the Judiciary. (1993). *Terrorism in America.* Washington, DC: U.S. Government Printing Office.

EXTERNAL COMBUSTION ENGINES

See: Engines

F

FARADAY, MICHAEL (1791–1867)

Michael Faraday has been called the "patron saint of electrical engineering." He produced the first electric motor and the first electric generator and is considered the greatest experimental scientist of the nineteenth century. Faraday came from humble beginnings. He was born in a village that is now part of London, and his father was a migrant blacksmith who was often ill and unable to support his family. Faraday often went hungry as a child and his only formal education was at a Sandemanian Church Sunday school. (The Sandemanians were a small fundamentalist Christian sect, and Faraday later became an elder of the church.) At age thirteen, he was apprenticed to a bookbinder for seven years. In addition to binding the books, he read them voraciously. Although he completed the apprenticeship, he subsequently sought a way out of a trade that he considered selfish and vicious.

Faraday's great opportunity came when a friend offered him a ticket to attend the lectures on chemistry given by Sir Humphrey Day, the director of the Royal Institution in London. After attending the lectures, Faraday sent Davy a neatly bound copy of his notes and asked for employment. In 1812 Davy did require a new assistant and, remembering the notes, hired Faraday. Davy was a leading scientist of his time and discovered several chemical elements, but it has been said that Faraday was his greatest discovery. Faraday was given quarters at the Royal Institution where he was to remain for forty-five years (staying on even after his marriage). Davy and his wife took Faraday with them as secretary to Europe on a grand tour in 1813. Despite the hostilities between France and England, they received Napoleon's permission to meet with French scientists in Paris. During this time Faraday's talent began to be recognized internationally.

In 1820 Faraday finished his apprenticeship under Davy and in the following year married and settled into the Royal Institution. Faraday's early reputation as a chemist was so great that in 1824 he was elected to the Royal Society. In 1825 Davy recommended that Faraday succeed him as director of the Royal Institution. The appointment paid only a hundred pounds a year, but Faraday soon received some adjunct academic appointments that enabled him to give up all other professional work and devote himself full-time to research. Faraday's scientific output was enormous, and at the end of his career, his laboratory notebooks, which covered most of his years at the Royal Institution, contained more than sixteen thousand neatly inscribed entries, bound in volumes by Faraday himself.

In his early work, Faraday was primarily a chemist. He liquefied several gases previously considered incapable of liquefaction, discovered benzene, prepared new compounds of carbon and chlorine, worked on new alloys of steel, and discovered the laws of electrolysis that bear his name. The latter discovery became the basis for the electroplating industry that developed in England during the early nineteenth century. In 1821, however, Hans Christian Oersted's discovery that an electric current could produce a magnetic field led Faraday away from chemistry for a while. With a better understanding of electric and magnetic fields, Faraday succeeded in building the first elementary electric motor.

Faraday returned to chemistry, but after 1830, his investigations again concentrated on electric and magnetic phenomena. He had become convinced that the reverse action to the phenomenon discovered by Oersted was also possible, that a magnetic

Michael Faraday. (Corbis Corporation)

field could produce an electric current. He also believed that he could induce a current in a circuit using an electromagnet like the one invented by Sturgeon in 1824. Wrapping two wires many times around opposite sides of an iron ring, Faraday discovered that when an electric current in one wire was turned on or off a current appeared in the other wire. Thus Faraday discovered the law of electromagnetic induction that bears his name: The electromotive force (voltage) induced in a circuit is equal to the (negative) time rate-of-change of the magnetic flux through the circuit. This law was not only the basis for the first elementary electric generator that Faraday produced, but also for all subsequent electric power dynamos that employ coils rotating in a magnetic field to produce electric power. Although Faraday is credited with the discovery because he was the first to publish his work, it was later learned that induction had been discovered shortly before by Joseph Henry, then an instructor in an obscure school in Albany, New York.

In the decade that followed this discovery, Faraday continued to make fundamental discoveries about electricity and magnetism. He showed that the electricity obtained from various sources was the same, analyzed the effects of dielectrics on electrostatics, and studied electric discharges in gases. At the end of 1839, however, his health broke down. There is reason to believe that Faraday may have suffered from mercury poisoning, a common affliction in that period. He did not return to work completely until 1845, when he discovered the phenomena of magnetically induced birefringence in glass and of diamagnetism. In 1846 he published a paper in which he suggested that space was a medium that bore electric and magnetic strains and that these strains were associated with the propagation of light. Later scientists recognized that these ideas were the forerunners of the modern theory of electromagnetic propagation and optical fields.

Faraday worked alone; he had no students, just ordinary assistants. Although devoted to laboratory work, he was also a brilliant public lecturer. Personally, he was invariably described as a gentle and modest person. He never forgot his humble beginnings, and he had a clear view of his own worth and a disdain of the class system. In 1858, when Queen Victoria, in view of his lifetime of great achievement, offered him a knighthood and the use of a house, Faraday accepted the cottage but refused the knighthood, stating that he preferred to remain "plain Mr. Faraday."

Faraday's activity slowed after 1850, and, in 1865, a progressive loss of memory forced his complete retirement. He died in 1867 and he was buried, not in Westminster Abbey, but perhaps more befitting his egalitarian ideals, in Highgate Cemetery, London.

Leonard S. Taylor

See also: Electric Motor Systems; Electric Power, Generation of; Magnetism and Magnets; Oersted, Hans Christian.

BIBLIOGRAPHY

Atherton, W. A. (1984). *From Compass to Computer.* San Francisco: San Francisco Press.

Segrè, E. (1984). *From Falling Bodies to Radio Waves.* New York: W. H. Freeman.

Meyer, H. W. (1971). *A History of Electricity and Magnetism.* Cambridge, MA: MIT Press.

FEDERAL ENERGY REGULATORY COMMISSION

See: Government Agencies

FERMI, ENRICO (1901–1954)

Enrico Fermi was both a brilliant theorist and an unusually gifted experimentalist — a combination of talents seldom found among twentieth-century physicists. Born in 1901 in Rome, Fermi obtained his doctor's degree in physics magna cum laude from the University of Pisa at the age of 21, with a dissertation on x-rays.

After two years of post-doctoral research at Max Born's Institute in Göttingen, and then with Paul Ehrenfest in Leiden, Fermi taught for two years at the University of Florence, where he soon established his reputation by developing what are now known as the Fermi-Dirac statistics. In 1926 he was appointed to a full professorship in physics at the University of Rome, where he quickly gathered around him a group of talented young faculty members and students, who helped him make a name for Rome in the fields of nuclear physics and quantum mechanics. His theoretical work culminated in a 1933 theory of nuclear beta decay that caused a great stir in world physics circles, and is still of major importance today.

Fermi had been fascinated by the discovery of the neutron by James Chadwick in 1932. He gradually switched his research interests to the use of neutrons to produce new types of nuclear reactions, in the hope of discovering new chemical elements or new isotopes of known elements. He had seen at once that the uncharged neutron would not be repelled by the positively-charged atomic nucleus. For that reason the uncharged neutron could penetrate much closer to a nucleus without the need for high-energy particle accelerators. He discovered that slow neutrons could be produced by passing a neutron beam through water or paraffin, since the neutron mass was almost equal to that of a hydrogen atom, and the consequent large energy loss in collisions with hydrogen slowed the neutrons down very quickly. Hence these "slow" or "thermal" neutrons would stay near a nucleus a longer fraction of a second and would therefore be more easily absorbed by the nucleus under investigation. Using this technique, Fermi discovered forty new artificial radioactive isotopes.

In 1934 Fermi decided to bombard uranium with neutrons in an attempt to produce "transuranic" elements, that is, elements beyond uranium, which is number 92 in the periodic table. He thought for a while that he had succeeded, since unstable atoms were produced that did not seem to correspond to any known radioactive isotope. He was wrong in this conjecture, but the research itself would eventually turn out to be of momentous importance both for physics and for world history, and worthy of the 1938 Nobel Prize in Physics.

Fermi's wife, Laura, was Jewish, and as Hitler's influence over Mussolini intensified, anti-Jewish laws were passed that made Laura's remaining in Italy precarious. After accepting his Nobel Prize in Stockholm, Fermi and his wife took a ship directly to the United States, where they would spend the rest of their lives. Enrico taught at Columbia University in New York City from 1939 to 1942, and at the University of Chicago from 1942 until his death in 1954.

In 1938 Niels Bohr had brought the astounding news from Europe that the radiochemists Otto Hahn and Fritz Strassmann in Berlin had conclusively demonstrated that one of the products of the bombardment of uranium by neutrons was barium, with atomic number 56, in the middle of the periodic table of elements. He also announced that in Stockholm Lise Meitner and her nephew Otto Frisch had proposed a theory to explain what they called "nuclear fission," the splitting of a uranium nucleus under neutron bombardment into two pieces, each with a mass roughly equal to half the mass of the uranium nucleus. The products of Fermi's neutron bombardment of uranium back in Rome had therefore not been transuranic elements, but radioactive isotopes of known elements from the middle of the periodic table.

Fermi and another European refugee, Leo Szilard, discussed the impact nuclear fission would have on physics and on the very unstable state of the world

Enrico Fermi. (Library of Congress)

itself in 1938. The efforts of Szilard in 1939 persuaded Albert Einstein to send his famous letter to President Franklin D. Roosevelt, which resulted in the creation of the Manhattan Project to contruct a nuclear bomb. Fermi was put in charge of the first attempt to construct a self-sustaining chain reaction, in which neutrons emitted by a fissioning nucleus would, in turn, produce one or more fission reactions in other uranium nuclei. The number of fissions produced, if controlled, might lead to a useful new source of energy; if uncontrolled, the result might be a nuclear bomb of incredible destructive power.

Fermi began to assemble a "nuclear pile" in a squash court under the football stands at the University of Chicago. This was really the first nuclear power reactor, in which a controlled, self-sustaining series of fission processes occurred. The controls consisted of cadmium rods inserted to absorb neutrons and keep the reactor from going

"critical." Gradually the rods were pulled out one by one, until the multiplication ratio of neutrons produced to neutrons absorbed was exactly one. Then the chain reaction was self-sustaining. To proceed further would run the risk of a major explosion. Fermi had the reactor shut down at exactly 3:45 P.M. on December 2, 1942, the day that is known in history as the beginning of nuclear energy and nuclear bombs.

Fermi lived only a little more than a decade after his hour of triumph. He spent most of this time at the University of Chicago, where, as in Rome, he surrounded himself with a group of outstanding graduate students, many of whom also later received Nobel Prizes. Fermi died of stomach cancer in 1954, but his name remains attached to many of the important contributions he made to physics. For example, element 100 is now called Fermium.

Fermi's overall impact on physics is well summarized by the nuclear physicist Otto Frisch (1979,

p. 22): "But occasionally one gets a man like Enrico Fermi, the Italian genius who rose to fame in 1927 as a theoretician and then surprised us all by the breathtaking results of his experiments with neutrons and finally by engineering the first nuclear reactor. On December the second, 1942, he started the first self-sustaining nuclear chain reaction initiated by man and thus became the Prometheus of the Atomic Age."

Joseph F. Mulligan

BIBLIOGRAPHY

Allison, S. K., et al. (1955). "Memorial Symposium in Honor of Enrico Fermi held on April 29, 1955 at meeting of the American Physical Society in Washington, D.C." *Reviews of Modern Physics* 27:249–275.

Fermi, E. (1962–1965). *Collected Papers*, 3 vols., ed. E. Segrè, et al. Chicago: University of Chicago Press.

Fermi, L. (1954). *Atoms in the Family: My Life with Enrico Fermi.* Chicago: University of Chicago Press.

Frisch, O. (1979). *What Little I Remember.* Cambridge: Cambridge University Press.

Segrè, E. (1970). *Enrico Fermi, Physicist.* Chicago: University of Chicago Press.

Segrè, E. (1971). "Fermi, Enrico." In *Dictionary of Scientific Biography,* ed. Charles Coulston Gillispie, Vol. 4, pp. 576–583. New York: Scribner.

Smyth, H. D. (1945). *Atomic Energy for Military Purposes.* Washington, DC: U.S. Government Printing Office.

Wattenberg, A. (1993). "The Birth of the Nuclear Age." *Physics Today* 46(January):44–51.

FERMI NATIONAL ACCELERATOR LABORATORY

See: Government Agencies

FIELD ENERGY

See: Waves

FISSION, NUCLEAR

See: Nuclear Fission

FISSION BOMB

See: Nuclear Fission

FLYWHEELS

Flywheels store kinetic energy (energy of motion) by mechanically confining motion of a mass to a circular trajectory. The functional elements of the flywheel are the mass storing the energy, the mechanism supporting the rotating assembly, and the means through which energy is deposited in the flywheel or retrieved from it.

Energy can be stored in rings, disks, or discrete weights, with spokes or hubs connecting the storage elements to shafts, and bearings supporting the assembly and allowing it to rotate. Energy may be transferred into or out of the wheel mechanically, hydraulically, aerodynamically, or electrically.

HISTORY

Ubiquitous in rotating machinery, flywheels have been used as a component of manufacturing equipment since their application in potters' wheels before 2000 B.C.E.

Flywheels attained broad use during the Industrial Revolution. In the embodiment of this era, flywheels used heavy rims built from cast iron to damp pulsations in engines, punches, shears, and presses. Often the pulsations to be damped arose from reciprocating motive forces or reciprocating end processes. The conversion of reciprocation into rotation enabled formatting of the flow of this energy. The most important types of formatting were transportation of energy by shafts, conversion of torque and speed by gears, and damping by flywheels.

A 23-ton flywheel in the Palace of Engineering at the 1924 British Empire Exhibition. (Corbis Corporation)

Flywheels are found in internal-combustion engines, where they damp out torque pulses caused by the periodic firing of cylinders. In this application, energy is stored very briefly before it is used— for less than one revolution of the wheel itself.

The evolution of flywheel materials and components and a systemic approach to design have led to the development of stand-alone flywheel energy storage systems. In these systems the rotating element of the flywheel transfers energy to the application electrically and is not directly connected to the load through shafting. These systems typically store energy that is released over many revolutions of the wheel, and as a system may be used in place of electrochemical energy storage in many applications. By being separate and distinct from the process it supports, the stand-alone flywheel system may use materials and components optimally. Of the various flywheel types, stand-alone systems will typically have the highest energy and power density as well as the highest rim speed and rotation rate.

TECHNOLOGY

The kinetic energy stored in the flywheel rotor is proportional to the mass of the rotor and the square of its linear velocity. Transformed into a cylindrical system, the stored kinetic energy, KE, is

$$KE = \tfrac{1}{2} J \omega^2 = \tfrac{1}{2} mr^2\omega^2$$

where ω is the rate of rotation in radians per second and J is the moment of inertia about the axis of rotation in kilogram-meter^2. For the special case of a radially thin ring, the moment of inertia is equal to its mass, m, multiplied by the square of its radius, r. This radius is also known as the radius of gyration.

Stress in the rim is proportional to the square of linear velocity at the tip. When rotor speed is dictated

Applications	Types	Unique Attributes
Stationary engines (historic), damp pulsations	Spoked hub, steel rim	Massive, low-speed rotors, belt or shaft mechanical connection to application
Automobile engine, damp out torque for	Solid metal rotor	Mounted on engine shaft, very low cost
Satellite stabilization and energy storage	Control moment gyro/reaction wheel	Lightweight, extremely long life, and high reliability
Stationary UPS system	Steel or composite rotor in vacuum	Electrical connection to application; high power density (relatively high power generator)
Energy storage for hybrid propulsion	Composite rotor in vacuum	Electrical connection to application; high energy density (lightweight rotor)

Table 1.
Flywheel Applications, Types, and Unique Attributes

by material considerations, the linear velocity of the tip is set, and rotation rate becomes a dependent parameter inversely proportional to rotor diameter. For example, a rotor with a tip speed of 1,000 meters per second and a diameter of 0.3 meter would have a rotation rate of about 63,700 rpm. If the diameter were made 0.6 meters instead, for this material the rotation rate would be about 31,850 rpm.

To maximize stored energy, the designer seeks to spin the rotor at the highest speed allowed by the strength of the materials used. There is a trade-off between heavier, lower-strength materials and stronger, lighter materials. For a thin rim, the relationship between rim stress and specific energy or energy stored per unit mass of rim is given by

$$KE/m = \sigma_h/2\rho$$

where σ_h is the hoop stress experienced by the ring in newtons per square meter, and ρ is the density of the ring material in kilograms per meter^3. Thus high specific energy corresponds directly to high specific strength: $\sigma sh/\rho$ and rotors made from carbon composite may be expected to store more energy per unit weight than metal rotors.

Since energy is proportional to the square of speed, high performance will be attained at high tip speed. Rims produced from carbon fiber have attained top speeds in excess of 1,400 meters per second and must be housed in an evacuated chamber to avoid severe aerodynamic heating.

The flywheel rim is connected to a shaft by spokes or a hub. The rotor experiences high centrifugal force and will tend to grow in size while the shaft will not. The spoke or hub assembly must span the gap between the shaft and the rim, allowing this differential growth while supporting the rim securely. High-speed composite rims may change dimension by more than 1 percent in normal operation. This large strain or relative growth makes hub design especially challenging for composite flywheels.

Bearings support the shaft and allow the flywheel assembly to rotate freely in applications where the flywheel is a component in a more extensive rotating machine, the rim and hub are supported by the shafting of the machine, and no dedicated bearings are used. In stand-alone systems, flywheel rotors are typically supported with hydrodynamic or rolling element bearings, although magnetic bearings are sometimes used either to support part of the weight of the rotor or to levitate it entirely.

Compact, high-rim-speed, stand-alone flywheel systems may require that the bearings run continuously for years at many tens of thousands of revolutions per minute. Smaller rotors will operate at higher rotation rates.

Historically, flywheels have stored and discharged energy through direct mechanical connection to the load. The flywheel may be affixed to the load or may communicate with the load through gears, belts, or shafts. Stand-alone flywheel systems may convert electrical energy to kinetic energy through a motor, or convert kinetic energy to electricity through a generator. In these systems the flow of energy into and out of the flywheel may be regulated electronically using active inverters controlled by microprocessors or digital signal processors.

FUTURE ADVANCES

The flywheel has become an integral energy storage element in a broad range of applications. New designs and innovations in current designs and diverse opportunities for bettering flywheel performance continue to emerge.

The modern stand-alone flywheel embodies a number of sophisticated technologies, and advances in these technologies will yield further improvement to this class of flywheel. Flywheel progress will track development in power electronics, bearings, and composite materials. Power electronics such as active inverters and digital signal processors will continue to become power dense, reliable, and inexpensive. Rolling element and magnetic bearings will mature to a point where decades-long operating life is considered routine.

The energy stored in a flywheel depends on the strength of the rotor material. Carbon fiber tensile strength remains well below theoretical limits. Expected increases in strength along with reduction in cost as the use of this material expands will translate into more energy dense, less expensive rotors.

Progress, which is likely to be incremental, is certain to improve the performance and energy efficiency of all applications.

Donald A. Bender

See also: Bearings; Conservation of Energy; Heat and Heating; Kinetic Energy; Materials; Potential Energy; Storage; Storage Technology; Tribology.

BIBLIOGRAPHY

Genta, G. (1985). *Kinetic Energy Storage.* London: Butterworths.

Post, R. F.; Fowler, T. K.; and Post, S. F. (1993). "A High-Efficiency Electromechanical Battery." *Proceedings of the IEEE* 81(3).

Post, R. F.; and Post, S. F. (1973). "Flywheels." *Scientific American* 229(Dec.):17.

U.S. Department of Energy. (1995). *Flywheel Energy Storage Workshop.* Springfield, VA: U.S. Department of Commerce.

FOSSIL FUELS

Fossil fuel is a general term for any hydrocarbon or carbonaceous rock that may be used for fuel: chiefly petroleum, natural gas, and coal. These energy sources are considered to be the lifeblood of the world economy. Nearly all fossil fuels are derived from organic matter, commonly buried plant or animal fossil remains, although a small amount of natural gas is inorganic in origin. Organic matter that has long been deeply buried is converted by increasing heat and pressure from peat into coal or from kerogen to petroleum (oil) or natural gas or liquids associated with natural gas (called natural gas liquids). Considerable time, commonly millions of years, is required to generate fossil fuels, and although there continues to be generation of coal, oil and natural gas today, they are being consumed at much greater rates than they are being generated. Fossil fuels are thus considered nonrenewable resources.

This article provides a brief historical perspective on fossil energy, focusing on the past several decades, and discusses significant energy shifts in a complex world of constantly changing energy supply, demand, policies and regulations. United States and world energy supplies are closely intertwined (Figure 1). World supplies of oil, gas and coal, are less extensively developed than those of the United States and the extent of remaining resources is extensively debated. Fossil fuels such as coal, natural gas, crude oil and natural gas liquids currently account for 81 percent of the energy use of the United States, and in 1997 were worth about $108 billion. For equivalency discussions and to allow comparisons among energy commodities, values are given in quadrillion British thermal units (Btus). For an idea of the magnitude of this unit, it is worth knowing that 153 quadrillion Btus of energy contains about a cubic mile of oil. The world consumes the equivalent of about 2.5 cubic miles of oil energy per year, of which 1 cubic mile is oil, 0.6 cubic mile is coal, 0.4 cubic mile is gas, and 0.5 cubic

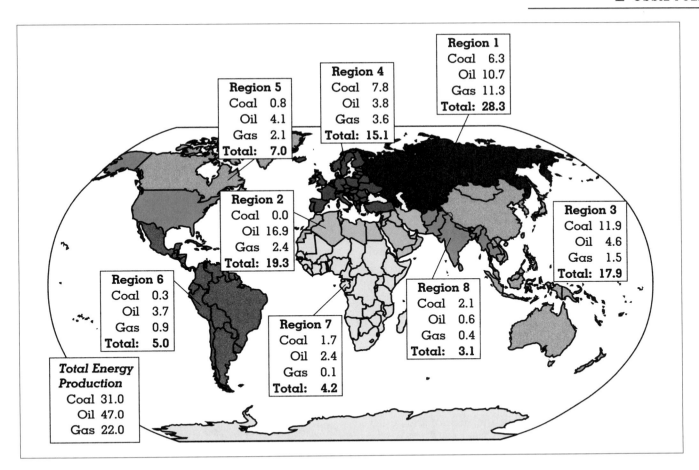

Figure 1.
Percentage comparisons of fossil fuel production for coal, oil, and natural gas for the world exclusive of the
United States.

mile is all other forms of energy. Total oil reserves are about 32 cubic miles of oil and total energy reserves are equivalent to about 90 cubic miles of oil. Quadrillion Btus can also be converted to billion barrel oil equivalents (BBOE) at the ratio of about 5:1 (i.e., 5 quadrillion Btu to 1 BBOE).

Both energy consumption and production more than doubled between 1960 and 2000, reflecting the United States's increasing need for energy resources (Figure 1). In 2000 the United States was at its historically highest level of both fossil energy consumption and production. However, significant shifts in domestic energy consumption and production occurred between 1980 and 2000. Production of oil and natural gas in the United States did not meet consumption between 1970 and 2000.

Surprisingly, coal is the largest energy source in the

United States and the world (Figure 1), despite perceptions that it has been replaced by other sources. In 1997 production of both coal (23.2 quadrillion Btus, or about 4.6 billion barrels of oil) and natural gas (19.5 quadrillion Btus, or about 3.9 billion barrels of oil) on an energy equivalent basis exceeded U.S. domestic oil production (13.6 quadrillion Btus, equivalent to about 2.7 billion barrels, or 3.1 billion barrels of oil if natural gas liquids are included). Coal production in the United States nearly doubled from 1970 to 2000 (from about 600 million tons to about 1 billion tons produced annually). Meanwhile, petroleum consumption at 18.6 million barrels of oil per day is near the all-time high of 18.8 million barrels of oil per day in 1978. Net U.S. petroleum imports (8.9 million barrels of oil per day) in 1997 were worth $67 billion and exceeded U.S. petroleum production (8.3 million barrels of oil per day). Concerns about petroleum supplies, specifically oil shortages, caused the United States to build a strategic petroleum reserve in 1977 that currently holds 563 million barrels of oil or about sixty-three days worth of net imported petroleum. This strategic supply is about half of the 1985 high when the reserve would have provided 115 days worth of net imported petroleum. Although U.S. oil and gas production generally declined from 1970 to 1993, natural gas production has been increasing since the mid-1980s and energy equivalent production from natural gas exceeded U.S. oil production in the late 1980s. Natural gas is expected to play an increasing role in the United States in response to both environmental concerns and anticipated major contributions from unconventional (or continuous) natural gas sources requiring reductions in carbon emissions, particularly if the Kyoto Protocol is passed by the United States.

Due to factors such as extreme fluctuations in commodity prices, particularly oil prices, wasteful oil and gas field practices, and perceived national needs, U.S. government involvement in energy markets has been part of U.S. history from the turn of the century. The U.S. began importing oil in 1948. The 1973 oil embargo by the Organization of Petroleum Exporting Countries (OPEC) made a strong impact on U.S. policy makers who responded by developing regulations designed to encourage new domestic oil production. A two-tiered oil pricing system was introduced that changed less for old oil and more for new oil. The prospect of higher prices for new oil produced record high drilling levels, focusing on oil development, in the late 1970s and early 1980s. Domestic production fell dramatically in the early 1980s following oil price deregulation that permitted world market forces to control oil prices. Since that time, the United States has returned to a period of reliance on oil supplied by the OPEC comparable to early 1970s levels.

Regulations also have a strong impact on natural gas supply, demand and prices. Exploration for and development of natural gas historically have been secondary to oil because of the high costs of transportation, as well as a complex transportation and marketing system that allowed for U.S.-federally regulated interstate gas pipelines but essentially unregulated intrastate pipelines. The natural gas supply shortfalls in 1977 and 1978 resulted in The Natural

Gas Policy Act of 1978, which was designed to deregulate natural gas on a 10-year schedule. The Act also extended U.S. federal regulation to all pipelines and gave incentives to explore for and develop certain classes of resources such as unconventional gas. Section 29 of the 1980 Windfall Profits Tax provided tax credits for coal-bed methane, deep-basin gas, and tight-gas reservoirs. These tax credits were graduated and substantial; for example, the tax credit on coal-bed methane was 90 cents per million Btus and 52 cents per million Btus for tight-gas reservoirs by the end of 1992 when the credit was terminated. The price of coal-bed methane during the years the credit was in effect ranged from $1 to $3 per million Btus. Drilling for Section 29 gas wells increased during the late 1980s and early 1990s, allowing drillers to establish production prior to 1993 and thus take advantage of the tax incentive, which remains in effect for gas produced from these wells through 2003.

By contrast, U.S. coal resources are not restricted by supply; however, the environmental consequences of coal use have had a major impact on coal development. The Power Plant and Industrial Fuel Use Act of 1978 was developed in response to perceived natural gas shortages; it prohibited not only the switching from oil to gas in power generation plants, but also the use of oil and gas as primary fuel in newly built large plants. However, coal remains the least expensive source of energy; consequently, coal has soared from 1980 to 2000. The most significant change in the use of coal reflects compliance with the Clean Air Act Amendments of 1990 (CAAA 90), which have stringent sulfur dioxide emission restrictions. Production of coal that complies with this Act has caused a shift from production east of the Mississippi to west of the Mississippi. Many western states have substantial coal resources, particularly low-sulfur coal resources, such as those in the Powder River Basin of Wyoming. In fact, Wyoming has been the largest coal-producing state since 1988. CAAA 90 allows utilities to bring coal-fired generating units into compliance, for example, by replacing coal-fired units with natural gas, or by using renewable or low-sulfur coals. Conversion to natural gas from coal in power plants, made feasible by the relatively low costs of natural gas generation in the late 1980s and early 1990s has eroded coal's 1990 share of 53 percent of domestic electricity generation. Clearly policy and regulations such as CAAA

90 impact U.S. coal quality issues and promote low-sulfur coal and natural gas usage.

Many of the resource additions to natural gas will come from unconventional accumulations, also called continuous deposits, such as tight-gas reservoirs and coal-bed methane. Since deregulation of the oil and gas industry in the 1980s, policy decisions have increasingly affected the energy industry (particularly of natural gas) by providing tax incentives to produce unconventional or continuous hydrocarbon accumulations. However, there are associated costs. For example, coal-bed methane is an important and growing natural gas resource, but the disposal cost of waste water generated by coal-bed methane production is thirty-eight times greater than the cost of disposing waste water generated by an onshore conventional gas well. Similarly, the electricity generated by alternative energy sources, such as windmills near San Francisco, is three times more expensive than electricity produced by conventional electric generators.

Increased energy use both for coal and natural gas likely will have the greatest impact on western U.S. federal lands because major unconventional or continuous natural gas deposits are known to exist in Wyoming, Utah, Montana, and New Mexico. The U.S. Geological Survey, in cooperation with the Department of Energy, estimates that an in-place natural gas resource in the Green River Basin in Wyoming is more than ten times larger than a recent estimate of the entire recoverable conventional natural gas resources of the United States. However, recoverable resources make up only a small percentage of this large in-place resource. These continuous deposits are distributed over a wide area as opposed to conventional resources that are localized in fields. The majority of conventional undiscovered oil and natural gas resources will likely be found on U.S. federally-managed lands, and thus the federal government and policies will continue to play an increasing role in energy development.

Implementation of the 1998 Kyoto Protocol, which is designed to reduce global carbon emissions, will have dramatic effects on fossil fuel usage worldwide. The Kyoto Protocol mostly affects delivered prices for coal and conversion of plants to natural gas, nuclear and/or renewable resources. However, as pointed out by the International Energy Agency, increased natural gas consumption in the United States may likely have the effect of increased reliance on imported oil in the short-term to fulfill transportation sector needs. New technologies, such as gas to liquid conversion, have the potential to create revolutionary industrywide change. This change is contingent upon sufficient commodity prices to support the necessary infrastructure to develop underutilized gas resources worldwide.

Some economists argue that civilization moves toward increasingly efficient energy resources, moving from sources such as wood to coal to oil to natural gas and ultimately to non-carbon based energy sources in 100 to 200 year cycles. Others argue for a far more complicated process involving demand, supply and regulations. Fossil fuels will remain critical resources well into the next century. In the meantime, their abundance and potential shortages are debated.

Thomas S. Ahlbrandt

See also: Geography and Energy Use; Reserves and Resources.

BIBLIOGRAPHY

Ahlbrandt, T. S. (1997). "The World Energy Program." *U.S. Geological Survey Fact Sheet.* FS 007–97.

Ahlbrandt, T. S., and Taylor, D.J. (1993). "Domestic Conventional Natural Gas Reserves—Can They Be Increased by the Year 2010?" In *The Future of Energy Gases,* ed. David B. Howell. U.S. Geological Survey Professional Paper 1570, 527–546.

Campbell, C. J. (1997). *The Coming Oil Crisis.* Brentwood, United.Kingdom: Multi-Science Publishing Company.

Edwards, J. D. (1997). "Crude Oil and Alternate Energy Production Forecasts for the Twenty-first Century: The End of the Hydrocarbon Era." *American Association of Petroleum Geologists Bulletin* 81:1292–1305.

Gautier, D. L.; Dolton, G. L.; Attanasi, E. D. (1998). *1995 National Oil and Gas Assessment of Onshore Federal Lands.* U.S. Geological Survey Open-File Report 95–75.

International Energy Agency. (1998). *World Energy Outlook.* Paris, France: International Energy Agency/OECD.

Jackson, J., ed. (1997). *Glossary of Geology,* 4th ed. Alexandria, Virginia: American Geological Institute.

Klett, T. R.; Ahlbrandt, T. S.; Schmoker, J. W.; and Dolton, G. L. (1997). *Ranking of the World's Oil and Gas Provinces by Known Petroleum Volumes.* U.S. Geological Survey Open File Report 97–463, CD-ROM.

Law, B. E., and Spencer, C. W. (1993). "Gas in Tight Reservoirs—An Emerging Major Source of Energy." In *The Future of Energy Gases,* ed. David B. Howell. U.S. Geological Survey Professional Paper 1570, 233–252.

Masters, C. D.; Attanasi, E. D; and Root, D. H. (1994). "World Petroleum Assessment and Analysis." *Proceedings* of the 14th World Petroleum Congress. London: John Wiley and Sons.

Munk, N. (1994). "Mandate Power." Forbes, August, 41–42.

Nakicenovic, N. (1993). "The Methane Age and Beyond." In The Future of Energy Gases, ed. David B. Howell. U.S. Geological Survey Professional Paper 1570, 661–676.

National Petroleum Council. (1992). The Potential for Natural Gas in the United States. Washington, DC, National Petroleum Council.

U.S. Department of Energy. (1998). Comprehensive National Energy Strategy. Washington, DC: Author.

U.S. Energy Information Administration. (1998). Annual Energy Review 1997. Washington, DC: Department of Energy/Energy Information Administration.

U.S. Energy Information Administration. (1998). International Energy Outlook 1998. Washington, DC: Department of Energy/Energy Information Administration.

U.S. Energy Information Administration. (1998). What Does the Kyoto Protocal Mean to U.S. Energy Markets and the U.S. Economy?. Washington, DC: Department of Energy/Energy Information Administration.

U.S. Geological Survey. (1993). The Future of Energy Gases. U.S. Geological Survey Circular 1115.

FOURIER, JEAN BAPTISTE JOSEPH (1768–1830)

Every physical scientist knows the name Fourier; the Series, Integral, and Transform that bear his name are essential mathematical tools. Joseph Fourier's great achievement was to state the equation for the diffusion of heat. The Fourier Series is a series of origonometric terms which converges to a periodic function over one period. The Fourier Integral is the limiting form of a Fourier Series when the period of the periodic function tends to infinity. The Fourier Transform is an analytic tool derived from the coefficients of the integral expansion of a Fourier Series. He pioneered the use of Fourier Series and Integrals because he needed them to solve a range of problems related to the flow of heat. Fourier was not the first person to realize that certain trigonometric expressions could be used to represent certain functions but he did develop the systematic use of such expressions to represent *arbitrary* functions.

BIOGRAPHY

The son of a tailor, Joseph Fourier was a member of a large family. Both of his parents died by the time he was nine. His education began at a local, church-run, military school, where he quickly showed talent in his studies and especially in mathematics. His school persuaded him to train as a priest. While preparing to take holy orders he taught his fellow novices mathematics. Fourier may well have entered the priesthood, but due to the French Revolution new priests were banned from taking holy orders. Instead he returned to his home town of Auxerre and taught at the military school. His friend and mathematics teacher, Bonard, encouraged him to develop his mathematical research, and at the end of 1789 Fourier travelled to Paris to report on this research to the Academie des Sciences.

Inspired by what he had experienced in Paris, Fourier joined the local Popular party on his return to Auxerre. In later years Fourier's involvement in local politics would lead to his arrest. He was arrested twice but each time was granted clemency.

France lost many of its teachers during the first years of the Revolution. One of the solutions to the shortage of teachers was the establishment of the Ecole Normale in Paris. Fourier, as a teacher and an active member of the Popular Society in Auxerre, was invited to attend in 1795. His attendance at the short-lived Ecole gave him the opportunity to meet and study with the brightest French scientists. Fourier's own talent gained him a position as assistant to the lecturers at the Ecole Normale.

The next phase of his career was sparked by his association with one of the lecturers, Gaspard Monge. When the French ruling council, the Directorate, ordered a campaign in Egypt, Monge was invited to participate. Fourier was included in Monge's Legion of Culture, which was to accompany the troops of the young general Napoleon Bonaparte (even then a national hero due to his successful campaigns in Italy).

The Egyptian campaign failed, but Fourier along with his fellow scientists managed to return to France. The general Fourier had accompanied to Egypt was now First Consul. Fourier had intended to return to Paris but Napoleon appointed Fourier as prefect of Isère. The prefecture gave Fourier the resources he needed to begin research into heat propagation but thwarted his ambition to be near the capital.

Jean Baptiste Joseph Fourier. (Library of Congress)

THEORIES OF HEAT PROPAGATION

The prevailing theory of heat, popularized by Simeon-Denis Poisson, Antoine Lavoisier and others, was a theory of heat as a substance, "caloric." Different materials were said to contain different quantities of caloric. Fourier had been interested in the phenomenon of heat from as early as 1802. Fourier's approach was pragmatic; he studied only the *flow* of heat and did not trouble himself with the vexing question of what the heat actually was.

The results from 1802–1803 were not satisfactory, since his model did not include any terms that described why heat was conducted at all. It was only in 1804, when Jean-Baptiste Biot, a friend of Poisson, visited Fourier in Grenoble that progress was made. Fourier realized that Biot's approach to heat propagation could be generalized and renewed his efforts. By December of 1807 Fourier was reading a long memoir on "the propagation of heat in solids" before the Class of the Institut de France.

The last section of the 1807 memoir was a description of the various experiments which Fourier had undertaken. It concentrated on heat diffusion between discrete masses and certain special cases of continuous bodies (bar, ring, sphere, cylinder, rectangular prism, and cube). The memoir was never published, since one of the examiners, Lagrange, denounced his use of the Fourier Series to express the initial temperature distribution. Fourier was not able to persuade the examiners that it was acceptable to use the Fourier Series to express a function which had a completely different form.

In 1810, the Institut de France announced that the Grand Prize in Mathematics for the following year was to be on "the propagation of heat in solid bodies." Fourier's essay reiterated the derivations from his earlier works, while correcting many of the errors. In 1812, he was awarded the prize and the sizeable honorarium that came with it.

Though he won the prize he did not win the outright acclaim of his referees. They accepted that Fourier had formulated heat flow correctly but felt that his methods were not without their difficulties. The use of the Fourier Series was still controversial. It was only when he had returned to Paris for good (around 1818) that he could get his work published in his seminal book, *The Analytical Theory of Heat.*

THE FOURIER LEGACY

Fourier was not without rivals, notably Biot and Poisson, but his work and the resulting book greatly influenced the later generations of mathematicians and physicists.

Fourier formulated the theory of heat flow in such a way that it could be solved and then went on to thoroughly investigate the necessary analytical tools for solving the problem. So thorough was his research that he left few problems in the analysis of heat flow for later physicists to investigate and little controversy once the case for the rigour of the mathematics was resolved. To the physical sciences Fourier left a practical theory of heat flow which agreed with experiment. He invented and demonstrated the usefulness of the Fourier Series and the Fourier integral—major tools of every physical scientist. His book may be seen both as a record of his pioneering work on heat propagation and as a mathematical primer for physicists. Lord Kelvin described the *Theory of Heat* as "a great mathematical poem."

Fourier's own attitude to his work is illustrated by

509

the "Preliminary Discourse" he wrote to introduce his book. As a confirmed positivist, he stated that whatever the causes of physical phenomena, the laws governing them are simple and fixed—and so could be discovered by observation and experiment. He was at pains to point out that his work had application to subjects outside the physical sciences, especially to the economy and to the arts. He had intended to write a companion volume to his *Theory of Heat* that would cover his experimental work, problems of terrestrial heat, and practical matters (such as the efficient heating of houses); but it was never completed.

His talents were many: an intuitive grasp of mathematics, a remarkable memory and an original approach. Fourier was a man of great common sense, a utilitarian, and a positivist.

David A. Keston

See also: Heat and Heating.

BIBLIOGRAPHY

Bracewell, R. N. (1989). "The Fourier Transform." *Scientific American* 6:62.

Grattam-Guiness, I. (1972). *Joseph Fourier (1768–1830): A Survey of His Life and Work.* Cambridge, MA: MIT Press.

Herivel, J. (1975). *Joseph Fourier: The Man and the Physicist.* New York, NY: Clarendon Press.

FREIGHT MOVEMENT

The history of freight movement is the history of continuing improvements in speed and efficiency. Customers demand speed to meet a schedule; shippers demand efficiency to lower costs and improve profitability. Efficiency for the shipper can be broken down into three major components: equipment utilization, labor productivity, and energy efficiency. Dramatic strides in improving labor productivity, equipment innovations, and better utilization of that equipment have been made since the 1970s. It has been estimated that freight industry energy efficiency gains have saved the United States more than 15 percent (3.8 quadrillion Btus) to 20 percent between 1972 and 1992. All sectors of the freight transportation industry—air, water, rail, and truck—have been able to pass these savings on to their customers

through lower prices and faster and more dependable service.

THE NATURE OF FREIGHT AND ENERGY

Movement of freight accounts for one-third of all U.S. transportation energy consumption. But if the U.S. share of the export/import cargo shipping market was included, the figure would be even higher.

The U.S. economy has changed drastically since the early 1970s. There has been a great change in shipping as well as a vast change in what is being shipped. Though the need to ship raw materials such as coal and minerals and basic commodities such as farm products has grown, the shipment of components and finished products has expanded much more rapidly. Awareness of shifting trends is important because each cargo industry segment has unique needs (see Table 1).

For the majority of shipped goods, the scope and influence of freight on an economy are often underappreciated because much of freight movement is "hidden." By the time a typical automobile is fully assembled and delivered, it has gone through thousands of transportation steps. It consists of numerous basic materials and thousands of components, all of which need to be fabricated, assembled, and sent to the automobile manufacturer for final assembly.

From the raw materials to the shipment of the finished product—including all the costs associated with shipments of various components to manufacture a product—the total cost for shipping is approximately only 2 percent of the average cost of the product in the United States. The shipping of the finished product, which usually adds another 1 to 3 percent to the selling price, is highly variable because of the tremendous variations in the mass and volume of the freight being shipped.

Of course, these costs can change with fluctuations in the price of energy. When the energy crisis of 1973 drove up oil prices, transportation costs were more affected than manufacturing costs. And the bigger the product and the greater the number of components needed for it, the greater was the run-up in freight costs. The transportation industry passed along these higher energy costs to the customers, making the industry a prime mover in pushing up inflation. Likewise, as energy prices stabilized or fell

Freight	Mass to Volume Ratio	Delivery Speed and Why	Usual Mode and Why
Commodities (coal, grain, raw materials)	Great	Slowest. Speed is not necessary. Because customers use large quantities of a commodity, customers want a reliable steady supply.	Waterways and Railways. This is the least expensive means of delivery.
Perishable Food (meat, fish and produce)	Moderate	Fast. Refrigerated transport is very expensive.	Refrigerated Truck/Container or air for Fresh Fish.
Flowers	Low	Fast. Flowers are perishables with a very short shelf life.	Air from South and Central America to the United States.
Manufactured Components	Varies	Fast. The just in time requirements of manufactures demand quick dependable deliveries.	Truck.
Finished Products	Varies	Fairly Fast. Delivery as quick as possible for payment as soon as possible.	Truck and Rail.
Spare Parts	Varies	Usually fast. It is extremely important to keep equipment fully utilized.	Air and Truck.
High Value Goods (gold, diamonds, jewels)	Moderate	Speed and the prevention of pilferage is foremost.	Air. Low volumes and greater security precautions.
Documents and Packages for Service Economy	Low	Very fast. The service economy requires quick communication.	Air.

Table 1.
Cargo and How It Is Moved
The wide variety of freight and the wide variety of needs favors a wide variety of different transportation modes. Although the railways and waterways move greater tonnage, trucking commands more than 80 percent of freight revenue.

through most of the 1980s and 1990s, the cost of freight transportation stabilized or fell as well.

The railways and waterways should continue to secure a larger share of the freight transportation market due to the energy efficiency advantages they command. Studies have found that railways can move competitive traffic at a fuel saving typically in the range of 65 to 70 percent compared with trucks (Blevins and Gibson, 1991) (see Table 2).

Although freight modes differ in propulsion sys-

Mode	Typical Capacity	Propulsion System	Customer Price/Speed for Shipping (40 x 8 x 8 container size)	Typical Speed in United States	Mileage (ton-miles /gallon)		Energy Consumption (Btu/ton-mile)	
					1975	2000	1975	2000
Container Ship	82,000 tons	Large Diesel	$500 Port-to-Port charge for container transported 2000 miles in 4 days	24 knots	182	230 est.	550	400
Container Train	7,000 tons	Large Diesel	$700 Door-to-Door Intermodal Container, Los Angeles to Chicago in 50 hours	60 to 70 mph (a)	185	380	685	370
Truck	35 tons	Diesel	$1,500 Door-to-Door, Los Angeles to Chicago in 40 hours	65 mph	44	44	2,800	2,800
747-400 Cargo Aircraft	100 tons	Jet Engine	$66 for 20 lb. package overnight	500 mph	3	6 est.	42,000	19,150

Table 2.
Energy Requirements for Freight Transportation
Much higher energy costs for the air and trucking industries are passed along in the form of higher rates, rates customers willingly pay for faster service.
   However, when crude oil prices more than doubled between 1998 and 1999 (fuel costs increased by about $200 million from 1998–1999 to 1999–2000 for Federal Express), both air freight and trucking operations resisted raising rates fearing competition from waterborne shipping and railroads, which have been closing the faster-services advantage.
Note: (a) The less vibration resulting from welded rail makes higher speed travel possible.

tems, resistance encountered, and speeds traveled, waterborne shipping and railroads will always have an inherent energy efficiency advantage over trucking for several reasons. First, propulsion becomes more efficient the longer the ship, train, or truck. When a truck adds a second or a third trailer (where allowed by law), it can double or triple the tonnage transported for only an incremental increase in fuel consumption. Thus, on a ton-miles-per-gallon basis, a large truck will be more energy-efficient than a small truck, a large ship more than a small ship, and a large plane more than a small plane. Second, trucks must possess the reserve power to overcome gravity; the need to negotiate grades that are usually limited to 4 percent, but that can reach as high as 12 percent. This "back-up" reserve power is more easily met by locomotives and is not needed for the gradeless waterways. Third, weight reduction for trains and trucks is more impor-

tant because it lowers the need for the reserve power needed to change speeds and climb grades. However, much more energy goes into propelling the non-freight component of the truckload than the trainload. And finally, trucks are at a distinct disadvantage in terms of air and rolling resistance. Locomotives and container ships move at slower speeds and present a greater frontal area in relation to the volume and mass of the load, so the amount of additional energy needed to overcome air resistance is only a fraction of what a truck encounters; moreover, rolling resistance is far less for a steel wheel rolling on a steel rail than for a truck tire on a paved road.

Some of trucking's disadvantage relative to shipment by water or rail can be explained by the type of product shipped. As more heavier freight is moved by rail, low-weight voluminous shipments continue to command a larger share of trucking freight. So although the volume of freight moved by trucks has increased, the energy intensity (ton-miles per gallon) of trucking has stayed the same. There have been technical improvements in trucking (e.g., better engines, less resistance), but they have been offset by shipment of less dense goods that fill up trailer space well before reaching the truck's weight limit; increased highway speeds, resulting in greater aerodynamic drag; and slow turnover among trucking fleets. Taking the differences of freight in mind, one study found that rail shipment achieved 1.4 to 9 times more ton-miles per gallon than competing truckload service (Department of Transportation, 1991). This estimate includes the fuel used in rail switching, terminal operations (e.g., loading and unloading a 55-foot trailer weighing approximately 65,000 pounds), and container drayage (e.g., trucking freight across town from one rail terminal to another).

Trucking also lacks the flexibility of rail shipment. If a railroad has to ship many lightly loaded containers, the railroad can simply add more train cars. Moreover, railroads can add train cars with only an incremental increase in fuel consumption. Constrained by regulations limiting trailer number, length, and weight, the same flexibility is not available to trucks.

THE ERA OF TRUCKING

Of all transportation options, trucking commands a lower share of ton-miles per gallon shipped than by rail or waterway, yet if considering only the volume of goods shipped, the percentage moved by truck would be much greater than by railway or waterway. In 1997 alone, trucks logged more than 190 billion miles, a 135 percent increase from 1975.

The ability of the trucking industry to thoroughly dominate transportation from 1960 until 1980 began to become apparent shortly after the conclusion of World War II because of four very favorable market conditions: (1) the shifting of freight away from military to commercial; (2) the building and completion of an extensive interstate highway system speeding delivery; (3) the decision of companies to expand and relocate facilities away from city centers and near interstate highway loops; and (4) the inability or unwillingness of railroads to adapt to these changes.

Trucking grew unabatedly until the energy crises of the 1970s. In an era of cheap energy and extensive interstate highways, trucking could offer speedy dependable service at a very competitive price. The only area in which trucking could not compete effectively with rail was for heavy commodities such as coal and grains. Railroads had a huge energy advantage, yet for many reasons—some beyond their control—could not adapt to the more competitive environment.

By the 1980s it was apparent that the real threat to trucking dominance was not the railroads but congestion. Highway congestion started to significantly worsen during the late 1970s and early 1980s, eroding much of the speed advantage and the accentuating the energy and labor productivity disadvantages of the railroads. Congestion continued to increase through the 1980s and 1990s, and it promises to worsen in the future. It has been estimated that growing traffic congestion might lower the on-road fuel economy of trucking by up to 15 percent by 2010 (compared to 1990), and the labor productivity of trucking will drop as well, as more time is spent stuck in traffic rather than moving freight.

Air quality suffers from this increasing congestion as well. Motor vehicles create the majority of air quality problems in urban areas, so for cities to comply with stringent ambient air quality standards, they will have to reduce motor vehicle emissions. Trucking accounts for only 4 percent of the U.S. motor vehicle fleet, yet can easily be responsible for 30 to 40 percent of the air quality problems because, in comparison to automobiles, the fleet is far older, is driven far more miles each year, and the emissions per vehicle are far greater.

Another congestion-related problem facing trucking is safety. Although crashes per mile decreased almost 50 percent from the 1970s to 2000, each year more than 5,000 people die in accidents involving trucks. Much of the reason for these accidents has been attributed to unsafe trucks, high speed limits, and driver fatigue. In 1997, trucks were involved in for 22 percent of the motor vehicle fatalities (excluding single-car accidents) in the United States.

For the economy as a whole, traffic jams in the ten most congested U.S. cities cost more than $34 billion a year, according to Texas Transportation Institute. In recognition of the part that trucking plays in this congestion problem, Congress instituted the Congestion Mitigation and Air Quality Improvement Program as part of the Internodal

Regulation or Subsidy	Distortion
Federal regulation of railroad rates to ensure that railroads did not unfairly use their monopoly power.	Allowed trucking companies to cherry-pick the most desirable traffic, leaving more volume of the less desirable traffic to the railroads. Repealed 1981.
Property tax abatements encouraging companies to locate and expand in suburban and rural areas.	If they did not make the siting decision for property tax saving reasons, efficient transportation would have been a higher priority. Started in 1960s and more prevalent now than ever.
Public funding for road and highway construction and reconstruction. Federal highway funding alone was $217 billion in 1998.	Although truckers pay road and fuel taxes, it pays only a small fraction of the cost of safety inspections and road damage from trucking. The taxes paid/benefits received imbalance has only grown through the years.
City and state subsidies in the bidding war for the next generation of ports.	The ship lines will make redevelopment decisions based more on who offers the bigge subsidy rather than the best rail links.
Anti-trust law that discourages railroad communication and cooperation necessary to make intermodal transportation work.	Without any coast-to-coast railroads, the railroads must cooperate to deliver containers across country.

Table 3.
Regulations and Subsidies That Have Distorted Energy Use in the Freight Transportation Market

Surface Transportation Efficiency Act of 1991 (ISTEA). A major goal of this program is to move more of the truck traffic to rail as a means of reducing congestion, improving air quality, and saving energy.

The trucking industry has been trying to make up for the growing labor and energy-efficiency advantages of the railroads by pushing for favorable legislation to raise the maximum workweek from sixty to seventy-two hours, gain higher weight limits, and allow more multi-trailers. These dynamics are moving the industry into what economist Michael Belzer calls "sweatshops on wheels."

Yet despite all the problems associated with the 190 billion miles logged by truckers each year, trucks are vital because they offer an unmatchable service: door-to-door delivery. Companies are located everywhere from city centers to distant suburban locations and rural areas, so the door-to-door railway delivery is not even an option for most businesses. Even if the United States made massive investments in railway infrastructure and raised the taxes of the trucking industry to fully account for its contribution to road and environmental damage, the mismatch of railroads to customer needs ensures that there always will be a great need for trucking. Except for the commodities and basic-materials industries, few other companies since the 1950s gave rail transportation a high enough priority in expansion and relocation plans. The corporate exodus from the cities long ago cemented many freight transportation and energy use patterns that are nearly impossible to change.

INTERMODALISM

It is a tedious, difficult debate to compare the advantages of one freight mode over another. Each mode has its distinct advantages. The important thing for most nations is to try to make each mode as efficient as possible. If each transportation mode can become equally efficient, and if government policy does not distort the market, the market will fairly dictate the most efficient way to move freight (see Table 3). All else being equal, it usually will be the most energy efficient choice too.

As the freight transportation industry moves into the twenty-first century, the greatest potential for improving efficiency is found along the railways because rail is the centerpiece in the evolution toward intermodalism. The intermodal industry consists of all modes of transportation working together to ship goods, whether it be via rail carrier, sea vessel, air transporter, or truck. Each mode has one thing in common: the delivery of a consignee's goods directly to the final destination, regardless of how the goods were originally shipped.

The first piggyback move, now called intermodal, came about when the Barnum & Bailey Circus went from state to state and town to town on a special train of flat railroad cars in 1872. Today this is referred to as a dedicated train. The cars were loaded with the tents, animals were put in caged containers, and passenger cars were hooked behind them carrying all of the circus personnel and wares. When the "circus train" arrived at the desired destination, the circus personnel would then untie the cables that secured the containers to the cars, back up a large portable ramp that was ten feet wide and approximately sixty feet long with a twenty degree angle, and attach it to the train where the locomotive was disconnected. The container cage would have wheels mounted under the container so that a tractor could back up to the ramp and attach to the container, pulling the container down for unloading. The same operation would also be used for loading.

Intermodal Evolution

To capitalize on the energy-efficiency advantages of intermodal freight movement, and also to come closer to matching the speed and reliability of trucking, the railroads had to overcome the equipment and infrastructure problems plaguing the industry, the aversion to cooperate in developing intraindustry and interindustry standards to achieve coast-to-coast delivery, price inflexibility due to regulation, and the reluctance to cooperate and customize equipment and processes to meet the needs of customers.

Prior to the containerization movement, work at ports and railyards was very slow and labor-intensive because much of the freight being moved entailed unprotected palleted cargo from ships and less-than-carload freight moved in boxcars. It took a lot of

Equipment like the Travelift, which is customized for container movements for rail terminal or harbor operations, is a more efficient means of loading and unloading cargo than labor-intensive circus ramping or forklift and pallet operations. (Mi-Jack Products, Inc.)

human energy to load and unload cargo in this manner and the upstart trucking industry had huge labor productivity advantages over the railroads.

The most important step in making intermodal shipment possible was the development of container standards and lifting equipment to meet the needs of ship lines and railroads. When Mi-Jack and the Santa Fe Railroad introduced a reliable overhead rubber-tired crane in 1963 to the piggyback (intermodal) industry, it started to take shipping out of the dark ages as loading or unloading times went from 45 minutes per load (circus ramping) to 2.5 to 3 minutes (rail terminal operators in 2000 guarantee the railroads 1 to 1.5 minutes per lift). All the associated operating costs and personnel requirements of circus ramping were eliminated as the new lift equipment, whether it be a sideloader (forklift truck) or a mobile gantry crane, greatly reduced energy requirements.

Although the containerization of cargo dates to the 1920s and earlier, the movement really accelerated during World War II, when the U.S. Armed Forces started to use containers to ship top-priority supplies because the cargo could be kept secret, handled less, and loaded and unloaded faster from truck to rail to sea vessel. Prior to containerization, pilferage was high because most freight was exposed and stacked on a 4-foot-square pallet for shipment.

After World War II, Malcolm McLean recognized the huge potential of containerization for the com-

Mode	Cost
Double-Stack Container	30 cents per mile
Piggyback Trailers on Spines	40 cents per mile
Truckload Highway Carrier	75 cents per mile

Table 4.
An example of costs for 1,000 mile haul
Although rail costs are route specific and vary tremendously, shipping by double-stack container is usually the most economical option, largely reflecting the energy savings. The energy-related advantage of the double-stack over the piggyback trailer comes from more cargo weight per rail car ("leaving the wheels behind," stacking, and lighter cars). As a result, a train can carry more cargo without increasing length (more cargo per 6,000 ft. of train). Trains also suffer less air resistance since double-stack sits lower. However, when drayage (local trucking) is expensive (up to $4.00 per mile in some areas), much of the economic advantage of intermodal shipping using double-stack containers is negated.

mercial market. McLean, who was in the trucking industry at that time, purchased two surplus sea vessels and started his own shipping company, SeaLand, first shipping fifty-eight containers from New York to Houston on April 26, 1956. Shipping would never be the same. SeaLand became very successful quickly, and eventually other shipping lines, such as Matson, also began to ship containers, in 1958. By the early 1960s every shipping line in the world was handling containers to ship cargo.

By using railroads to transport their containers, the shipping lines discovered in the 1960s that they could bypass the costly Panama Canal in going from Japan to Europe. The container could be transported from the Pacific coast port to the Atlantic coast port and placed on an awaiting ship, improving in-transit time and vastly reducing the cost of shipping and fuel. This is called the landbridge, which differs from the micro-bridge, wherein the container is loaded on a ship in Japan, unloaded onto a Santa Fe container car in San Francisco for delivery in Chicago, and then unloaded to a chassis for trucking to its final destination.

However, the containerization movement still had a major problem: standardization of equipment. It was very difficult to get the various shipping lines to agree on a standard container corner casting that any spreader on the crane could handle to facilitate the loading and unloading of the container. Matson and SeaLand were reluctant to change their design to the 1965 standards of the International Standard Organization and the American Standard Association (ISO-ASA). It took a major effort of the lifting equip-

ment manufacturers Drott and Mi-Jack, with the help of railroads, to finally start converting all containers to the new standards a few years later. During the interim, Mi-Jack designed a special cobra-head latch that could accommodate all three corner castings (each required a different picking point to lift).

Another major innovation was the introduction of the double-stack car by Southern Pacific in the late 1970s. The shipping lines realized the economic advantages of shipping two or three containers on a double-stack car. Because it was in essence a skeleton frame, it weighed much less than a conventional container on flat car (COFC). In the process, significant weight reduction of the container itself was achieved. Lighter containers carried on lighter double-stack cars, with a better aerodynamic profile than piggyback trailers, dramatically improved railroad energy efficiency.

Charlie Kaye, the President of XTRA, wisely took advantage of railroads shortage of capital by having XTRA lease containers and trailers to the railroad on a per-diem basis in the early 1960s. For the railroads, it was a good arrangement. By using a leasing company for trailers and container equipment, it freed up capital to invest in new locomotives and track improvements.

In the early 1980s, Don Oris of American President Lines followed Kaye's lead. This shipper purchased its own double-stack railroad cars and did not rely on railroads to furnish equipment for shipping containers from the West Coast. Because it was now involved in both sea and rail transport, American President Lines saw the benefit of modifying the double-stack car design to fit all lift equipment at port and rail terminals.

Leasing became so popular that by the 1990s, railroads could bid among many leasing companies for the various types of equipment required. This leasing movement, and the start of the boom in imports, marked the beginning of railroads shipping more containers than trailers. The 1980 balance of 40 percent containers and 60 percent trailers began to change, shifting to 60 percent containers and 40 percent piggyback trailers by 1998.

Largely due to stifling regulations, the railroads had no other choice but to lease equipment. In the early 1970s the rate of return for the rail industry was in the 1 to 2 percent range, and bankruptcy was commonplace. By the mid-1970s, 25 percent of the nation's rail miles had to be operated at reduced speed because of dangerous conditions resulting from lack of investment in infrastructure.

It took the combination of the Staggers Act of 1980 deregulating the railroads and the boom in imports starting in the 1980s for the railroads to be confident that they could recoup their investments in terminals, lifting equipment, new lightweight trailers on flat cars (TOFC) and double-stack cars, and energy-efficient locomotives. New terminals were designed for handling TOFC-COFC as well as very sophisticated lift equipment for loading trailers and containers weighing 40,000 to 60,000 pounds. This equipment had to operate every 1.5 to 2 minutes, five to seven days a week, working eighteen to twenty hours per day depending on the volume at the terminal.

Equally impressive was the introduction of new locomotive technology. For approximately the same fuel economy, a top-of-the-line locomotive in 1997 could generate twice the horsepower, pull more than twice the load, and reach a top speed 5 mph faster (75 mph) than a top-of-the-line locomotive in 1972. The better locomotives and lighter double-stack cars of the 1990s could replace as many as 280 trucks from the roadways for every 6,000 ft. train.

As intermodal shipping increased in volume, each mode of transportation realized that it needed to cooperate with competitors to increase the volume of freight. Mike Haverty, president of the Santa Fe Railroad in 1989, convinced two major truck carriers, J. B. Hunt and Schneider International, which previously were major competitors of the railroads, to become their shipping partners. Both agreed to put piggyback trailers on TOFC cars for trips greater than 500 miles for customers who required delivery within ten to fourteen hours. By 1992 Hunt realized the benefits of leaving the wheels behind, and began a transition to container and double-stack operations.

Railroads also entered into contracts with shipping lines to deliver containers after they were unloaded at ports. Again, truckers were working with shipping lines and airlines, as well as railroads, to deliver containers or trailers to their final destination. Each of the modes of transportation needed the other to complete the seamless delivery of freight from point of origin to final destination.

All these favorable dynamics were responsible for intermodal volume growing from 3 million containers and trailers in 1980 to nearly 9 million in 1998, as railroad customer rates fell from approximately 3.2 cents to 2.5 cents per ton-mile (1.5 cents if adjusted for inflation).

Trucks line North Capitol Street in Washington, D.C., as they convoy toward the U.S. Capitol on February 22, 2000. Hundreds of trucks entered the district to protest a steep rise in diesel fuel prices. (Corbis Corporation)

Future of Freight: Reducing Bottlenecks

If energy prices remain stable, customer rates will continue to fall as the freight transportation industry gets bigger and faster. On the equipment front, the trend is for bigger container ships, more powerful locomotives, and more multi-trailered trucks, regulations permitting. On the logistics front, there is certain to be more consolidation and cooperative efforts of railroad, shipping lines and truck carriers to reduce energy requirements and expedite cargo delivery. However, expansion and consolidation are not going to solve all freight transportation problems. To handle ever greater and faster-moving volume, the freight industry will have to find solutions to bottleneck problems.

The benefits of eliminating bottlenecks are twofold: energy efficiency and speedier deliveries. In the air freight area, hub-and-spoke systems are very efficient because large carriers such as Federal Express

A tugboat escorts a cargo ship in the Panama Canal. (Corbis Corporation)

have the volume and an efficient hub facility to quickly move freight through the terminal. The same is true for some of the major interstate trucking firms. Intermodal rail and shipping industries are a bit different: it takes a great deal of cooperation for different companies to share facilities and coordinate activities.

Going big is certain to help ton miles per gallon fall for sea vessels. The new container sea vessel *March Regima* can ship more than 8,000 TEUs (one TEU is equivalent to one 20-foot container), and even larger vessels are planned to handle 10,000 TEUs. These new vessels will be more than twice the size of the average medium-size to large-size container sea vessel of the 1990s, such as the Ckass C10, which can ship 4,500 TEUs, with 45,000 hp to travel at 22 knots to 25 knots, or the Class 11 container sea vessel, which can ship 4,800 TEUs requiring approximately 66,000 hp to travel at 25 knots to 30 knots.

To realize even greater efficiency with these large vessels, new port terminals are being designed to eliminate storage areas there and to provide faster loading and unloading sequences that will enable a ship to depart as quickly as possible. Depending on the size of the vessel, the cost to be in port can be as low as $5,000 per hour or as high as $15,000 per hour. These new ports will do away with the need to have ship-to-shore cranes unload containers on chassis, to be driven to a storage area at the port for pickup. Instead, loading and unloading of containers directly from vessels and railcars will eliminate extra steps in the process, dramatically cutting the time a larger-container vessel needs to be in port. Ports will be handling much more traffic in much less space.

Joint ventures such as the revitalization of the Panama Canal Railway—a 47-mile railroad running parallel to the Panama Canal and connecting ports on the Atlantic and Pacific Oceans—should be more common. Kansas City Southern Railroad and Mi-

Jack Products, a U.S. Class 1 railroad and North America's leading independent intermodal terminal operator, respectively, have invested $60 million for start-up expenses to revive and modernize the 143-year old first transcontinental railroad of the Americas. Fully operational in 2000, the revitalized railroad provides an efficient intermodal link for world commerce and complements the existing transportation provided by the canal, the Colón Free Trade Zone, and the port terminals.

This landbridge for the Americas reduces truck traffic on the major highway between Balboa and Colón by operating continuously, with a capacity for ten trains running in each direction every twenty-four hours. The new line accommodates locomotives and rolling stock capable of traveling at speeds of up to 65 to 95 km/hr, reducing energy requirements as well as emissions. The fine-tuned coordination of this port-to-port rail transport system, which allows railcars to pull up alongside vessels, helps the two coastal ports work like one enormous hub. Since the Panama Canal cannot handle the growing fleet of larger-container ships, this giant port-to-port terminal could eventually handle the majority of container traffic.

In the United States, the most important step being taken by railroads to relieve congestion bottlenecks is a new major hub terminal that would eliminate crosstown drayage. The through-port terminal will have a rail-mounted crane straddling ten or more tracks with the capacity to immediately select any group of containers to transfer from one train to another. For example, a westbound and an eastbound train arrive at the through-port terminal. Designated containers going east are on the westbound train. They will be transferred to the eastbound train for final destination. The same procedure will be performed for north and south, northeast and southeast, etc. The interchange, which normally would have taken one to three hours to unload a container or trailer and deliver it across town to an eastbound terminal from a westbound terminal, will now take only minutes using the through-port interchange operation.

Another alternative being considered is the in-line terminal design. This design consists of a trackside operation and storage area along trackside, in-line with the ramping and deramping operation of handling containers and trailers. It will also include one-way traffic on each side of the storage area. Truck

THE WAYS OF MAJOR EXPORTERS

Unlike the United States, countries such as Japan and China are far greater exporters than importers, and they site manufacturing facilities with easy access to waterways. Many of the ports operate as effectively as U.S. ports, but the intermodal operations needed for fast, dependable overland service lag well behind U.S. operations.

drivers will be in and out of the terminal with their cargo in fewer than ten minutes, whether they are picking up or delivering containers or trailers. The gate dispatcher will now act as a traffic controller, similar to those used in the airline industry, and will be responsible for guiding the truck driver and freight to the delivery site of the trailer or container as well as the yard section or storage area location. With present terminal design, it takes approximately thirty minutes to two hours to pick up or deliver a trailer or container, from the arrival at the entrance gate to the delivery of the container and the driver's exit of the terminal. With an in-line terminal design, an intermodal truck driver's time will be more productive than that of the nonintermodal counterpart, who usually spends approximately 25 percent of the time waiting to load and unload.

Railroads also will need to do a better job of catering to customer needs so that customers choose intermodal transport for all shipping greater than 500 miles and not just for longer distances. United Parcel Service (UPS) was an early intermodal pioneer using piggyback to ship freight in the 1960s, and has continually encouraged railroads to invest in lifting equipment, railroad cars, and terminal infrastructure to improve ramping and deramping. To this day, it is still one of the largest customers shipping trailers in the intermodal industry. However, UPS is in the transportation business. For intermodal to expand significantly, the industry has to convince major industry of intermodal's benefits. "How close is the nearest intermodal terminal?" will need to become a top industry priority for expansion plans. More companies need to adopt a variation on the mine-mouth-to-boiler thinking of utilities for constructing new coal-fired power

NOVEL ENERGY SAVINGS: QUILTING INSTEAD OF REFRIGERATING

"Don't cool it, cover it" is the latest trend in transportation. Shipping temperature-sensitive refrigerated cargo has always been one of most expensive means to move freight. Just to run a compressor to refrigerate an 800 cubic foot trailer for three days can take more than 70 gallons of diesel fuel. Aside from fuel, refrigerated units must have regularly scheduled maintenance performed as well, which is why the shipping price to move a refrigerated trailer cross-country can be three or more times that of using a dry trailer. For companies in competitive marketplaces, this is a burdensome cost to absorb.

The Cargo Quilt was developed to offer an alternative, to drastically lower the cost of moving temperature-sensitive cargo by avoiding refrigeration. After the freight is loaded on a dry trailer or container, the Cargo Quilt, which works similarly to a thermos bottle, is draped over the cargo. Depending on the bulk of the shipment, it can maintain the temperature—whether hot or cold—from five to thirty days. And because there are no mechanical parts that can malfunction (e.g., thermostat or compressor), there is no chance of spoilage.

Some ultra-temperature-sensitive cargo will always need refrigeration, yet the manufacturer of the Cargo Quilt estimates that up to 50 percent of temperature-sensitive freight currently serviced by refrigerated trailers and containers could use a Cargo Quilt in a dry trailer at substantial savings. And as each mode of freight transportation lowers its guaranteed in-transit time, this percentage will continue to rise.

Quilts also reap savings by allowing the intermixing of chilled, frozen and freezables within the same mechanical trailer. Oscar Meyer Company uses a chilled unit for the majority of its meat products, yet needs to keep four to six pallets of turkey nuggets frozen as well. It achieves a dual-temperature zone container by covering the frozen turkey nuggets in a PalletQuilt. All the perishables arrive as loaded: some chilled, some frozen. Coors, Nestlé and Anheuser-Busch are just a few of the other major corporations that have switched to the Cargo Quilt and dry trailers.

Who would have ever thought that the Cargo Quilt, which has nothing to do with improving propulsion or reducing resistance, would turn out to be one of the most promising energy-conserving transportation innovations of the 1990s?

---

plants. The closer the intermodal terminal, the less of a concern is trucking and traffic congestion.

Railroads are expected to become more aggressive in moving away from their middleman status as intermediary to trucking and shipping. However, the mountainous debt taken on due to railroad mergers and acquisitions in the late 1990s brings into question railroads' ability to make future investments in infrastructure and marketing channels.

Intermodal service must improve for corporations that adopted just-in-time inventory in the 1980s and 1990s. Many companies feel that the benefits of keeping minimal stock are worth the premium price paid for faster door-to-door delivery by truck, which is a major reason why truck freight revenues remain at more than 80 percent of the country's total freight revenues.

There is a grave transportation risk in relying on just-in-time production methods. Any freight transportation system breakdown can be catastrophic for companies and nations. If key parts get stuck at the border, or if a natural disaster destroys a highway or a rail line, some factories might have to close down, others write off permanent losses. And if it is a component for an essential product or service, the ripple effect could cost the economy billions of dollars.

Automobile companies are major users of just-in-time methods, coordinating the delivery of thousands of parts from many different suppliers. A transportation delay of a few hours for one part could shut down an assembly line for half a day.

For the intermodal equipment and infrastructure investments to fully pay off, and to grab a larger share of just-in-time freight, will require further computerization improvements in tracking and processing equipment for the industry, as well as by the U.S. Customs Service. Although the 1980s transition from pen and paper to computerization by Customs has helped speed intermodal traffic, the tripling of imports from 1985 to 2000 put tremendous strain on the computer system used to process container shipments. This system, which is being upgraded, has created costly shipment delays for companies relying on just-in-time production methods. And there remains the fear of a major computer system breakdown that might take weeks to resolve and that would cost many companies billions of dollars as goods are not transported and as assembly plants are idled waiting for parts.

The intermodal industry seems to be addressing all these challenges and is highly likely to continue to grow and shift more freight toward rail because of the large advantages over roadways in safety, congestion, pollution, noise, land use, and energy consumption. Nothing is more energy-efficient than moving goods intermodally, and nobody does it better than the United States. Government and business people from around the world come to America to watch, learn, and marvel at the way the different modes cooperate to deliver cargo so efficiently.

John Zumerchik
Jack Lanigan, Sr.

See also: Air Travel; Government Approaches to Energy Regulation; Government and the Energy Marketplace; Locomotive Technology; Propulsion; Ships; Tires; Traffic Flow Management; Transportation, Evolution of Energy Use and.

BIBLIOGRAPHY

Abacus Technology Corp. (1991). *Rail vs. Truck Fuel Efficiency: The Relative Fuel Efficiency of Truck Competitive Rail Freight and Truck Operations Compared in A Range of Corridors.* Washington, DC: U.S. Department of Transportation, Federal Railroad Administration, Office of Policy.

Belzer, M. H. (2000). *Sweatshops on Wheels : Winners and Losers in Trucking Deregulation.* New York: Oxford University Press.

Berk, G. (1994). *Alternative Tracks: The Constitution of American Industrial Order, 1865–1917.* Baltimore: John Hopkins University Press.

Blevins, W. G., and Gibson, A. W. (1991). "Comparison of Emissions and Energy Use for Truck and Rail."

Transportation Association of Canada Conference, Vol. 4, Conference Proceedings.

Davis, S. C. (1997). *Transportation Energy Databook 17,* ORNL-6919. Oak Ridge, TN: Oak Ridge National Laboratory.

DeBoer, D. J. (1992). *Piggyback and Containers: A History of Rail Intermodal on America's Steel Highways.* San Marino, CA: Golden West Books.

Greene, D. L. (1996). *Transportation and Energy.* Washington, DC: Eno Transportation Foundation, Inc.

Kahn, A. M. (1991). *Energy and Environmental Factors in Freight Transportation.* Ottawa: A. K. SocioTechnical Consultants for Transport Canada, Economic Research Branch.

FRICTION

See: Tribology

FUEL ADDITIVE

See: Gasoline and Additives

FUEL CELLS

A fuel cell is equivalent to a generator: it converts a fuel's chemical energy directly into electricity. The main difference between these energy conversion devices is that the fuel cell acccomplishes this directly, without the two additional intermediate steps, heat release and mechanical motion.

A fuel cell has two basic elements: a fuel delivery system and an electro-chemical cell that converts the delivered fuel into useful electricity. It is this unique combination that enables fuel cells to potentially offer the best features of both heat engines and batteries. Like batteries, the cell generates a dc electric output and is quiet, clean, and shape-flexible, and may be manufactured using similar plate and film-rolling processes. By contrast, the fuel delivery system ensures that fuel cells, like heat engines, can be

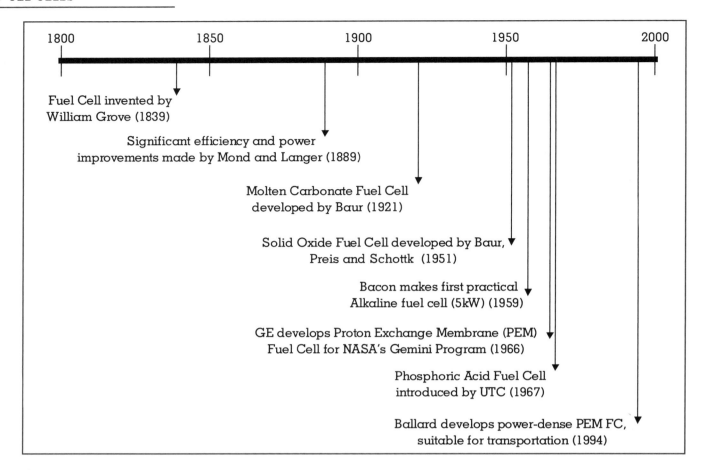

Figure 1.
Timeline of major fuel cell milestones.

quickly refueled and operate for long periods between stoppages.

"Fuel cell" is an ambiguous term because, although the conversion occurs inside a fuel cell, these cells need to be stacked together, in a fuel cell *stack*, to produce useful output. In addition, various ancillary devices are required to operate the stack properly, and these components make up the rest of the fuel cell *system*. In this article, fuel cell will be taken to mean *fuel cell system* (i.e., a complete stand-alone device that generates net power).

HISTORICAL INTEREST IN FUEL CELLS

Although fuel cells were invented over 150 years ago, Figure 1 reveals that there have been only a few key milestones in fuel cell development. For this reason they have only recently attracted significant and

widespread interest from governments, research laboratories and major corporations. Two developments are behind this shift: impressive recent technology advances, and growing concern over the state of the environment.

For many years after their invention in 1839 by an English lawyer, Sir William Grove, fuel cells were little more than a laboratory curiosity because their performance was unreliable and few uses could be found for them. With rapid developments in electricity during the late 1800s it is surprising that fuel cells did not manage to compete with electrochemical batteries or generators as a source of electricity. In retrospect, the invention of the automobile in 1885 could have stimulated fuel cell development because a battery's limited energy storage makes it unsatisfactory for transportation. However, the internal combustion engine was introduced soon after the fuel cell, managed to improve at a faster rate than all alternatives, and has remained the prime mover of choice.

It was not until the 1960s that fuel cells successfully filled a niche that the battery or heat engine could not. Fuel cells were the logical choice for NASA's Gemini and Apollo Programs because they could use the same fuel and oxidant that was already available for rocket propulsion, and could generate high-quality electricity and drinking water in a relatively lightweight system. Although this application enabled a small fuel cell industry to emerge, the requirements were so specific, and NASA's cost objectives were so lenient, that fuel cells remained, and still remain, a minor power source.

This situation is beginning to change because recent years have seen impressive progress in reducing the size and cost of fuel cell systems to the point where they are now considered one of the most promising "engines" for the future. Simultaneously, increased concerns over climate change and air quality have stimulated many organizations to fund and develop technologies that offer significant environmental benefits. Their high efficiency and low emissions make fuel cells a prime candidate for research funding in many major industries, particularly as major growth is expected in developing nations, where energy is currently produced with low efficiency and with few emissions controls. In fact, if fuel cells are given the right fuel, they can produce zero emissions with twice the efficiency of heat engines.

Although environmental trends are helping to drive fuel cell development, they are not the only drivers. Another is the shift toward decentralized power, where many small power sources replace one large powerplant; this favors fuel cells since their costs tend to be proportional to power output, whereas the cost per kilowatt ($/kW) for gas turbines increases as they are shrunk. Moreover, the premium on high-quality electricity is likely to increase in the future as the cost of power outages rises. Fuel cells that generate electricity on-site (in hotels, hospitals, financial institutions, etc.) from natural gas promise to generate electricity without interruption and, unlike generators, they can produce this electricity very quietly and cleanly.

At the other end of the power spectrum, there is increasing interest in fuel cells for small electronic appliances such as laptop computers, since a high value can be placed on extending the time period between power outages. It is possible that a small ambient-pressure fuel cell mounted permanently in the appliance would allow longer-lasting hydrogen cartridges or even methanol ampoules to replace battery packs.

ELECTRICITY PRODUCTION BY A FUEL CELL

Fuel Cell Stack

As with a battery, chemical energy is converted directly into electrical energy. However, unlike a battery, the chemical energy is not contained in the electrolyte, but is continuously fed from an external source of hydrogen.

In general, a fuel cell converts gaseous hydrogen and oxygen into water, electricity (and, inevitably, some heat) via the following mechanism, shown in Figure 2:

i. The anode (positive pole) is made of a material that readily strips the electron from the hydrogen molecules. (This step explains why hydrogen is so important for fuel cell operation; with other fuels, it is difficult to generate an exchange current because multiple chemical bonds must be broken before discrete atoms can be ionized).

ii. Free electrons pass through an external load toward the cathode—this is dc electric current—while the hydrogen ions (protons) migrate through the electrolyte toward the cathode.

iii. At the cathode, oxygen is ionized by incoming electrons, and then these oxygen anions combine with protons to form water.

Since a typical voltage output from one cell is around 0.4–0.8 V, many cells must be connected together in series to build up a practical voltage (e.g., 200 V). A bipolar plate performs this cell-connecting function and also helps to distribute reactant and product gases to maximize power output.

Fuel cell stack voltage varies with external load. During low current operation, the cathode's activation overpotential slows the reaction, and this reduces the voltage. At high power, there is a limitation on how quickly the various fluids can enter and

---

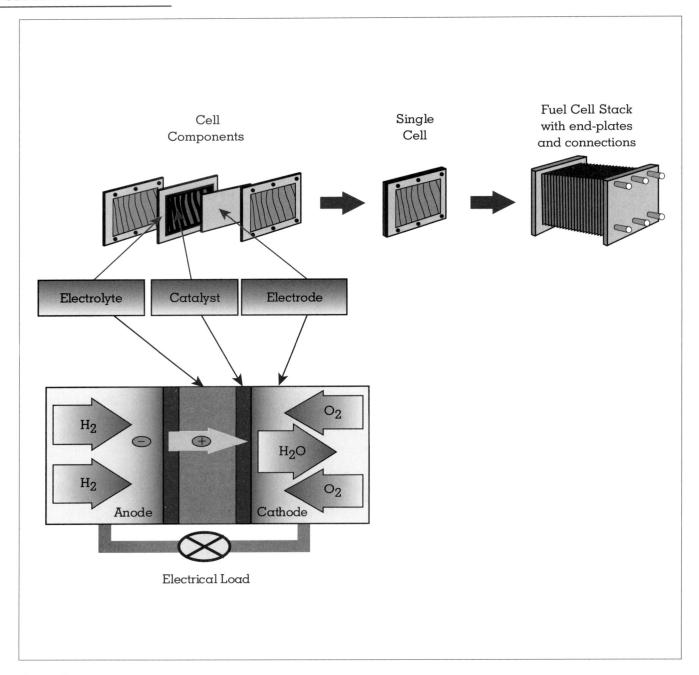

Figure 2.
Fuel cell stack.

exit the cell, and this limits the current that can be produced. In most of the operating range, however, it is ohmic polarization, caused by various electrical resistances (e.g., inside the electrolyte, the electrolyte-electrode interfaces, etc.) that dominates cell behavior. Continued research into superior elec-

trodes and electrolytes promises to reduce all three types of losses.

Because of these losses, fuel cells generate significant heat, and this places a limit on the maximum power available because it is very difficult to provide adequate cooling to avoid formation of potentially dangerous "hot spots." This self-heating, however, helps to warm up the system from cold-start (mainly a concern for transportation applications).

A complete fuel cell system, even when operating on pure hydrogen, is quite complex because, like most engines, a fuel cell stack cannot produce power without functioning air, fuel, thermal, and electrical systems. Figure 3 illustrates the major elements of a complete system. It is important to understand that the sub-systems are not only critical from an operational standpoint, but also have a major effect on system economics since they account for the majority of the fuel cell system cost.

Air Sub-System

The reaction kinetics on the cathode (air) side are inherently slow because oxygen's dissociation and subsequent multi-electron ionization is a more complex sequence of events than at the (hydrogen) anode. In order to overcome this activation barrier (and hence increase power output) it is necessary to raise the oxygen pressure. This is typically accomplished by compressing the incoming air to 2–3 atmospheres. Further compression is self-defeating because, as pressurization increases, the power consumed by compression more than offsets the increase in stack power. In addition, air compression consumes about 10–15 percent of the fuel cell stack output, and this parasitic load causes the fuel cell system efficiency to drop rapidly at low power, even though the stack efficiency, itself, increases under these conditions. One way to recover some of this energy penalty is to use some of the energy of the hot exhaust gases to drive an expansion turbine mounted on the shaft compressor. However, this substantilly raises system cost and weight.

Alternatively, the fuel cell stack can be operated at ambient pressure. Although this simplifies the system considerably and raises overall efficiency, it does reduce stack power and increase thermal management challenges.

Fuel Sub-System

On the fuel side, the issues are even more complex. Hydrogen, although currently it is made in relatively large amounts inside oil refineries for upgrading petroleum products and for making many bulk chemicals (e.g., ammonia), it is not currently distributed like conventional fuels.

Moreover, although there are many ways to store hydrogen, none is particularly cost effective because hydrogen has an inherently low energy density. Very high pressures (e.g., 345 bar, or 5,000 psi) are required to store practical amounts in gaseous form, and such vessels are heavy and expensive. Liquid hydrogen is more energy-dense and may be preferred for transportation, but it requires storage at –253°C (–423°F) and this creates handling and venting issues. Moreover, hydrogen is mostly made by steam reforming natural gas, and liquefying is so energy-intensive that roughly half of the energy contained in the natural gas is lost by the time it is converted into liquid hydrogen. A third approach is to absorb hydrogen into alloys that "trap" it safely and at low pressure. Unfortunately, metal hydrides are heavy and are prohibitively expensive.

Because it is so difficult to economically transport and store hydrogen, there is interest in generating hydrogen on-demand by passing conventional fuels through catalysts; such an approach is called fuel processing. The challenge for fuel cell application is to convert available, relatively impure, fossil fuels into a hydrogen-rich gas (hydrogen-feed) without contaminating the various catalysts used in the fuel processor and fuel cell stack. Depending on the type of fuel cell stack and its operating temperature, there can be as many as four sequential stages involved in fuel processing:

i. fuel pre-treatment (such as desulfurization and vaporization)

ii. reforming (conversion of fuel into hydrogen-rich feed, sometimes called synthesis gas, or syngas)

iii. water-gas shift for converting most (>95%) of the carbon monoxide byproduct into carbon dioxide

iv. preferential oxidation for final removal of the remaining carbon monoxide.

Most of the differentiation between various fuel processor strategies comes in the second stage. One reforming method uses steam produced by the fuel cell stack reaction to reform the fuel into hydrogen. This process, steam reforming, is endothermic (absorbs heat), may require high temperatures (depending on the fuel) and, because the catalysts that enable the reaction are selective, different fuels cannot be used in the same catalytic reactor. However, a major advantage is that the product is

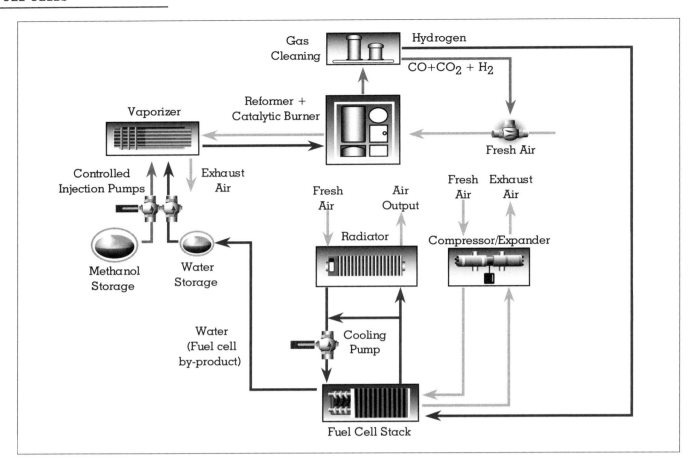

Figure 3.
Fuel Cell System.

hydrogen-rich; for example, when methane is steam reformed, the hydrogen concentration is ~75% ($CH_4 + H_2O \rightarrow 3H_2 + CO$). This process is favored for stationary applications, where a single fuel and steady-state operation are typical.

In contrast to steam reforming, partial oxidation (POX) uses air instead of steam and, as its name implies, burns the fuel in restricted amounts of air so that it generates partially combusted products, including hydrogen. POX generates heat and can, therefore, potentially respond faster than a steam reformer. This is beneficial for load-following applications (e.g., transportation).

Moreover, because all fuels burn, POX does not demand a catalyst, although advanced designs often use one to lower flame temperatures, which helps to relax materials requirements and to improve efficiency and emissions. The hydrogen concentration, however, is considerably lower (~40%) because

none comes from steam and there is about 80 percent nitrogen diluent in the air ($CH_4 + \frac{1}{2}O_2 + 2N_2 \rightarrow 2H_2 + CO + 2N_2$).

Autothermal reforming, ATR, combines steam reforming and POX. Since the heat released from POX is consumed by steam reforming, the reactor can be adiabatic (or autothermal). For some applications, ATR may offer the best of both worlds: fuel-flexibility, by partly breaking the fuel down into small HC fragments using air, and relatively high hydrogen yield, by steam reforming these HC fragments.

Thermal Sub-System

Cooling strongly depends on fuel cell operating temperature and also depends on the fuel cell's external environment. For low temperature fuel cells, cooling imposes a significant energy debit because pumps need to force coolant out to a heat

	Operating Temperature (°C)	Advantages	Disadvantages	Potential Application
Alkaline	25-100	• Mature technology • No precious metals	• Must use pure hydrogen	• Space
Proton Exchange Membrane	0-85	• Can operate at ambient temperature • High power density	• Sensitive to CO-poisoning • Need for humidification	• Transportation • Distributed Power
Phosphoric Acid	170-220	• Mature • Reformate-intolerant	• Bulky • Cannot start from ambient	• Heavy-duty transportation • Distributed Power
Molten Carbonate	~650	• Some fuel flexibility • High-grade waste heat	• Fragile electrolyte matrix • Electrode sintering	• Distribute power • Utilities
Solid Oxide	800-1000	• Maximum fuel flexibility • Highest co-generation efficiency	• Exotic materials • Sealing and cracking issues	• Distribute power • Utilities

Table 1.
Comparison of various fuel cells.

exchanger, from which heat must be rejected to the air. Operating the fuel cell at maximum efficiency reduces heat loads, but also reduces power output, forcing an increase in fuel cell stack size and cost. For high-temperature fuel cells, however, waste heat can be utilized by expanding the off gases through a turbine to generate additional electricity; such co-generation efficiencies can reach 80 percent. In some applications, even the remaining 20 percent can provide value (e.g., by warming the building's interior).

Heat rejection is only one aspect of thermal management. Thermal integration is vital for optimizing fuel cell system efficiency, cost, volume and weight. Other critical tasks, depending on the fuel cell, are water recovery (from fuel cell stack to fuel processor) and freeze-thaw management.

Electrical Sub-System

Electrical management, or power conditioning, of fuel cell output is often essential because the fuel cell voltage is always dc and may not be at a suitable level. For stationary applications, an inverter is needed for conversion to ac, while in cases where dc voltage is acceptable, a dc-dc converter may be needed to adjust to the load voltage. In electric vehicles, for example, a combination of dc-dc conversion followed by inversion may be necessary to interface the fuel cell stack to a 300 V ac motor.

TYPES OF FUEL CELLS

There are five classes of fuel cells. Like batteries, they differ in the electrolyte, which can be either liquid (alkaline or acidic), polymer film, molten salt, or ceramic. As Table 1 shows, each type has specific advantages and disadvantages that make it suitable for different applications. Ultimately, however, the fuel cells that win the commercialization race will be those that are the most economical.

The first fuel cell to become practical was the alkaline fuel cell (AFC). In space applications, liquid hydrogen and liquid oxygen are already available to provide rocket propulsion, and so consumption in the AFC, to create on-board electricity and potable water for the crew, is an elegant synergy. It therefore found application during the 1960s on the Gemini manned spacecraft in place of heavier batteries. Their high cost ($400,000/kW) could be tolerated because weight reduction is extremely valuable; For example, it could allow additional experimental equipment to be carried on-board.

The AFC has some attractive features, such as relatively high efficiency (due to low internal resistance and high electrochemical activity), rapid start-up, low corrosion characteristics, and few precious metal requirements.

However, the AFC's corrosive environment demands that it uses some rather exotic materials, and the alkaline (potassium hydroxide solution) concentration must be tightly controlled because it has poor tolerance to deviations. Critically, the alkali is readily neutralized by acidic gases, so both the incoming fuel and air need carbon dioxide clean-up. This limits AFC applications to those in which pure hydrogen is used as the fuel, since a fuel processor generates large amounts of carbon dioxide. The small amount of carbon dioxide in air (~0.03%) can be handled using an alkaline trap upstream of the fuel cell and, consequently, is not as much of a problem.

Because of this extreme sensitivity, attention shifted to an acidic system, the phosphoric acid fuel cell (PAFC), for other applications. Although it is tolerant to CO_2, the need for liquid water to be present to facilitate proton migration adds complexity to the system. It is now a relatively mature technology, having been developed extensively for stationary power usage, and 200 kW units (designed for co-generation) are currently for sale and have demonstrated 40,000 hours of operation. An 11 MW model has also been tested.

In contrast with the AFC, the PAFC can demonstrate reliable operation with 40 percent to 50 percent system efficiency even when operating on low quality fuels, such as waste residues. This fuel flexibility is enabled by higher temperature operation (200°C vs. 100°C for the AFC) since this raises electro-catalyst tolerance toward impurities. However, the PAFC is still too heavy and lacks the rapid start-up that is necessary for vehicle applications because it needs preheating to 100°C before it can draw a current. This is unfortunate because the PAFC's operating temperature would allow it to thermally integrate better with a methanol reformer.

The PAFC is, however, suitable for stationary power generation, but faces several direct fuel cell competitors. One is the molten carbonate fuel cell (MCFC), which operates at ~650°C and uses an electrolyte made from molten potassium and lithium carbonate salts. High-temperature operation is ideal for stationary applications because the waste heat can enable co-generation; it also allows fossil fuels to be reformed directly within the cells, and this reduces system size and complexity. Systems providing up to 2 MW have been demonstrated.

On the negative side, the MCFC suffers from sealing and cathode corrosion problems induced by its high-temperature molten electrolyte. Thermal cycling is also limited because once the electrolyte solidifies it is prone to develop cracks during reheat-ing. Other issues include anode sintering and elution of the oxidized nickel cathode into the electrolyte.

These problems have led to recent interest in another alternative to PAFC, the solid oxide fuel cell (SOFC). As its name suggests, the electrolyte is a solid oxide ceramic. In order to mobilize solid oxide ions, this cell must operate at temperatures as high as 1,000°C. This ensures rapid diffusion of gases into the porous electrodes and subsequent electrode reaction, and also eliminates the need for external reforming. Therefore, in addition to hydrogen and carbon monoxide fuels, the solid oxide fuel cell can even reform methane directly. Consequently, this fuel cell has attractive specific power, and cogeneration efficiencies greater than 80 percent may be achievable. Moreover, the SOFC can be air-cooled, simplifying the cooling system, although the need to preheat air demands additional heat exchangers. During the 1980s and 1990s, 20–25 kW "seal-less" tubular SOFC modules were developed and tested for producing electricity in Japan. Systems producing as much as 100 kW have recently been demonstrated.

Because this design has relatively low power density, recent work has focused on a "monolithic" SOFC, since this could have faster cell chemistry kinetics. The very high temperatures do, however, present sealing and cracking problems between the electrochemically active area and the gas manifolds.

Conceptually elegant, the SOFC nonetheless contains inherently expensive materials, such as an electrolyte made from zirconium dioxide stabilized with yttrium oxide, a strontium-doped lanthanum manganite cathode, and a nickel-doped stabilized zirconia anode. Moreover, no low-cost fabrication methods have yet been devised.

The most promising fuel cell for transportation purposes was initially developed in the 1960s and is called the proton-exchange membrane fuel cell (PEMFC). Compared with the PAFC, it has much greater power density; state-of-the-art PEMFC stacks can produce in excess of 1 kW/l. It is also potentially less expensive and, because it uses a thin solid polymer electrolyte sheet, it has relatively few sealing and corrosion issues and no problems associated with electrolyte dilution by the product water.

Since it can operate at ambient temperatures, the PEMFC can startup quickly, but it does have two significant disadvantages: lower efficiency and more stringent purity requirements. The lower efficiency is due to the difficulty in recovering waste heat, whereas the catalyzed electrode's tolerance toward impurities drops significantly as the temperature falls. For example, whereas a PAFC operating at 200°C (390°F) can tolerate 1 percent CO, the PEMFC, operating at 80°C (175°F) can tolerate only ~0.01% (100 ppm) CO. The membrane (electrolyte) requires constant humidification to maintain a vapor pressure of at least 400 mmHg (~0.5 bar), since failure to do so produces a catastrophic increase in resistance. Operation at temperatures above 100°C would greatly simplify the system, but existing membranes are not sufficiently durable at higher temperatures and will require further development.

Fuel cells can run on fuels other than hydrogen. In the direct methanol fuel cell (DMFC), a dilute methanol solution (~3%) is fed directly into the anode, and a multistep process causes the liberation of protons and electrons together with conversion to water and carbon dioxide. Because no fuel processor is required, the system is conceptually very attractive. However, the multistep process is understandably less rapid than the simpler hydrogen reaction, and this causes the direct methanol fuel cell stack to produce less power and to need more catalyst.

FUTURE CHALLENGES

The biggest commercial challenge facing fuel cells is cost, and mass production alone is insufficient to drive costs down to competitive levels. In the stationary power market, fuel cell systems currently cost ~$3,000/kW and this can be split into three roughly equal parts: fuel cell stack, fuel processor, and power conditioning. With mass production, this cost might fall to below $1,500/kW but this barely makes it competitive with advanced gas turbines. Moreover, for automotive applications, costs of under $100/kW are necessary to compete with the internal combustion engine. Bringing the cost down to these levels will require the development of novel system designs, materials, and manufacturing processes, in addition to mass production. However, even if PEM fuel cells fail to reach the stringent automotive target, they are still likely to be far less expensive than fuel cells designed specifically for other applications.

Even in a "simple" hydrogen fuel cell system, capital cost reduction requires improvements in many diverse areas, such as catalyst loadings, air pressuriza-

tion, cell thermal management, and sealing. Compounding the challenge is the need for durability and reliability. For example, electrodes and seals must be resistant to corrosion, stress, temperature fluctuations, and fuel impurities. Unfortunately, stable materials tend to be more expensive, and so a trade-off between life and cost can be expected as the technology nears the market stage.

This trade-off may not even occur in some cases. Membranes used in the PEMFC have been developed for the chlor-alkali industry and have 40,000-hour durability (shutdowns are prohibitively expensive in stationary applications), require only 5,000-hour durability (corresponding to 100,000 miles) for automotive applications. Hence, it may be possible to develop less expensive membranes that still meet automotive requirements.

Operating costs, in contrast, are more straightforward to determine because they depend on system efficiency, which, in turn, is related to voltage and current density (the current generated per unit area of electrolyte). Fuel savings are expected since the fuel cell operates more efficiently than a heat engine, and there may be lower maintenance and repair costs because fuel cells have fewer moving parts to wear out.

In addition to cost, a major technical challenge is in the fuel processor sub-system. Improvements are still needed in reducing size, weight, and, for transportation applications, cold-starting. Vaporizers also rely on heat generated from unused fuel leaving the fuel cell, and this combustion must be emission-free if the fuel cell system is to be environmentally attractive. This leads to the use of catalytic burners and, although such burners reduce NO_x emissions, they have yet to demonstrate sufficient durability. For the PEMFC, the final CO clean-up stage, PROX, uses precious metal catalysts and requires very fine temperature control in order to maintain the catalyst's selectivity toward oxidizing a small amount (1%) of CO in the presence of large amounts (>40%) of hydrogen. Such precision is difficult to achieve under conditions where the load varies continuously.

Full system integration is a major challenge since system design often must accommodate contradictory objectives. For example, it is relatively straightforward to design a fuel cell for high efficiency by maximizing thermal integration, but this is likely to increase complexity and degrade dynamic response. It may also increase cost and, given the dollar value of

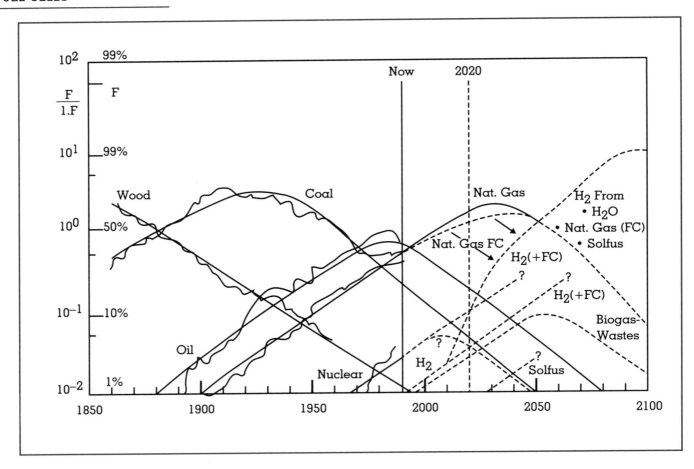

Figure 5.
Shift towards a hydrogen economy?

each efficiency percentage gain, this may not be economically justifiable.

System integration involves numerous miscellaneous development activities, such as control software to address system start-up, shut-down and transient operation, and thermal sub-systems to accomplish heat recovery, heat rejection and water recovery within the constraints of weight, size, capital and operating costs, reliability, and so on. Depending on the application, there will be additional key issues; automotive applications, for example, demand robustness to vibrations, impact, and cold temperatures, since if the water freezes it will halt fuel cell operation.

CONCLUSION

The same environmental drivers that are stimulating fuel cell development are also causing increased inter-est in alternatives. For example, in stationary power applications, microturbines are being developed that might compete with fuel cells for distributed power generation. In transportation, there is renewed interest in diesel engines, hybrid propulsion systems, and alternative-fueled vehicles. Advances in solar cells may also eliminate potential markets for fuel cells. Despite the strong, entrenched competition, there are reasons to believe that fuel cell commercialization is inevitable. Perhaps the strongest energy trend is the gradual shift toward renewable hydrogen. For example, wind-generated electricity can be used for electrolyzing water and, despite the extra step, the *potential* advantage of hydrogen over electricity is its easier transmission and storage. As Figure 5 indicates, the process of using fuels with ever-increasing H:C ratio has been developing for over 100 years. Recently, such a shift has been reinforced by environmental arguments. Should hydrogen evolve to be the fuel of the future, there is a compelling case to convert this into end-use electricity using fuel cells.

It must be recognized, however, that hydrogen usage for transportation will always create a trade-off between energy efficiency (fuel economy) and energy density (range), whereas this trade-off is non-existent for other applications.

In summary, fuel cell development is being accelerated both by the wide variety of applications and by the search for cleaner and more efficient utilization of primary energy and, ultimately, renewable energy. Because these forces for change are unlikely to disappear, it is quite likely that fuel cells will emerge as one of the most important and pervasive power sources for the future.

Christopher E. Borroni-Bird

BIBLIOGRAPHY

Appleby, A. J., and Foulkes, F. R. (1989). *Fuel Cell Handbook.* New York: Van Nostrand Reinhold Co.

FUEL CELL VEHICLES

ENVIRONMENTAL CONCERNS CONVERGE ON TECHNOLOGY PROGRESS

International concerns about the release of greenhouse gas emissions, deteriorating air quality, and reliance on oil sources in the Middle East oil continue to place pressure on the automobile industry to develop cleaner and more efficient vehicles. In practice there are three main ways this can be achieved: vehicle weight reduction, deployment of more efficient propulsion systems, and cleaner fuels. Fuel cell technology, by virtue of its unique means of operation, is well placed to address the power train limitations of both conventional internal combustion (IC), engines (i.e., nonzero emissions and relatively low efficiency) and batteries (i.e., inadequate range and long "recharging" time).

Many of the world's major automakers, prompted by both this consumer demand and progress in reducing the inherent cost and size of fuel cells, are now committed to developing and commercializing fuel cell vehicles.

FUEL CELL BASICS

A fuel cell creates electricity directly from a fuel, usually hydrogen. The hydrogen is fed into one side of the fuel cell (the anode) where electrons are stripped off it to produce hydrogen ions (protons). In a vehicle, these electrons energize a motor to turn the wheels, and then they return to the cathode to combine with incoming oxygen from the air to produce oxygen anions. Meanwhile, the protons pass through the electrolyte and link up with the oxygen ions to produce water. The net effect is identical to the combustion of hydrogen in air except that the transfer of electrons has been channeled directly into making electricity, as opposed to heat.

As with batteries, differences in electrolytes create several types of fuel cells. The automobile's demanding requirements for compactness and fast start-up have led to the Proton Exchange Membrane (PEM) fuel cell being the preferred type. This fuel cell has an electrolyte made of a solid polymer.

CHALLENGES TO FUEL CELL VEHICLE COMMERCIALIZATION

The biggest challenge facing fuel cells is cost. Current fuel cell systems probably cost about ten times as much as the $25/kW target set by the IC engine. Novel materials and manufacturing processes must be developed because mass production alone will not drive fuel cell stack (series of fuel cells combines to generate a useable voltage) costs down to the required level. Inexpensive bipolar plate materials, more effective catalyst utilization, and refined flowfields for efficiently channelling the input gases and exhaust water will contribute to the cost reduction.

It is expected that the fuel cell should be able to compete with an IC engine in terms of size and weight. As an added advantage, many fuel cell components can be configured into a relatively wide array of shapes to take advantage of space onboard the vehicle.

A fuel cell system also needs ancillaries to support the stack, just as an IC engine has many of the same type of ancillary subsystems. Major subsystems are needed for providing adequate humidification and cooling, and for supplying fuel and oxidant (air) with the correct purity and appropriate quantity.

These subsystems profoundly affect the fuel cell system performance. As an example, the inherently slow air (oxygen) electrode reaction must be accelerated using an air compressor, but this creates noise and exacts high power consumption, typically 10 to 15 percent of the fuel cell stack output. Fuel cell *system* efficiency, therefore, drops drastically when the fuel cell is operating at light load (below 10% rated power) even though the *stack* efficiency increases under these conditions. Low-cost, highly efficient air compressors that exploit the hot compressed exhaust gases may have more impact on increasing system efficiency than stack improvements would have.

Although the fuel cell is more efficient than an IC engine, it does produce waste heat at low temperature, and this poses problems from a heat rejection standpoint. Because there are vehicle styling constraints on how large the radiator can be, there is significant research into raising fuel cell stack operating temperatures by up to 10°C. If this can be achieved without degrading the electrolyte's durability, then it may almost halve the temperature differential between waste heat and ambient air in extremely hot conditions, with corresponding reductions in radiator size.

An automobile must also operate reliably in freezing conditions, which means that the humidifying, de-ionized water may need draining from the fuel cell system at key-off and reinjection during start-up. The coolant must also be freeze tolerant. These requirements for immediate power will probably force the vehicle to contain an energy storage component, such as a battery, flywheel, or ultracapacitor. This hybridization should also boost fuel economy because it enables regenerative braking and reduces the time that the fuel cell system spends below 10% rated load (where it is least efficient and where the driven car spends much of its time).

The vehicle also puts constraints on the choice of fuel. A combination of factors, such as the anticipated fuel and vehicle cost, both vehicular and "well-to-wheels" emissions and efficiency, perceived or real safety, and infrastructure will determine the choice of optimum fuel.

FUEL SELECTION AFFECTS PERFORMANCE AND COMMERCIALIZATION

Environmentally benign hydrogen is clearly the ideal fuel for use with fuel cells. The question is whether to store the hydrogen on-board or to generate it on board from liquid fuels that are easier to store and distribute.

The problem is that hydrogen, even at 10,000 psi (or 690 bar), requires five to ten times the volume of today's gasoline tank, depending on the fuel cell vehicle's real world efficiency. Packaging volume is compromised even further because pressurized tanks require thick carbon fiber walls and are, therefore, nonconformable. Moreover, they may cost several thousand dollars more than a conventional gasoline tank.

Although liquifying the hydrogen to 20K (–253°C) does increase energy density, it also creates new problems, such as cryogenic handling and venting (as ambient heat passes through the walls of the insulated vessel and evaporates the liquid). In addition, liquefaction is highly energy intense.

The third method of storing hydrogen is to absorb it into a metal. The only metal hydrides that can currently liberate their absorbed hydrogen at 80°C (allowing exploitation of the waste heat from the PEM fuel cell) are low-temperature ones and these can only store around 2% hydrogen by weight. Since 5 kg (11 lb) of hydrogen may be required to provide 380 miles (610 km) range in an 80 mpg (3 l/100 km) vehicle, the hydride will weigh more than 250 kg (550 lb). This weight not only reduces fuel economy but adds significant cost.

Hydrogen can also be adsorbed onto activated carbons; storage occurs in both gaseous and adsorbed phases. Existing carbon adsorbents outperform compressed gas storage only at relatively low pressures, when most of the gas is stored in the adsorbed phase. At pressures above 3,000 psi (207 bar), which are needed to store significant amounts, the adsorbent tends to be counterproductive since it blocks free space. Carbon adsorbents and metal hydrides introduce several control issues, such as risk of poisoning, and heat management during refueling and discharging.

Improved hydrogen storage is the key to a hydrogen economy. Without adequate storage, hydrogen may remain a niche transportation fuel which in turn, could limit the development of a hydrogen infrastructure. Of course, hydrogen could still play a leading role in *stationary* power generation without the need for dramatic improvements in storage.

Another concern with hydrogen is safety. Reputable organizations with extensive experience in using hydrogen have concluded that hydrogen is often safer than gasoline because it has a tendency to rapidly disperse upwards. However, its invisible flame and wide flammability range cause legitimate concerns. Reversing hydrogen's image requires a combination of public education and real world vehicle demonstrations. Automakers must design safe ways to store hydrogen on board. Despite these storage and safety concerns, hydrogen is considered the most attractive fuel for fuel cell buses because central refueling is possible and there is adequate space for hydrogen storage on the roof, which is also probably the safest location. However, for cars and light-duty trucks used by individuals, there is great interest in generating hydrogen on board the vehicle from fuels that are easier to store and transport.

Natural gas (CH_4) is also bulky to store, but reacting it with steam generates methanol (CH_3OH), an ambient-temperature liquid fuel. Having already been manufactured from natural gas outside the vehicle and containing no C-C bonds, methanol can generate hydrogen relatively easily and efficiently on board the vehicle using the steam produced by the fuel cell ($CH_3OH + H_2O \rightarrow 3H_2 + CO_2$). This hydrogen is then fed into the fuel cell to generate electricity. Unlike alkaline fuel cells used in the Apollo space missions, the carbon dioxide diluent can be fed into the PEM fuel cell without harm. Another attraction of methanol is the potential to ultimately eliminate the complexity of a reformer by using a direct methanol fuel cell where methanol is converted into hydrogen inside the cell. Because of these advantages, methanol is currently the leading candidate to propel mainstream fuel cell vehicles, absent a hydrogen storage breakthrough.

However, there are several issues with widespread methanol usage. Methanol production from natural gas is relatively inefficient (~67%), and this largely offsets the vehicular improvement in efficiency and carbon dioxide reduction (since gasoline can be made with ~85% efficiency from oil). Additionally, the PEM fuel cell demands very pure methanol, which is difficult to deliver using existing oil pipelines and may require a new fuel distribution infrastructure.

Compared with methanol, gasoline is more difficult to reform into hydrogen because it contains C-C (carbon-carbon) bonds. Breaking these down to generate hydrogen requires small amounts of air (called partial oxidation or POX) and this lowers the efficiency and makes heat integration more difficult. Also, gasoline is not a homogenous compound like methanol. Its sulfur poisons the fuel processor and fuel cell stack catalysts, and its aromatics have a tendency to form soot. Finally, gasoline's hydrogen-carbon (H:C) ratio is under two whereas methanol's is four. Gasoline, therefore, it yields less hydrogen and this reduces fuel cell stack output and efficiency. The net effect is that gasoline fuel cell vehicles are likely to be less efficient and more complex than methanol fuel cell vehicles. However, use of gasoline keeps the refueling infrastructure intact, and even gasoline fuel cell vehicles offer potential to be cleaner and more efficient than gasoline engines.

A fuel closely related to gasoline is naphtha, which is also a potential fuel cell fuel. Naphtha is already produced in large quantities at refineries and is a cheaper fuel than gasoline, which must have octane-boosting additives blended into it. Unlike methanol, naphtha can be distributed in the same pipelines as gasoline. From the fuel cell's perspective, it has a higher H:C ratio and lower sulfur and aromatics content than gasoline.

Given the technical challenges facing gasoline reforming, it is possible that gasoline fuel cells may be introduced to vehicles as an Auxiliary Power Unit (APU) feature. A 5 kW APU that provides electricity to power air conditioning and other electrical loads can operate with less demanding transient performance than is necessary for propulsion. Moreover, lower efficiency can be tolerated since it competes with a gasoline engine-alternator combination at idle and not with a gasoline engine directly.

The transition from the IC engine to fuel cells will not alleviate energy supply insecurity. All of these fuel options—hydrogen, methanol, gasoline, and naphtha will be generated most economically from fossil fuels (natural gas or oil) in the near to mid-term. However, fuel cells are still expected to reduce carbon dioxide emissions since propulsion is not dependent on combustion. With respect to local air quality, fuel cell vehicles should significantly reduce pollutants, particularly if they use hydrogen. The only emissions from a hydrogen fuel cell vehicle would be pure water, thus allowing it to qualify as a Zero Emission Vehicle. In addition, fuel cell vehicles should stimulate development of cleaner fuels more quickly than IC engines because they are less tolerant of fuel impurities and benefit more from hydrogen-rich fuels.

Christopher Borroni-Bird

See also: Fuel Cells.

533

BIBLIOGRAPHY

Borroni-Bird, C. E. (1996). "Fuel Cell Commercialization Issues for Light-Duty Vehicle Applications." *Journal of Power Sources* 61:33–48.

FUEL ECONOMY

See: Automobile Performance

FUEL OIL

See: Residual Fuels

FULLER, R. BUCKMINSTER, JR. (1895–1983)

Richard Buckminster Fuller, Jr., best known as the architect of Houston's Astrodome and other geodesic structures, enjoyed a long and varied career as a structural engineer and unconventional humanistic thinker. A colorful and gregarious individual, Fuller was first embraced by government officials for his innovative designs and later cherished by the 1960s counterculture. He patented more than twenty new inventions, authored twenty-five books and dozens of articles, lectured globally on energy issues and the wise use of world resources, and dabbled in both art and science. Never one to be modest, Fuller called himself "an engineer, inventor, mathematician, architect, cartographer, philosopher, poet, cosmologist, comprehensive designer and choreographer." He especially liked the self-description "anticipatory comprehensive design scientist," because he saw himself as a scientist who anticipated human needs and answered them with technology in the most energy-efficient way.

Fuller's "more with less" philosophy first gained the attention of Americans in the 1920s with his "Dymaxion" inventions. Fuller employed this term—a combination of "dynamic," "maximum," and "ion"—to describe inventions that do the most with the least expenditure of energy. His Dymaxion house was self-sufficient in that it generated its own power, recycled water, and converted wastes into useable energy. It also featured air conditioning and built-in laborsaving utilities including an automatic laundry machine, a dishwasher that cleaned, dried, and reshelved the dishes, and compressed air and vacuum units. The Dymaxion bathroom was a one-piece aluminum unit containing a "fog gun"—an economical showerhead using a mixture of 90 percent air and 10 percent water. Fuller's Dymaxion car ran on three wheels, had front-wheel drive with rear-wheel steering, and registered a reported forty to fifty miles per gallon in 1933. None of these projects were mass-produced, but his futuristic designs, and later his geodesic buildings, World Game workshops, and startling questions and proposals about natural resources and human survival stimulated imaginations and encouraged others to explore ways to create a more energy-efficient environment.

Born in Milton, Massachusetts, in 1895 to Richard and Caroline Fuller, Bucky, as most knew him, spent his summers at the family retreat, Bear Island, in Penobscot Bay, Maine. As Fuller remembered it, his interest in building better "instruments, tools, or other devices" to increase the "technical advantage of man over environmental circumstance" began there. One of his tasks each day as a young boy was to row a boat four miles round trip to another island for the mail. To expedite this trip, he constructed his "first teleologic design invention," a "mechanical jelly fish." Noting the structure of the jellyfish and attending to its movement through the water, Fuller copied nature and produced a boat of greater speed and ease. Observing natural phenomena remained his lifelong source of inspiration which was not surprising, given that Fuller was the grandnephew of transcendentalist Margaret Fuller.

Following his Brahmin family's tradition, Fuller went to Harvard, but instead of graduating as a lawyer or Unitarian minister as his ancestors had, he was expelled twice for cutting classes, failing grades, and raucous living. He never did gain a bachelor's degree. Instead, Fuller tutored himself in the arts and

Buckminster Fuller, sitting beside a model of his "Dymaxion" house, seen in the 1930 World's Fair in Chicago. (Corbis Corporation)

sciences, and the Navy and apprenticeships at a cotton mill machinery plant and a meat-packing factory provided him with a practical education. Eventually, Fuller garnered multiple honorary doctorates and awards, including a Presidential Medal of Freedom shortly before his death in 1983.

In 1917, Fuller married Anne Hewlett, the daughter of the respected New York architect James M. Hewlett. A year later their daughter Alexandra was born, only to die a few years later from infantile paralysis. From 1922 to 1926, Fuller and his father-in-law founded and ran the Stockade Building System that produced lightweight construction materials. Fuller failed miserably at the business and, in 1927, stung by the death of his daughter, years of carousing, and financial failure, he considered suicide. Impoverished and living in the gangster region of Chicago with his wife and newborn daughter, Allegra, Fuller walked to the shore of Lake Michigan

with the intent of throwing himself in. Instead, he had a revelation: "You do not have the right to eliminate yourself, you do not belong to you. You belong to the universe. . . apply yourself to converting all your experience to the highest advantage of others." At this point, Fuller made it his ambition to design "tools for living," believing that human life would improve if the built environment was transformed.

Fuller approached his mission both philosophically and practically. He explored mathematics in search of "nature's coordinate system" and invented his own "energetic-synergetic geometry" as his expression of the underlying order he saw. He also studied philosophy and physics for understanding of time and motion. But his philosophical musings always had a practical bent. For example, Fuller, taken with Albert Einstein's theories about time and motion, sought to employ physics in his design initiatives.

Fuller devoted his early years to the problem of

building energy-efficient and affordable housing, but his projects and ideas were largely confined to a few students in architectural schools, executives of small corporations, and readers of *Shelter* and *Fortune* magazines. His most successful enterprise was the manufacture of Dymaxion Deployment Units for use by the Army in World War II. Fuller tried to convert these military units into civilian housing, but failed to raise production funds.

In 1949, Fuller's luck changed with the construction of his first geodesic dome at Black Mountain College. Motivated by a desire to create an energy-efficient building that covered a large space with a minimal amount of material, Fuller fashioned a sphere-shaped structure out of triangular pyramids. He joined these tetrahedrons together by pulling them tight in tension rather than relying on compression to raise the dome. In doing so, Fuller reconceptualized dome engineering and created a practical, efficient, cost-effective, strong, and easy-to-transport-and-assemble building.

Several hundred thousand geodesics function today worldwide as auditoriums, aviaries, banks, churches, exposition halls, greenhouses, homes, industrial plants, military sheds, planetariums, playground equipment, and sports arenas. Some of these geodesics, including the clear, thin-skinned "Skybreak" and "Garden-of-Eden" structures, were experiments in using renewable energy sources. These transparent spheres, primarily functioning as greenhouses or as experimental biodomes by environmentalists before the much-publicized Biosphere projects, relied on solar power to regulate temperature. Fuller also had more fantastic ideas to reduce energy losses in summer cooling and winter heating: in 1950, for example, he proposed building a dome over Manhattan to regulate the environment.

Fuller's vision was more expansive than creating better shelter systems, however. His scope was global and his ideals utopian. At the same time that he was creating "machines for living," he was surveying Earth's resources, designing maps, and plotting strategies for an equitable distribution of goods and services. In 1927 Fuller started his "Inventory of World Resources, Human Trends and Needs." In the 1930s he began his Dymaxion map projects to gain a global perspective, and by the 1960s, in the heart of the Cold War, Fuller was busy devising ways to ensure the survival of the earth.

For Earth to continue functioning for the maximum gain of people everywhere, Fuller believed that resources must be used wisely and shared equally. He scorned reliance on fossil fuels and encouraged the development of renewable energy sources, including solar, wind, and water power. To convey his ideas, Fuller employed the metaphor of Earth as a spaceship. This spacecraft, he explained in his best-selling book, *Operating Manual for Spaceship Earth*, was finite and in need of careful management. With limited materials on board, "earthians" must work out an equitable and just distribution to keep the spaceship operating smoothly and efficiently.

Intent upon channeling human energy and resources into projects for "livingry," as opposed to "weaponry," Fuller participated in cultural exchanges between the Soviet Union and the United States, worked on United Nations projects, and traveled the world to communicate his vision. He conducted marathon "thinking-aloud" sessions before large audiences, taking on a cult status for some and the role of crackpot for others. He criticized political solutions to world problems and promoted technological design to reallocate wealth, labor, and resources.

The culmination of these thoughts can be found in Fuller's strategic global planning organization, the World Game Institute. Counter to military war games, World Game was both a tool for disseminating information and an exercise to engage others in problem-solving. Through a set of simulated exercises, participants used Fuller's inventory of resources, synergetic geometry, and Dymaxion maps to plot strategies to "make the world work for 100% of humanity in the shortest possible time, through spontaneous cooperation, and without ecological offense or the disadvantage of anyone." Begun in small college classrooms in the 1960s, World Game attracted a dedicated following. In the 1990s, the World Game Organization established itself on the Internet to facilitate innovative thinking about world resources.

Fuller's lifelong engagement in global energy issues and his lasting contributions in engineering design make him a noteworthy study in twentieth century debates on energy supply and use. Fuller believed that world problems of war, poverty, and energy allocation could be eradicated by cooperation between nations and through technological innovation. His optimistic celebration of technology pitted him against individuals such as historian Lewis Mumford who questioned the salutary effects of technology, and placed him at odds with Paul Ehrlich and other environmentalists who believed that Earth's resources could not support the world's growing population. Fuller, a modernist at heart, maintained that the trouble was in distribution, and that continual employment of technology would not destroy but save the planet.

Linda Sargent Wood

BIBLIOGRAPHY

Baldwin, J. (1996). *BuckyWorks: Buckminster Fuller's Ideas for Today.* New York: Wiley.

Fuller, R. B. (1963). *Ideas and Integrities: A Spontaneous Autobiographical Disclosure*, edited by R. W. Marks. Englewood Cliffs, NJ: Prentice-Hall.

Fuller, R. B. (1963). *Operating Manual for Spaceship Earth.* New York: E. P. Dutton.

Fuller, R. B. (1969). *Utopia or Oblivion: The Prospects for Humanity.* New York: Bantam.

Fuller, R. B. (1975). *Synergetics: Explorations in the Geometry of Thinking.* New York: Macmillan Publishing Company.

Gabel, M. (1975). *Energy Earth and Everyone: A Global Energy Strategy for Spaceship Earth.* San Francisco: Straight Arrow Books.

McHale, J. (1962). *R. Buckminster Fuller.* New York: George Brazillier, Inc.

Sieden, L. S. (1989). *Buckminster Fuller's Universe: An Appreciation.* New York: Plenum Press.

FULTON, ROBERT (1767–1815)

Robert Fulton was unique among inventors of his time, since he was born in America, apprenticed in Europe and returned to his native country to perfect his greatest invention, the steamboat. Fulton was born in 1765 of a respectable family in Little Britain, Pennsylvania, in Lancaster County. Unlike many inventors, he was not motivated by paternal leadership since he grew up fatherless after the age of three. His mother is credited for his soaring interest in the world of painting. Despite, or perhaps because of, the loss of his father, Robert's free time was spent with local mechanics or alone with his drawing pencil. By age seventeen, he was accomplished enough as a landscape and portrait artist that he earned income from it in the city of Philadelphia for four years. He earned enough to build a home for his mother in Washington County before receiving a generous

Robert Fulton's "Clermont." (Corbis Corporation)

offer to take his artistic talents to England in 1786, at age 21, where they were equally well received.

Over the next seven years, he made numerous acquaintances, including two who were instrumental in his transition from artist to civil engineer—the Duke of Bridgewater, famous for work with canals, and Lord Stanhope, who specialized in works of science and mechanical arts. By 1793, Robert was experimenting with inland navigation, an area that remained of interest throughout his life. A year later, he filed a patent in Britain for a double inclined plane, and spent several years in Birmingham, the birthplace of the Industrial Revolution that came about thanks to the powerful engines built by James Watt as early as 1763.

Remarkable for someone so talented in the arts, Fulton was able to use his accomplished drawing skills to express his designs. As a now noted draftsman, he invented numerous pieces of equipment such as tools to saw marble, spin flax, make rope and excavate earth. Sadly, he once lost many of his original manuscripts during a shipping accident.

By the beginning of the nineteenth century, Fulton turned his attention to his obsession with submarines and steamships. He made no secret of his goal for submarines—he intended to build them in order to destroy all ships of war so that appropriate attention could be devoted by society to the fields of education, industry and free speech. In 1801, he managed to stay under water for four hours and twenty minutes in one of his devices, and in 1805 he demonstrated the ability to utilize a torpedo to blow up a well built ship of two hundred tons. Unfortunately for Robert, neither the French nor British government was particularly impressed with the unpredictable success, nor the importance of his innovations; so he packed his bags and returned to America in December of 1806.

By the time Robert Fulton arrived in America, he had been studying steam navigation for thirteen years. Five years earlier, he had met Chancellor Livingston, who was partly successful in building a steam vessel, but yielded to Fulton's skill by allowing him to take out a patent in the latter's name to begin his improvements. By 1807, there was much skepticism and derision expressed along the banks of the East River in New York when Fulton prepared to launch his steamship, "The Clermont," from the ship yard of Charles Brown. A few hours later, there were nothing

but loud applause and admiration at the spectacular achievement that had been termed "Fulton's Folly." Finding his own design flaws, Fulton modified the water wheels and soon launched the ship on a stunning three-hundred-mile round-trip run to Albany. The total sailing time was sixty-two hours, a remarkable endurance record for so new an invention. Soon there were regular runs being made on the Hudson River and by rival ships around the country, with many disputes over the patent rights of different vessels in different states. But the industry was clearly set in motion by Fulton's uncanny designs.

By 1812, he added a double-hulled ferryboat to his credits, designing floating docks to receive them. In 1814, the citizens of New York vindicated his years of war vessel experiments by demanding a protective ship in defense of their New York Harbor. In March of that year, the President of the United States, James Madison, was authorized to enlist Fulton to design a steam-powered frigate, with full explosive battery, as a military defense weapon. He also empowered Fulton to complete a design for the defensive submarine Fulton longed to finish. Unfortunately, just three months before the completion of the steam frigate, Fulton fell victim to severe winter exposure on the river, dying on February 24, 1815, in New York City at the age of 50.

The public mourning was said to be equal to that expressed only for those who held public office of considerable acclaim. But by this time, the word 'folly' had long been dropped from his name. Just as George Stephenson followed the inventors of locomotives to gain credit for creating Britain's railways, Robert Fulton was not the inventor of the steam boat. What he *was*, however, was the brilliant creator of such an improved version that it made possible the many advancements of steamships in succeeding years. He took theory out of the experimentation process and designed machinery that opened the door to unlimited practical use from the technology available at the time.

Dennis R. Diehl

BIBLIOGRAPHY
Howe, H. (1858). *Memoirs of the Most Eminent American Mechanics: Also, Lives of Distinguished European Mechanics.* New York: Harper & Brothers.

FURNACES AND BOILERS

Furnaces and boilers are devices that burn fuel to space heat homes, offices, and industrial facilities. Natural gas, liquefied petroleum gas, and heating oil are the dominant fuels used for furnaces and boilers. In the United States, furnaces and boilers burning gas and oil take care of over 75 percent of all space heating.

EARLY HISTORY

The first oil burning devices for heating appeared in the oil-rich Caucasus region of Russia as early as 1861. But because of the remoteness of this region, those devices remained in obscurity, out of the flow of marketable goods. An event almost 115 years ago marks the birth of the modern oil burner. On August 11, 1885, the U.S. Patent office granted a patent to David H. Burrell of Little Falls, New York, for a "furnace apparatus for combining and utilizing oleaginous matters." As noted in the June 1985 centennial issue of *Fuel Oil News*, Burrell's invention "...was the forerunner of today's modern oil burner, and is generally accepted as the one that started the oil heat industry."

Gas burners are different from oil burners in that they control the mixing of air and gas and are referred to as *aerated burners*. This type of burner was invented around 1855 by Robert Wilhelm von Bunsen and is often called a *Bunsen burner* or *blue flame burner*. Natural gas is mostly methane and typically found underground in pockets. Liquified petroleum gas (LPG) occurs in "wet" natural gas and crude oil and must be extracted and refined before use.

GAS AND OIL HEAT BASICS

Liquid fuels, including heating oil after it is refined from crude oil, are transported by pipelines, oil tankers, barges, railroads, and highway tanker trucks to local bulk storage facilities, and delivered by tank trucks to the homeowner's tank, which is located either underground or in the basement, garage, or utility room. Figure 1 shows a typical residential oil heating system. These tanks typically store 275 to 500 gallons of fuel that can supply heat for a typical house for a month, and usually several oil tank truck deliveries to the home are needed during the year. LPG is

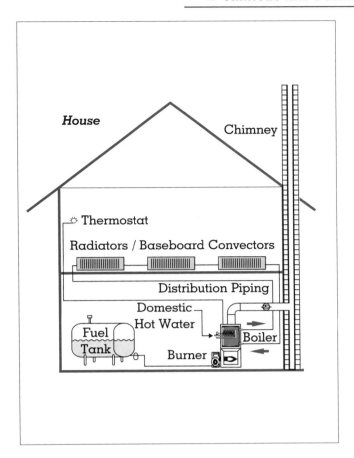

Figure 1.
Oil heating system.

similarly transported under pressure as a liquid by tanker trucks to the homeowner's above-ground tank. LPG may also be delivered in portable tanks typically 100-250 lbs, or as small as the 20-lb. tanks, used for portable gas grills. LPG is heavier than air and should not be stored inside a house for safety reasons.

Natural gas is transmitted in pipelines first across the country under high pressure in transcontinental pipe lines to local gas utility companies. The gas utility company reduces the pressure and distributes the gas through underground mains, and again reduces pressure from the street main into the home.

Gas or oil fuel is burned in a furnace or boiler (heat exchanger) to heat air or water, or to make steam that carries the heat absorbed from the combustion process to the rooms in the house, as shown in Figure 2. This is accomplished by circulating the warm air through ducts directly to the rooms, or by circulating hot water or steam through pipes to baseboard hot water convector units or radiators in the room. The boiler loop is completed with warm water returned through pipes to the boiler, or return air through the return register to air ducts back to the furnace to be reheated. In recent years, radiant floor heating systems that also rely on hot water circulation have come back into use. Boilers may also serve the additional function of heating water for showers and baths. A water heating coil immersed within the boiler is often used to heat this water as needed at a fixed rate. Another option is to have the coil immersed in a very well insulated water tank. The tank can provide a large supply of hot water at a very high rate of use. This is accomplished by exchanging heat from boiler water that passes through the inside of the coil to the domestic hot water stored in the tank. This method can provide for more hot water since a full tank of hot water is standing by for use when needed. The boiler operates only when the tank temperature drops below a certain set point. This results in fewer burner start-ups and is a more efficient way to heat water during the non-heating season months, enhancing year-round efficiency.

A recent alternative to the furnace or boiler for space heating is to use the hot water heater to heat the house. A separate air handler is sold along with a pump that pumps hot water from the water heater to the air handler where hot water flows through a finned tube heat exchanger coil. A blower in the air handler pushes air over the hot water coil and then through ducts to the rooms. Water out of the coil returns back to the water heater to be reheated. This way the water heater does double duty: it operates as a combination space heater and water heater. Gas water heaters can be used this way in mild climates and for homes or apartments with smaller heating requirements. Oil-fueled water heaters can also be used this way to handle homes in colder climates because they have higher firing rates and re-heat water faster. The installation cost is much lower for this approach in new construction, but units with high efficiency are not available for the gas water heaters as they are for gas furnaces or boilers.

Furnaces and boilers sold today must by law have annual fuel utilization efficiency of at least 78 to 80 percent. Gas water heaters operating this way as space heaters are equivalent to the efficiency of pre-1992 furnaces and boilers which had space heating efficiencies typically in the mid-60 percent range. However, the combined efficiency for space and

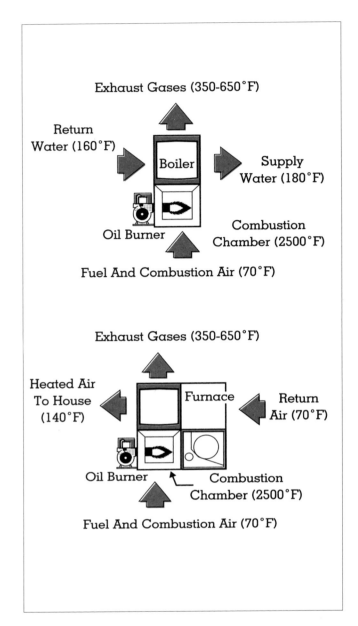

Figure 2.
Typical boiler or furnace operation
NOTE: Temperatures can vary.

water heating together may partly make up for the inefficiency of the water heater alone. There is also a concern regarding the life expectancy of water heaters, typically 5 to 10 years as compared to a boiler that will last anywhere from 20 to 40 years (or longer) depending on its design.

Both gas and oil-fired heating systems consist of several subsystems. The oil burner pump draws fuel from the tank; the fuel is atomized and mixed with just the right amount of air for clean combustion; an electric spark provides the ignition of the fuel/air mixture; and the flame produces the heat in a combustion chamber. The heat is then released to the heat exchanger to heat air in the furnace and water in the boiler or to make steam. The process is similar for gas that is supplied under regulated gas pressure into the house piping to the furnace. The sequence of operation for the system starts at the wall thermostat. When the room temperature drops below the set point, a switch closes allowing the gas valve to send gas to the burners. With almost all new gas furnaces today, a spark or glowing hot wire igniter typically lights a gas pilot, or a spark system ignites the oil spray. The presence of the ignition source or pilot is proven, and the gas valve opens, sending gas mixed with air to the burner for ignition. The flame sensing system must be satisfied that the flame ignites or else it will quickly shut the burner off to prevent significant amounts of unburned fuel from accumulating in the heating appliance. The flue pipe connects the unit to a chimney to exhaust the products of combustion from the furnace into the outside ambient environment. A side wall vented flue pipe made of a plastic is used for some high efficiency condensing furnaces. The combustion of natural gas and fuel oils, which are primarily hydrocarbon molecules, results primarily in the formation of water vapor and carbon dioxide. The process also results in very minor amounts of sulfur dioxide (which is proportionate to the sulfur content of fuel oil, typically less than one half of one percent by weight), and trace amounts of other sulfur oxides, nitrogen oxides, and particulate matter (unburned hydrocarbon-based materials) in the range of parts per million to parts per billion or less.

The furnace blower or boiler circulator pump starts up to send heated room air or hot water through the ducts or hot water pipes to the steam radiators, or hot water baseboard units, or individual room air registers located throughout the house. When the thermostat is satisfied that the room temperature has reached the set point, the burner shuts off. In furnaces, the blower continues to run a few seconds until the air temperature drops to about 90° F, then the blower also shuts off. The furnace blower may come on again before the next burner start-up to purge heat out of the furnace, particularly if the fan has a low turn-on set point. The cycle repeats when the room temperature drops below the set point. In boilers, heat may be purged from the boiler by running the circulator for a controlled amount of additional time, delivering either more heat to the living space or hot water to a storage tank. Heat will continue to be supplied to the room from the radiators or hot water baseboard units until they cool down to room temperature.

U.S. HOUSEHOLD ENERGY CONSUMPTION AND EXPENDITURES

There were 10.8 million U.S. households that used fuel oil for space and/or water heating in 1993. The average household using fuel oil typically comsumed 684 gallons a year for space heating. In addition, 3.6 million U.S. homes used kerosene. Together they represented 14.4 million homes or 15 percent of the 96.6 million households in the United States. These households consumed a total of 7.38 billion gallons of heating fuel and 340 million gallons of kerosene.

There were 52.6 million U.S. households that used natural gas for space and/or water heating and/ or use in other appliances (mainly cooking ranges) in 1993. In addition there were 5.6 million U.S. homes that used LPG. Together they represented 58.2 million homes or 60 percent of the 96.6 million households in the United States. These households consumed a total of 4,954 billion cubic feet of natural gas and 3.84 billion gallons of LPG for space and water heating. The average household using natural gas typically uses about 70 million Btu a year for space heating.

Whereas heated homes are mostly located in the colder regions of the United States like the north and northeast, natural gas and LPG are used for heating homes through out the United States in both warmer and colder climates.

EFFICIENCY

Most new gas and oil-fueled furnaces and boilers have similar efficiencies. The range of efficiency has narrowed with the introduction of minimum efficiency standards for new products sold since 1992. New gas and oil heating equipment currently available in the marketplace have Annual Fuel Utilization Efficiency (AFUE) ratings of at least 78 to 80 percent. AFUE is a measure of how efficient a furnace operates on an annual basis and takes into account cycling losses of the furnace or boiler. It does not include the

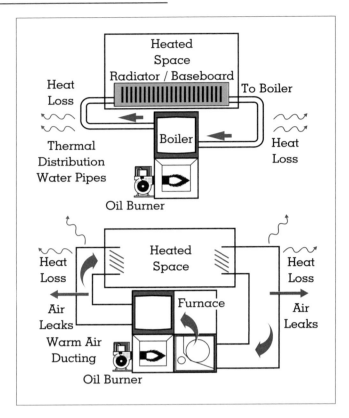

Figure 3.
Heat loss from distribution pipes and air ducts.

distribution losses in the duct system, which can be as much as 30 percent or more, especially when ducts are located in unheated areas such as the attic or unheated crawl space (see Figure 3). The AFUE also does not include the electricity to operate the blowers and any other electrical motors, controls or ignitors. The AFUE for gas furnaces are clustered in two groups: those in the range of 78 to 82 percent AFUE and those in the range of 90 to 96 percent. Few if any gas furnaces are available between these ranges because the upper range is a condensing furnace that requires an additional heat exchanger of a special alloy steel, a condensate pump and drain, and a special flue pipe material.

Because of the price diffferential between low- and high-efficiency condensing furnaces, only 22 percent of gas furnaces sold in the mid-1990's were high-efficiency condensing-type furnaces. Condensate is water that forms as a result of the combustion process. When the hydrogen in the fuel combines with oxygen from the combustion air, it forms water

vapor (steam) that is cooled and condensed in a condensing furnace during the heat exchange process. Carbon dioxide in the flue gases makes this water act like soda water and becomes acidic. Fuel oil contains less hydrogen than natural gas and is therefore less likely to condense water vapor in the flue gases when operated in the mid-80 percent range. Therefore oil furnaces with up to 87 percent AFUE are available without condensing. The availability of condensing oil-fueled furnaces is limited since there is little efficiency benefit to be gained from the additional cost of a special alloy heat exchanger.

Some new features of furnaces available today include variable speed blowers, which deliver warm air more slowly and more quietly when less heat is needed, and variable heat output from the burner, which when combined with the variable speed blower allows for more continuous heating than the typical fixed firing rate. Distribution system features can be sophisticated with zoned heating which employs a number of thermostats, a sophisticated central controller, and a series of valves or dampers that direct airflow or water to different parts of the home only when needed in those areas.

The average AFUE of all installed furnaces is 65 to 75 percent, much lower than post-1992 efficiency standards due to the different vintages of furnaces and boilers. Systems that are 40 years old or older are even less efficient (55-65%), but these represent a very small fraction of furnaces operating in the United States today.

The U.S. Department of Energy develops test procedures for efficiency measurements and sets minimum efficiency standards for furnaces, boilers, and water heaters. Information on energy efficiency of buildings and equipment is available from the DOE.

Esher Kweller
Roger J. McDonald

BIBLIOGRAPHY

American Council for an Energy Efficient Economy. (1989). *The Most Energy Efficient Appliances.* Washington, DC: Author.

American Council for an Energy Efficient Economy. (1989). *Oil and Gas Heating Systems.* Washington, DC: Author.

Beckett Corporation. (1979). *The Professional Serviceman's Guide to Oil Heat.* Elyria, OH: Author.

Energy Efficiency and Renewable Energy Network (EREN) Home Page. Washington DC: U.S. Department of

Energy, Energy Efficiency and Renewable Energy Network. August 4, 2000 <http://www.eren.doe.gov/>.

Lanthier, George. (1995). *Combustion and Oil Burning Equipment: An Advanced Primer*. Arlington, MA: Firedragon Enterprises.

Petroleum Marketer Association of America. (1995). *Advanced Oil Heat: A Guide to Improved Efficiency*. Arlington, VA: Kendall/Hunt Publishing.

Trethwey, Richard. (1994). *This Old House: Heating, Ventilation, and Air Conditioning*. Boston: Little, Brown.

Weinberger, Herbert. (1992). *Residential Oil Burners*. Albany, NY: Delmar Publishers.

FUSES

See: Electricity

FUSION, BOMB

See: Nuclear Fusion

FUSION, POWER

See: Nuclear Fusion

FUTURES

OVERVIEW

A futures contract is an agreement that calls for a seller to deliver to a buyer a specified quantity and quality of an identified commodity, at a fixed time in the future, at a price agreed to when the contract is made. An option on a commodity futures contract gives the buyer of the option the right to convert the option into a futures contract. Energy futures and options contracts are used by energy producers, petroleum refin-

ers, traders, industrial and commercial consumers, and institutional investors across the world to manage their inherent price risk, to speculate on price changes in energy, or to balance their portfolio risk exposure.

With very limited exceptions, futures and options must be executed on the floor of a commodity exchange through persons and firms registered with regulatory authorities. The contracts are traded either by open outcry, where traders physically transact deals face to face in specified trading areas called pits or rings, or electronically via computerized networks. The futures market provides a standardized trading environment such that all users know exactly what they are trading and where their obligations and risks lie. By entering into a standard futures and/or options contract, a certain amount of price assurance can be introduced into a world of uncertainty and price volatility.

Most futures contracts assume that actual delivery of the commodity can take place to fulfill the contract. However, some futures contracts require cash settlement instead of delivery. Futures contracts can be terminated by an offsetting transaction (i.e., an equal and opposite transaction to the one that opened the position) executed at any time prior to the contract's expiration. The vast majority of futures contracts are terminated by offset or a final cash payment rather than by physical delivery.

HISTORICAL PERSPECTIVE

Futures contracts for agricultural commodities have been traded in the United States since the nineteenth century and have been under federal regulation since the 1920s. Starting in the late 1970s, futures trading has expanded rapidly into many new markets, beyond the domain of traditional physical and agricultural commodities such as metals and grains. Futures and options are now offered on many energy commodities such as crude oil, gasoline, heating oil, natural gas, and electricity, as well as on a vast array of other commodities and financial instruments, including foreign currencies, government securities, and stock indices.

TERMS AND CONDITIONS

A typical futures contract might call for the delivery of 1,000 barrels (42,000 U.S. gallons) of unleaded gasoline meeting defined specifications at petroleum product terminals in New York Harbor during the next twelve months at an agreed price in dollars and cents per gallon. All terms and conditions other than the price are standardized. Gasoline is sold through hundreds of wholesale distributors and thousands of retail outlets, and is the largest single volume refined product sold in the United States. It accounts for almost half of national oil consumption. A market that diverse is often subject to intense competition, which in turn breeds price volatility and the need for reliable risk management instruments for energy producers and users.

OPTIONS

There are two types of options — call options and put options. A call option on a futures contract gives the buyer the right, but not the obligation, to purchase the underlying contract at a specified price (the strike or exercise price) during the life of the option. A put option gives the buyer the right to sell the underlying contract at the strike or exercise price before the option expires. The cost of obtaining this right to buy or sell is known as the option's "premium." This is the price that is bid and offered in the exchange pit or via the exchange's computerized trading system. As with futures, exchange-traded option positions can be closed out by offset.

HOW FUTURES AND OPTIONS DIFFER

The major difference between futures and options arises from the different obligations of buyer and seller. A futures contract obligates both buyer and seller to perform the contract, either by an offsetting transaction or by delivery. Both parties to a futures contract derive a profit or loss equal to the difference between the price when the contract was initiated and when it was terminated. In contrast, an option buyer is not obliged to fulfill the option contract. Buying an options contract is similar to buying insurance. The buyer is typically paying a premium to remove risk, while the seller earns the premium and takes on risk. The option buyer's loss is limited to the premium paid, but in order for the buyer to make a profit, the price must increase above (call option) or decrease below (put option) the option's strike price by more than the amount of the premium paid. In turn, the option seller (writer or grantor), in exchange for the premium received, must fulfill the option contract if the buyer so chooses. Thus, the

option's exercise takes place if the option has value (is "in the money") before it expires.

FEATURES NEEDED FOR A WELL-FUNCTIONING MARKET

Whatever the item, such as crude oil, underlying the futures or options contract, every market needs certain ingredients to flourish. These include

- Risk-shifting potential—the contract must provide the ability for those with price risk in the underlying item to shift that risk to a market participant willing to accept it. In the energy world, commercial producers, traders, refiners, distributors and consumers need to be able to plan ahead, and frequently enter into commitments to buy or sell energy commodities many months in advance.
- Price volatility—the price of the underlying item must change enough to warrant the need for shifting price risk. Energy prices are subject to significant variance due to factors affecting supply and demand such as level of economic activity, weather, environmental regulations, political turmoil, and war. Market psychology also plays its part.
- Cash market competition—the underlying cash (or physicals) market for the item must be broad enough to allow for healthy competition, which creates a need to manage price risk and decreases the likelihood of market corners, squeezes, or manipulation. The physical market in energy commodities is the largest such market in the world.
- Trading liquidity—active trading is needed so that sizable orders can be executed rapidly and inexpensively. Popular markets such as crude oil, heating oil, and unleaded gasoline have thousands of contracts traded daily.
- Standardized underlying entity—the commodity or other item underlying the futures contract must be standardized and/or capable of being graded so that it is clear what is being bought and sold. Energy commodities are fungible interchangeable goods sold in accordance with strict specifications and grades.

Energy is considered to be a well-functioning market because it satisfies these criteria. The existence of such a market has a significant modifying effect on short-term price volatility, and will temper the impact of any future disruptions such as those that occurred in the pre-futures market 1970s. However, even a well-functioning futures market cannot be expected to eliminate the economic risks of a massive physical supply interruption.

MARKET PARTICIPANTS

Most of the participants in the energy futures and option markets are commercial or institutional energy producers, such as petroleum producers, refiners, and electric utilities; traders; or users, such as industrial and transportation companies. The energy producers and traders, most of whom are called "hedgers," want the value of their products to increase and also want to limit, if possible, any loss in value. Energy users, who are also hedgers, want to protect themselves from cost increases arising from increases in energy prices. Hedgers may use the commodity markets to take a position that will reduce their risk of financial loss due to a change in price. Other participants are "speculators" who hope to profit from changes in the price of the futures or option contract. It is important to note that hedgers typically do not try to make a killing in the market. They use futures to help stabilize their revenues or their costs. Speculators, on the other hand, try to profit by taking a position in the futures market and hoping the market moves in their favor. Hedgers hold offsetting positions in the cash market for the physical commodity but speculators do not.

THE MECHANICS OF TRADING

The mechanics of futures and options trading are straightforward. Typically, customers who wish to trade futures and options contracts do so through a broker. Both, buyers and sellers, deposit funds—traditionally called margin, but more correctly characterized as a performance bond or good-faith deposit—with a brokerage firm. This amount is typically a small percentage—less than 10 percent—of the total value of the item underlying the contract.

The New York Mercantile Exchange (NYMEX) is the largest physical commodity futures exchange in the world. The exchange pioneered the concept of risk management for the energy industry with the launch of heating oil futures in 1978, followed by options and/or futures for sweet and sour crude oil, unleaded gasoline, heating oil, propane, natural gas, and electricity. The NYMEX is owned by its members and is governed by an elected board of directors. Members must

be approved by the board and must meet strict business integrity and financial solvency standards.

The federal government has long recognized the unique economic benefit futures trading provides for price discovery and offsetting price risk. In 1974, Congress created the Commodity Futures Trading Commission (CFTC), replacing the previous Commodity Exchange Authority, which had limited jurisdiction over agricultural and livestock commodities. The CFTC was given extensive authority to regulate commodity futures and related trading in the United States. A primary function of the CFTC is to ensure the economic utility of futures markets as hedging and price discovery vehicles.

The London-based International Petroleum Exchange (IPE) is the second largest energy futures exchange in the world, listing futures contracts that represent the pricing benchmarks for two-thirds of the world's crude oil and the majority of middle distillate traded in Europe. IPE natural gas futures may also develop into an international benchmark as the European market develops larger sales volume.

Besides NYMEX and IPE, there are a number of other exchanges offering trading opportunities in energy futures. These include the Singapore International Monetary Exchange (North Sea Brent crude), The Chicago Board of Trade (electricity), Kansas City Board of Trade (western U.S. natural gas), and the Minneapolis Grain Exchange (Twin Cities' electricity). Domestic energy futures trading opportunities have arisen due to deregulation of the electricity and natural gas industries introducing many new competitors prepared to compete on the basis of price.

There has been discussion of the possibility of a futures market in emission credits arising from domestic regulations or international treaties to reduce energy use-related greenhouse gas emissions. These so-called pollution credits would be generated when Country A (or Corporation A) reduces its emissions below a specific goal, thereby earning credits for the extra reductions. At the same time, Country/Corporation B decides that emission controls are too expensive, so it purchases A's emission reduction credits. A declining cap on allowable emissions would reduce the available number of credits over time. The controversial theory is that market forces would thereby reduce emissions. Although there is some U.S. experience with the private sale and barter of such emission credits (a cash or "physicals" market), it remains to be seen if a true exchange-traded futures market in emission credits will arise.

ENERGY AND FUTURES PRICES

In addition to providing some control of price risk, futures and options markets are also very useful mechanisms for price discovery and for gauging market sentiment. There is a world-wide need for accurate, real-time information about the prices established through futures and options trading, that is, a need for price transparency. Exchange prices are simultaneously transmitted around the world via a network of information vendors' terminal services directly to clients, thereby allowing users to follow the market in real time wherever they may be. Energy futures prices are also widely reported in the financial press. These markets thus enable an open, equitable and competitive environment.

Frank R. Power

BIBLIOGRAPHY

Battley, N. (1991). *Introduction to Commodity Futures and Options*. New York: McGraw-Hill.

Brown, S. L., and Errera, S. (1987). *Trading Energy Futures: A Manual for Energy Industry Professionals*. New York: Quorum Books.

Clubley, S. (1998). *Trading in Energy Futures and Options*. Cambridge, Eng.: Woodhead.

Commodity Futures Trading Commission. (1997). *Futures and Options: What You Should Know Before You Trade*. Washington, DC: Author.

Commodity Futures Trading Commission. (1997). *Glossary: Guide to the Language of the Futures Industry*. Washington, DC: Author.

Futures Industry Institute. (1998). *Futures and Options Fact Book*. Washington, DC: Author.

Hull, J. (1995). *Introduction to Futures and Options Markets*. Englewood Cliffs, NJ: Prentice-Hall.

Kolb, R. W. (1991). *Understanding Futures Markets*. New York: New York Institute of Finance.

Kozoil, J. D. (1987). *A Handbook for Professional Futures and Options Traders*. New York: John Wiley.

Razavi, H., and Fesharaki, F. (1991). *Fundamentals of Petroleum Trading*. New York: Praeger.

GASOLINE AND ADDITIVES

Gasoline is the primary product made from petroleum. There are a number of distinct classes or grades of gasoline. Straight-run gasoline is that part of the gasoline pool obtained purely through distillation of crude oil. The major portion of the gasoline used in automotive and aviation is cracked gasoline obtained through the thermal or catalytic cracking of the heavier oil fractions (e.g., gas oil). A wide variety of gasoline types are made by mixing straight-run gasoline, cracked gasoline, reformed and synthesized gasolines, and additives.

Motor fuels account for about one-quarter of all energy use in the United States. The energy content of gasoline varies seasonally. The average energy content of regular gasoline stands at 114,000 Btu/gal in the summer and about 112,500 Btu/gal in the winter. The energy content of conventional gasolines also varies widely from batch to batch and station to station by as much as 3 to 5 percent between the minimum and maximum energy values.

Gasoline can be made from coal as well as petroleum. In the 1930s and 1940s, Germany and other European countries produced significant quantities of gasoline from the high-pressure hydrogenation of coal. But to convert the solid coal into liquid motor fuels is a much more complex and expensive process. It could not compete with the widely available and easily refined petroleum-based motor fuels.

In the United States, all gasoline is produced by private commercial companies. In many cases, they are vertically integrated so that they drill for, find, and transport oil, process the oil into gasoline and other products and then sell the gasoline to a network of retailers who specialize in certain brand name gasolines and gasoline blends. These large refiners may also market these products to other refiners, wholesalers, and selected service stations.

GROWING DEMAND FOR GASOLINE

At the end of the nineteenth century, virtually all of the gasoline produced (around 6 million barrels) was used as a solvent by industry, including chemical and metallurgical plants and dry cleaning establishments, and as kerosene for domestic stoves and space heaters. But by 1919, when the United States produced 87.5 million barrels of gasoline, 85 percent was consumed by the internal combustion engine (in automobiles, trucks, tractors, and motorboats).

Between 1899 and 1919, as demand for gasoline grew, the price increased more than 135 percent, from 10.8 cents/gal to 25.4 cents/gal. From 1929 to 1941, gasoline use by passenger cars increased from 256.7 million barrels to 291.5 million barrels. Consumption of aviation fuel went from only 753,000 barrels in 1929 to over 6.4 million barrels at the start of World War II. By 1941, gasoline accounted for over one-half of petroleum products with 90 percent of gasoline output used as fuel for automotive and aircraft engines.

Between 1948 and 1975, per capita consumption of gasoline in the United States increased from about 150 gal/yr to a little less than 500 gal/yr. A growing trend after the war was the increasing use of jet fuel for aircraft and the decline in use of aviation gasoline. After 1945, oil production increased in other parts of the world, especially the Middle East and Latin America. By the 1970s, the Middle East became a dominant oil producing region. The cartel formed by the major Middle Eastern oil producing countries, known as OPEC, became a major force in setting oil prices internationally through the control of oil production.

Since the mid-1970s, the rate of growth of per capita gasoline consumption has slowed. An important factor in causing this moderation in demand was the trend to improve automobile fuel economy that was initiated by worldwide fuel shortages. Fuel economy hovered around 14.1 mpg between 1955 and 1975; it rose sharply over the next 15 years, reaching around 28.2 mpg in 1990.

An aging population and continued improvements in engine technology and fuel economy may slow U.S. gasoline demand in the early part of the twenty-first century from the 2 percent annual growth rate of the 1990s.

KNOCKING AND OCTANE RATING

The process of knocking has been studied extensively by chemists and mechanical engineers. Knocking is rapid and premature burning (i.e., preignition) of the fuel vapors in the cylinders of the engine while the pistons are still in the compression mode. Research on knocking was carried out prior to World War I, but it was only with the increase in the size and power of automotive engines after 1920 that significant attempts were made to deal with the problem on a commercial basis.

Knocking, which has a distinctive metallic "ping," results in loss of power and efficiencies and over time causes damage to the engine. Knocking is a great energy waster because it forces the automobile to consume greater quantities of gasoline per mile than do engines that are functioning properly. The problem of engine knocking was an important factor in the U.S. push for a gasoline rating system. Around the time of World War I, there was no single standard specification or measure of gasoline performance. Many states developed their own specifications, often conflicting with those promulgated by the automotive and petroleum industries and the federal government. Even the various branches of government had their own specifications. The specifications might be based on the boiling point of the gasoline fraction, miles allowed per gallon of fuel, or the chemical composition of the gasoline.

The octane numbering system was developed in the late 1920s and was closely linked to the federal government's program of measurement standards, designed jointly by the Department of the Army and the National Bureau of Standards.

The octane number of a fuel is a measure of the tendency of the fuel to knock. The octane scale has a minimum and maximum based on the performance of reference fuels. In the laboratory, these are burned under specific and preset conditions. One reference fuel is normal heptane. This is a very poor fuel and is given an octane rating of zero. On the opposite end of the scale is iso-octane (2,2,4 trimethyl pentane). Iso-octane is a superior fuel and is given a rating of 100.

The octane rating of fuels is derived by simple laboratory procedures. The fuel being tested is burned to determine and measure its degree of "knocking." Then the two reference fuels are blended together until a reference gasoline is formed that knocks to the same degree as the tested fuel. The proportion of iso-octane present in the reference fuel is then the octane number of the tested gasoline. Some compounds, like methanol and toluene, perform better than iso-octane and, by extrapolation, their octane numbers are over 100. A higher octane number is important from a very practical consideration: it gives better engine performance in the form of more miles per gallon of gasoline.

From the 1920s to the 1940s, catalytic cracking processes were developed that not only increased processing efficiencies, but progressively raised the octane number of gasoline. In 1913, prior to the devising of the octane scale, the commercialization of the Burton Process, a noncontinuous thermal technology, produced gasoline with an estimated octane number of between 50 and 60. Continuous thermal cracking, first operated in the early 1920s, produced gasoline with an octane number of close to 75. With the first catalytic process in the form of the Houdry technology (1938), cracked gasoline reached the unprecedented octane level in the high 80s. Fluid catalytic cracking, the culmination of the cracking art that came on line in 1943, pushed the quality of gasoline to an octane level of 95.

While octane rating provided an objective and verifiable measure of performance across all grades of gasoline, it did not immediately lead to unified standards. It was not until the 1930s, when both the octane rating and new types of octane-boosting additives entered the industry, that automotive fuel began to center around two major types of gasoline—regular and premium—each operating within its own octane range. Over the next 60 years, octane rating of gasoline increased due to improved refining practices and the use of additives. In the 1970s and 1980s, the use of additives became increasingly tied to environmental concerns (i.e., clean air), as well as higher octane ratings. Gasoline has come a long way since the Model T, and it is important to note that, in terms of constant dollars, it is cheaper today than it was in 1920.

Table 1 shows the major gasoline additives that were introduced from the 1920s through the 1980s. The increase in octane number of gasoline with use of these additives is shown.

TETRAETHYLEAD: THE FIRST GASOLINE ADDITIVE

One source of knocking was related to the vehicle engine. All else being equal, an automobile engine with a higher compression ratio, advanced spark schedule, or inefficient combustion is more likely to experience knocking. Within the United States, research into knocking has focused on the chemical aspects of gasoline, which is a complex hydrocarbon mixture of paraffins, naphthenes, and aromatics.

Chemical additives first entered the industry in the first decade of the century. These additives served a number of uses. For example, they lessened the capacity of gasoline to vaporize out of the gas tank or to polymerize (i.e., produce gummy residues) in the engine. In the early 1920s, the most important application for these substances was to eliminate knocking. Tetraethylead (TEL) was the first major gasoline additive to be commercialized for this purpose.

Charles F. Kettering, the inventor of the self starter, the Delco battery, and other major components of modern automotive engineering, started to work on the problem in 1916 at his Dayton Engineering Laboratories Company (DELCO). Kettering was induced into this research not by the problems faced by the automobile but by gasoline-powered electric lighting systems for farms. These systems employed generators utilizing internal combustion engines. These engines, which burned kerosene and not gasoline, knocked badly. Kettering and his team addressed this nonautomotive concern as a profitable research project and one of potentially great benefit to the agricultural sector.

Kettering hired Thomas Midgley Jr., a mechanical engineer from Cornell, to work with him on the

Additive	Octane Number
Tetraethyl lead (TEL)	100
Methanol	107
Ethanol	108
Methyl-t-butyl ether (MTBE)	116
Ethyl-t-butyl ether (ETBE)	118

Table 1.
Octane Numbers of Common Gas Additives

project. Studying the combustion process in more detail, Kettering and Midgley determined that low volatility in the fuel caused knocking to occur. This conclusion led them to search for metallic and chemical agents to blend with the gasoline to increase volatility and reduce knocking.

A promising line of research led Midgley to the halogen group of chemicals and specifically iodine and its compounds. General Motors purchased DELCO Labs in 1919 and the search for an antiknock agent came under GM management, with Kettering and Midgley remaining on board to continue the work, but with the focus now on automotive application.

Using iodine as their starting point, they experimented with a series of compounds including the anilines and a series of metals near the bottom of the periodic table. Lead turned out to be the most effective of the additives tested. But lead alone caused a number of problems, including the accumulation of its oxide in engine components, and particularly the cylinders, valves, and spark plugs.

Experiments continued to find an appropriate form of lead that could at the same time prevent the formation of oxide deposits. Ethylene was found to combine with lead to form tetraethyllead (TEL), a stable compound that satisfied this requirement.

Kettering and Midgley were the first to identify it as a prime antiknock agent, though the compound had been known since 1852. They estimated that only a very small amount of TEL—a few parts per thousand—would result in a 25 percent increase in horsepower as well as fuel efficiency.

The next stage of development was to design a production process to link the ethyl group to lead. GM attempted to make TEL from ethyl iodide. They built an experimental plant, but the process proved too expensive to commercialize.

An alternative source of the ethyl component was ethyl bromide, a less expensive material. It was at this point that GM called upon DuPont to take over process development. DuPont was the largest U.S. chemical company at the time. It had extensive experience in the scale-up of complex chemical operations, including explosives and high-pressure synthesis. The manufacturing process was undertaken by DuPont's premier department, the Organic Chemical section. GM contracted with DuPont to build a 1,300 pound per day plant. The first commercial quantities of TEL were sold in February 1923 in the form of ethyl premium gasoline.

In 1923, GM set up a special chemical division, the GM Chemical Co., to market the new additive. However, GM became dissatisfied with DuPont's progress at the plant. In order to augment its TEL supply, and to push DuPont into accelerating its pace of production, GM called upon the Standard Oil Company of New Jersey (later Esso/Exxon) to set up its own process independently of DuPont. In fact, Jersey Standard had obtained the rights to an ethyl chloride route to TEL. This turned out to be a far cheaper process than the bromide technology. By the mid-1920s, both DuPont and Jersey were producing TEL.

GM brought Jersey in as a partner in the TEL process through the formation of the Ethyl Corporation, each party receiving a 50 percent share in the new company. All operations related to the production, licensing, and selling of TEL from both DuPont and Jersey were centralized in this company.

Soon after production began, TEL was held responsible for a high incidence of illness and deaths among production workers at both the DuPont and Jersey Standard plants. The substance penetrated the skin to cause lead poisoning. Starting in late 1924, there were forty-five cases of lead poisoning and four fatalities at Jersey Standard's Bayway production plant. Additional deaths occurred at the DuPont Plant and at the Dayton Laboratory. This forced the suspension of the sale of TEL in 1925 and the first half of 1926.

These incidents compelled the U.S. Surgeon General to investigate the health effects of TEL. The industry itself moved rapidly to deal with the crisis by instituting a series of safety measures. Now, ethyl fluid was blended at distribution centers and not at service stations (it had been done on the spot and increased the chances of lead poisoning to service sta-

tion attendants). Also, ethyl gasoline was dyed red to distinguish it from regular grade gasoline. DuPont and Jersey placed tighter controls over the production process. The federal government placed its own set of restrictions on TEL. It set the maximum limit of 3 cc of TEL per gallon of gasoline. By 1926, TEL was once again being sold commercially.

Ironically, this episode proved beneficial to DuPont. DuPont became the dominant source of TEL after the mid-1920s because they perfected the chloride process and were far more experienced than Jersey Standard in producing and handling toxic substances.

The Ethyl Corp. and DuPont held the TEL patent, and controlled the TEL monopoly. The company held the sole right to the only known material that could eliminate automotive knocking. And it used its influence in the gasoline market to manipulate prices. Over the next few years, the company wielded its monopoly power to maintain a 3–5 cent differential between its "ethyl" gasoline and the regular, unleaded gasoline sold by the rest of the industry.

Throughout the 1930s TEL proved itself a profitable product for DuPont, which remained virtually the only TEL producer into the post–World War II period. With no advantage to be gained in further collaboration, DuPont severed its ties with Ethyl Corp. in 1948 and continued to manufacture TEL independently.

COMPETING AGAINST TEL: ALTERNATIVE ANTIKNOCK TECHNOLOGY

As the automotive industry continued to introduce higher compression engines during the 1920s and 1930s, refiners increasingly relied on TEL to meet gasoline quality. By 1929, fifty refiners in the United States had contracted with the Ethyl Corp. to incorporate TEL in their high test gasoline.

TEL was not the only way to increase octane number. Those few companies who did not wish to do business with Jersey Standard, sought other means to produce a viable premium gasoline. TEL represented the most serious threat to the traditional gasoline product. It was cheap, very effective, and only 0.1 percent of TEL was required to increase the octane number 10 to 15 points. In contrast, between 50 to 100 times this concentration was required of alternative octane enhancers to achieve the same effect.

Benzol and other alcohol-based additives improved octane number, up to a point. Experiments using alcohol (ethanol, methanol) as a replacement for gasoline began as early as 1906. In 1915, Henry Ford announced a plan to extract alcohol from grain to power his new Fordson tractor, an idea that never achieved commercial success.

The shortage of petroleum after World War I induced an intense search for a gasoline substitute in the form of alcohol. The trade press felt alcohol would definitely replace gasoline as a fuel at some point. The advantages of alcohol cited in the technical press included greater power and elimination of knocking.

The push to use alcohol as a fuel surfaced at various times coinciding with real or perceived gasoline shortages and often directed by the farm lobby during periods of low grain prices. The great discoveries of oil in the Mid-Continental fields in the1920s reduced the incentive for the use of alcohol as a fuel. But in the 1930s the severe agricultural crisis brought back interest in alcohol. Alcohol distillers, farmers, and Midwest legislators unsuccessfully attempted to regulate the blending into gasoline of between 5–25 percent ethanol. It took the oil supply disruptions of the 1970s for farm state legislators to pass legislation to highly subsidize ethanol. The subsidies, which remain in effect today, are the reason ethanol continues to play a notable role as a fuel additive.

As experiments at Sun Oil Co. in the early 1930s indicated, there were serious disadvantages associated with alcohol. While alcohol did in fact appear to increase the octane number, it left large amounts of deposits in the engine. Alcohol also vaporized out of the gas tank and engine at rapid rates. And the combustion temperature of the alcohol group is lower than for hydrocarbons because it is already partially oxidized.

The most effective competitive approach for the more independent refiners was in developing new types of cracking technologies. Companies like Sun Oil, one of the few companies who remained independent of Jersey and Ethyl, continued to expand the limits of thermal cracking, notably by employing higher pressures and temperatures. Sun's gasoline reached octane levels close to those achieved by gasolines spiked with TEL (i.e., between 73 and 75). Sun Oil continued to compete with additives purely through advanced cracking technology, a path that

would lead by the late 1930s to the first catalytic cracking process (i.e., the Houdry Process). But by this time, more advanced refining processes were coming on line and competing with the Houdry Process. By the early 1940s, Jersey Standard developed fluidized bed catalytic cracking technology. Fluidized cracking proved superior to Houdry's fixed bed process with respect to both production economies and the quality of the product (i.e., octane rating of the gasoline). Fluidized cracking quickly displaced Houdry's catalytic cracking technology as the process of choice.

Competition did not center on quality alone. Price and packaging were called into play as weapons against the onslaught of TEL. For example, as a marketing tool, Sun Oil dyed its gasoline blue to more easily identify it as a high premium fuel (customers actually saw the gasoline being pumped in a large clear glass reservoir on top of the gas pump). Sun then competed aggressively on price. Whereas TEL-using refiners sold two grades of gasoline, regular and premium, Sun marketed only its premium "blue Sunoco" at regular grade prices. Sun could do this because it was not burdened, as TEL was, with such additional costs as blending and distribution expenses that cut into profit margins.

By the late 1920s and into the 1940s, with the use of either TEL, other additives, or advanced cracking technology, a number of premium grade gasolines appeared on the market. In addition to Sun's premium, there was Gulf's "NoNox," Sinclair's premium "H.C.'s" gasoline, and Roxana Petroleum's "Super-Shell." The use of TEL has plummeted since the government's mandate in 1975 to install catalytic converters for reducing the carbon monoxide and unburned hydrocarbons in automotive exhaust gases. This is because lead poisons the noble metal (chiefly platinum) catalysts used. In addition the lead bearing particulates in the emissions from engines burning leaded fuel are toxic in their own rights.

POSTWAR DEVELOPMENTS: GASOLINE AND THE ENVIRONMENT

Additives and the blending process became an increasingly important part of gasoline manufacture after World War II. Refiners had to balance such factors as customer specifications, regulatory requirements, and probable storage (i.e., nonuse) time. The industry became more precise in how, when, and how many of components should be added to gasolines. The large, modern refinery increasingly incorporated complex computer programs to help plan and effect blending requirements. Critical factors that had to be factored into these calculations included seasonal adjustments, current and anticipated demand, regulatory levels, and supply schedules of the various components.

Since the 1950s, an increasing portion of a refiner's R&D has gone into new and improved additives. Beyond their role as antiknock agents, additives and blending agents have taken on an ever broadening range of functions to improve the performance of fuels in automotive and aircraft engines.

Sulfur and Gasoline

In recent years, there has been a greater understanding of the role of automotive emissions as environmental pollutants. Sulfur dioxide, nitrogen oxides, and carbon monoxide degrade the earth's atmosphere and are health hazards. Carbon dioxide adds to the atmospheric buildup of greenhouse gases and in turn accelerates the process of global warming.

Sulfur is a particular problem as an environmental hazard. It occurs naturally in various concentrations in petroleum, and it is difficult and costly to remove all of it. Distillation and cracking removes some, but small amounts survive the distillation and cracking processes and enter into the gasoline. The average level of sulfur in gasoline has not changed much since 1970, remaining at 300 parts per million (ppm) with a range between 30 and 1,000 ppm.

High levels of sulfur not only form dangerous oxides, but they also tend to poison the catalyst in the catalytic converter. As it flows over the catalyst in the exhaust system, the sulfur decreases conversion efficiency and limits the catalyst's oxygen storage capacity. With the converter working at less than maximum efficiency, the exhaust entering the atmosphere contains increased concentrations, not only of the sulfur oxides but also, of hydrocarbons, nitrogen oxides, carbon monoxides, toxic metals, and particulate matter.

In the 1990s, the EPA began controlling sulfur through its reformulated gasoline program. It developed regulations in 1999 that would sharply reduce the sulfur content in gasoline from 300 ppm to a maximum of 80 ppm.

The new regulations, scheduled to go into effect in 2004, are compelling certain refiners to purchase low-sulfur content ("sweet") crude oil. This is the strategy being pursued by Japanese refiners. However, the Japanese are not major oil producers but import oil from other producing countries. U.S. refiners, in contrast, consume oil from a wide range of ("sour") petroleum sources that have a high-sulfur content, including Venezuela, California, and parts of the Gulf Region. U.S. companies own and operate oil producing infrastructures (i.e., derricks, pipelines), within the United States and overseas. They are committed to working these oil fields, even if producing high-sulfur oil. U.S. refineries thus need to continue dealing with high-sulfur crude oil. Imported crude from the Middle East, while historically low in sulfur, is also becoming increasingly less sweet.

Petroleum refiners will have to reduce sulfur content at the refineries. This will require the costly retooling of some of their plant operations in order to achieve a suitable fuel mix. Removing additional amounts of sulfur at the refinery will entail installation of separate catalyst-based process such as hydrosulfurization. Another possible approach is the removal of sulfur in liquid oil or gasoline by the use of both organic and inorganic scavenger agents added to the oil or gasoline to seek out, combine with, and precipitate out sulfur and its compounds.

Reformulated Gasoline and MTBE

Prior to the Clean Air Act of 1990, environmental regulations were aimed at reducing emissions as they left the exhaust system. The catalytic converter has been the primary means of attacking air pollution in this way. After 1990, regulations for the first time undertook to alter the composition of the fuel itself. Reformulated gasoline applies to gasoline that is sold in the nine metropolitan areas designated by the EPA with the highest level of ozone pollution. About 48 million people reside in areas where ozone concentrations exceed federal standards.

Reformulation refers to the transformation of gasoline to make it cleaner with respect to emissions. Beginning in 1995, specifications for reformulated gasoline included a 2 percent minimum oxygen content and a maximum content of various organic and inorganic pollutants. In addition, heavy metal additives in gasoline are prohibited. A disadvantage of

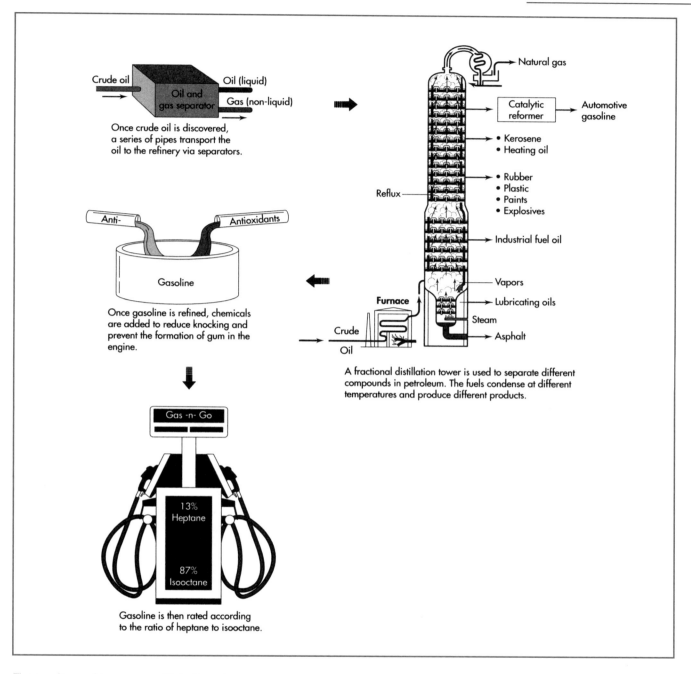

The gasoline-making process. (Gale Group)

reformulated gasoline is that it contains 1 to 3 percent less energy per gallon than traditional gasoline.

Many reformulated gasolines use oxygenated compounds as additives. Clean Air regulations specify the need for oxygenated fuel in 39 metropolitan areas with high carbon monoxide concentrations. The regulations for oxygenated fuel are seasonal: during the winter season, gasoline must contain a minimum of 27 percent oxygen. The oxygen helps engines to burn the fuel more completely which, in turn, reduces monoxide emissions. The major additive to supply the additional oxygen to reformulate gasoline to satisfy these requirements is the methanol derivative, methyl tertiary butyl ether (MTBE).

Currently, this additive is used in over 30 percent of U.S. gasoline.

MTBE was first used as a fuel additive in the 1940s and was a popular additive in Europe in the 1970s and 1980s. In the late 1970s, MTBE began replacing lead in this country to enhance octane number. In the late 1980s, California led the way in the United States for its use as an oxygenate for cleaner burning fuel. The consumption of MTBE in the United States increased rapidly between 1990 and 1995 with the passage of the Clean Air Act and, a few years later the implementation of the federal reformulated gasoline program. Currently, MTBE is produced at 50 U.S. plants located in 14 states. About 3.3 billion gallons of MTBE, requiring 1.3 billion gallons of methanol feedstock, are blended annually into reformulated gasoline.

In the late 1990s, MTBE came under serious attack on grounds of both efficacy and safety. A report by the National Research Council (1999) stated that the addition of oxygen additives in gasoline, including MTBE and ethanol, are far less important in controlling pollution than emission control equipment and technical improvement to vehicle engines and exhaust systems.

Moreover, MTBE has been found in groundwater, lakes and reservoirs used for drinking water, and it has been linked to possible serious disease. The probable occurrence of cancerous tumors in laboratory rats injected with MTBE alerted federal agencies as to its possible health hazards. In 1999, the EPA reversed itself, recommending the phasing out of MTBE as an additive to gasoline.

During the first half of 2000, MBTE production in the Unites States averaged 215,000 barrels per day. In the same six-month period, the average production of fuel ethanol was 106,000 barrels per day. In light of the EPA's 1999 recommendation, ethanol will most likely replace MTBE as an effective oxygenate additive. In addition to its use as an oxygenate, ethanol enhances octane ratings and dilutes contaminants found in regular gasoline.

New and Emerging Gasoline Additives

The development and blending of additives is undertaken for the most part by the petroleum refining industry. Additives are essential to the economic well-being of the industry because they tend to boost sales for gasoline and diesel fuel. In most cases, additives do not differ in price by more than three to four cents a gallon. The recently developed additives do

not necessarily sacrifice fuel efficiency for higher octane numbers. They are multifunctional. In addition to boosting octane ratings they may also clean the engine, which, in turn, leads to greater fuel efficiency.

Beyond their role in enhancing octane numbers and reducing emissions, the group of more recent fuel additives performs a growing range of functions: antioxidants extend the storage life of gasoline by increasing its chemical stability; corrosion inhibitors prevent damage to tanks, pipes, and vessels by hindering the growth of deposits in the engine and dissolving existing deposits; demulsifiers or surface active compounds prevent the formation of emulsions and the dirt and rust entrained in them that can foul the engine and its components.

Beginning in the 1970s, gasoline additives increasingly took on the role of antipollutant agent in the face of government attempts to reduce automotive emissions into the atmosphere. Despite the advances made in cracking and reforming technologies and in the development and blending of additives (not to mention enhancements in the engine itself), the use of automotive gasoline has increased the level of air pollution. This is so because modern distillates, blends of straight run and cracked or chemically transformed product, tend to have a higher aromatic content. The result is longer ignition delays and an incomplete combustion process that fouls the engine and its components and increases particulate and oxide emissions.

Continued implementation of clean air legislation, especially within the United States, is expected to accelerate the consumption of fuel additives. In 1999, the EPA proposed wide-ranging standards that would effectively reformulate all gasoline sold in the United States and significantly reduce tailpipe emissions from trucks and sports utility vehicles. These regulations require potentially expensive sulfur-reducing initiatives from both the oil industry and the automakers. For refiners, it will require significant redesign and retooling of plant equipment and processes will be required in order to achieve suitable changes in the fuel mix because the U.S. oil industry is committed to continue development of its sour petroleum reserves. The DOE expects that the more complex processing methods will add six cents to the cost of a gallon of gasoline between 1999 and 2020.

In the United States alone, the demand for fuel

additives is expected to reach over 51 billion pounds by 2002. Oxygenates are anticipated to dominate the market, both within the United States and internationally. Nonpremium gasoline and diesel fuel represent the fastest growing markets for fuel additives.

A recently marketed fuel additive is MMT (methylcyclopentadienylmanganese tricarbonyl). MMT was first developed by the Ethyl Corporation in 1957 as an octane enhancing agent and has experienced a growth in demand in the 1990s. MMT was Ethyl Corporation's first major new antiknock compound since TEL.

However, in 1997, the EPA blocked the manufacture of MMT. The Agency took this action for two reasons. It determined that MMT had the potential for being hazardous to humans, and in particular to children. The EPA is especially concerned about the toxic effects of the manganese contained in MMT. Also, the EPA discovered that MMT was likely interfering with the performance of the catalytic converters in automobiles and in turn causing an increase in exhaust emissions in the air. In 1998, the EPA decision was overturned by a federal appeals court in Washington. The court's decision allows the Ethyl Corp. to test MMT while it is selling the additive. The decision set no deadline for the completion of tests. In addition to its use in the United States, MMT is consumed as an additive in unleaded gasoline in Canada.

In addition to MTBE and MMT, other kinds of additives are being developed. Some of these are derivatives of alcohol. Variations of MTBE are also being used, especially the ether-derived ETBE (ethyl-t-butyl ether).

A new generation of additives specifically designed for aircraft gasoline are also being developed. These additives address such problems as carbon buildup, burned and warped valves, excessive cylinder head temperatures, stuck valves and piston rings, clogged injectors, rough idle, and detonation. Aviation fuel additives often act as detergents (to remove deposits), octane enhancers, and moisture eliminators.

Competition from Alternative Fuels

Most alternative fuel vehicles on the road today were originally designed for gasoline, but converted for use with an alternative fuel. Because the petroleum industry has successfully responded to the competitive threats of alternative fuels by developing reformulated gasolines that burn much cleaner, the conversions are typically performed more for economic reasons (when the alternative fuel is less expensive, which has occurred with propane) rather than environmental reasons. It is likely that technical advances will continue to permit petroleum refiners to meet the increasingly more stringent environmental regulations imposed on gasoline with only minor increases in the retail price. And since petroleum reserves will be abundant at least through 2020, gasoline promises to dominate automotive transportation for the foreseeable future.

However, fuel cell vehicles, which are designed to generate their power from hydrogen, pose a major long-term threat to the preeminence of gasoline. Automakers believe the best solution is to extract hydrogen from a liquid source because hydrogen has a low energy density and is expensive to transport and store. All the major automakers are developing fuel cell vehicles powered by hydrogen extracted from methanol because reforming gasoline into hydrogen requires additional reaction steps, and a higher operating temperature for the reformer. Both requirements are likely to make the gasoline reformer larger and more expensive than the methanol reformer. Moreover, the sulfur content of gasoline is another major reason that automakers are leery of developing gasoline reformers for fuel cell vehicles. Quantities as low as a few parts per million can be a poison to the fuel cell stack. There are no gasoline reformer fuel cell vehicles in operation, so an acceptable level of sulfur has not been determined. If it is determined that an ultralow-sulfur gasoline blend can be developed specifically for fuel cell vehicles, it would be a far less expensive solution than developing the fuel production, delivery and storage infrastructure that would be needed for methanol-powered fuel cell vehicles.

Sanford L. Moskowitz

See also: Efficiency of Energy Use, Economic Concerns and; Engines; Fuel Cells; Fuel Cell Vehicles; Hydrogen; Methanol; Synthetic Fuel.

BIBLIOGRAPHY

Alexander, D. E., and Fairbridge, R. W., eds. (1999). *The Chapman & Hall Encyclopedia of Environmental Science.* Boston, MA: Kluwer Academic.

Buell, P., and Girard, J. (1994). *Chemistry: An Environmental Perspective.* Englewood Cliffs, NJ: Prentice-Hall.

Encyclopedia of Chemical Technology, 4th ed. (1994). New York: John Wiley & Sons, Inc.

Enos, J. (1962). *Petroleum, Progress and Profits: A History of Process Innovation*. Cambridge, MA: MIT Press.

Environmental Protection Agency. (1998). "Press Release: EPA Announces Blue-Ribbon Panel to Review the Use of MTBE and Other Oxygenates in Gasoline." Washington, DC: Author.

Environmental Protection Agency. (1999). "Press Release: Statement by the U.S. EPA Administrator on the Findings of EPA's Blue Ribbon MTBE Panel." Washington DC: Author.

"Ford Expects Output of Flexible Fuel Vehicles to Represent 10% of its U.S. Automotive Production." *Purchasing.* 125(9):9.

Giebelhaus, A. W. (1980). *Business and Government in the Oil Industry: A Case Study of Sun Oil*, 1876–1945. Greenwich, CT: JAI Press.

Hounshell, D. A., and Smith, J. K. (1988). *Science and Corporate Strategy: DuPont R&D, 1902–1980*. New York: Cambridge University Press.

Larson, H. M., et al. (1971). *History of Standard Oil Company (New Jersey): New Horizons, 1927–1950*. New York: Harper and Row.

"MTBE Phaseout Will Spur Ethanol Demand." *Purchasing.* 127(3):32C1.

National Research Council. (1990). *Fuels to Drive Our Future*. Washington, DC: National Academy Press.

"New Fuel May Meet Emissions Rules." *Purchasing.* 118(10):67.

Speight, J. G., ed. (1990). *Fuel Science and Technology Handbook*. New York: Marcel Dekker.

Spitz, P. (1988). *Petrochemicals: The Rise of an Industry*. New York: John Wiley and Sons.

U.S. International Trade Commission (ITC), Office of Industries. (1998). "Industry and Trade Summary: Refined Petroleum Products," USITC Publication 3147. Washington, DC: Author.

Williamson, H. F., et al. (1963). *The American Petroleum Industry, Vol. II: The Age of Energy, 1899–1959*. Evanston, IL: Northwestern University Press.

GASOLINE ENGINES

The gasoline engine is a device to convert the chemical energy stored in gasoline into mechanical energy to do work—to mow a lawn; chainsaw a tree; propel a car, boat, or airplane; or to perform myriad other tasks. The energy in the gasoline is transformed into heat within the engine through combustion, so the gasoline engine is an internal combustion engine.

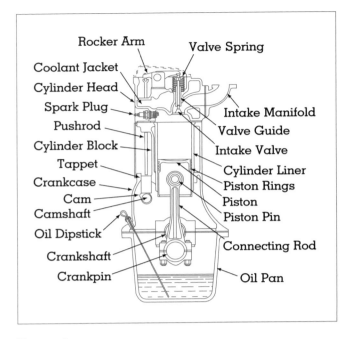

Figure 1.

Cross section through gasoline engine using push-rod valve actuation.

Because combustion is normally initiated by an electric spark, the gasoline engine is also frequently known as a spark-ignition engine.

A number of different kinematic mechanisms have been used to extract mechanical work from the heated products of combustion. The preferred option is the slider-crank mechanism, which is incorporated into the gasoline-engine cross section of figure 1. In the slider-crank mechanism, the piston reciprocates up and down within a cylinder, alternately doing work on and extracting work from the gas enclosed by the piston, cylinder walls, and cylinder head. A poppet-type intake valve in the cylinder head opens during part of the engine cycle to admit a fresh charge of air and fuel. A spark plug ignites the mixture at the appropriate time in the cycle. A poppet-type exhaust valve (hidden behind the intake valve in Figure 1) opens later in the cycle to allow the burned products of combustion to escape the cylinder.

The reciprocating motion of the piston is transformed into rotary motion on the crankshaft by two of the links in the slider-crank mechanism—the connecting rod and the crank (hidden from view in this cross section). The connecting rod joins the piston pin to the crank pin. The crank connects the crank

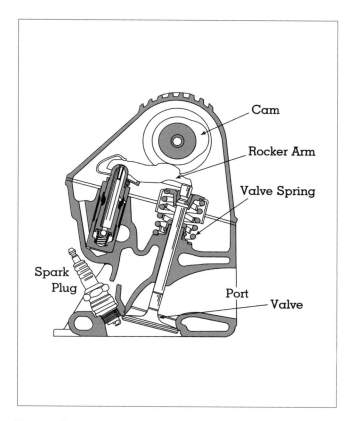

Figure 2.
Overhead-cam valve actuation.

pin to the crankshaft. The crank and the crank pin are usually integral parts of the crankshaft.

In Figure 1, the intake and exhaust valves are actuated by a camshaft chain, or a gear-driven from the crankshaft at half engine speed. The camshaft, operating through a tappet and push rod, tilts one end of a pivoted rocker arm at the appropriate time during the cycle. The opposite end of the rocker arm, working against the force of a valve spring, opens and closes a poppet valve. Each intake and exhaust valve has its own such mechanism.

This arrangement of valves defines a push-rod engine. In the alternative approach of Figure 2, the camshaft is moved to a position above the cylinder head, eliminating the push rod. This configuration defines an overhead-cam engine. The overhead camshaft is driven from the crankshaft by either a belt or a chain.

Other essential elements of a gasoline engine are also evident in Figure 1. The oil pan, fastened to the bottom of the crankcase, contains a reservoir of oil

that splashes up during engine operation to lubricate the interface between the cylinder wall and the piston and piston rings. These rings seal the gas within the space above the piston. The dipstick used to verify an adequate oil supply is evident. Liquid in the coolant jacket seen in the cylinder head and surrounding the cylinder wall maintains the engine parts exposed to combustion gases at an acceptable temperature.

HISTORICAL BACKGROUND

The first practical internal-combustion engine is attributed to the Frenchman Jean Lenoir in 1860. His single-cylinder engine completed its cycle in but two strokes of the piston. On the first stroke, the piston drew a fresh charge of air and a gaseous fuel into the cylinder. Near midstroke, the intake valve closed, a spark ignited the trapped charge, the ensuing combustion quickly raised cylinder pressure, and the remainder of the piston stroke involved expansion of the combustion products against the piston to produce useful output work for transmission to the crankshaft. On the return stroke of the piston, the combustion products were expelled from the cylinder through an exhaust valve.

Four-Stroke Piston Engine

In France in 1862, Beau de Rochas outlined the principles of the four-stroke engine so common today. However, he never transformed those principles into hardware. Among the improvements proposed by de Rochas was compression of the charge prior to combustion. In contrast, the charge in the Lenoir engine was essentially at atmospheric pressure when combustion was initiated.

The operating principles of the four-stroke cycle are illustrated in Figure 3. In Figure 3a the descending piston draws a fresh charge into the cylinder through the open intake valve during the intake stroke. In Figure 3b, the intake valve has been closed, and the ascending piston compresses the trapped charge. As the piston approaches top dead center (TDC) on this compression stroke, the spark plug ignites the mixture. This initiates a flame front that sweeps across the chamber above the piston. Combustion is normally completed during the expansion stroke (see Figure 3c) well before the piston reaches bottom dead center (BDC). Before BDC is reached, the exhaust valve begins to open, releasing pressurized combustion products from the cylinder.

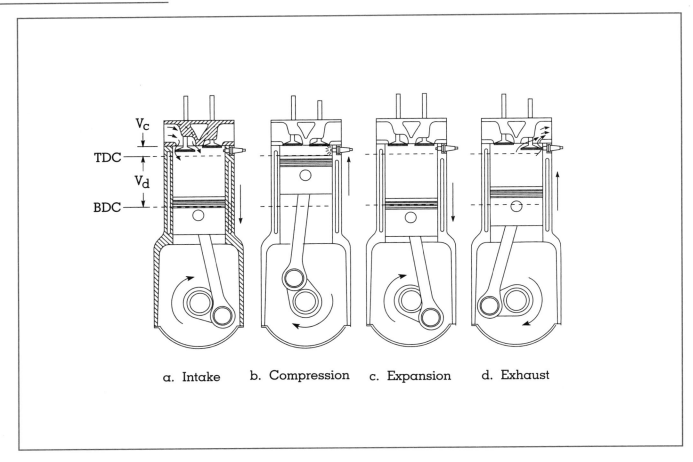

Figure 3.
Four-stroke cycle. TDC and BDC = top dead center and bottom dead center positions of the piston, respectively. V_d = displacement. V_c = clearance volume. Compression ratio = $(V_d + V_c)/V_c$.

During the exhaust stroke (see Figure 3d) the ascending piston expels most of the remaining products through the open exhaust valve, in preparation for a repetition of the cycle.

In 1876 in Germany, Nikolaus Otto built the first four-stroke engine, even though he was apparently unaware of the proposals of de Rochas. An idealized version of the cycle on which his 1876 engine operated is represented on coordinates of cylinder pressure versus volume in Figure 4. From 1 to 2, the piston expends work in compressing the fresh charge as it moves from BDC to TDC. At 2, combustion releases chemical energy stored in the fuel, raising cylinder pressure to 3. From 3 to 4, the products expand as the piston returns to BDC, producing useful work in the process. From 4 to 5, the burned gas is expelled from the cylinder as pressure drops to its initial value. These four events comprise what has come to be known in engineering thermodynamics as the Otto cycle.

The horizontal line at the bottom of the pressure-volume diagram of Figure 4 traces the other two strokes of the four-stroke cycle. On the exhaust stroke, from 5 to 6, the rising piston expels most of the remaining combustion products from the cylinder. On the intake stroke, from 6 to 7 (= 1), the descending piston inducts a fresh charge for repetition of the cycle. The net thermodynamic work developed in this cycle is proportional to the area enclosed by the pressure-volume diagram. In the ideal case, both the exhaust and intake strokes occur at atmospheric pressure, so they have no effect on the net output work. That justifies their exclusion from the thermodynamic representation of the ideal Otto

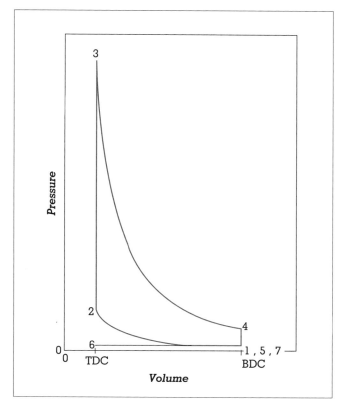

Figure 4.

Pressure-volume diagram for ideal Otto cycle (1-2-3-4-5), with exhaust and intake of four-stroke cycle (5-6-7) added .

cycle. The pressure-volume diagram for an actual operating engine deviates somewhat from this idealization because of such factors as noninstantaneous combustion, heat loss from the cylinder, and pressure loss across the valves during intake and exhaust.

Two-Stroke Piston Engine

Toward the end of the nineteenth century, successful two-stroke engines operating on the Otto cycle were developed by Dugald Clerk, James Robson, Karl Benz, and James Day. In this engine, the intake, combustion, expansion, and exhaust events all occur with but two piston strokes, or one crankshaft revolution. In principle this should double the output of a four-stroke engine of equal piston displacement. However, instead of the intake and exhaust events taking place during sequential strokes of the piston, they occur concurrently while the piston is near BDC. This impairs the ability of the engine to induct and retain as much of the fresh charge as in an equivalent four-stroke engine, with its separate intake and exhaust strokes. Thus the power delivered per unit of piston displacement is somewhat less than twice that of a comparable four-stroke engine, but still translates into a size and weight advantage for the two-stroke engine.

To deliver a fresh charge to the cylinder in the absence of an intake stroke, the two-stroke engine requires that the incoming charge be pressurized slightly. This is often accomplished by using the underside of the piston as a compressor, as illustrated in Figure 5. In Figure 5a, the piston is rising to compress the charge trapped in the cylinder. This creates a subatmospheric pressure in the crankcase, opening a spring-loaded inlet valve to admit a fresh charge. In Figure 5b, the mixture has been ignited and burned as the piston descends on the expansion stroke to extract work from the products. Later during the piston downstroke, Figure 5c, the top of the piston uncovers exhaust ports in the cylinder wall, allowing combustion products to escape. During descent of the piston, Figures 5b and c, the air inlet valve has been closed by its spring, and the underside of the piston is compressing the charge in the crankcase. In Figure 5d, the piston is just past BDC. Intake ports in the cylinder wall have been uncovered, and the piston, which has just completed its descent, has transferred the fresh charge from the crankcase into the cylinder through a transfer passage.

Note in Figure 5 that with the piston near BDC, both intake and exhaust ports are open concurrently. This provides a pathway whereby some of the incoming charge can "short-circuit" the cycle and exit with the exhaust gas. If the engine uses an upstream carburetor to mix fuel into the air before the charge enters the crankcase, then a fraction of the fuel leaves with the exhaust gas. That penalizes fuel economy and increases exhaust emissions. This escape path for unburned fuel can be eliminated by injecting fuel directly into the cylinder after both ports are closed, but at the cost of increased complexity.

If the crankcase compression illustrated in Figure 5 is used, the reservoir of lubricating oil normally contained in the crankcase of a four-stroke engine (see Figure 1) must be eliminated. Cylinder lubrication is then usually accomplished by mixing a small quantity of oil into the fuel. This increases oil consumption. An alternative allowing use of the

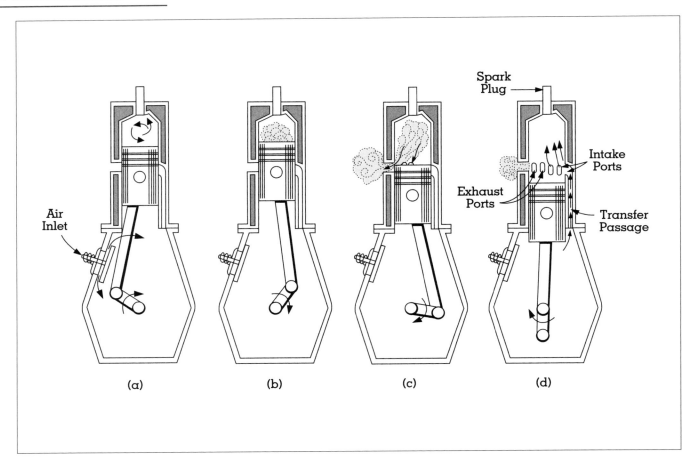

Figure 5.
Two-stroke cycle with crankcase compression.

crankcase as an oil reservoir is to eliminate the transfer passage and add an engine-driven blower to provide the pressurized fresh charge directly through the intake ports. Again this increases engine complexity.

Rotary Engine

The four-stroke and two-stroke engines described above both use the slider-crank mechanism to transform piston work into crankshaft torque, but other intermittent-combustion engines have been conceived that use different kinematic arrangements to achieve this end. The only one that has realized significant commercial success is the rotary engine first demonstrated successfully in Germany by Felix Wankel in 1957.

Illustrated in Figure 6, this engine incorporates a flat three-sided rotor captured between parallel end walls. The rotor orbits and rotates around the central shaft axis, and within a stationary housing that is specially shaped so that the three apexes of the rotor always remain in close proximity to the inner wall of the housing. Linear apex seals separate three chambers, each enclosed by a rotor flank, the stationary housing, and the two end walls. The chambers are further sealed by rotor side seals that rub against the end walls. As the rotor orbits and rotates, the volume of each chamber periodically increases and decreases, just as the cylinder volume above a piston of the slider-crank mechanism changes throughout the engine cycle. A crankshaft-mounted eccentric transmits the output work from the rotor to the engine output shaft.

The engine cycle can be understood by following the flank AB in Figure 6. In Figure 6a, the fresh charge is entering the chamber through a peripheral inlet port. As the rotor rotates clockwise, in Figure 6b the volume of the chamber is decreasing to compress the charge. In Figure 6c, the chamber volume is near its minimum and the spark plug has ignited the mix-

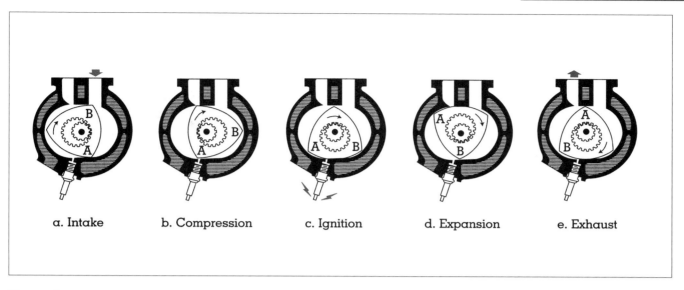

a. Intake   b. Compression   c. Ignition   d. Expansion   e. Exhaust

Figure 6.
Wankel rotary combustion engine.

ture. In Figure 6d, the burned products are expanding. In Figure 6e, apex A has uncovered the peripheral exhaust port to release the expanded products. Thus each chamber mimics the four-stroke cycle of the slider-crank mechanism.

Given the three flanks of the rotor, there are three power pulses per rotor revolution. The output shaft rotates at three times rotor speed. Thus there is one power pulse for each revolution of the output shaft, just as in a two-stroke piston engine. This, combined with the absence of connecting rods and the ability of the engine to run smoothly at high speeds, contributes to the compactness of the Wankel rotary engine. However, the segments of the housing exposed to the heat of the combustion and exhaust processes are never cooled by the incoming charge, as in the reciprocating piston engine. The high surface-to-volume ratio of the long, thin combustion chamber promotes high heat loss. The chambers have proved difficult to seal. Such factors penalize the fuel economy of the Wankel engine relative to its slider-crank counterpart, which is one of the factors that has impaired its broader acceptance.

MULTICYLINDER ENGINES

Otto's single-cylinder engine of 1876 had a nominal cylinder bore of half a foot and a piston stroke of a foot. Operating at 180 rpm, it developed about 3 horsepow-

er from its 6-liter displacement. Today a 6-liter automotive engine would likely have eight cylinders and deliver more than a hundred times as much power.

The most common arrangements for multicylinder engines are illustrated in Figure 7. Configurations employing four cylinders in line (I-4), and vee arrangements of either six (V-6) or eight (V-8) cylinders, currently dominate the automotive field. However, I-3, I-5, I-6, V-10, V-12, and six-cylinder horizontally opposed (H-6) arrangements are used as well.

Also represented in Figure 7 is the radial arrangement of cylinders common in large aircraft engines before the advent of jet propulsion. Five, seven, or nine cylinders were arranged in a bank around the crankshaft. Larger engines used two banks, one behind the other. One of the last, most powerful radial aircraft engines employed twenty-eight cylinders in four banks of seven cylinders each. These radial aircraft engines were air-cooled.

Individual-cylinder piston displacement in contemporary gasoline engines generally ranges from 0.15 to 0.85 liter, with the bore/stroke ratio varying between 0.8 and 1.3. Mean piston speed, which is twice the product of stroke and crankshaft rotational speed, generally falls between 8 and 15 meters/per second. Thus engines with larger cylinders are designed for slower rotational speeds. Choosing the number of cylinders

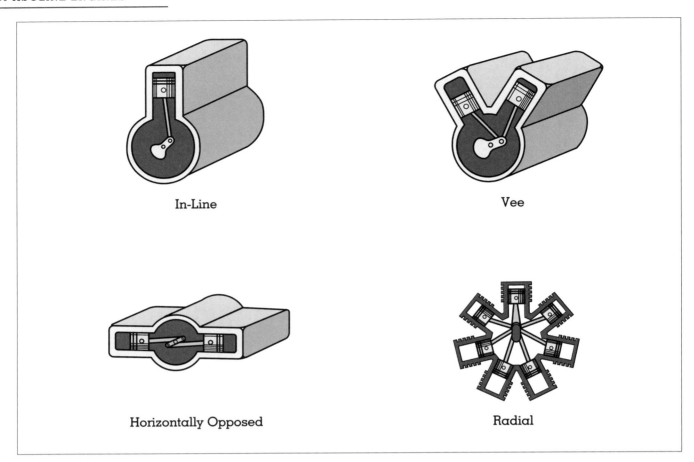

Figure 7.
Some multicylinder engine arrangements.

and their physical arrangement in engine design are often influenced significantly by such issues as the dimensions of the engine compartment or the bore-centerline spacing built into existing tooling.

ENGINE EFFICIENCY

It is common practice to measure the effectiveness of the engine as an energy conversion device in terms of the fraction of the fuel energy consumed that is actually delivered as useful output work. This fraction is customarily termed "thermal efficiency." In vehicular use, fuel economy (e.g., miles/per gallon) is directly proportional to thermal efficiency, although vehicle fuel economy also depends on such parameters as vehicle weight, aerodynamic drag, rolling resistance, and the energy content of a gallon of fuel.

The efficiency of the 1860 Lenoir engine was no more than 5 percent. An efficiency of 14 percent was claimed for Otto's original four-stroke engine of 1876. The best efficiency of modern automobile engines burning gasoline is in the 30 to 35 percent range. Small engines are generally somewhat less efficient than large ones. As discussed above, the simplicity of the crankcase-scavenged two-stroke engine, commonly employed in such low-power engines as those used in gardening equipment, carries an additional efficiency penalty. At the other extreme, large, slow-speed marine engines have demonstrated efficiencies in the 50 percent range, but these are compression-ignition engines operated on diesel fuel rather than gasoline.

The best efficiency attainable from a gasoline engine of specified power rating depends heavily on four parameters: compression ratio, air/fuel ratio, spark timing, and the fraction of the mechanical energy developed in the cylinder or cylinders devoted to overcoming parasitic losses. The compression ratio is the ratio of cylinder volume at BDC to volume at TDC. The air/fuel ratio is simply the mass of air inducted by the cylinders divided by the mass of fuel added to it. The spark timing is the number of degrees of crankshaft rotation before TDC at which the ignition spark is discharged. Parasitic losses include (1) engine friction, (2) the energy consumed by such essential components as the oil pump, fuel pump, coolant pump, generator used for charging the battery, and radiator cooling fan, and (3) the pumping loss associated with drawing the fresh charge into the cylinder during intake and expelling the combustion products from the cylinder during the exhaust.

Compression Ratio

Raising the compression ratio generally increases the thermal efficiency of the gasoline engine, as can be demonstrated by thermodynamic analysis of the engine cycle. The simplest such analysis is of the air-standard Otto cycle (see Figure 3, path 1-2-3-4-5). This technique assumes that the cylinder is filled with air behaving as a perfect gas (i.e., with constant specific heat), that from 2 to 3 heat is added to the air from an external source (no internal combustion), that from 4 to 5 heat is rejected from the air to the surroundings (no exhaust process), and that during compression (1 to 2) and expansion (3 to 4) the air exchanges no heat with its surroundings. This constitutes a thermodynamic cycle because the composition, pressure, and temperature of the cylinder contents are the same at the end of the cycle, point 5, as at the beginning, point 1. The thermodynamic efficiency (η) of this air-standard Otto cycle is given by $\eta = 1R^a$ where R = compression ratio $\alpha = 0.4$.

A thermodynamically closer approximation to the actual engine is provided by fuel-air cycle analysis. In it the cylinder contents during compression are assumed to consist of air, fuel, and residual burned gas from the previous cycle. The specific heat of the cylinder gas is varied with temperature and composition. Constant-volume combustion is assumed, with the products in chemical equilibrium. At the end of the expansion stroke and during the exhaust stroke, the combustion products are allowed to escape from the cylinder, being replaced by a fresh charge during the intake stroke. No exchange of heat between the cylinder gas and its surroundings is considered.

In contrast to the air-standard cycle, the fuel-air cycle is a cycle in the mechanical sense only. It is not a true thermodynamic cycle because neither the composition nor the temperature of the cylinder contents returns to its initial state at point 1. This, of course, is also true of a real internal-combustion engine. With the more realistic assumptions of the fuel-air cycle, exponent $\alpha \equiv 0.28$ in $\eta = 1R^a$ for the normal range of air-fuel ratios used with a gasoline engine. This relationship defines an upper limit for efficiency of an otto-cycle engine burning premixed air and gasoline.

According to $\eta = 1R^a$, the efficiency of the ideal Otto cycle increases indefinitely with increasing compression ratio. Actual engine experiments, which inherently include the real effects of incomplete combustion, heat loss, and finite combustion time neglected in fuel-air cycle analysis, indicate an efficiency that is less than that given by $\eta = 1R^a$ when $\alpha = 0.28$. Furthermore, measured experimental efficiency reached a maximum at a compression ratio of about 17 in large-displacement automotive cylinders but at a somewhat lower compression ratio in smaller cylinders.

Otto's engine of 1876 had a compression ratio of 2.5. By the beginning of the twentieth century, the efficiency advantage of higher compression ratios was recognized, but liquid-fueled engines built with higher compression ratios often experienced an unexplained, intolerable, explosive combustion noise. Over a couple of decades, researchers came to realize that the noise was caused by autoignition in the unburned mixture ahead of the normal flame front as it advanced from the spark plug. This autoignition is caused by the increase in pressure and temperature within the cylinder once normal combustion has been initiated. The phenomenon can be likened to the compression ignition of fuel in the diesel engine. Experience taught that a fuel such as kerosene was more prone to this "combustion knock" than more volatile petroleum products, such as gasoline.

Thomas Midgley, working with a group of researchers under Charles F. Kettering, discovered a fuel-additive, tetraethyl lead that enhanced the knock resistance of existing fuels. By the time this additive entered commercial use in 1923, the average compression ratio of new U.S. cars had advanced to 4.3.

An octane rating scale was devised for fuels to quantify their knock resistance. Further research led to cataloguing the antiknock qualities of the myriad individual hydrocarbon species found in gasoline.

Fuel refiners learned how to rearrange the atoms in fuel molecules to enhance the octane number of the fuel available at gasoline stations. Studying the combustion process led to combustion-chamber designs with lower octane requirements. All of these activities fostered a nearly continuous increase in the average compression ratio of the new U.S. car, as follows: 4.3 in 1923; 5.1 in 1930; 6.3 in 1940; 7.0 in 1950; 9.0 in 1960; 9.8 in 1970; and 8.6 in 1980. The drop from 1970 to 1980 resulted from the imposition of increasingly severe exhaust emission standards in the United States, which included the need to design engines to run on unleaded gasoline. This was necessitated by the introduction of the catalytic converter to the exhaust system. The catalyst is poisoned by lead in the fuel.

Since 1980, further improvements in engine design have allowed the average compression ratio to creep back up to the range of 9 to 10, despite the absence of tetraethyl lead. Barring an unanticipated increase in the octane number of commercial gasoline, a further major increase in the average compression ratio of the contemporary gasoline engine is unlikely. A variable compression ratio engine that would retain the currently acceptable ratio at high engine loads but increase it at light load, where combustion knock is less likely, is a concept that has long existed but has yet to be implemented commercially on a significant scale.

Air-Fuel Ratio

The second important parameter affecting efficiency is air/fuel ratio. For every hydrocarbon fuel, there is an air/fuel ratio that, in principle, causes all the hydrogen in the fuel to burn to water vapor and all the carbon in the fuel to burn to carbon dioxide. This chemically correct proportion is called the stoichiometric ratio.

If the engine is fed a mixture containing more fuel than the stoichiometric amount, the mixture is said to be rich, and carbon monoxide (CO) and hydrogen are added to the combustion products. Because these two gases are fuels themselves, their presence in the exhaust signifies incomplete combustion and wasted energy.

If the engine is fed a mixture containing more air than the stoichiometric amount, the mixture is said to be lean, and unconsumed oxygen appears in the combustion products. This signifies that the full power-producing potential of the air inducted by the engine has not been utilized.

Among the lesser constituents of engine exhaust are unburned hydrocarbons (HC) and oxides of nitrogen (NOx). Exhaust HC concentration increases rapidly as the mixture is enriched from stoichiometric. Exhaust NOx concentration peaks at a mixture ratio slightly leaner than stoichiometric and falls off as the mixture is made either richer or leaner. These two exhaust-gas constituents react in the atmosphere in the presence of sunlight to form ozone, a major ingredient of photochemical smog. Consequently their mass emissions, along with that of CO, are regulated for environmental reasons.

For conventional gasoline, the stoichiometric ratio is approximately 14.7. Its precise value varies slightly with the composition of the gasoline. Maximum power is achieved with a slightly rich air/fuel ratio—say, 12.5. Maximum efficiency is achieved with a slightly lean mixture—say, 16—although this best-economy mixture ratio is somewhat dependent on combustion quality.

For most of the nineteenth century, the fuel and air were mixed upstream of the engine in a carburetor. In the automobile, the carburetor contains an inlet throttle valve linked to the accelerator pedal. The throttle valve is fully opened when the accelerator pedal is depressed to the floorboard, but barely open when the pedal is released so that the engine runs at its idle condition. The intent of the carburetor is to supply the engine with its highest air-fuel ratio during midspeed cruise conditions. As the throttle approaches its wide-open position, the mixture is enriched to maximize engine power. The mixture prepared by the carburetor is also temporarily enriched when the throttle is opened suddenly, as during a sharp vehicle acceleration, because the resultant fuel-flow increase lags behind the increase in airflow. As engine idle is approached, the mixture is enriched to compensate for poor combustion quality at the low-speed, low-pressure idle condition. For cold starting, the carburetor includes a choke valve that enriches the mixture so that enough of the fuel is vaporized near the spark plug to ensure ignition.

In early days, the mixture supplied to the engine was often quite rich in order to ensure smooth engine operation. This, of course, wasted fuel energy through incomplete combustion. Fleet surveys in early years showed marked improvement in the percentage of fuel that was wasted: 15.5% in 1927, 7.5%

in 1933, and only 1.5% by 1940. (The vehicles were operated at 40 miles per hour, with gasoline having a stoichiometric air-fuel ratio of 15).

Following World War II, engine improvements included further leaning of the mixture, especially toward the leaner-than-stoichiometric best-economy ratio during cruise. Then in the early 1980s, U.S. emission standards became so stringent that the catalytic converter, which had been added in about 1975 to oxidize HC and CO in the exhaust, also had to reduce NOx. This has led to the widely used three-way catalyst that controls all three emissions, but only if the air/fuel ratio is kept in a narrow range about the stoichiometric ratio.

Such tight mixture control is beyond the capability of the traditional carburetor. Consequently, after sorting through a number of alternatives, industry has settled on closed-loop-controlled port-fuel injection. Typically, an electronically controlled fuel injector is mounted in the intake port to each cylinder. A sensor in the air intake system tells an on-board computer what the airflow rate is, and the computer tells the fuel injectors how much fuel to inject for a stoichiometric ratio. An oxygen sensor checks the oxygen content in the exhaust stream and tells the computer to make a correction if the air/fuel ratio has drifted outside the desired range. This closed-loop control avoids unnecessary use of an inefficient rich mixture during vehicle cruise.

An old concept for increasing thermal efficiency at part load that is the subject of renewed interest is the use of a much leaner average mixture ratio at part load. To burn such an overall-lean mixture in the time made available by the contemporary gasoline engine, this concept calls for stratifying the fresh charge. Thus a combustible richer-than-average cloud of air and fuel is segregated near the spark plug and surrounded by fuel-free air. Although gains in part-load efficiency have been demonstrated with this concept, emissionscontrol issues must be resolved before this concept is suitable for automotive use.

Ignition Timing

The third important parameter affecting engine efficiency is ignition timing. For each engine operating speed and throttle position, there is a spark timing giving best efficiency. In the early days of the automobile, timing was adjusted manually by the driver, so fuel economy was somewhat dependent on driver skill.

Eventually, automatic control of spark advance evolved. The conventional controller included two essential elements. Mechanically, spring-restrained rotating flyweights responding to centrifugal force advanced the spark as engine speed increased. Pneumatically, a diaphragm that sensed the increasing vacuum in the intake manifold as the intake throttle was closed added additional spark advance as the resulting reduction in cylinder pressure slowed flame speed.

With the advent of electronic engine controls, the on-board computer now manages ignition timing. It offers greater versatility in timing control, being able to integrate signals from sensors monitoring engine speed, airflow rate, throttle position, intake-manifold pressure, engine temperature, and ambient pressure.

Combustion knock can still be experienced in the engine near full throttle if the octane rating of the fuel is too low for the compression ratio, or vice versa. When ignition timing was controlled manually, the driver could retard the spark when knock was heard. In the era of mechanical/pneumatic control, the built-in timing schedule typically included slight retard from best-economy spark advance near full throttle to allow use of a slightly higher compression ratio for greater engine efficiency at part load. With the advent of electronic control, many engines are now equipped with knock sensors that detect incipient knock and maximize efficiency by retarding the spark just enough to avoid it. This technology has contributed to the modest increase in average compression ratio in automobiles since the 1980s.

Parasitic Losses

When parasitic losses are measured as a fraction of the power developed in the cylinder or cylinders, that fraction can be improved in two ways. Either the parasitic losses themselves can be reduced, or the power developed in the cylinder or cyinders can be increased. Over the years, both paths have been pursued.

Energy lost to mechanical friction has been decreased significantly through a large number of small, incremental improvements. In the typical modern engine, about half of the friction occurs between the cylinder wall and the piston/piston ring assembly. The sidewall contact surfaces of the piston have been specially shaped to minimize engine friction. Piston ring tension has been reduced, commensurate with adequate sealing of the combustion chamber. Oil viscosity has been chosen, and oil composition altered, to reduce friction. Analysis of the effects of operating pressure and temperature on engine dimensional changes has led to improved conformity between piston and cylinder during engine operation. Bearing dimensions and clearance have been selected to decrease friction. Reciprocating elements of the slider-crank mechanism have been lightened to decrease inertia loads on the bearings. In the valvetrain, sliding contact between adjacent members has often been replaced with rolling contact. Valves have been lightened, and valve-spring tension reduced.

Improved aerodynamic design of intake and exhaust ports has contributed to lower pumping losses. Paying greater attention to pressure drop in the catalytic converter also has helped. When recirculating exhaust gas into the intake system at part load to decrease NOx production, a common practice since 1973, the throttle has to be opened further to produce the same power at a given speed. This reduces intake pumping loss at a given part-load power.

Power requirements for oil, coolant, and fuel pumps are generally minor. Coolant-system power requirement increased when the thermo-siphon cooling system of early automobiles was superseded by an engine-driven coolant pump as engine power rating increased. Avoiding the use of unnecessarily high oil pressure has been beneficial. On the other hand, the higher pressure used in fuel injectors, compared to the fuel-pressure requirement of the carburetor, has not. In passenger cars, perhaps the largest reduction in auxiliary-power consumption has come by changing from an engine-driven cooling fan, running continuously, to an electrically driven fan that is inoperative as long as the radiator provides adequate cooling from ram air flowing through it as a result of vehicle forward motion.

The power that can be produced in each cylinder depends on the amount of air it can induct. Recent interest in replacing the traditional single-intake valve per cylinder with two smaller ones, and in a few cases even three, has increased total intake-valve area, and hence engine airflow at a given speed. At the same time, engine speed capability has been increased.

These changes have raised the air capacity of the engine, resulting in increased power delivery for a given displacement. If the engine displacement is then decreased to hold maximum power constant, the parasitic losses are smaller for the same output power capability. This can be translated into improved vehicle fuel economy.

Charles A. Amann

See also: Automobile Performance; Combustion; Diesel Cycle Engines; Drivetrains; Engines; Otto, Nikolaus August; Steam Engines; Stirling Engines; Tribology; Turbines, Gas; Turbines, Steam.

BIBLIOGRAPHY

Heywood, J. B. (1998). *Internal Combustion Engine Fundamentals.* New York: McGraw-Hill Book Company.

Obert, E. F. (1973). *Internal Combustion Engines and Air Pollution.* New York: Indext Educational Publishers.

Stone, R. (1999). *Introduction to Internal Combustion Engines,* 3rd ed. Warrendale, PA: Society of Automotive Engineers.

GEAR

See: Drivetrains; Mechanical Transmission of Energy

GEAR BOX

See: Drivetrains

GENERATOR, ELECTRICAL

See: Electric Motor Systems

GEOGRAPHY AND ENERGY USE

Geography looks for patterns in the distribution of diverse phenomena, and it tries to explain their variation by examining a wide range of underlying environmental and socioeconomic factors. Many fundamental realities of modern civilization cannot be fully appreciated without the basic understanding of the geography of energy use. There is a very high degree of inequality in the use of energy resources throughout the word: this is true for aggregate use of energy, for every individual fuel or renewable energy flow, as well as for the consumption of electricity. These consumption disparities are inevitably accompanied by large differences in energy self-sufficiency and trade.

A country's dependence on energy imports, its size and climate, its stage of economic development, and the average quality of life of its inhabitants are the key variables determining the patterns of fuel and electricity consumption among the world's 180 countries. Large countries also have distinct regional patterns of energy consumption, but the progressing homogenization of energy use has been reducing some of these differences. For example, in the United States sports utility vehicles are now favored by both urban and rural drivers, and, unlike a few decades ago, utilities in both southern and northern states have summer peak loads due to air conditioning.

ENERGY SELF-SUFFICIENCY AND IMPORTS

Energy self-sufficiency is not just a matter of possessing substantial domestic resources; it is also determined by the overall rate of consumption. The United States, following the sharp decline of Russian's output, is now the world's second largest producer of crude oil (after Saudi Arabia); it is also the world's largest importer of crude oil, buying more than 50 percent of it in order to satisfy its high demand for transportation energy. The only affluent economies self-sufficient in fossil fuels are Canada (due to its relatively small population and vast mineral resources), and the United Kingdom and Norway (thanks to the North Sea oil and gas fields).

Russia is the most notable case of a middle-income country with large surpluses of energy. Its huge oil and gas exports are now surpassed only by Saudi

Arabia, the largest OPEC oil producer. OPEC produces about 40 percent of the world's crude oil output and it supplies about 45 percent of all traded petroleum. In total, almost 60 percent of the world's crude oil extraction is exported from about forty-five hydrocarbon-producing countries—but the six largest exporters (Saudi Arabia, Iran, Russia, Norway, Kuwait, and the United Arab Emirates) sell just over 50 percent of the traded total. In contrast, more than 130 countries import crude oil and refined oil products; besides the United States, the largest buyers are Japan, Germany, France, and Italy.

About 20 percent of the world's natural gas production was exported during the late 1990s, three-quarters of it through pipelines, and the rest by LNG tankers. The former Soviet Union, Canada, the Netherlands, and Norway are the largest pipeline exporters, while Indonesia, Algeria, and Malaysia dominate the LNG trade. The largest importers of piped gas are the United States, Germany, Italy, and France; Japan and South Korea buy most of the LNG.

In comparison to hydrocarbons, coal trade is rather limited, with only about a tenth of annual extraction of hard coals and lignites exported, mainly from Australia, United States, South Africa, and Canada. Because of their lack of domestic resources and large iron and steel industries, Japan and South Korea are the two biggest buyers of steam and metallurgical coal. Although the development of high-voltage networks has led to rising exports of electricity, less than 4 percent of the global generation is traded internationally; France (mainly due to its large nuclear capacity), Canada, Russia, and Switzerland are the largest exporters. This trade will rise as the markets for electricity grow, and electricity transmission and distribution improves. Only 33 of the 180 countries are net energy exporters, and about 70 countries do not export any commercial energy. Self-sufficiency in energy supply was a major goal of many nations following the first "Energy Crisis" in 1973, but the collapse of OPEC's high crude oil prices in 1985 and the subsequent stabilization of the world's oil supply have greatly lessened these concerns during the 1990s.

AGGREGATE ENERGY USE AND ITS COMPOSITION

During the late 1990s annual consumption rates of commercial energy ranged from less than 25 kgoe (or

Source	Energy Consumption				
	1950	1973	1985	1990	1998
Solid fuels	1040	1700	2170	2330	2220
Liquid fuels	460	2450	2520	2790	3010
Natural gas	170	1100	1460	1720	2020
Primary Electricity	30	130	300	360	440
Total	1700	5380	6450	7200	7650

Table 1.
Global Energy Consumption
Note: All fuel conversions according to the UN rates; all primary electricity expressed in terms of its thermal equivalent.

less than 1 GJ) per capita in the poorest countries of sub-Saharan Africa to nearly 8 toe (or more than 300 GJ) per capita in the United States and Canada. The global mean was close to 1.5 toe (or 60 GJ) per capita—but the sharply bimodal distribution of the world's energy use reflecting the rich-poor divide meant that only a few countries (including Argentina and Portugal) were close to this level. Affluent countries outside North America averaged close to 3.5 toe per capita, while the mean for low-income economies was just 0.6 toe, close to the Chinese average. This huge gap in aggregate energy consumption has been narrowing slowly. In 1950 industrialized countries consumed about 93 percent of the world's commercial primary energy. Subsequent economic development in Asia and Latin America reduced this share, but by 1998 industrialized countries containing just one fifth of global population still consumed about 70 percent of all primary energy.

The United States alone, with less than 5 percent of the world population, claims about 25 percent of the world is total commercial energy use. Among the world's affluent countries only Canada has a similarly high per capita use of fossil fuels and primary electricity (about 8 t of crude oil equivalent per year). In spite of its huge fuel and electricity production, the United States imports more than a fifth of its total energy use, including more than half of its crude oil consumption. Almost two fifths of all commercial energy is used by industry, a quarter in transportation, a fifth by households, and a bit over one sixth goes into the commercial sector.

In contrast, during the late 1990s the poorest quarter of humanity—made up of about fifteen sub-Saharan African countries, Nepal, Bangladesh,

Indochina, and most of rural India—consumed a mere 2.5 percent of all commercial energy. The poorest people in the poorest countries, including mostly subsistence peasants but also millions of destitute people in large cities, do not directly consume any commercial fuels or electricity at all!

All of the world's major economies, as well as scores of smaller, low-income nations, rely mainly on hydrocarbons. Crude oil now supplies two-fifths of the world's primary energy (Table 1). There are distinct consumption patterns in the shares of light and heavy oil products: the United States burns more than 40 percent of all its liquid fuels as gasoline, Japan just a fifth; and the residual fuel oil accounts for nearly a third of Japanese use, but for less than 3 percent of the U.S. total. Small countries of the Persian Gulf have the highest per capita oil consumption (more than 5 t a year in the United Arab Emirates and in Qatar); the U.S. rate is more than 2.5 t a year; European means are around 1 t; China's mean is about 120 kg, and sub-Saharan Africa is well below 100 kg per capita.

Natural gas supplies nearly a quarter of the world's primary commercial energy, with regional shares ranging from one half in the former Soviet Union to less than 10 percent in the Asia Pacific. The United States and Russia are by far the largest consumers, followed by the United Kingdom, Germany, Canada, Ukraine, and Japan; leaving the small Persian Gulf emirates aside, Canada, the Netherlands, the United States, Russia, and Saudi Arabia have the highest per capita consumption. Coal still provides 30 percent of the world's primary commercial energy, but outside China and India—where it is still used widely for heating, cooking, and in transportation, and where it supplies, respectively, about three quarters and two

Global Electricity Production (in TWh)

Source	Year				
	1950	1973	1985	1990	1997
Fossil Fueled	620	4560	6260	7550	8290
Hydro	340	1320	2000	2210	2560
Nuclear	—	190	1450	1980	2270
Geothermal	—	—	30	40	50
Wind, Solar	—	—	—	50	80
Total	960	6070	9740	11,830	13,250

Table 2.
Global Electricity Production

thirds of commercial energy consumption—it has only two major markets: electricity generation and production of metallurgical coke.

When converted at its heat value (1 kWh = 3.6 MJ) primary electricity supplied about 6 percent of global commercial energy consumption during the late 1990s. Hydro and nuclear generation account for about 97 percent of all primary electricity, wind, geothermal, and solar—in that order—for the rest (Table 2). Canada, the United States, Russia, and China are the largest producers of hydroelectricity; the United States, Japan, and France lead in nuclear generation; and the United States, Mexico, and the Philippines are the world's largest producers of geo-thermal electricity.

ENERGY CONSUMPTION AND ECONOMIC DEVELOPMENT

On the global level the national per capita rates of energy consumption (Table 3) correlate highly (r ≥ 0.9) with per capita gross domestic product (GDP): the line of the best fit runs diagonally from Nepal (in the lower left corner of a scattergram) to the United States (in the upper right corner). This commonly used presentation has two serious shortcomings: the exclusion of biomass fuels and the use of exchange rates in calculating national GDPs in dollars. Omission of biomass energies substantially under-rates actual fuel use in low-income countries where wood and crop residues still supply large shares of total energy demand (more than 90% in the poorest regions of sub-Saharan Africa; about a fifth in China), and official exchange rates almost invariably undervalue the GDP of low-income countries.

Inclusion of biomass energies and comparison of

Countries	Energy consumption			
	1950	1973	1985	1995
World	700	1450	1330	1300
Developed countries				
U.S.	5120	8330	6680	7500
France	1340	3150	2810	2650
Japan	390	2760	2650	2330
Largest oil exporters				
Saudi Arabia	110	690	4390	4360
Russia	1120	3540	4350	4630
Developing countries				
Brazil	140	440	500	640
China	60	300	490	680
India	70	130	180	270
Poorest economies				
Bangladesh	10	20	40	70
Ethiopia	—	20	10	20

All fuel conversions according to the UN rates; all primary electricity expressed in terms of its thermal equivalent.

Table 3.
Per Capita Consumption of Commercial Energy

GDPs in terms of purchasing power parities (PPP) weaken the overall energy-economy correlation, and disaggregated analyses for more homogeneous regions show that energy-GDP correlations are masking very large differences at all levels of the economic spectrum. Absence of any strong energy-GDP correlation is perhaps most obvious in Europe: while France and Germany have very similar PPP-adjusted GDPs, Germany's per capita energy use is much higher; similar, or even larger, differences can be seen when comparing Switzerland and Denmark or Austria and Finland.

Another revealing look at the energy-economy link is to compare national energy/GDP intensities (i.e., how many joules are used, on the average, to produce a unit of GDP) expressed in constant monies. These rates follow a nearly universal pattern of historical changes, rising during the early stages of economic development and eventually commencing gradual declines as economies mature and become more efficient in their use of energy. This shared trend still leaves the economies at very different energy-intensity levels. The U.S. energy intensity fell by more than a third since the mid-1970s—but this

569

impressive decline has still left the country far behind Japan and the most affluent European countries. China cut its energy intensity by half since the early 1980s—but it still lags behind Japan.

Weak energy-GDP correlations for comparatively homogeneous groups of countries and substantial differences in energy intensities of similarly affluent economies have a number of causes. A country's size plays an obvious role: there are higher energy burdens in integrating larger, and often sparsely inhabited, territories by road and rail, and in affluent countries the need to span long distances promotes air travel and freight, the most energy-intensive form of transportation. Not surprisingly, Canadians fly two to three times more frequently than most Europeans do. Even within countries, there are wide consumption disparities. In the United States, gasoline and diesel fuel consumption by private cars is highest in Wyoming where an average car travels nearly 60 percent more miles annually than the national mean; Montana and Idaho are the other two thinly populated states with considerably longer average car travel.

Climate is another obvious determinant of a country's energy use. For example, Canada averages annually about 4,600 heating degree days compared to Japan's 1,800, and this large difference is reflected in a much higher level of household and institutional fuel consumption. But climate's effects are either partially negated or highly potentiated by different lifestyles and by prevailing affluence. The Japanese and British not only tolerate much lower indoor temperatures (below 15°C) than Americans and Canadians (typically above 20°C), but they also commonly heat only some rooms in the house or, in the case of Japan, merely parts of some rooms (using *kotatsu* foot warmers).

Larger, overheated and often poorly insulated American houses mean that the U.S., with the national annual mean of 2,600 heating degree days, uses relatively more fuel and electricity for heating than does Germany with its mean of 3,200 heating degree days. Space heating takes half of U.S. residential consumption, water heating (with about 20%) comes second, ahead of air conditioning. Widespread adoption of air conditioning erased most of the previously large differences in residential energy consumption between the U.S. snowbelt and the sunbelt: now Minnesota and Texas, or Nevada and Montana have nearly identical per capita averages of household energy use. The most notable outliers are Hawaii (40% below the national mean: no heating, limited air conditioning) and Maine (almost 30% above the mean due to heating).

Composition of the primary energy consumption makes a great deal of difference. Because coal combustion is inherently less efficient due to the presence of combustible ash than the burning of hydrocarbons, the economies that are more dependent on coal (China, United States) are handicapped in comparison with nations relying much more on liquid fuels, natural gas, and primary electricity (Japan, France). So are the energy exporting countries: energy self-sufficiency (be it in Russia or Saudi Arabia) is not conducive to efficient conversions, but high dependence on highly taxed imports (as in Japan or Italy) promotes frugality.

Differences in industrial structure are also important: Canada is the world's leading producer of energy-intensive aluminum—but Japan does not smelt the metal at all. And as the only remaining superpower, the United States still invests heavily in energy-intensive production of weapons and in the maintenance of military capacity. Annual energy consumption of all branches of the U.S. military averaged about 25 million t of oil equivalent during the 1990s (more than half of it as jet fuel): that is more than the total primary commercial consumption of nearly two-thirds of the world's countries! In contrast, the size of Japan's military forces is restricted by the country's constitution.

But no single factor is more responsible for variations in energy intensity among high-income countries than the level of private, and increasingly discretionary (as opposed to essential), energy use. During the late 1990s, the Japanese used only about 0.4 toe in their generally cramped and poorly heated apartments and houses, but U.S. residential consumption was about 1 toe. Ubiquitous air-conditioning, overheating of oversized houses, and heating of large volumes of water explain the difference. Refrigerator and washing machine ownership is nearly universal throughout the rich world, but the appliances are smaller outside of North America where clothes dryers, dishwashers, and freezers are also much rarer.

There are even greater disparities in energy used for transportation: North American commuters commonly consume three times as much gasoline per year as do their European counterparts who rely much more on energy-efficient trains; and Americans and

Canadians also take many more short-haul flights, the most energy-intensive form of transportation, in order to visit families and friends or to go for vacation. They also consume much more fuel during frequent pastime driving and in a still growing range of energy-intensive recreation machines (SUVs, RVs, ATVs, boats, snowmobiles, seadoos).

In total, cars and light trucks consume nearly three-fifths of all fuel used by U.S. transportation. In spite of a more than 50 percent increase in average fuel efficiency since 1973, the U.S. cars still consume between 25 to 55 percent more fuel per unit distance than the average in European countries (11.6 l per 100 kilometers compared to 9.1 l in Germany, and 7.4 l in Denmark in 1995). All forms of residential consumption and private transportation thus claim more than 2 toe per capita in the United States, compared to less than 1 toe in Europe and about 0.75 toe in Japan.

ENERGY USE AND THE QUALITY OF LIFE

But does the higher use of energy correlate closely with the higher quality of life? The answer is both yes and no. There are obvious links between per capita energy use and the physical quality of life characterized above all by adequate health care, nutrition, and housing. Life expectancy at birth and infant mortality are perhaps the two most revealing indicators of the physical quality of life. The first variable subsumes decades of nutritional, health-care and environmental effects and the second one finesses these factors for the most vulnerable age group. During the 1990s average national life expectancies above 70 years required generally annual per capita use of 40 to 50 GJ of primary energy—as did the infant mortality rate below 40 (per thousand newborn).

Increased commercial energy use beyond this range brought first rapidly diminishing improvements of the two variables and soon a levelling-off with hardly any additional gains. Best national achievements—combined male and female life expectancies of 75 years and infant mortalities below ten—can be sustained with energy use of 70 GJ per year, or roughly half of the current European mean, and less than a quarter of the North American average. Annual commercial energy consumption around 70 GJ per capita is also needed in order to provide ample opportunities for postsecondary schooling—and it appears to be the desirable minimum for any society striving to combine a decent physical quality of life with adequate opportunities for intellectual advancement.

On the other hand, many social and mental components of the quality of life—including such critical but intangible matters as political and religious freedoms, or satisfying pastimes—do not depend on high energy use. Reading, listening to music, hiking, sports, gardening and craft hobbies require only modest amounts of energy embodied in books, recordings, and appropriate equipment or tools—and they are surely no less rewarding than high-energy pastimes requiring combustion of liquid fuels.

It is salutary to recall that the free press and the ideas of fundamental personal freedoms and democratic institutions were introduced and codified by our ancestors at times when their energy use was a mere fraction of ours. As a result, contemporary suppression or cultivation of these freedoms has little to do with overall energy consumption: they thrive in energy-rich United States as they do in energy-poor India, and they are repressed in energy-rich Saudi Arabia as they are in energy-scarce North Korea. Public opinion polls also make it clear that higher energy use does not necessarily enhance feelings of personal and economic security, optimism about the future, and general satisfaction with national or family affairs.

The combination of abundant food energy supplies and of the widespread ownership of exertion-saving appliances has been a major contributor to an epidemic of obesity (being at least a 35% over ideal body weight) in North America. National health and nutrition surveys sin the United States how that during the 1990s every third adult was obese, and an astonishing three-quarters of all adults had body weights higher than the values associated with the lowest mortality for their height.

Vaclav Smil

See also: Agriculture; Energy Economics; Energy Intensity Trends; Reserves and Resources.

BIBLIOGRAPHY

Biesiot, W., and Noorman, K. J. (1999). "Energy Requirements in Household Consumption: A Case Study of the Netherlands." *Ecological Economics* 28:367–383.

BP Amoco. (2000). *BP Statistical Review of World Energy.* BP Amoco, London. <http://www.bpamoco.co.il/worldenergy>.

Darmstadter, J. (1977). *How Industrial Societies Use Energy: A Comparative Analysis*. Baltimore, MD: Johns Hopkins University Press.

Energy Information Administration. (2000). *Annual Energy Review*. EIA, Washington, DC. <http://www.eia.doe.gov>.

Energy Information Administration. (2000). *Annual Energy Outlook*. EIA, Washington, DC. <http://www.eia.doe.gov>.

Energy Information Administration. (2000). *State Energy Data Report, Consumption Estimates*. EIA, Washington, DC. <http://www.eia.doe.gov>.

Energy Information Administration. (2000). *International Energy Outlook*. EIA, Washington, DC. <http://www.eia.doe.gov>.

International Energy Agency. (2000). *Energy Statistics of OECD Countries*. EIA, Paris. <http://www.eia.doe.gov>.

International Energy Agency. (2000). *Energy Balances of OECD Countries*. EIA, Paris. <http://www.eia.doe.gov>.

International Energy Agency. (2000). *Energy Statistics and Balances of Non-OECD Countries*. IEA, Paris. <http://www.ei.doe.gov>.

International Energy Agency. (1997). *Indicators of Energy Use and Efficiency: Understanding the Link Between Energy and Human Activity*. Paris: OECD/EIA.

Meyers, S., and Schipper, L. (1992). "World Energy Use in the 1970s and 1980s: Exploring the Changes." *Annual Review of Energy* 17:463–505.

Sathaye, J., and Tyler, S. (1991). "Transitions in Household Energy Use in Urban China, India, the Philippines, Thailand, and Hong Kong." *Annual Review of Energy* 16:295–335.

Schipper, L. et al. (1989). "Linking Life-styles and Energy Use: A Matter of Time?" *Annual Review of Energy* 14:273–320.

Schipper, L., and Meyers, S. (1992). *Energy Efficiency and Human Activity*. New York: Cambridge University Press.

Smil, V. (1991). *General Energetics*. New York: John Wiley.

Smil, V. (1992). "Elusive Links: Energy, Value, Economic Growth and Quality of life." *OPEC Review* Spring 1992:1–21.

United Nations. (2000). *Yearbook of World Energy Statistics*. UNO, New York.

GEOTHERMAL ENERGY

Geothermal energy is heat energy that originates within Earth itself. The temperature at the core of our planet is 4,200°C (7,592°F), and heat flows outward to the cooler surface, where it can produce dramatic displays such as volcanoes, geysers, and hot springs, or be used to heat buildings, generate electricity, or perform other useful functions. This outward flow of heat is continually being maintained from within by the decay of radioactive elements such as uranium, thorium, and radium, which occur naturally in Earth. Because of its origin in radioactivity, geothermal energy can actually be thought of as being a form of natural nuclear energy.

The U.S. Department of Energy has estimated that the total usable geothermal energy resource in Earth's crust to a depth of 10 kilometers is about 100 million exajoules, which is 300,000 times the world's annual energy consumption. Unfortunately, only a tiny fraction of this energy is extractable at a price that is competitive in today's energy market.

In most areas of the world, geothermal energy is very diffuse—the average rate of geothermal heat transfer to Earth's surface is only about 0.06 watt per square meter. This is very small compared to, say, the solar radiation absorbed at the surface, which provides a global average of 110 watts per square meter. Geothermal energy can be readily exploited in regions where the rate of heat transfer to the surface is much higher than average, usually in seismic zones at continental-plate boundaries where plates are colliding or drifting apart. For example, the heat flux at the Wairakei thermal field in New Zealand is approximately 30 watts per square meter.

A related aspect of geothermal energy is the thermal gradient, which is the increase of temperature with depth below Earth's surface. The average thermal gradient is about 30°C (54°F) per kilometer, but it can be much higher at specific locations—for instance, in Iceland, where the increase is greater than 100°C (180°F) per kilometer in places.

TYPES OF GEOTHERMAL SOURCES

Geothermal sources are categorized into various types: hydrothermal reservoirs, geopressurized zones, hot dry rock, normal geothermal gradient, and magma.

Hydrothermal Reservoirs

Groundwater can seep down along faults in Earth's crust and become heated through contact with hot rocks below. Sometimes this hot water accumulates in an interconnected system of fractures and becomes a hydrothermal reservoir. The water might remain underground or might rise by convection through fractures to the surface, producing geysers and hot springs.

THERMAL GRADIENT MAP OF THE CONTERMINOUS U.S.

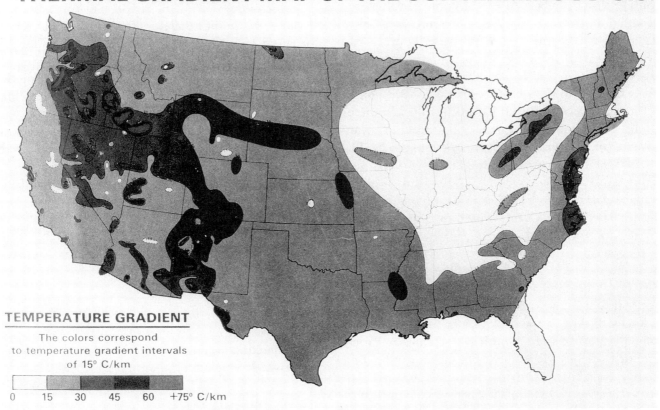

TEMPERATURE GRADIENT

The colors correspond to temperature gradient intervals of 15° C/km

0   15   30   45   60   +75° C/km

Map indicating regions of high thermal gradient where HDR geothermal techniques may be applied. (U.S. Department of Energy)

Geothermal power plant located in a lava field in Blue Lagoon, Iceland. (Corbis-Bettmann)

Hydrothermal reservoirs are the only geothermal sources that have been used for commercial energy production. Because of the high pressure deep below Earth's surface, the water in these sources can become heated well above the usual boiling temperature of 100°C (212°F). As the superheated water makes its way to the surface, either by convection or because a geothermal well has been drilled, some or all of the water will vaporize to become steam because of the lower pressures encountered. The most desirable geothermal sources have very high temperatures—above 300°C (572°F)—and all of the water vaporizes to produce dry steam (containing no liquid water), which can be used directly in steam-electric turbines.

Wet steam reservoirs are much more common than the simple dry type. Again, the field is full of very hot water, under such high pressure that it cannot boil. When a lower-pressure escape route is provided by drilling, some of the water suddenly evaporates (flashes) to steam, and it is a steam-water mixture that reaches the surface. The steam can be used to drive a turbine. The hot water also can be used to drive a second turbine in a binary cycle, described in the section "Electricity Generation" in this article.

Many geothermal reservoirs contain hot water at a temperature too low for electricity generation. However, the water can be used to heat buildings such as homes, greenhouses, and fish hatcheries. This heating can be either direct or through the use of heat pumps.

Geopressurized Zones

Geopressurized zones are regions where water from an ancient ocean or lake is trapped in place by impermeable layers of rock. The water is heated to temperatures between 100°C (212°F) and 200°C (392°C) by the normal flow of heat from Earth's core, and because of the overlying rock, the water is held under very high pressure as well. Thus energy is contained in the water because of both the temperature and the pressure, and can be used to generate electricity. Many geopressurized zones also contain additional energy in the form of methane from the decay of organic material that once lived in the water. The U.S. Geological Survey has estimated that about one third of the energy from geopressurized zones in the United States is available as methane.

Hot Dry Rock

In many regions of the world, hot rocks lie near Earth's surface, but there is little surrounding water. Attempts have been made to fracture such rocks and then pump water into them to extract the thermal energy, but the technical difficulties in fracturing the rocks have proven to be much more troublesome than anticipated, and there has been a problem with water losses. Consequently, progress in extracting energy from hot, dry rocks has been slow. However, experiments are ongoing in the United States, Japan, and Europe because the amount of energy available from hot, dry rocks is much greater than that from hydrothermal resources. The U.S. Department of Energy estimates that the total energy available from high-quality hot, dry rock areas is about 6,000 times the annual U.S. energy use.

Normal Geothermal Gradient

In principle the normal geothermal gradient produces a useful temperature difference anywhere on the globe. If a hole is drilled to a depth of 6 kilometers (which is feasible), a temperature difference of about 180°C (324°F) is available, but no technology has been developed to take advantage of this resource. At this depth, water is unlikely, and the problems of extracting the energy are similar to the difficulties encountered with hot, dry rocks near the surface.

Magma

Magma is subterranean molten rock, and although the potential thermal resource represented by magma pools and volcanoes is extremely large, it also presents an immense technological challenge. The high temperatures produce obvious problems with melting and deformation of equipment. The most promising candidates for heat extraction from magma are young volcanic calderas (less than a few million years old) that have magma relatively close to the surface. There have been some preliminary test wells drilled at the Long Valley caldera, located about 400 kilometers north of Los Angeles. The hope is that heat can be extracted in this area by drilling a well down to the magma level and pumping water into the well to solidify the magma at the bottom of the well. Then more water could be pumped into the well, become heated by contact with the solidified magma, and then returned to the surface to generate steam for electricity production.

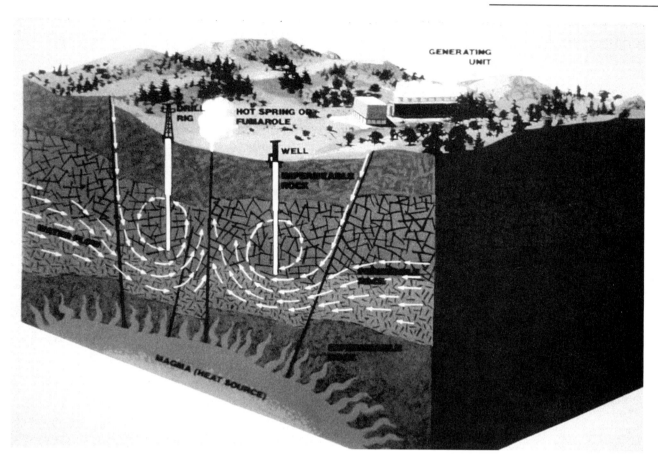

Cutaway drawing of the earth showing source of geothermal energy. (U.S. Department of Energy)

ELECTRICITY GENERATION

There are two general ways by which geothermal energy can be utilized: generation of electricity, and space heating. Production of electricity from a geothermal source was pioneered in an experimental program at the Larderello thermal field in Italy in 1904, and a 205-kilowatt generator began operation at this site in 1913. By 1998 the world's geothermal electrical generating capacity was about 8,240 megawatts, which represents only a small portion of the total electricity capacity of 2 million megawatts from all sources. However, the geothermal capacity has been growing steadily, and the 1998 amount represents a 40 percent increase above the 1990 value of 5,870 megawatts. The amount of electrical energy generated geothermally worldwide in 1998 was about 44 terawatt-hours, or 0.16 exajoule, representing approximately 0.4 percent of global electricity generation.

The United States is the world leader in geothermal electricity production, with about 2850 megawatts of capacity. As shown in Table 1, nine other countries each had more than 100 megawatts of electrical capacity in 1998, with this group being led by the Philippines, producing 1,850 megawatts of geothermal electrical power.

The world's most developed geothermal source is at the Geysers plant in California's Mayacamas Mountains, about ninety miles north of San Francisco. Electricity has been generated at this site since 1960, and as of 1999 the total installed generating capacity there was 1,224 megawatts. It has been demonstrated at the Geysers and at other geothermal electric plants that electricity from geothermal resources can be cost-competitive with other sources.

To produce electricity from a geothermal resource, wells are drilled into the reservoir, and as the hot, high-pressure water travels to Earth's surface, some of it vaporizes into steam as the pressure decreases. The hotter the original source, the greater the amount of dry steam produced. For dry-steam sources, the technology is basically the same as for electric plants that use the burning of fossil fuels to produce steam, except that the temperature and pressure of the geothermal steam are much lower. Dry-steam fields are being used in the United States, Italy, and Japan. In the Geysers geothermal area in California, the steam temperature is about 200°C (392°F) and the pressure about 700 kilopascals (7 atmospheres). Because geothermal steam is cooler than fossil-fuel steam, the efficiency of conversion of thermal energy to electricity is less than at a fossil-fuel plant; at the Geysers it is only 15 to 20 percent, compared to 40 percent for fossil fuels. The generating units are also smaller, ranging in size from 55 to 110 megawatts at the Geysers.

For geothermal reservoirs that are at lower temperatures, and hence produce less dry steam, working plants usually employ multiple-vaporization systems. The first vaporization ("flash") is conducted under some pressure, and the remaining pressurized hot water from the ground, along with the hot residual water from the turbine, can be flashed again to lower pressure, providing steam for a second turbine. Flashed power production is used in many countries, including the United States, the Philippines, Mexico, Italy, Japan, and New Zealand.

If the temperature of the original hot water is too low for effective flashing, the water can still be used to generate electricity in what is referred to as binary cycle (or organic cycle) electricity generation. The hot water is pumped to the surface under pressure to prevent evaporation, which would decrease the temperature, and its heat is transferred to an organic fluid such as isobutane, isopentane, Freon, or hexane, all of which have a boiling temperature lower than that of water. The fluid is vaporized by the heat from the water and acts as the working fluid in a turbine. It has been estimated that geothermal reservoirs with temperatures suitable for binary cycle generation are about fifty times as abundant as sources that provide pure dry steam.

Some geothermal power plants use a combination of flash and binary cycles to increase the efficiency of electricity production. An initial flash creates steam that drives a turbine; then the binary cycle is run, using either the hot water remaining after the initial flash or the hot exhaust from the turbine.

Geopressurized zones, discussed earlier, also are areas where electricity could be generated geother-

World Total	8240
U.S.A.	2850
Philippines	1850
Italy	770
Mexico	740
Indonesia	590
Japan	530
New Zealand	350
Iceland	140
Costa Rica	120
El Salvador	110
Nicaragua	70
Kenya	40
China	30
Turkey	20
Portugal (Azores)	10
Russia	10

Table 1.
Geothermal Electricity Generation Capacity in 1998 (in megawatts) Countries with fewer than 10 megawatts of capacity are not listed.

mally, not only from the hot water but also from the associated methane. Three geopressurized well sites in the United States have been developed experimentally for electricity production, but the cost of the electricity generated is considerably higher than that from conventional energy sources.

DIRECT GEOTHERMAL ENERGY USE

Geothermal sources at too low a temperature for electricity generation can be utilized for space heating, bathing, and other uses. Hot fluid from the reservoir is piped to the end-site location through insulated pipes to minimize loss of heat energy. Pumps are installed either in the geothermal well itself (if the temperature is low enough) or at the surface to drive the fluid through the piping system. To operate, these pumps require energy, usually supplied by electricity. The hot fluid can be used itself to provide the heating, or it can be pass through a heat exchanger where another working fluid is heated.

Direct geothermal energy is used for space heating of homes, greenhouses, livestock barns, and fish-farm ponds. As well, it is employed as a heat source in some industrial processes, such as paper production in New Zealand and drying diatomite in Iceland. Since the industrial applications usually require high-

	World	Japan	Iceland	China	U.S.A.
Space heating	33	21	77	17	10
Bathing, swimming, therapeutic use of baths	19	73	4	21	11
Greenhouses	14	2	4	7	5
Heat pumps (heating/cooling)	12	0	0	0	59
Fish farming	11	2	3	46	10
Industry	10	0	10	9	4
Snow melting	1	2	2	0	1

Table 2.
Types of Geothermal Direct Use Worldwide and in the Top Four Countries, 1995 (in %)

er temperatures than, say, space heating, it is advantageous to cascade the geothermal fluid, using it first at high temperature in industry and then afterward at lower temperature for another use.

Another way by which heat from the ground can be utilized is through the use of heat pumps. Pipes containing a fluid are buried in the ground, and heat can be extracted from the ground in the winter to heat a building, and dissipated in the ground in the summer to provide air conditioning. Essentially a heat pump acts like a refrigerator, extracting thermal energy from one area and moving it to another area. In the winter, the ground is cooled and the building is heated, and in the summer the heat pump runs in reverse, to cool the building and heat the ground.

An extensive summary of the various direct uses of geothermal energy has been compiled. Worldwide the most common use is for space heating (33%), followed by bathing, swimming, and therapeutic use of baths (19%). Table 2 shows the worldwide percentages for other uses, as well as percentages for the top four direct-use countries. As seen in this table, the specific uses of geothermal energy vary greatly from country to country.

The geothermal direct-use power capacity worldwide in 1997 was close to 10,000 megawatts, with China and the United States leading the way with 1910 megawatts each. Iceland and Japan also had more than 1,000 megawatts each, as shown in Table 3, and more than thirty countries in total were using geothermal heat. Total geothermal direct-use energy in 1997 was about 37 terawatt-hours, or 0.13 exajoule.

One of the countries that makes extensive use of direct geothermal heat is Iceland. This might seem surprising for a country with "ice" as part of its name, but Iceland lies in a region of continental-plate activity. Virtually all the homes and other buildings in Iceland are heated geothermally, and there is also a small geothermal electric plant. In Iceland's capital city, Reykjavik, geothermally heated water has been used for space heating since 1930, and the cost of this heating is less than half the cost if oil were used.

Other important examples of geothermal heating are in France, both near Paris and in the Southwest. During oil-exploration drilling in the 1950s, hot water was discovered in the Paris region, but exploitation did not begin until the 1970s as a result of rapidly increasing oil prices. In France, the equivalent of 200,000 homes are being provided with space heating and water heating from geothermal sources. One interesting feature of the French geothermal sources is that they do not occur in regions of elevated thermal gradient.

ENVIRONMENTAL EFFECTS OF GEOTHERMAL ENERGY

The most important potential environmental impacts of geothermal energy are water and air pollution. At the largest geothermal plants, thousands of tons of hot water and/or steam are extracted per hour, and these fluids contain a variety of dissolved pollutants. The hot geothermal water dissolves salts from surrounding rocks, and this salt produces severe corrosion and scale deposits in equipment. To prevent contamination of surface water, the geothermal brine must either be returned to its source or discarded carefully in another area. In addition to salts, the geothermal waters sometimes contain high concentrations of toxic elements such as arsenic, boron, lead,

Countries	
World total	**9960**
China	1910
U.S.A.	1910
Iceland	1440
Japan	1160
Hungary	750

Table 3.
Geothermal Direct-Use Capacity in 1997 (megawatts)—Top Eight
Countries and World Total

Located in Sonoma and Lake Counties in Calif., the geysers complex produces over 900,000 kW of electricity using steam from geothermal wells 7,000 to 10,000 feet below the surface. (U.S. Department of Energy)

and mercury. The 180-megawatt Wairakei geothermal electrical plant in New Zealand dumps arsenic and mercury into a neighbouring river at a rate four times as high as the rate from natural sources nearby.

Geothermal water often contains dissolved gases as well as salts and toxic elements. One of the gases that is often found in association with geothermal water and steam is hydrogen sulfide, which has an unpleasant odor—like rotten eggs—and is toxic in high concentrations. At the Geysers plant in California, attempts have been made to capture the hydrogen sulfide chemically, but this job has proven to be surprisingly difficult; a working hydrogen sulfide extractor is now in place, but it was expensive to develop and install, and its useful life under demanding operating conditions is questionable. Geothermal brines also are a source of the greenhouse gas carbon dioxide, which contributes to global warming. However, a typical geothermal electrical plant produces only about 5 percent of the carbon dioxide emitted by a fossil-fuel-fired plant generating the same amount of electricity. Many modern geothermal plants can capture the carbon dioxide (as well as the hydrogen sulfide) and reinject it into the geothermal source along with the used geothermal fluids. At these facilities, the carbon dioxide that escapes into the atmosphere is less than 0.1 percent of the emissions from a coal- or oil-fired plant of the same capacity.

Another potential problem, particularly if geothermal water is not returned to its source, is land subsidence. For example, there has been significant subsidence at the Wanaker field in New Zealand. Finally, an annoying difficulty with geothermal heat has been the noise produced by escaping steam and water. The shriek of the high-pressure fluids is intolerable, and is usually dissipated in towers in which the fluids are forced to swirl around and lose their kinetic energy to friction. However, the towers provide only partial relief, and the plants are still noisy.

GEOTHERMAL ENERGY—RENEWABLE OR NOT?

Most people tend to think of geothermal energy as being renewable, but in fact one of the major problems in choosing a geothermal energy site lies in estimating how long the energy can usefully be extracted. If heat is withdrawn from a geothermal source too rapidly for natural replenishment, then the temperature and pressure can drop so low that the source becomes unproductive. The Geysers plants have not been working at full capacity because the useful steam would be depleted too quickly. Since it is expensive to drill geothermal wells and construct power plants, a source should produce energy for at least thirty years to be an economically sound venture, and it is not an easy task to estimate the working lifetime beforehand.

THE FUTURE OF GEOTHERMAL ENERGY

The main advantage of geothermal energy is that it can be exploited easily and inexpensively in regions where it is abundantly available in hydrothermal reservoirs, whether it is used for electricity production or for direct-use heat. Geothermally produced electricity from dry-steam sources is very cheap, second only to hydroelectric power in cost. Electricity from liquid-dominated hydrothermal sources is cost-competitive with other types of electrical generation at only a few sites. It is unlikely that other types of geothermal sources, such as hot, dry rock and the normal geothermal gradient, will soon become economical. Hence, geothermal energy will provide only a small fraction of the world's energy in the foreseeable future.

An important feature of geothermal energy is that it has to be used locally, because steam or hot water cannot be piped great distances without excessive energy loss. Even if electricity is generated, losses also are incurred in its transmission over long distances. As a result, geothermal energy use has geographical limitations.

The future development of geothermal energy resources will depend on a number of factors, such as cost relative to other energy sources, environmental concerns, and government funding for energy replacements for fossil fuels. An important concern that many members of the public have about future developments is centered around whether resources such as hot springs and geysers should be exploited at all. Many of these sources—such as those in Yellowstone National Park in Wyoming—are unique natural phenomena that many people feel are important to protect for future generations.

Ernest L. McFarland

See also: Biofuels; Diesel Fuel; District Heating and Cooling; Economically Efficient Energy Choices; Electric Power, Generation of; Environmental Problems and Energy Use; Fossil Fuels; Heat and Heating; Heat Pumps; Heat Transfer; Hydroelectric Energy; Reserves and Resources; Seismic Energy; Thermal Energy; Thermal Energy, Historical Evolution of the Use of; Thermodynamics; Water Heating.

BIBLIOGRAPHY

Atomic Energy of Canada Limited. (1999). *Nuclear Sector Focus: A Summary of Energy, Electricity, and Nuclear Data*. Mississauga, Ont.: Author.

Fridleifsson, I. B. (1998). "Direct Use of Geothermal Energy Around the World." *Geo-Heat Center Quarterly Bulletin* 19(4):4–9.

Golub, R., and Brus, E. (1994). *The Almanac of Renewable Energy: The Complete Guide to Emerging Energy Technologies*. New York: Henry Holt.

Howes, R. (1991). "Geothermal Energy." In *The Energy Sourcebook*, ed. R. Howes and A. Fainberg. New York: American Institute of Physics.

Lund, J. W. (1996). "Lectures on Direct Utilization of Geothermal Energy." United Nations University Geothermal Training Programme Report 1996-1. Reykjavik, Iceland.

McFarland, E. L.; Hunt, J. L.; and Campbell, J. L. (1997). *Energy, Physics and the Environment*, 2nd ed. Guelph, Ont.: University of Guelph.

GIBBS, JOSIAH WILLARD (1839–1903)

Gibbs came from an academic family in New Haven, Connecticut. His father was a noted philologist, a graduate of Yale and professor of sacred literature there from 1826 until his death in 1861. The younger Gibbs grew up in New Haven and graduated from Yale College, having won a number of prizes in both Latin and mathematics. He continued at Yale as a student of engineering in the new graduate school and, in 1863, received one of the first Ph.D. degrees granted in the United States. After serving as a tutor in Yale College for three years, giving elementary instruction in Latin and physics, Gibbs left New Haven for further study in Europe. He spent a year each at the universities of Paris, Berlin, and Heidelberg, attending lectures in mathematics and physics and reading widely in both fields. He was never a student of any of the luminaries whose lectures he attended (the list includes Liouville and Kronecker in mathematics, and Kirchhoff and Helmholtz in physics) but these European studies, rather than his earlier engineering education, provided the foundation for his subsequent scientific work. A qualification was a life-long fondness for geometrical reasoning, evident in Gibbs's scientific writings, but first developed in his dissertation.

Gibbs returned to New Haven in 1869. He never again left America and seldom left New Haven except for annual summer holidays in northern New England and a very occasional journey to lecture or attend a meeting. Gibbs never married and lived all his life in the house in which he had grown up, less than a block from the college buildings. In 1871, two years before he published his first scientific paper, Gibbs was appointed professor of mathematical physics at Yale. He held that position without salary

Josiah Willard Gibbs. (Library of Congress)

In his first work on thermodynamics in 1873, Gibbs immediately combined the differential forms of the first and second laws of thermodynamics for the reversible processes of a system to obtain a single "fundamental equation":

$$dU = TdS - pdV$$

an expression containing only the state variables of the system in which U, T, S, p and V are the internal energy, temperature, entropy, pressure, and volume, respectively. Noteworthy here is the assumption, which Gibbs made at the outset but which was not common at the time, that entropy is an essential thermodynamic concept. At the same time, the importance of energy was also emphasized. As Gibbs wrote at the beginning of his great memoir, "On the Equilibrium of Heterogeneous Substances," whose first installment appeared in 1876: "The comprehension of the laws which govern any material system is greatly facilitated by considering the energy and entropy of the system in the various states of which it is capable." The reason, as he then went on to explain, is that these properties allow one to understand the interactions of a system with its surroundings and its conditions of equilibrium.

As was usual with him, Gibbs sought to resolve the problem in general terms before proceeding to applications. Again beginning with the differential forms of the first two laws (which, in effect, define the state functions U and S), but this time for any process, whether reversible or irreversible, he combined the two expressions to yield the general condition of equilibrium for any virtual change:

$$\delta U - T\delta S - \delta W = 0$$

where W is the external work. If a system is isolated, so that $\delta W = 0$, this condition becomes:

$$(\delta U)_s \geq 0 \text{ or } (\delta S)_U \leq 0$$

for constant S and U, respectively. The second inequality, which Gibbs showed to be equivalent to the first, immediately indicates that thermodynamic equilibrium is a natural generalization of mechanical equilibrium, both being characterized by minimum energy under appropriate conditions.

The first and probably most significant application of this approach was to the problem of chemical equilibri-

for nine years, living on inherited income. It was during this time that he wrote the memoirs on thermodynamics that, in most estimates, constitute his greatest contribution to science. Gibbs declined the offer of a paid appointment at Bowdoin College in 1873, but he was tempted to leave Yale in 1880 when he was invited to join the faculty of the newly-founded Johns Hopkins University in Baltimore. Only then did Yale provide Gibbs a salary, as tangible evidence of the high regard his colleagues had for him and of his importance to the University. Gibbs remained at Yale and continued to teach there until his death, after a brief illness, in 1903.

Gibbs worked on electromagnetism during the 1880s, concentrating on optics and particularly on James Clerk Maxwell's electromagnetic theory of light, and on statistical mechanics from at least the mid-1880s until his death. The latter research resulted in his seminal *Elementary Principles in Statistical Mechanics*, published in 1902. However, it seems more appropriate in this place to briefly describe Gibbs's memoirs on thermodynamics.

um. In a heterogeneous system composed of several homogeneous phases, the basic equilibrium condition leads to the requirement that temperature, pressure, and the chemical potential (a new concept introduced by Gibbs) of each independent chemical component have the same values throughout the system. From these general conditions, Gibbs derived the phase rule:

$$\delta = n+2-r,$$

that cornerstone of physical chemistry, which specifies the number of independent variations δ in a system of r different coexistent phases having n independent chemical components.

Among many other valuable results in his memoir on heterogeneous equilibrium is a formulation of the Gibbs free energy, also called the Gibbs function, which is defined by the equation:

$$G = H-TS,$$

where H is the enthalpy, that is, the sum of the internal energy of a body or system and the product of its pressure and volume. It is useful for specifying the conditions of chemical equilibrium at constant temperature and pressure, when G is a minimum. More generally, Gibbs's memoir greatly extended the domain covered by thermodynamics, including chemical, electric, surface, electromagnetic, and electrochemical phenomena into one integrated system.

Gibbs's thermodynamic writings were not as widely read—much less appreciated—as they deserved to be in the decade following their appearance. One reason is that they were published in the obscure *Transactions of the Connecticut Academy of Sciences*. Gibbs sent offprints of his memoirs to many scientists, but only Maxwell seems to have recognized their importance. That changed after Wilhelm Ostwald published a German translation in 1892. In the meantime, continental scientists such as Helmholtz and Planck independently developed Gibbs's methods and results, unaware of his prior work.

Another reason for lack of interest is the severity of Gibbs's style. Austere and logically demanding, it was a challenge even for mathematicians as distinguished as Poincaré. The same severity extended to Gibbs's lectures, which his few students found clear and well-organized, but not easy to understand, owing to their great generality and meticulous precision.

A third reason is that Gibbs made no effort to promote or popularize his results. He seems to have been a solitary, self-contained, and self-sufficient thinker, confident in his ability, who worked at his own unhurried pace, neither needing nor wanting feedback from others. An attitude of detachment from the work of his students plus his own solitary habit of work is undoubtedly responsible for the fact that Gibbs founded no "school" or group of students to develop his ideas and exploit his discoveries.

Robert J. Deltete

BIBLIOGRAPHY

Deltete, R. J. (1996). "Josiah Willard Gibbs and Wilhelm Ostwald: A Contrast in Scientifc Style." *Journal of Chemical Education* 73(4):289–295.

Klein, M. J. (1972). "Gibbs, Josiah Willard," in *Dictionary of Scientific Biography*, Vol. 5, ed. C. G. Gillispie. New York: Scribner.

Klein, M. J. (1989). "The Physics of J. Willard Gibbs in His Time." In *Proceedings of the Gibbs Symposium. Yale University, May 15–17, 1989*, ed. D. G. Caldi and G. D. Mostow. New York: American Mathematical Society.

GLASS TECHNOLOGY

See: Windows

GLOBAL WARMING

See: Climatic Effects

GOVERNMENT AGENCIES

Energy is the economic lifeblood of all economies. It is an essential gear for the economies of the developed world, and of ever growing importance to developing nations. It is of concern to the largest of nations as well as the smallest. And it is as great a concern of local governments as national ones.

Government action or inaction at all levels in energy policy has varied tremendously. Prior to 1930, involvement in energy issues at all levels of government was minimal in the United States because of relatively moderate demand for energy consuming technology, and an abundant and relatively cheap supply of fossil fuels. The government's approach to energy changed with the Depression and the New Deal. The United States began to subsidize energy by building hydroelectric stations (the Tennessee Valley Authority), supporting rural electrification, and subsidizing of nuclear research. Much of the developing world, as well as the Soviet Union, began subsidizing energy early in the twentieth century as a means to buy votes and strengthen political support. Because of World War II, the strategic role of energy to economic growth and national security became much more apparent. Almost all governments began to take a more active role in energy markets as world energy consumption accelerated, as spending was boosted on all aspects of the energy puzzle—exploration, production, distribution, consumption, and the energy statistics to track all of the above.

Traditionally, the developed world has paid the greatest attention to petroleum (reserves, resources, security, price), and devoted the most resources to nuclear energy research (both fission and fusion), a source that proponents in the 1950s believed would someday turn out to be "too cheap to meter." Planning, funding, and development of renewables and other energy resources has fluctuated much more wildly through the years, usually inversely to the real and perceived availability of oil. During and following an oil crisis, where there is great supply uncertainty and prices skyrocket, planning, funding and development grow, and correspondingly decrease when oil supplies are more secure and stable.

WORLD AGENCIES

The pooling of resources to meet shared objectives have been the factor most responsible for the establishment and growth of international agencies concerned with energy issues. The three most important issues bringing about cooperation have been the need to coordinate production, the pooling of resources for research and development of energy technology, and the coordinating of activities to secure supply.

Organization of Petroleum Exporting Countries (OPEC)

The OPEC cartel was founded by Iraq, Iran, Saudia Arabia, Kuwait, and Venezuela in September 1960 as a way to coordinate petroleum production and pricing among member countries. It was not until the 1970s that the cartel tried to become an effective monopoly.

As of 2000, membership has expanded to thirteen, accounting for over 60 percent of all production. The reserves controlled by Member Countries were much higher in the 1970s, and consequently so was the cartel's monopoly power in controlling prices. But because of exploration discoveries, advances in technology (enhanced recovery of existing wells, improved offshore equipment), and the lack of cooperation from non-OPEC producers, the power of the cartel to control oil production and prices has steadily diminished. During the 1970s, an announcement of an OPEC meeting would be the major news story of the day, but by the 1990s, low oil prices and a secure supply caused the media to only superficially cover OPEC meetings. As most economists predicted, the ability of the cartel to control the price of petroleum did not last long. There will always be an incentive for member states—especially the smaller producers—to "cheat" on their quotas, and the free market will always react swiftly to higher prices by pumping up other sources of oil production and securing other energy resources.

As OPEC's share of the world oil supply market continued to fall in the 1990s, they began taking steps to better coordinate production with non-OPEC producers such as Mexico and other members of the Independent Petroleum Exporting Countries (IPEC). By exchanging information, and undertaking joint studies of issues of common interest, the hope was to stabilize prices and improve the economic outlook for all oil producers. This collaboration between OPEC and major non-OPEC producers helped raise oil prices to over $27 a barrel in 1999 from a low of less than $13 in 1998.

International Energy Agency (IEA)

In response to the Arab embargo, and the attempt of OPEC to dictate the supply and price of petroleum, the Organization for Economic Cooperation and Development (OECD) established the International Energy Agency (IEA) in November 1974. IEA membership included all of Western Europe, Canada, the

United States, Australia, New Zealand, and Japan. Besides compiling useful consumption and production statistics, its mandate was for cooperation and coordination of activities to secure an oil supply in times of supply disruption. IEA's first action to insulate member countries from the effects of supply disruptions was the Emergency Sharing System (ESS). This system was to be put into effect only in cases of serious disruptions, an actual or anticipated loss of 7 percent of expected supply. The System consisted of three parts: a building up of supplies, a reduction of consumption during periods of short supply, and a complex sharing system that would attempt to distribute the loss equitably. During the 1979–1981 oil supply disruption, the system was never really tested because the loss of supplies never reached a level to trigger the ESS. Nevertheless, because of the economic hardship felt by many nations during the 1979–1981 disruption, in 1984 the Coordinated Emergency Response Measures (CERM) was adopted. The CERM were intended as a means to reach rapid agreement on oil stockpile drawdown and demand restraint during oil supply disruptions of less than 7 percent. Because of ESS and CERM, and the agreement of other OPEC Member States to boost production, the supply disruption caused by the Iraqi occupation of Kuwait, and the subsequent United Nations embargo of all oil exports from Iraq and Kuwait, did not have such a large effect.

Twenty-five years after its establishment, the focus of the IEA has changed and expanded significantly. Whereas in 1974, the IEA looked at coal and nuclear energy as the two most promising alternatives to oil, the call for more environmentally acceptable energy sources has pushed to the forefront the development of renewable energy, other nonfossil fuel resources, and the more clean and efficient use of fossil fuels. To avoid duplication of effort and better research and development results, the IEA coordinates cooperation among members to more efficiently use resources, equipment, and research personnel.

United Nations (UN)

The United Nations officially came into existence at the end of World War II on October 24, 1945 to help stabilize international relations and better secure peace. Through the years, the UN has greatly expanded its mission to include other issues such as development and protecting the environment. Because energy is such a key piece of the develop-

ment puzzle, in 1947 the UN established a statistics division that began tracking energy consumption and production throughout the world in 1992.

The divisions established to take an active role in energy and development issues were the United Nations Development Program (UNDP) and the Commission on Sustainable Development in 1993. Funded by voluntary contributions from member states, with funding directed toward countries with annual per-capita GNP of $750 or less, UNDP focuses on preventing unsustainable production and consumption patterns, and ways of curbing pollution and slowing the rate of resource degradation—economic growth with environmental protection and conservation. The burden of environmental protection is far greater for poorer nations because it diverts resources away from more pressing problems such as poverty, low levels of social development, inadequate energy infrastructure, and a lack of capital for infrastructure. Despite emissions of industrial countries dropping from 1980 to 2000, many emissions, most notably toxic substances, greenhouse gases, and waste volumes are continuing to increase in the developing world.

Global warming is another energy-related problem that the UN has attempted to address through its Intergovernmental Panel of Climate Change (IPCC). The IPCC was very successful in negotiating an agreement to solve the problem of chlorofluorocarbons (January 1987 Montreal Protocol on Substances That Deplete the Ozone Layer), yet the task of getting member countries to agree on carbon emission reductions to combat the perceived threat of global warming has proven to be a much more daunting task. The science is more uncertain, there are no easy solutions (the combustion that provides the majority of the world's energy needs always entails carbon emissions), and to achieve the cuts that climate modelers feel would be necessary to fully address the problem will require a complete restructuring of current civilization. Nevertheless, based on the work of the IPCC, the United Nations Framework Convention on Climate Change adopted the Kyoto Protocol on December 11, 1997, as a first step in the process. The Kyoto Protocol calls for most of the developed nations to reduce carbon emissions by 10 percent of 1990 levels by the year 2010. Most nations ratified the treaty despite no reduction commitments from the developing nations. The poorer nations strongly objected to universal emission reductions. They countered that energy drives development, and the exploitation of cheap and abundant energy, often at the expense of the poorer nations, is how the richer nations achieved their higher standard of living. Burdensome environmental regulations of energy would severely retard development; thus, the poorer nations feel it is only equitable that the richer nations, who created the majority of carbon emissions in the last 150 years, first make the effort to curtail emissions.

The Kyoto negotiations show the tremendous friction energy issues can cause between the richest and poorest countries in the world. Addressing future energy-related conflicts—whether it be wars over oil fields or disputes over carbon emission quotas—will remain a major role of the UN.

The International Atomic Energy Agency (IAEA)

To address the technology, waste, safety and security issues concerning nuclear energy, the UN established the International Atomic Energy Agency (IAEA) in 1957, a few years after U.S. President Dwight D. Eisenhower's famous "Atoms for Peace" speech before the United Nations General Assembly.

As the political, economic, and technological realities of the world evolved, so did IAEA. The scope of products, services, and programs of the IAEA has expanded well beyond its original function as the world's central intergovernmental forum for scientific and technical nuclear cooperation, and the watchdog for civilian nuclear power programs. In the 1990s, the IAEA helped countries carry out comparative cost effective assessments of how to expand electric power generation capacity such as the potential roles of renewable energy, and the different options and costs for reducing atmospheric and greenhouse gas emissions. They also conduct research and provide outreach through the Agency's laboratories. The IAEA tries to go beyond technology transfer and energy capacity building to help provide solutions to problems of sustainable human development.

International Institute for Applied Systems Analysis

Founded in 1972 by the United States, the Soviet Union, and ten other countries, the International Institute for Applied Systems Analysis is a research organization located in Laxenburg, Austria that conducts scientific studies in many energy-related areas of global consequences such as transboundary air pollution, sustainable forest resources, climate change, and environmentally compatable energy strategies.

World Bank

Established in 1946, with a subscribed capital fund of $7.67 million, the primary mission of the World Bank is to combat poverty by securing low cost funding for sustainable development. The largest shareholder is the U.S., followed by the United Kingdom, Japan, Germany, and France.

Although the World Bank provides loans for a variety of purposes, energy-related infrastructure receives over 7 percent, which primarily goes toward electric power infrastructure (hydroelectric, fossil fuel, nuclear) but also oil and natural gas exploration, production and distribution. Other energy-related areas making up a considerable share of the World Bank's loan portfolio are the transportation and agricultural sector. Loans to improve the energy and logistic efficiency of the transportation sector account for 11 percent of funding, followed by 10 percent going toward agriculture to provide assistance in expanding the amount of food energy produced.

By 1999 the World Bank loan portfolio for the energy and mining sectors had grown to $4.1 billion. Its strategy for energy development is to reform and restructure markets to attract private investment to build energy infrastructure, and expand energy access to the rural and low-income populations, and promote an environmentally responsible energy supply and use. One priority is to get more of the 2 billion plus poor people access to modern energy for cooking and lighting. Without modern energy, the world's poor rely on wood, crop residues, and other biofuels that result in environmentally damaging deforestation. Deforestation is of interest to all nations because it is not only a habitat destroyer, but also contributes to the increasing concentrations of carbon dioxide, a gas long suspected of creating global warming.

The World Bank grants financing for fossil fuel electricity generation, yet finances only facilities that have advanced emission control equipment. And although the World Bank has never financed a nuclear power plant, a zero carbon emitter, it is very active in evaluating hydropower projects, helping to establish the World Commission on Large Dams.

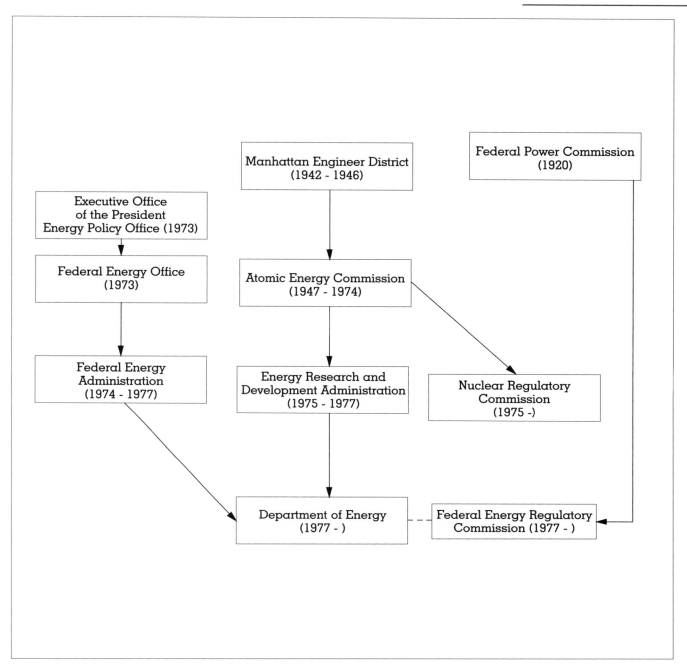

Figure 1.
Precursors and agencies related to the U.S. Department of Energy.

The World Bank also looks for energy efficiency improvement opportunities by pushing for the elimination of fossil fuel subsidies (estimated at over $200 billion a year in the developing world), improving demand-side efficiency, and making the energy supply system more cost conscious.

UNITED STATES AGENCIES

The focus of U.S. energy policy is with the Department of Energy, but because energy issues touch almost every sector of the economy, the Department of Commerce, the Department of Transportation, Department of Agriculture,

Department of Defense, and the Environmental Protection Agency also devote a considerable amount of resources toward energy issues.

The Department of Energy (DOE)

The Department of Energy was established on June 3, 1977, unifying offices, laboratories, and staffs from other federal agencies (see Figure 1). Besides replacing and taking on all the responsibilities of the Federal Energy Administration, the Energy Research and Development Administration, and the Atomic Energy Commission, limited functions were transferred from the Departments of Agriculture, Commerce, Housing and Urban Development, and Transportation. The newly formed agency had around 20,000 employees and a budget of $10.4 billion.

Prior to the energy crisis of 1973, the federal government took a limited role in formulating a national energy policy. Markets operated freely, long-range planning was left up to the private sector, and oversight controlled by state, local, and regional authorities. Americans felt comfortable with private industry controlling production, distribution, marketing, and pricing. But in the case of natural monopolies, such as in the interstate transportation of natural gas and electricity, it was generally acknowledged that the Federal Power Commission (established in 1920) was necessary to ensure fair prices.

The regulatory agency that the DOE established for oversight of natural monopolies was the Federal Energy Regulatory Commission (FERC). The five-member commission was set up to control the regulation and licensing of hydroelectric power facilities, the regulation of the transmission and sale of electric power, the transportation and sale of natural gas, and the operation of natural gas and oil pipelines.

After the Oil Embargo, the Economic Regulatory Administration within DOE administered oil pricing, energy import programs, the importing and exporting of natural gas, programs to curtail natural gas consumption, and to supervise the conversion of electric power production from natural gas and oil to coal. These command and control programs continued through the Carter Administration, but ended early in the Reagan Administration, a strong free market advocate.

To provide long-term energy trends to the Department, President, Congress and the public, the DOE set up a statistics division called the Energy Information Administration. The federal government had been gathering and publishing energy statistics since 1949, but under the centralized control of the Energy Information Administration, the scope of data gathering expanded and continued to expand through the 1990s to encompass related areas of interest like carbon emissions.

During the Carter Administration, funding increased to pay for the greater focus on conserving energy, greater oil production including increasing the production of oil, natural gas and synthetic fuel from coal and shale oil reserves, and speeding the development and implementation of solar power. The Carter Administration took a very activist approach to energy policy, and believed that the United States could be getting 20 percent of its energy from the sun or other renewable energy sources by the year 2000. Toward this end, the Carter Administration pushed through Congress very generous subsidies for residential and commercial solar and wind installations.

The Reagan Administration (1981–1989) came in with a completely different vision of the federal role in the energy field. Reagan wanted to abolish the DOE, but with that being politically impossible, he set about restructuring it, letting private industry and the free marketplace set energy priorities. By ending government regulations and price controls, which were detrimental to domestic oil and natural gas production, he felt a free marketplace would prevent or limit the impact of future energy crises. And although Reagan reduced and eliminated subsidies for energy conservation and energy technologies like solar—preferring private capital to demonstrate commercial viability of technology—he continued to support long-term energy research and development, such as fusion, that private industry felt too risky to undertake. Despite these cutbacks, the DOE budget grew during the Reagan era mainly because of the emphasis on defense-related spending and the Strategic Defense Initiative designed to stop incoming nuclear warhead missiles.

The Bush Administration (1989–1993) had a similar free marketplace philosophy as Reagan, but faced the daunting task of having to start directing billions toward cleaning up after forty years of neglect at the contaminated weapons complex, particularly the federal facilities at Savannah River South Carolina, Hanford Washington, and Rocky Flats Colorado. The cleanup plan was fourfold: characterize and prioritize all waste cleanups at departmental sites, confine and correct immediate problems, establish long-term cleanup plans, and mandate compliance with all applicable laws. By the end of the Bush Administration, environmental management of defense-related nuclear waste consumed nearly a third of the budget, which reflected the dauntingly difficult and expensive nature of the cleanup.

Perhaps the most noteworthy energy-related accomplishment of the Bush Administration was the handling of the Persian Gulf Crisis of 1970 and the January 15, 1991 Operation Desert Storm—a military action that many categorized as an energy war. Because Iraqi and Kuwaiti production constituted around 4.3 million barrels per day, or 9 percent of the world total, the role of the DOE during the crisis was to reassure the public and press about oil issues, improve energy coordination with other countries primarily through the IEA, and promote energy conservation and increase energy production. Rapidly rising spot market prices for crude oil as well as gasoline prices did occur, but because of these efforts, along with the short duration of the crisis, the impact on world economies was minimized.

The Clinton Administration (1993–2000) that followed was far more inclined to embrace environmental activism than Reagan or Bush, and far more likely to propose command and control solutions to energy and environmental problems. However, the Clinton Administration also realized the need to allow markets to work, to do otherwise would result in some of the disastrous consequences of intervention policies used in the 1970s.

To pay for boosted spending for conservation grants, energy research and development, and spending on defense waste management, the Administration dramatically cut defense research and development. Many energy policy decisions of the Administration were directly linked to the quality and health of the environment. And unlike the Reagan and Bush Administrations that remained skeptical of global warming, the Clinton Administration made action on the global warming threat a priority. On the first Earth Day of the Administration, Clinton promised to stabilize greenhouse gas emissions at 1990 levels by the year 2000, an extremely ambitious goal. The method proposed was the Btu tax, a 25.7 cents per million Btus tax (4 %) on all forms of energy except solar, geothermal, and wind, and an additional 34.2 cents per million Btus tax for gasoline (4 cents a gallon) and other refined petroleum products. Although projected to

reduce emissions by 25 million metric tons, the proposal was rejected by Congress primarily because it was widely viewed as a tax increase to finance increased social program spending. In December 1997, the Clinton Administration signed the Kyoto Protocol calling for dramatic carbon emission reductions, but never sent it to the Senate for ratification because of certain defeat. Almost all Senators felt there should be future carbon emission targets for the developing world, and the inequity of the cuts would give an unfair competitive advantage to the developing economies of the world, such as China, that did not agree to emission reductions.

By the end of the Clinton Administration in 2000, the budget for the agency had nearly doubled, and the number of employees exceeded 170,000, a figure that includes all those employed at DOE's national laboratories, cleanup sites, and other facilities. Despite organizational reshuffling and shifts in funding, much of the DOE mission—energy security, developing new energy sources, and collecting statistics about energy production and consumption—remains the same. The 1997 Budget allocated 38 percent toward nuclear security, followed by nuclear cleanup at 36 percent, basic science 15 percent, and energy subsidies 11 percent. Many critics feel the $6 billion spent annually for nuclear cleanup of DOE facilities is extremely wasteful. Instead of trying to return these sites to pristine conditions, critic feel DOE should renegotiate the cleanup plan with the EPA, and state and local authorities so that the emphasis can be shifted toward containment and neutralization of waste.

As the DOE moves into the twenty-first century, the Department faces significant challenges in defining a mission to warrant its generous funding. As long as fossil fuel supplies remain cheap and abundant, and the public remains skeptical of global warming, there will remain a desire to cut funding.

The Environmental Protection Agency (EPA)

To consolidate the environmental protection activities of the federal government, the Environmental Protection Agency was formed on December 2, 1970 from three federal Departments, three Administrations, three Bureaus, and several other offices. The mission of EPA was to collect data on pollution, conduct research on the adverse effects of pollution, and through grants and technical assistance, develop methods for controlling it.

Much of the EPA's regulatory activity centers around the energy business, and in many ways, its actions have a greater impact on altering production and consumption patterns in the energy sector than the DOE's actions. Oil drilling, production, transportation (pipelines and tankers), and refining are all heavily regulated segments of the energy sector. Refining, in particular, is subjected to periodic EPA audits for hazardous waste by-products and airborne emissions released during the refining process. The stringency of EPA refinery regulations, and the uncertainty about future regulations, partly explains why the industry has not built any new major refineries in the United States since the mid-1970s.

Power plant and auto emissions are two other energy areas touched by EPA regulations. These are regulated through the Clean Air Act of 1970, and the amendments in 1977 and 1990 that set guidelines for acceptable emission levels. The Clean Air Act has been widely praised for improving air quality, significantly lowering emissions of sulfur dioxide, nitrous oxide, and other particulate matter. However, most economists feel that the benefits from the EPA administered safety investments (including nonenergy aspects) are far less cost effective than the safety investments of other regulatory agencies (see Table 1).

The 1990 amendments, which gave the Agency broad new power to revise Clean Air standards if EPA felt the results of studies warranted changes, has been particularly troublesome for the energy sector. Much of the EPA's air standard actions taken during the 1990s came under intense criticism for not being warranted by science. In particular, the EPA severely toughened new Clean Air particulate matter standards for power plant emissions in 1996, claiming it would save 15,000 lives a year, and reduce hospital admissions and respiratory illness. This was followed, in 1999, by a new lower sulfur content gasoline standard for refiners (effective in 2004) that will add 2 to 6 cents to the price of a gallon of gasoline. The EPA claims an additional 2,400 deaths will be prevented every year from this new standard. It was impossible to dispute the merits of either new standard because the EPA prevented public review of the data upon which these regulations are based (the "Pope" study) as required by the Freedom of Information Act (taxpayer-funded scientific data used to support federal regulations must be made available). Without access

Regulatory Agency	Median Cost/Life-Year Saved
Federal Aviation Administration	$23,000
Consumer Product Safety Commission	$68,000
National Highway Traffic Safety Administration	$78,000
Occupational Safety and Health Administration	$88,000
Environmental Protection Agency	$7,629,000

Table 1.
Median Value of Cost/Life-Year Saved For Five Regulatory Agencies
SOURCE: Tengs, 1995

to the data, there can be no confirmation that the new standards will save any lives.

On the nuclear energy front, the EPA was instrumental in the late 1980s for bringing DOE nuclear facilities into compliance with the Nuclear Waste Policy Act of 1982 and 1987 (high-level radioactive wastes), the Low-Level Radioactive Waste Policy Act of 1980 and 1985, the Uranium Mill Tailings Radiation Control Act of 1978, and the Superfund statute. The most egregious sites needing cleanup are the nuclear weapons production sites located at Savannah River South Carolina, Rocky Flats Colorado, and Hanford Washington, that are all going to take decades to cleanup and cost billions of dollars. The EPA action at these DOE sites was touted as a sign that the federal government can police itself, and that the environmental laws that apply in the private sector apply equally to the public sector.

To combat the dangers of global warming by reducing carbon emissions, the Clinton Administration formed the Climate Protection Division (CPD), formerly the Atmospheric Pollution Prevention Division. This Division has been directed to find nonregulatory ways to reduce greenhouse gases through energy-efficiency improvements in all sectors of the economy. In collaboration with DOE, the EPA established the Energy Star Labeling Program as a way to develop voluntary energy-efficiency specifications for products such as office equipment, heating and cooling equipment, residential appliances, and computers. This Program allows manufacturers to prominently place an Energy Star label on qualifying products. The hope is for consumers to learn to recognize the label as the symbol for energy-efficiency, and become accustomed to buying only energy efficient products with the Energy Star label.

The Department of Commerce (DOC)

The commercial and industrial sectors account for more than fifty percent of all energy consumption in the United States. The Department of Commerce (DOC), whose mission is to improve the overall competitiveness of the commercial and industrial sector, is very concerned with reducing consumption through conservation and energy efficiency, and in improving the competitiveness of the energy businesses on the production issues.

On the energy use side, the DOC helps promote the DOE and the EPA energy efficiency programs (Energy Star) for the industrial and commercial sectors of the economy. To aid the energy businesses on the production side, the International Trade Administration and the Office of Energy, Infrastructure and Machinery assist all the energy fuel industries in improving their market competiveness and ability to participate in international trade. One of the more aggressive actions of the Department under the Clinton Administration was to sponsor trade missions to promote the export of U.S. electric power production technology.

Another major function of the DOC is the management of energy-related research through the National Institute of Standards and Technologies Laboratories in Gaithersburg, Maryland and Boulder, Colorado and the research laboratories of the National Oceanic and Atmospheric Administration.

The National Institute of Standards and Technology (NIST), formerly the National Bureau of Standards (NBS), was established by Congress in 1901. Its mission is to assist industry in the development of technology needed to improve product quality, to modernize manufacturing processes, to ensureproduct reliability, and to facilitate rapid commercialization of products based on new scientific discoveries. For example, it provides measurements that support the nations' standards for lighting and electric power usage. In 1988, NIST added three major new programs to its measurement andstandards laboratories: The Advanced Technology Program (ATP) which in partnership with industry accelerates innovative, enabling technologies with strong potential to deliver large payoff for the nation; the Manufacturing Extension Partnership (MEP), a

network of local centers offering technical and business assistance to smaller manufacturers; and the Malcomb Baldridge National Quality Award that recognizes business performance excellence. Programs the ATP is funding in the energy arena include the development of rapid thermal processing to produce low-cost solar cells with Solarex (a business unit of Amoco/Enron Solar), and ultrathin silicon ribbon for high-efficiency solar cells with Evergreen Solar, Inc.

The Office of Oceanic and Atmospheric Research (OAR) is the division of NOAA that conducts and directs oceanic and atmospheric research. Since carbon dioxide is a greenhouse gas and fossil fuels are the leading generator of carbon dioxide, the work of the twelve Environmental Research Laboratories and eleven Joint Institutes of OAR to describe, monitor, and assess climate trends are of great interest to all parties interested in the affect of energy use on climate change.

The Department of the Interior

The Department of the Interior is the home of the Bureau of Land Management, the Minerals Management Service, the Office of Surface Mining Reclamation and Enforcement, and the U.S Geological Survey (USGS).

The Bureau of Land Management is responsible for the leasing of land for coal, oil, and natural gas exploration and production, and the Minerals Management Service does likewise for offshore leasing. Both divisions rely on resource evaluation information provided by the USGS in negotiating leases. The main responsibilities are the inspection and enforcement of leases, and the collection of royalty payments and other revenues due the Federal Government from the leasing and extraction of energy resources. Besides providing the statistics for the other Department of the Interior divisions, the USGS compiles databases and geologic maps of energy reserves and resources.

The mining industry has taken great steps in limiting environmental problems in developing energy resources, yet much of the mining industry prior to the 1980s abandoned sites leaving huge environmental problems. The Office of Surface Mining Reclamation and Enforcement was established to formulate policy and working plans for the Abandoned Mine Land reclamation programs. This was necessary because the Department cannot sue many of the offending mining companies for cleanup costs because most no longer exist.

The Department of Agriculture (USDA)

Energy issues are a major, yet indirect, focus of the Department of Agriculture (USDA). Energy is involved in every step of agriculture: it takes energy to produce the inputs of nitrogen, phosphate, seeds and pest/weed control, energy to run the machinery to plant, maintain, harvest and process crops, and energy to transport crops by truck, train, barge and container ships. And the even the end result, food itself, is essentially energy. Food is as essential to humans as gasoline is to the automobile.

The USDA has an inherent conflict of interest: promote eating more dairy products and meats to relieve agriculture surpluses, while promote good eating with the ubiquitous USDA Food Pyramid. Out of a budget of over $65 billion in 2000, about two-thirds of the USDA budget goes toward nutritional programs and social programs for the poor, such as food stamps, to ensure that all Americans can afford an ample supply of food energy. For the middle and upper class of society, food prices are artificially low because of the subsidies to producers. An overabundance of food, that is more affordable to more of the population than ever before, has turned the United States into an equal opportunity obesity society. Over 20 percent of the population is obese (over 50 percent overweight), and obesity can be found in large percentages at all income levels. This is unlike most of the world where obesity is far less prevalent and mostly found among the affluent.

Another function of the USDA is the promotion of biofuels like ethanol as an important market for the nation's farm products. Ethanol is an alcohol fuel produced from corn and is blended with gasoline to enhance octane and reduce automobile emissions of pollutants. From 1980 through 2000, ethanol producers have received a subsidy of nearly $10 billion. To prop up the price of corn and help the ethanol industry, in 1986 the USDA Office of Energy started to give away free corn for all ethanol producers, which was extended to even the very largest and most profitable ethanol producers like Archer Daniels Midland Corporation. These huge subsidies are very controversial. Proponents claim ethanol as a fuel provides an additional market for corn farmers, and also claim the fuel is better for the environment and helps reduce imports of foreign oil. Critics counter that the energy content of ethanol is one-third that of gasoline. Moreover, the DOE and Congressional Research Service found it was not better for the environment, and would retail at a price much higher than gasoline if not for the heavy subsidies. And because it takes considerable energy to convert corn to ethanol, there can be a net loss in energy in producing ethanol (a gallon of ethanol contains around 76,000 British Thermal Units (Btus) of energy and estimates of the energy to produce that gallon range from around 60,000 to 90,000 Btus).

Another biofuel of importance is wood. The Forest Service, which is part of USDA, administers national forest lands for the sale of wood for wood fuel. Besides determining the quantity of wood fuel to bring to market by collecting and analyzing statistics on woody biomass supply and use, the Forest Service sponsors forest biomass energy-related research in conjunction with federal and state agencies, as well as universities.

The Department of Transportation (DOT)

Because every means of transportation requires energy for propulsion, how energy is used in transportation is something that is carefully tracked by the Office of Transportation Policy Development within the Department of Transportation (DOT). The transportation sector felt the greatest impact from the oil supply disruptions in the 1970s because it was, and continues to be, the sector most dependent on oil. It is also the sector with the least flexibility to switch fuels. (see also Consumption)

National Aeronautic and Space Administration (NASA)

The National Aeronautic and Space Administration is involved in every aspect of atmospheric and space science. Because of the special energy requirements of spacecraft and satellites, NASA has been the proving grounds for many emerging energy technologies such as fuel cells and photovoltaics. In the area of transportation, the Jet Propulsion Laboratory in Pasadena California is a leading center for bettering jet and rocket engines and developing new technologies such as ion propulsion. NASA is also responsible for building, launching, and collecting the data from satellites trying to detect global warming.

The Department of Defense (DOD)

The DOD consumes more energy than most nations of the world, and four times more energy than all the other federal agencies combined. The DOD took an interest in energy long before there was an energy crisis. Energy in the form of petroleum is what fuels military technology, and almost every military strategy involves, directly or indirectly, energy. The DOD is not only interested in the logistics of petroleum planning (through the Defense Fuel Supply Center), but also in developing an array of nontraditional energy technologies, everything from better nuclear propulsion for ships and submarines to solar photovoltaic applications and battery technologies to power mobile communication technologies.

Nuclear Regulatory Commission

The Nuclear Regulatory Commission (NRC) is an independent federal agency that licenses and decommissions commercial nuclear power plants and other nuclear facilities. NRC inspections and investigations are designed to assure compliance with the Agency's regulations, most notably the construction and operation of facilities, the management of high-level and low-level nuclear wastes, radiation control in mining, and the packaging of radioactive wastes for transportation.

STATE AND LOCAL AGENCIES

Almost every state in the nation has an energy policy and planning agency charged with ensuring a reliable and affordable energy supply. Duties of these agencies include forecasting future energy needs, siting and licensing power plants, promoting energy efficiency, and planning for energy emergencies. Instead of leaving all energy decisions entirely up to the free market, many states actively promote changes in production and consumption. Faced with some of the worst air quality problems, the California Energy Commission is one of the most active state agencies in implementing demand side management and market transformation strategies.

In an effort to coordinate policy, exchange information, and to convey to the federal government the specific energy priorities and concerns of the states, the state agencies formed the National Association of State Energy Officials (NASEO) in 1986. To better ensure that appropriate policy was being implemented, NASEO set up the Association of Energy Research and Technology Transfer Institute to track the successes and failures of different programs.

Because the federal government has taken a more hands-off approach toward energy since the 1970s, this trend of state and local governments becoming more active in making energy and environmental decisions about electricity production and transportation issues is likely to continue, especially for the more populace areas of the country like California and the Northeast.

John Zumerchik

See also: Agriculture; Air Pollution; Biofuels; Climatic Effects; Culture and Energy Usage; Demand-Side Management; Emission Control, Power Plant; Emission Control, Vehicle; Geography and Energy Use; Government and the Energy Marketplace; Market Transformation; Military Energy Use, Historical Aspects of; National Energy Laboratories; Nuclear Waste; Propulsion; Spacecraft Energy Systems.

BIBLIOGRAPHY

Cochrane, R. C. (1966). *Measures for Progress: A History of the National Bureau of Standards.* Gaithersburg, MD: U.S. Department of Commerce.

Energy Information Administration. (1998). *Energy Information Directory 1998.* Washington, DC: Author

Fehner, T., and Holl, J. M. (1994). *Department of Energy 1977–1994 A Summary History.* Oak Ridge, TN: Department of Energy.

Government Printing Office. (1980). *Cutting Energy Costs: The 1980 Yearbook of Agriculture.* Washington, DC: Author.

Greene, D. L. (1996). *Transportation and Energy.* Landsdowne, VA: Eno Transportation Foundation, Inc.

Hewlett, R. G., and Anderson, O. E., Jr. (1991). *History of the United States Atomic Energy Commission,* Vol. 1: 1939–1946. Berkeley, University of California Press.

Hewlett, R. G., and Duncan, F. (1991). *Atomic Shield,* Vol. 2: 1947–1952. Berkeley: University of California Press.

Hewlett, R. G., and Holl, J. (1991). *Atoms for Peace and War: Eisenhower and the Atomic Energy Commission,* Vol. 3: 1953–1961. Berkeley: University of California Press.

Landy, M. K.; Roberts, M. J.; Thomas, S. R.; and Lansy, M. K. (1994). *The Environmental Protection Agency: Asking the Wrong Questions: From Nixon to Clinton.* New York: Oxford University Press.

Scheinman, L. (1998). *International Atomic Energy Agency and World Nuclear Order.* Washington, DC: Resources for the Future.

Scott, R. (1994). *The History of the International Energy Agency 1974–1994,* Vol. 1: *Origins and Structure,* Vol. 2: *Major Policies and Actions.* Paris, France: OECD/IEA.

Tengs, R. O., et al. (1995). "Five-Hundred Life-Saving Interventions and Their Cost-Effectiveness." *Risk Analysis* 15(3):369–389.

Tetreault, M. A. (1981). *The Organizations of Arab Petroleum Exporting Countries.* Westport, CT: Greenwood Publishing Group.

United Nations. (1998). *Basic Facts About the United Nations.* New York: Author.

GOVERNMENT AND THE ENERGY MARKETPLACE

Government intervention in energy markets involves either government ownership of industry resources or, more commonly, planning and regulation of privately held resources. The proper role of government planning and regulation has been a hotly debated issue among economists and politicians for many years. Through the twentieth century, governments of the United States and of most of the rest of the developed world have become increasingly larger and more involved in taking action that affect the outcomes of how energy is produced, transported, and consumed. Some believe government intervention in markets is the best solution; others believe government intervention should be avoided. Given the extensive number of energy markets and the extensive use of government intervention in those markets, it is not surprising that the record is mixed. Markets have proved to be reasonably efficient, yet also can be harsh; government action has provided important protections, yet also has made some good situations bad and bad situations worse.

THE VIRTUES OF THE MARKETPLACE

Markets decide what energy resources shall be produced, how they shall be produced, and who will receive the benefits of the production processes. With millions of different activity options, and millions of individuals making individual and collective decisions, it is an overwhelmingly complex process. At the same time, it is usually very efficient because self-interested market players communicate through the price system. A self-interested rational individual will make decisions based on true preferences, and follow those preferences in a way that will provide the greatest satisfaction. Choices will be made under certain constraints that apply to all consumers: income, energy prices, and the ability to switch fuels or fuel providers.

Economic theory holds that individuals will make decisions based on what is in their best interests. Government intervention distorts these capital investment decisions; it alters constraints by favoring or giving preferences or incentives to one choice or type of behavior over another. Government intervention tends to subtly manipulate supply and demand of energy, which usually increases costs to users, reduces supply, and leads to higher prices. Political revolt to government intervention is rare since the consumer often is unable to separate the interventionist component of price (the price premium attributable to regulation and taxation) from what the price would be in the free market.

REASONS FOR GOVERNMENT INTERVENTION

Government intervention is justified as a way to correct the shortcomings of the marketplace. Proponents of intervention do not necessarily believe that markets in general do not work, but rather that there are dynamics going on in certain markets that require intervention to cause them to work in a more socially desirable manner. First, intervention is justified in the energy arena because of problems arising from the laws regarding the ownership and exchange of property rights and the purchase and sale of energy rights. Property rights are far more troublesome in the oil and gas production business than most other businesses for two reasons: the commodity of value sits below the surface, and the commodity is a liquid or a gas not a solid, which means it can migrate many miles. Since energy resources know no boundaries (oil and gas deposits often straddle the property of several owners) a system is needed to assign energy rights.

Second, intervention is justified because social costs may exceed private costs as well as private benefits. For example, when an individual chooses to take a personal automobile to work instead of mass transit, the individual driver receives the short-term benefits (privacy, comfort, speed, and convenience) while the negative social costs (greater air pollution, highway construction, traffic jams, and resource depletion) are shared by all. Intervention usually is an attempt to lower the social costs. However, the problem is that social costs may be easily identifiable in theory, but much more difficult to accurately quantify in practice.

A third function of intervention is the oversight of energy utility monopolies for electricity and natural gas. Utilities have been considered natural monopolies because it was generally cost effective to have a sole generator, transmitter or distributor of electricity or natural gas for a local area due to the large infrastructure required and economies of scale. Without competitors, government historically felt a duty to protect customers from unfair price gouging. But new technology is challenging the concept of natural monopoly in the utilities, and has resulted in the deregulation of energy generation, transmission and distribution in the 1990s. Government's role is shifting to establishing the rules, guidelines, and procedures that attempt to be equitable to all parties (industry and customers) and minimize the social costs to society by ensuring these markets are also socially desirable.

Finally, government intervention is often called for to establish standards by essentially reducing informational market barriers. This is important because the rational individual can only make a decision in his best interest when the information is at hand to make that decision.

TYPES OF GOVERNMENT INTERVENTION

Governments intervene in all energy markets—exploration, production, distribution, and consumption—and carry out intervention in many different ways. While some impacts of intervention are intended, other impacts occur indirectly as an unintended consequence. Examples of a direct impact are the price-lowering effect of subsidies for the biofuel ethanol, and the increased desirability of conservation resulting from a tax on energy use. Examples of unintended indirect impacts are the encouragement of single passenger driving over mass transit resulting from the provision of free parking, or a dramatic increase in the price of coal-generated electricity resulting from clean air regulations to reduce air pollution.

Taxes

Governments levy taxes to raise revenue and to discourage consumption of what is taxed. The primary purpose of United States federal and state gasoline taxes is to raise revenue for transportation infrastructure, particularly highway construction. In Europe and Japan, where gasoline taxes are many times greater (gasoline retails for over twice as much as in the United States), the purpose is also to discourage consumption. However, when taxes are implemented, rarely are drops in transportation energy consumption immediate. Even very steep increases in gasoline taxes usually take time to result in a reduction in consumption. People who drive fifty miles to work still need to get to work by driving in the short term. Only over the long term can people decide whether to move closer to work, change jobs, rearrange schedules to use mass transit or to carpool, purchase a more fuel efficient car, or accept higher energy costs by maintaining the same lifestyle.

Unlike the gasoline tax that only impacts the transportation sector, carbon taxes affect all sectors of the economy. Implemented by some European countries and proposed in the United States by the Clinton Administration in 1993, the carbon tax makes consumption of fossil fuels more expensive for the energy user. The goals of a carbon tax are to reduce the consumption of energy and to make non-carbon emitting sources like wind and hydroelectric more cost-competitive with fossil fuels.

Subsidies

Subsidies are transfer payments from governments to business interests and individuals. Governments have been involved in the subsidizing of energy since the early part of the twentieth century. For democratic governments, subsidies have been often used as a means to buy votes, and for authoritarian governments, as an important means of placating the masses. Price supports for ethanol, tax credits for renewable energy, depletion allowances for oil and gas production, and grants for energy conservation measures for low income families fit the narrow definition of energy subsidies. But if subsidies are considered in a broader context, trade barriers, regulations, and U.S. military spending to ensure the flow of oil from the Persian Gulf can be considered subsidies as well.

Regulation

Regulations are rules set by governments to protect consumers and to control or direct conduct in the marketplace. Production, transportation, consumption, and prices have all been affected by energy regulations.

Allocation control regulations affecting oil prices were issued in the 1970s in response to what was perceived as a number of market inefficiencies. The primary motivation was fairness to low-income individuals and families, yet price controls ended up exacerbating the inefficiencies—worsening the problem it was supposed to solve. Since prices convey critical information to buyers and sellers of energy, government intervention that tries to alter the message can distort market signals so that everyone ultimately ends up less satisfied. Price controls send the wrong signal by artificially stimulating consumption (more demand at a lower price) and reducing the quantity supplied (producers are willing to produce less at the lower price).

The most-well known energy regulations are probably the ones covering the production, transmission, and distribution of electricity and natural gas. To avoid the inefficiency of having multiple electric cables or gas lines run down each street and into every home, in the past governments awarded the right to serve an area to one firm—a "natural monopoly." Unlike most countries, where the government owns and operates electric and gas monopolies, in the United States most of the power industry remains privately-owned and the government regulates it by setting prices, the amount that can be earned, and the quality of service provided. Although Congress develops legislation that authorizes regulation, identifying and enforcing of the rules for specific cases are carried out by the Federal Energy Regulatory Commission and state public service commissions.

Criticism of the way public utility monopolies have been regulated usually centers around the built-in biases that may be favorable to the utility and the customer but are also socially undesirable. Electric utility regulation was structured so that companies earned more by building more facilities, selling more power, encouraging consumption, and discouraging conservation. In the late 1970s and early 1980s, states instituted demand-side management programs that attempted to shift the incentive away from consumption to conservation.

In the late 1970s, public policies began to focus on opening up power generation markets so that non-utility generators and independent power producers could have access to transmission systems and electricity markets. By the early 1990s, much more aggressive public policies were being considered and adopted. These public policies envisioned giving customers in all customer classes the opportunity to select their electricity suppliers. It was recognized that allowing customer choice would require not only providing open access to transmission systems but also the total "unbundling" of power production from power delivery and customer service functions to prevent discriminatory transactions and self dealing.

While unbundling was seen as a necessary step for deregulation, utility companies pointed out that unbundling could create "stranded assets." The creation of stranded assets could be devastating to those utility companies that had made investments in nuclear power plants and other assets that would be uneconomic or unable to compete in open markets. To address the stranded assets issue, government intervention sought to create transition plans elevating rates above market values and allowing the recovery of stranded costs. In the longer term, it is likely that many electric power marketers will be buying power from many different electricity producers, and then turning around and selling the power to industry, business and the residential customers.

"Cross subsidies" have traditionally been used in establishing rates. Industrial customers pay a higher rate for their electricity than residential customers and, in effect, subsidize the power supply for residential customers. In a different context, "cross subsidies" can be created to select power generation technologies based on their environmental impact. There is political pressure from the environmental interest groups to include ratepayer cross-subsidies so that the less desirable power generation technologies (e.g., coal-fired steam turbines) subsidize the more desirable power generation technologies (e.g., wind turbines). This is being considered as part of the deregulation process. Pollution costs fall largely on third parties. In the case of fossil fuel electric power plants, only a small fraction of air pollution costs are borne by the beneficiaries of the power production. Proponents view cross subsidies as an effective way to raise the cost of fossil fuel energy production, lower the cost of renewable energy, and support energy efficiency and conservation programs. Programs like "Plant a tree whenever you build a new fossil fuel power plant" are, in effect, cross subsidy programs because they balance environmental objectives with economic objectives in power generation planning.

Compared to regulation, cross subsidies are a more market-based approach for discouraging air pollution emissions from energy production. Economists generally agree that economic approaches to pollution control (taxes, subsidies, pollution allowances, and trading programs) have been far more efficient and equitable solutions than "command and control" approaches to regulations that have mandated specific emission levels and specific technology.

Since there is an environmental impact with almost all energy production, environmental regulations and policy are indirectly having a greater impact on the energy industry than direct energy policy. Clean air regulations continue to make it more expensive to burn fossil fuels, especially coal; clean water regulations present expensive challenges for the transportation (oil tankers) and storage (underground fuel tanks) of petroleum products; carbon dioxide reduction policies to prevent global warming favor the growth of renewable energy over fossil fuels.

Safety is another major regulatory area. The Occupational Safety and Health Administration (OSHA) enforces many worker safety regulations covering the operation of refineries and electric power plants, and the Nuclear Regulatory Commission oversees the regulations ensuring the safety of nuclear power plants. There are also numerous safety regulations covering the operations of oil and natural gas pipelines and oil tankers. These protective programs have unquestionably provided major social benefits. However, critics contend that these programs have gone too far, have been too expensive, and have placed too much emphasis on government as the interpreter of safety.

When regulation fails to achieve the intended objective, the question arises of whether it was a failure solely of administration or of theory too. If advocates of regulation can prove failure to be largely administrative, it is easier to make the case for new regulations under improved administration.

Information

There have been considerable efforts made at moral persuasion aimed at the social conscience to convince those involved in high consumption behavior to behave more responsibly. During the 1970s, the United States government embarked on a campaign to persuade the public to conserve energy by, among other things, driving 55 mph, keeping car tires inflated, and turning down thermostats in buildings to 65 degrees. In the 1980s, energy use appliance labels were mandated so that rational consumers could compare the efficiency of models and purchase the more energy efficient models. (This was followed by Energy Star Labeling, a joint action of the EPA and DOE in the 1990s that recognizes products such as computers or windows that meet a given energy efficiency standard. Early in the twenty-first century, an emerging target of moral persuasion is the sports utility vehicles (SUV). Compared to the average automobile, the average SUV consumes about 30 percent more fuel per mile and generates more air pollutants and far greater carbon dioxide emissions. The message is that it is unpatriotic to purchase these energy-guzzling, high-emission vehicles. The combination of moral persuasion and the reclassification of SUVs as automobiles (to meet the tougher emission and Corporate Average Fuel Economy standard for automobiles) could be an effective means of curtailing SUV purchases.

Moral persuasion has been most effective in times of short-term emergencies, like the energy shortages of the 1970s, but is far less effective in the long-term and for problems not universally viewed as problems. The benefits of such efforts are the speed, the inexpensiveness of the approach, and the implicit threat behind the effort: stop the undesirable activity by choice or society will take direct measures to curtail it.

Research and development

Europe, Russia, Japan, and the United States have spent billions for energy research and development since the 1970s. The United States alone spent over $60 billion from 1978 to 1996. Nations liberally spend on energy research to try to expand supply options, to develop indigenous resources for national security reasons, and to speed invention and innovation that will benefit society. When the energy research and development budget was significantly boosted in the mid-1970s, the hope was that nuclear fusion, solar energy and other renewable energy sources could develop into the primary energy sources of the next century. But because there have been no major breakthroughs, and fossil fuels have remained relatively inexpensive, renewables' share of energy production has actually fallen. Fossil fuels research and development have shown better results: cleaner transportation fuels and vehicles, more efficient natural gas turbines, and the clean coal program, which raised efficiency and reduced the harmful emissions from coal burning power plants. Largely because of the disappointing overall results of federal spending on energy research and development—the poor track record of picking winners, the inequitable funding based more on politics than scientific merit, and the draining of research and development capital away from the private sector—the amount of money spent on energy R&D has been falling throughout the 1980s and 1990s as a percent of total U.S. research and development. It is uncertain whether this trend will continue. Many energy experts still advocate government spending because industrial goals are usually short-term and do not necessarily reflect national goals such as energy security and environmental quality.

THE NEXT ENERGY CRISIS

Historically government intervention has been primarily crisis-driven. During the Energy Crisis of the 1970s, it was easy for government leaders to gather popular support for government policies that subsidized alternative energy sources (boosting supply) and promised to lower the price of energy. Yet once the crisis passed, gathering support for new initiatives became much more difficult. That is largely why there have been only minor new energy initiatives since the end of the Carter Administration in 1980.

Freer and more diverse world energy markets make another major energy crisis less likely. If one does occur, it is likely to again trigger strong cries for government action. But because of the valuable economic lessons learned from government actions in the 1970s, any crisis-related intervention should achieve better results; the ineffective and harmful mistakes of the past are unlikely to be repeated. Moreover, because the total value of crude oil to the economy is much smaller, any oil supply disruption is unlikely to be as severe for the American economy as the oil price hikes of 1973 and 1979. The economy can more easily substitute coal and natural gas for petroleum, and the very energy-dependent industries like aluminum, steel and petrochemicals have grown more slowly and make up a smaller share of the economy than high growth industries like pharmaceuticals, computer technology and the service economy.

John Zumerchik

See also: Appliances; Behavior; Capital Investment Decisions; Demand-Side Management; Efficiency of Energy Use, Economic Concerns and; Efficiency of Energy Use, Labeling of; Environmental Economics; Environmental Problems and Energy Use; Government Agencies; Green Energy; Market Imperfections; National Energy Laboratories; Property Rights; Regulation and Rates for Electricity; Subsidies and Energy Costs; Taxation of Energy; True Energy Costs

BIBLIOGRAPHY

Asch, P., and Seneca, R. S. (1988). *Government and the Marketplace*. Fort Worth, TX: HBJ School.

Bradley, R. L. (1996). *Oil, Gas and Government*. Lanham, MD: Rowman & Littlefield Publishers, Inc.

Yergin, D., and Stanislaw, J. (Contributor) (1998). *The Commanding Heights: The Battle Between Government and the Marketplace That Is Remaking the Modern World*. New York: Simon & Schuster.

Weidenbaum, M. L. (1995). *Business and Government in the Global Marketplace*. Englewood Cliffs, NJ: Prentice-Hall, Inc.

GRAVITATIONAL ENERGY

Earth's moon is held in orbit by an attractive gravitational force between Earth and the moon. Tides on Earth are due mainly to the gravitational pull of the moon on Earth. Any two masses whether or not Earth and the moon, experience a mutual gravitational force that tends to pull them together. A mass in a position to be pulled to another position by a gravitational force has gravitational potential energy. Anything—water, a book, a parachutist, a molecule in the atmosphere, etc.—has gravitational energy if it is in a position to move closer to the center of Earth. Ordinarily, something has to do work to get the object to the elevated position. A book on the floor of a room has no gravitational energy if it cannot move lower than the floor. But if you lift the book, and place it on top of a table, it has gravitational energy because it is in a position to fall to the floor. The act of lifting the book and doing work produces the gravitational energy of the book. Water atop a dam has gravitational potential energy, but the water did not get into its position at the top of a dam without some agent doing work. The mechanism elevating the water could be a mechanical pump or the natural processes of water evaporating, rising in the atmosphere, condensing to liquid, and falling as rain. Ocean water pulled into a natural basin by the gravitational pulls of the moon and the sun has gravitational potential energy that we refer to as tidal energy.

Electric energy produced by a hydroelectric power plant is derived from gravitational energy. The process starts with gravitational forces pulling the water to the bottom of the dam, where the gravitational energy is converted to kinetic energy (energy due to matter in motion). When the moving water impinges on the blades of a turbine, some of the kinetic energy is converted to rotational energy, causing the turbine to rotate. The turbine is coupled to an electric generator, where the rotational energy of the turbine is converted to electric energy. The electric generator is connected to transmission lines that deliver the electric energy to consumers.

Electricity supplied to consumers is produced pri-marily by massive generators driven by steam turbines. The system is most efficient and most economical when the generators produce electricity at a constant rate. However, the problem is that demand varies widely by time of day. At night, when consumer demand is low, it would be seem that a utility could keep their most efficient generators running and store electric energy for times when needed. Unfortunately, there is no practical way of storing the quantities of electric energy produced by a large generator. It is possible, however, to convert the electric energy to some other form and store it. Then when electricity is neededm the secondary form of energy is converted back again to electricity. One way of doing this is to keep efficient generators running at night, when the demand is low, and use them to operate electrically driven pumps that store the water as gravitational energy in an elevated reservoir.

The Paraná River meets the Itaipu Dam in Foz Do Iguacu, Brazil. (Corbis Corporation)

When consumer demand picks up during the day, the water is allowed to flow to a lower level to a hydroelectric unit, where the gravitational energy is converted to electric energy. A system like this is called a *pumped-storage facility*. The electric energy recovered in the hydroelectric unit is always less than the electric energy used to store the water as gravitational energy. However, the former is economical because electricity produced at night is relatively cheap. There are a number of these systems throughout the world, the most prominent in the United States being the Robert Moses Plant, near the base of Niagara Falls.

Joseph Priest

See also: Conservation of Energy; Kinetic Energy; Nuclear Energy; Potential Energy.

BIBLIOGRAPHY

Hobson, A. (1995). *Physics: Concepts and Connections.* Englewood Cliffs, NJ: Prentice-Hall.

Priest, J. (2000). *Energy: Principles, Problems, Alternatives,* 5th ed. Dubuque, IA: Kendall/Hunt Publishing Co.

Serway, R. A. (1998). *Principles of Physics*, 2nd ed. Fort Worth, TX: Saunders College Publishing.

GREEN ENERGY

Green pricing for electricity allows consumers to voluntarily pay a premium for more expensive energy sources that are considered better for the environment. The initiative presumes that the buying public believes: (1) the existing corpus of environmental regulation has not fully corrected the negative effects of producing electricity from hydrocarbons (oil, natural gas, orimulsion, and coal); (2) nuclear power is not environmentally friendly despite the absence of air emissions; and (3) certain renewable technologies are relatively benign for the environment.

OPEN-ACCESS RESTRUCTURING: DRIVER OF GREEN PRICING

Electricity typically has been purchased from a single franchised monopolist or municipality. The physical commodity (electrons) and the transmission service were bundled together into one price. The power could be generated (in terms of market share in the United States in 1998) from coal (56%), nuclear (20%), natural gas (11%), hydro (8%), oil (3%), biomass (1.5%), geothermal (0.2%), wind (0.1%), and solar (0.02%). Recently, wholesale and some retail markets have been unbundled, allowing competitors to sell electrons with the monopoly utility or municipality providing the transmission service. Open-access restructuring gives customers choices and creates a commodity market in which the lowest-cost electricity wins market share at the expense of higher-cost alternatives.

The prospect and reality of open-access restructuring is responsible for the current interest in green pricing. Monopoly suppliers became interested in green pricing as a marketing strategy to retain their customer base given the prospect of open-access competition from new commodity suppliers. Independent suppliers operating in open-access states embraced green pricing as an important form of product differentiation. At the same time, environmentalist interest groups campaigned for consumers to use their newfound buying power to absorb a price premium to support those energies perceived to have lower social costs than private costs due to their environmental advantages.

GREEN POWER SUBSIDIES

Electricity generated from biomass, geothermal, solar, and wind facilities (non hydro renewables) have been subsidized by government policies since the 1970s with research and development grants, accelerated depreciation, tax credits, ratepayer cross-subsidies, and must-buy provisions. On an energy production basis, these subsidies have been substantially greater than for conventional generating sources. Without government subsidies today, natural gas technologies would easily dominate the new capacity market in the United States due to their combined economic and environmental characteristics relative to other generation technologies, both conventional and nonconventional.

CHALLENGES FOR GREEN PRICING

The ability of consumers to voluntarily pay a premium to purchase environmentally preferred electric generation would seem to be a happy middle ground between free-market energy proponents and "clean energy" advocates. Yet, on closer inspection there are numerous issues that concern a maturing green-pricing market.

Government Codependence.

Green pricing is not stand-alone philanthropy, but a voluntary addition to mandated subsidies from the entire ratepayer and taxpayer class for the same renewables. Since "green" energy is often intermittent (the sun does not always shine nor the wind always blow), conventional energy with round-the-clock reliability sets the foundation for green pricing. Yet, conventional supply is not compensated for providing the reliability for intermittent "green" energy. Green pricing attempts to get uneconomic renewables "over the top," given upstream subsidies that alone might not be enough to make the renewables economic in a competitive market.

In a free market, the absence of these subsidies might make the green pricing premium prohibitive for consumers. In California, for example, a 1.5-cent per kilowatt hour (kWh) subsidy for green-pricing customers has led to over 100,000 customers for "green" power. There are serious questions about what will happen when the subsidy is reduced or expires. This subsidy is over and above other favors that have led to the construction of high-cost renewable capacity in the last decade—subsidies that may or may not continue in the future.

The codependency of green pricing on government favor is apparent in another way. Green pricing is a short-term consumer commitment, yet needed capacity is capital-intensive and long-lived, requires long lead times, and generates power that may be too expensive to profitably sell in the spot market. To secure financing for such projects, mandates and financial subsidies are required unless qualifying "green" capacity already exists (which predominately has been the case to date). Noneconomic projects built on the strength of short-term green-pricing commitments could become stranded or abandoned assets if the month-to-month consumer premiums cease. This is a reason why ten states as of 1999 have guaranteed a minimum percentage of new capacity to be renewables-based in their electricity restructuring programs. A federal Renewable Portfolio Standard is being sought as well to mandate (nonhydro) renewable quotas nationally.

Subjective Qualifications.

Green pricing assumes that an objective environmental standard exists for consumers to compare competing energies where, in fact, very complicated and subjective tradeoffs exist. A list of issues can be constructed.

Should the programs apply to existing generation, new generation, or both? Applying green pricing to existing projects is a windfall for owners since it is over and above the sustainable support the market originally provided. Many environmentalists have been critical of green pricing programs that utilize existing capacity since no incremental environmental benefits are secured in return for the premium that customers pay. "Greener" for some customers automatically makes the remainder "browner."

Should hydropower be part of the green portfolio since it is renewable and does not emit air pollution, albeit conventional? A green energy primer from the Renewable Energy Policy Project segments hydro into "cleaner" and "polluting" based on river system impacts. Some green pricing programs include projects below 30 megawatts, while other programs do not allow hydro at all.

Should wind projects, like hydro, be bifurcated into "low" and "high" impact categories depending on avian impacts and other nonair emission considerations? Prominent environmental groups have raised concerns over certain bird-sensitive wind sites without generating opposition from other environmental groups pushing windpower as a strategy to reduce air emissions.

Should biomass be included as a "green" energy, given that it has direct air emissions and may not be renewable in its most common application—waste burning?

Should geothermal be included as a "green" energy, given that projects deplete over time and release toxics in many, if not most, applications?

Is nuclear "green" since it produces no air emissions—and in this sense is the cleanest of all energy sources in the aggregate—or should waste disposal and the hazards of plant failure be controlling?

Should "green" energies be blended with "non-green" sources for genres of green pricing, and if so, in what amounts?

Should power generation from natural gas be included as a "green" energy since it is the cleanest of the hydrocarbons and may compare favorably with some major renewable options in as many as five environmental categories: front-end (embedded) air emissions, wildlife disturbance, land usage, noise, and visual blight? The Renewable Energy Policy

Project delineates between "cleaner" new gas technologies and "polluting" old natural gas technologies—and reserves a third "cleanest" category for energy efficiency/conservation, solar, wind, and geothermal (but not hydro or biomass).

Should "clean" coal plants meeting more stringent new source review standards be differentiated from older plants "grandfathered" to lesser requirements in an energy-environmental matrix? Restated, should operational distinctions used for other renewable sources and natural gas also be made for coal plants for the purpose of environmentally based price differentiation?

Should carbon dioxide (CO_2) emissions from hydrocarbons be rewarded or penalized in green-pricing programs? CO_2 fertilization plays a positive role in plant growth and agricultural productivity. Agricultural economists and scientists have calculated net benefits to a moderately warmer, wetter, and CO_2-enriched world. Other analysis has concluded that unregulated CO_2 emissions, potentially elevating global temperatures enough to destabilize the climate, should be penalized.

Can solar technologies be part of green-pricing programs since they are two or three times more expensive than other renewable options?

Who decides? The above issues point toward a potentially intricate color-coded system differentiating energy not only according to its general type, but also according to operational, site-specific attributes. The categorizations would also have to be revised for new scientific data and pollution-control developments. Can green-pricing programs accommodate this complexity and subjectivity short of resorting to government intervention to define what is "green?"

The leading green-pricing arbiter in the early U.S. experience is the nonprofit Center for Resource Solutions. Their Green-e Certification Seal sets an "objective" standard for green energy under the following criteria:

- Fifty percent or more of the volumetric offering must be from newly constructed, but not mandated, renewable capacity from wind, solar, biomass, geothermal, ocean, or small hydro.
- The remaining 50 percent or less of supply must have an average emission rate below the system average for fossil-fuel-emitted sulfur dioxide, nitrogen oxide, and carbon dioxide.

- No new nuclear supply can be added to the generation mix of a blended product.

State regulators across the country have approved the Green-e labeling program, although marketers do not have to receive the seal to operate. Of greater concern for free green choice is an interest by the Federal Trade Commission and the National Association of Attorneys General to set green power guidelines.

Tracking and Labeling Issues

How should "green" electrons be documented for premium-paying consumers? The industry has proposed that green electrons be certified upon generation, creating a paper record that can be traded and ultimately balanced downstream between sales and purchases. Certain environmentalist interest groups have proposed instead that each "green" electron be physically tracked in the transaction chain so that the final consumer gets actual "green" electrons. Industry groups have resisted this proposal, believing any tracking attempt would reduce flexibility and increase transaction costs when a paper system would result in the same impetus for "green" supply. Physics dictates that electrons flow freely; supplies by necessity are commingled and balanced as they move downstream. In other words, the person buying "green" electrons will not necessarily receive those electrons.

As part of green pricing, environmentalists have proposed a labeling system in which all batches of electricity enter the system with specific information about their generation mix and air-emission characteristics. Industry has proposed that labeling just be made for a positive assertion of green power rather than a blanket reporting of mix and emissions—again for cost and flexibility reasons.

OTHER CONCERNS

The environmental community has raised concerns that green pricing could be construed by lawmakers and the public as a substitute for taxpayer and ratepayer subsidies. A major consumer group has complained that green pricing is only for the affluent, and nonparticipants get to "free ride" on the more environmentally conscious users. Independent power marketers have worried about utility green-pricing programs increasing barriers to entry when the market opens to competition.

CONCLUSION

Green pricing, although nominally a free-market support program for consumer environmental preferences, is closely intertwined with government intervention in energy markets and prone to special interest politics. To pass a free market test, regulators and providers would have to establish open, stand-alone green-pricing programs that do not contain upstream subsidies or require downstream quotas on market participants. In such a world, voluntary labeling standards and common law protections against fraudulent misrepresentation can ensure the integrity of the program.

Green pricing in a true market would have to respect the heterogeneity and fluidity of consumer preferences. Consumer opinion about what is "green" can be expected to evolve and mature, particularly if environmental programs reduce or remove the perceived negative externalities of less expensive energy, relative economics substantially change between energies, or new information becomes available about environmental characteristics of competing energy choices.

Robert L. Bradley, Jr.

BIBLIOGRAPHY

Bradley, R. L. (1999). "The Increasing Sustainability of Conventional Energy." In *Advances in the Economics of Energy and Resources, Vol. 11: Fuels for the Future*, ed. J. Moroney. Stamford, CT: JAI Press.

Holt, E. A. (1997). "Disclosure and Certification: Truth and Labeling for Electric Power." Issue Brief no. 5. College Park, MD: Renewable Energy Policy Project.

Intergovernmental Panel on Climate Change. (1996). *Climate Change 1995: The Science of Climate Change.* Cambridge, UK: Cambridge University Press.

Mendelsohn, R., and Newman, J., eds. (1999). *The Impact of Climate Change on the United States Economy.* Cambridge, UK: Cambridge University Press.

Morris, J. (1997). *Green Goods? Consumers, Product Labels and the Environment.* London: Institute for Economic Affairs.

Rader, N. (1998). *Green Buyers Beware: A Critical Review of "Green Electricity" Products.* Washington, DC: Public Citizen.

Rader, N., and Short, W. (1998). "Competitive Retail Markets: Tenuous Ground for Renewable Energy." *Electricity Journal* 11(3):72–80.

Renewable Energy Policy Project. (1999). "Your Green Guide to Electricity Choices" (pamphlet).

Wiser, R.; Fans, J.; Porter, K.; and Houston, A. (1999). "Green Power Marketing in Retail Competition: An Early Assessment." NREL/TP.620.25939. Golden, CO: National Renewable Energy Laboratory.

Wittwer, S. (1995). *Food, Climate, and Carbon Dioxide: The Global Environment and World Food Production.* New York: Lewis Publishers.

GREENHOUSE EFFECT

See: Air Pollution; Environmental Problems and Energy Use

GUNPOWDER

See: Explosives and Propellants

HEAT AND HEATING

MAINTAINING CONSTANT TEMPERATURE

The purpose of a heating system is to maintain a constant temperature. To do this, all heating systems—whether it is for heating a small room or a domed stadium—must compensate for heat losses with heat gains. Even though the human body continually loses heat through conduction with the surrounding air and evaporation of water from the surface of the skin, the temperature of the human body is remarkably constant. For every joule of heat flowing from the body, there is one joule produced by mechanisms that convert energy from food. If heat loss quickens (e.g., while one is swimming), the heat regulators in the body convert more energy from food to compensate.

It is the same with a room in a building. Heat is lost through many tiny openings to the outside and through conduction through walls and windows. A heating system makes up this loss keeping the temperature constant. The greater the temperature difference between the inside and outside of a building, the greater is the heat loss. This is why the heating system must provide more heat when you want to maintain a higher temperature as well as when the outside temperature is extremely cold.

FORCED-AIR HEATING SYSTEM

For any substance, the total thermal energy is the sum of the energies of all the molecules. In other words, thermal energy is equal to the energy per molecule multiplied by the number of molecules. A volume of air may have a high temperature, but does not necessarily have a lot of thermal energy because the number of molecules might be small. Conversely, a volume of air may have low temperature and significant thermal energy if many molecules are involved. When a gas is mixed with one at a higher temperature, the mixture will come to a temperature somewhere between the temperatures of the gases being mixed. A forced-air heating system works on this principle by mixing warmer air with the air existing in the room. Sometimes the mixing is uneven in a room producing temperature variations that are annoying to the occupants.

HOT-WATER HEATING SYSTEM

A hot-water heating system forces water into pipes, or arrangements of pipes called registers that warm from contact with warm water. Air in the room warms from contact with the pipes. Usually, the pipes are on the floor of a room so that warmer, less dense air around the pipes rises somewhat like a helium-filled balloon rises in air. The warmer air cools as it mixes with cooler air near the ceiling and falls as its density increases. This process is called convection and the moving air is referred to as convection current. The process of convection described here is pipe-to-air and usually does a better job of heating evenly than in an air-to-air convection system—the circulation of air by fans as in a forced-air heating system.

ELECTRICAL RESISTANCE HEATING

Rather than have pipes warmed by water, some heating systems employ coils of wire or other configurations of metal that are warmed by electric currents. An electric current is a flow of electric charge that may be either positive, negative or a combination of positive and negative. Both positive and negative electric currents are involved in the battery of an

The first electric welding transformer, a device which uses electricity to weld joints of metal using resistance heating. (Corbis Corporation)

automobile. But in a metal the electric current is due to the flow of electrons, which have negative charge. The electrons are components of atoms making up the structure of the metal. In a very similar way, electrons are forced to move through a wire by electrical pressure in somewhat the same way that water molecules are forced to move through a pipe by water pressure. Water molecules encounter resistance as they flow through the pipe and electrons encounter resistance as they flow through a metal wire. The resistance of both the pipe and the wire depends on the dimensions of and obstructions in the pipe and wire. The resistance increases as the length increases but decreases as the diameter increases.

The material selected makes a difference too. For the same length and diameter, a wire made of copper has less resistance than a wire made of aluminum. A current-carrying wire warms because of electrical resistance.

The resistance (R) of a conductor is measured in ohms and the rate of producing heat (power P) in watts. When the conductor is connected to a constant voltage source, as in a household electric outlet, the thermal power produced is given by $P = V^2/R$. Note that if the resistance halves, the power doubles. Some electric toasters produce heat from current-carrying wires that you usually can see. Because the voltage is fixed, typically 120 volts in a house, a desired thermal power is attained by tailoring the resistance of the heating element. The wire chosen must also be able to withstand the high temperature without melting. Typically, the wires are nichrome (an alloy of nickel and chromium) and the electrical resistance is about 15 ohms, making the thermal power about 1,000 watts. One thousand watts means 1,000 joules of heat are produced each second the toaster is operating.

A current-carrying wire converts electric energy to thermal energy with an efficiency of nearly 100 percent. In this sense, electric heating is very efficient. However, the electricity for producing the current in the heater was likely produced by burning coal in a large electric power plant. The efficiency for delivering electric energy to a consumer is about 32 percent due to energy losses in the plant and in transmission to the house. Therefore, the efficiency for converting the thermal energy from burning coal to thermal energy delivered by an electric heater is very low compared to the 80–90 percent efficiency for modern forced-air heating systems burning natural gas. Electric resistance heat is clean and convenient and a good backup for other systems, but as the main source of heat, the economics are very unfavorable.

UNWANTED HEAT PRODUCED IN ELECTRONIC COMPONENTS

Resistors are components of any electronic device such as a computer. If there is need for a resistance of 1,000 ohms, a designer can purchase a resistor having 1,000 ohms resistance. Heat will be produced in the resistor when a current flows through it. If this heat is not transferred away, the temperature will rise and damage the resistor. Usually the designer relies on heat being conducted to surrounding air to keep the resistor from overheating, and the most efficient transfer of heat is by increasing the surface area of the resistor. The physical dimensions of the resistor will be chosen so that the thermal power can be dissipated without damaging the resistor. For example, a 1,000-ohm resistor designed to dissipate 1 watt of heat will be about four times larger than a 1,000-ohm resistor designed to dissipate ¼ watt. In some cases the transfer of heat by direct conduction to the air is not sufficient and the resistor will be mounted onto a metal housing having substantial mass and surface area.

In addition to resistors, the electronic world of computers, calculators, television sets, radios, and high-fidelity systems involves a multitude of electrical components with names like diode, transistor, integrated circuit, and capacitor. Unwanted heat is generated in these components because of ever-present electrical resistance. If the heat is not transferred away, the temperature of the device will increase and, ultimately, be damaged. The heat generated is especially large in transistors in the final stage of a high-fidelity system where the speakers are connected. These so-called power transistors are mounted on a metal structure having substantial mass and surface area to disperse heat transferred to it by the transistor.

Scientists are forever searching for electronic components like transistors that will perform their function with minimum production of heat. In early versions of personal computers the heat produced by the components could not be dissipated by natural means and fans were used to force the heat to the surroundings. Most contemporary personal computers, including laptops, rely on removing heat by natural means and do not require fans.

TRANSMISSION OF ELECTRIC POWER

An electric generator in a modern coal-burning or nuclear-fueled electric power plant converts mechanical energy to electric energy at an impressive efficiency, about 99 percent. The 1 percent of mechanical energy not going into electric energy is lost as heat. When an electrical device is connected to the generator, the power delivered depends on the voltage of the generator and the current in the device. Specifically, electric power (watts) is equal to voltage (volts) times current (amperes). Typically, the voltage is 10,000 volts and the generator is capable of delivering around 1 billion watts of electric power. If the generator were connected directly to wires that transmitted electric power to consumers, the current would be 1,000,000,000 watts per 10,000 volts, which equals 100,000 amperes. The wires through which this current must flow can be tens or hundreds of miles long and have non-negligible electrical resistance. Therefore, heat will be produced and lost. Even if the resistance of the wires was as low as 1 ohm, the energy produced by the generator would be completely lost as heat. The power company uses a device called a transformer to reduce the current in the wires and thereby reduce the heat loss. The transformer is a device that accepts electric power at some voltage and current and then delivers essentially the same power at a different voltage and current. At the power plant the voltage is stepped up from 10,000 volts to around 1,000,000 volts. A voltage of 10,000 volts delivering 100,000 amperes would be changed to a voltage of 1,000,000 volts and current of 1,000 amperes. Because the power lost as heat depends on the square of the electric current, the lower current reduces the heat loss significantly. When the power arrives at a distribution center for consumers, the voltage is stepped down from its transmission value of 1,000,000 volts. For household use this voltage is 120 volts for lights and small appliances and 240 volts for electric stoves, electric clothes dryers, and electric water heaters.

A superconductor is a material having zero electrical resistance. An electric current in a superconductor would produce no heat. To achieve zero resistance the superconductor must be cooled. This temperature has gotten progressively higher over the years. At this writing it is about 100 K, which is about −173°C. As research progresses on superconducting materials there is hope that they can be used in the transmission of electric energy. Even though there would be an energy expenditure involved in cooling the transmission lines, it would be more than offset by the savings accruing from not having power lost to heat generated in the transmission lines.

COOLING TOWERS AND LOW-GRADE HEAT

You can do work by rubbing your hands together and all the work will be converted to heat. A food mixer can churn through water and do work and all the work will be converted to heat. But a heat engine can only convert a portion of the heat it takes in to work. Nowhere is this more evident than in the steam turbines driving electric generators in electric power plants. The clouds seen emanating from the giant towers so noticeable at electric power plants are graphic evidence of heat being transferred to the environment.

High temperature steam pounding on the blades of a turbine causes the blades to rotate. After passing by the blades the cooler steam condenses to liquid water. Condensation liberates energy and if this energy is not removed, the condenser part of the system warms up. The heat liberated at the condenser is removed by circulating water in pipes in contact with the condenser. Each second the water must remove more than a billion joules of heat, so there is a substantial amount of thermal energy absorbed by the cooling water. The purpose of a cooling tower is to transfer the thermal energy in the cooling water to the air environment by splashing the warm water on the floor of the tower. A portion of the water evaporates and is carried into the air environment by a natural draft created by the chimney-like structure of the tower. The vapor cloud seen at the top of a cooling tower is from the water that has been evaporated.

Although there is significant thermal energy in the water circulating from the condenser of a turbine, it is only 10–15°C higher than the input water. It is a case of having a lot of energy by having a lot of molecules. This relatively low-temperature water is not suitable for heating buildings. Furthermore, the electric power plant is usually several miles removed from buildings where the warm water could be used. By the time that the heated water would reach the building, most of the heat energy would be lost to the environment. Locating industries and communities near plants would make it feasible. One suggestion for using the warm water is to stimulate plant growth in nearby greenhouses.

Remember that heat is lost to the environment in virtually all energy conversion processes. The more heat that is lost the less desirable the form of energy and the less efficient the process. A 100-watt fluorescent lightbulb is more efficient than a 100-watt incandescent bulb because more of the electric energy goes into radiant energy and less into heat. Modern electronics using transistor technology is enormously more energy efficient than the original vacuum tube technology. The total elimination of heat loss can never be achieved, but scientists and engineers work diligently to find better ways of minimizing heat loss and successes have been achieved on many fronts.

Joseph Priest

See also: Conservation of Energy; District Heating and Cooling; Furnaces and Boilers; Heat Transfer.

BIBLIOGRAPHY

Incropera, F. P. (1996). *Fundamentals of Heat and Mass Transfer*, 4th ed. New York: John Wiley and Sons.

Killinger, J. (1999). *Heating and Cooling Essentials*. Tinley Park, IL: Goodheart-Willcox Co.

Priest, J. (2000). *Energy: Principles, Problems, Alternatives*, 5th ed. Dubuque, IA: Kendall/Hunt Publishing Company.

Ristinen, R. A., and Kraushaar, J. J. (1999). *Energy and the Environment*. New York: John Wiley and Sons, Inc.

Stevenson, W. (1982). *Elements of Power System Analysis*, 4th ed. New York: McGraw-Hill Publishers.

Woodson, R. D. (1971). *Scientific American* 224(12):70–78.

HEATING OIL

See: Residual Fuels

HEAT PUMPS

A heat pump is a thermodynamic heating/refrigerating system used to transfer heat. Cooling and heating heat pumps are designed to utilize the heat extracted at a low temperature and the heat rejected at a higher temperature for cooling and heating functions, respectively.

The household refrigerator can provide a simple analogy. A refrigerator is actually a one-way heat pump that transfers heat from the food storage compartment to the room outside. In so doing, the inside of the refrigerator becomes progressively cooler as heat is taken out. In a closed room, the heat coming out of the refrigerator would make the room warmer. The larger the refrigerator is, the greater the potential amount of heat there is to transfer out. If the refrigerator door were open to the outdoors, there would be an almost unlimited amount of heat that could be transferred to the inside of a dwelling. Thus it is possible to design a refrigeration system to transfer heat from the cold outdoors (or any other cold reservoir, such as water or the ground) to the insides of a building (or any medium that it is desired to heat).

Heat pumps can move heat energy between any form of matter, but they are typically designed to utilize the common heating and cooling media, i.e., air or water. Heat pumps are identified and termed by transfer media. Thus terms air-to-air, water-to-air, or air-to-water are commonly used. Most heat pumps are designed to transfer heat between outside air and inside air, or, in the case of so-called geothermal or ground-source heat pumps, between the ground or well water and air. The principle application of heat pumps is ambient heating and cooling in buildings.

There are three types of heat pump applications: heating, cooling or heating and cooling. "Heating only" heat pumps are designed to transfer heat in one direction, from a cold source, such as the outdoors, to the inside of a building, or to a domestic hot water plumbing system, or to some industrial process. There is no term "cooling only heat pump," as the term would simply be describing what is commonly referred to as an air conditioner or cooling system.

"Heating and cooling" heat pumps have refrigeration systems that are reversible, permitting them to operate as heating or cooling systems. An analogy would be a window air conditioner that was turned around in winter, so that it was cooling the outside and blowing hot air inside. Most heat pumps in use are of the heating and cooling type, and are used to heat residences, or commercial and industrial buildings. Residences consume the bulk of those sold: in 1999 about a quarter of new single-family homes were equipped with a heating and cooling heat pump.

Most heat pumps utilize a vapor-compression refrigeration system to transfer the heat. Such systems employ a cycle in which a volatile liquid, the refrigerant, is vaporized, compressed, liquefied, and expanded continuously in an enclosed system. A compressor serves as a pump, pressurizing the refrigerant and circulating it through the system. Pressurized refrigerant is liquefied in a condenser, liberating heat. Liquid refrigerant passes through an expansion device into an evaporator where it boils and expands into a vapor, absorbing heat in the process. Two heat exchangers are used, one as a condenser, and one as an evaporator. In the case of an air-to-air heat pump, one heat exchanger is placed outside, and one inside. In a ground-source heat pump, the outdoor heat exchanger is placed in contact with well water, a pond, or the ground itself. In both cases, the refrigerant flow is made reversible so that each heat exchanger can be used as an evaporator or as a condenser, depending on whether heating or cooling is needed. The entire system is electrically controlled. (See Figures 1 and 2.)

Some heat pumps, called thermoelectric heat pumps, employ the Peltier effect, using thermocouples. The Peltier effect refers to the evolution or absorption of heat produced by an electric current passing across junctions of two suitable, dissimilar metals, alloys, or semiconductors. Presently, thermoelectric heat pumps are used only in some specialized applications. They have not been developed to a point to make them practical for general heating and cooling of buildings.

The energy efficiency of heat pumps is measured by calculating their coefficient of performance (COP), the ratio of the heat energy obtained to the energy input. The capacity of modern heat pumps in the United States is rated in British thermal units per hour (Btu/h). The COP at a given operating point can be calculated by dividing the Btu/h output of the system by the energy input in Btu/h. Since system input for most heat pumps is electricity measured in watt-hours, the watt figure is multiplied by a conver-

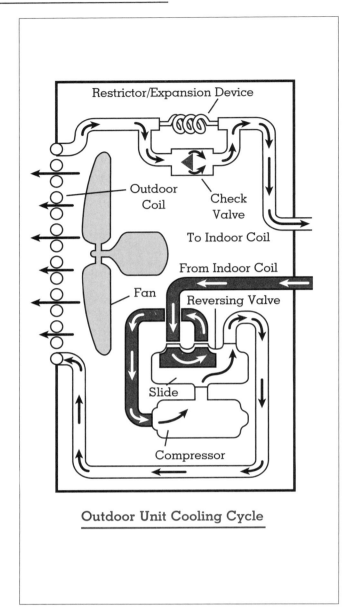

Figure 1.
Schematic of cooling cycle for a heat pump system.
SOURCE: Air Conditioning and Refrigeration Institute

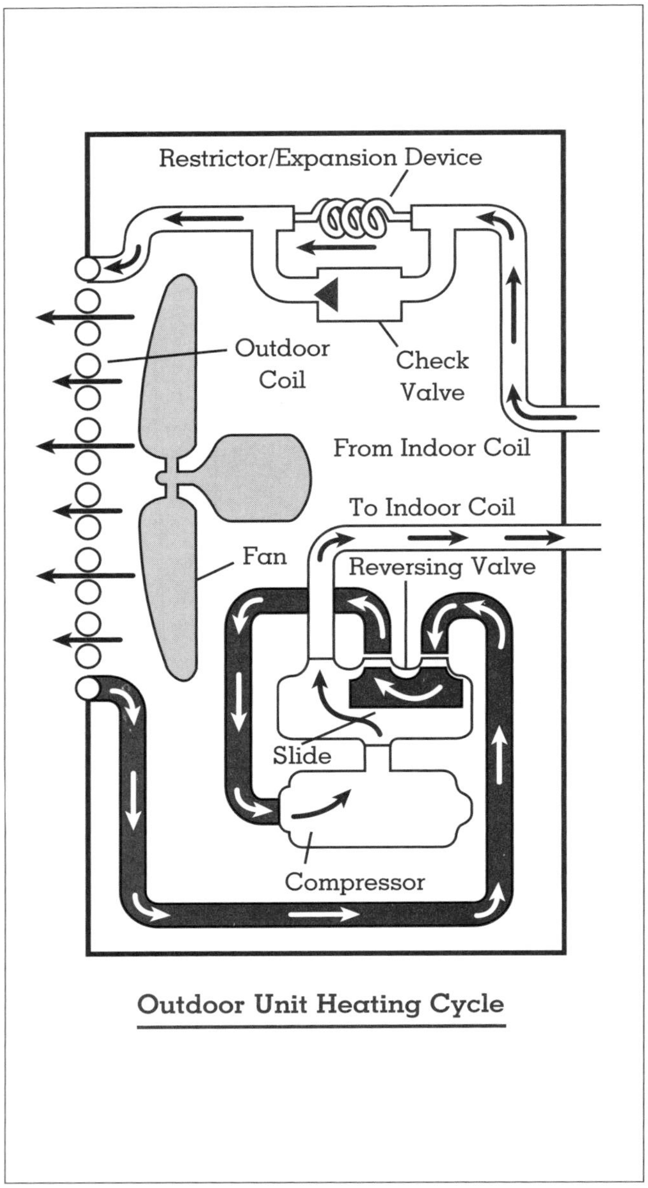

Figure 2.
Schematic of an outdoor heat pump's heating cycle.
SOURCE: Air Conditioning and Refrigeration Institute

sion factor of 3.412 to obtain the energy input in Btu/h.

There are two major factors that impact the COP: temperature difference and system component efficiency. A heat pump requires energy to move heat from a lower temperature to a higher one. As the difference in the two temperatures increases, more energy is required. The COP of a heat pump is high-er when the temperature difference is less, and less energy is consumed to transfer a given amount of heat. Thus the COP of a heat pump varies during its operation, and is relative to the system operating point, the temperature of the medium heat is transferred from, and the temperature of the medium it is transferred to. Comparison of the COP of two different systems is meaningless unless the operating points are the same. For this reason, although heat pump efficiency continues to improve, specific historical comparisons of heat pump efficiencies are difficult because system operation data was recorded at different operating points.

Air-to-air heat pumps are particularly prone to varying COPs, due to fluctuating outdoor weather conditions. In the winter, the COP decreases as the outdoor temperature decreases, due to the need for the system to transfer heat energy at greater temperature differences. The heat output decreases as the outdoor temperature drops at the same time that the heating needs of the building are increasing. In fact, the COP may drop so low that the heat pump cannot meet the thermal needs of a building. For this reason an auxiliary heat source such as electric resistance heaters or a fossil-fueled furnace is needed in geographic areas that have cold winters. The COP of air-to-air heat pumps is further reduced by the use of defrost cycles needed to clear the outdoor heat exchanger of frost buildup when the air temperature drops below about 40°F. During defrost, the refrigeration system is reversed into comfort cooling mode, heating the outdoor heat exchanger to melt the frost. Since the defrost mode is cooling the indoor air, supplemental heat is usually necessary to maintain comfort levels, reducing the overall COP of the system. The reduction varies with the duration of the defrost.

In the summer, the COP of an air-to-air heat pump decreases as the outdoor temperature rises, reducing the cooling capacity. Normally the thermal needs of the building are met since it is common practice to size a heat pump so that it will deliver adequate cooling capacity in all but the most extreme summer conditions. The winter heating capacity of the system is then determined by this tradeoff, and if the heating capacity is inadequate, supplemental electric or fossil fuel heat is required.

Because deep ground temperature has little change, geothermal heat pumps using well water function at a fairly stable COP. Geothermal types using the ground for thermal mass will see some variance in COP depending on the dryness of the soil. Wet ground conducts more heat than dry and theoretical soil conductivity may vary up to 1,000 percent. For this reason the ground loop heat exchanger should be buried deep enough to minimize soil moisture fluctuations.

System components can affect the COP because their design and performance can vary. In vapor-compression systems, the transfer efficiency of the heat exchangers and the energy efficiency of the compressor, and how these components are matched, help determine the operating COP. Compressors are performance-optimized for narrow operating ranges. If the optimization is done for heating, cooling operation may suffer. The opposite is also true.

The operating condition of the components also affects the COP. For example, if the heat exchangers become clogged or corroded, or if a homeowner fails to change a furnace filter, the operating COP of a heat pump will decrease. Any heating and cooling system will suffer some performance degradation once it has been put into use. The amount of degradation depends on the conditions under which the system has to operate, as well as how carefully system components are maintained.

Most heat pumps for residences are unitary systems; that is, they consist of one or more factory-built modules. Larger buildings may require built-up heat pumps, made of various system components that require engineering design tailored for the specific building. Sometimes, multiple unitary units are used in large buildings for ease of zone control; the building is divided up into smaller areas that have their own thermostats.

HISTORY

Oliver Evans proposed the closed vapor-refrigeration cycle in 1805 in *The Young Steam Engineer's Guide*. Evans noted: "Thus it appears possible to extract the latent heat from cold water and to apply it to boil other water." By 1852 William Thompson (Lord Kelvin) had proposed that a refrigeration system be used to either cool or heat the air in buildings, and outlined the design of such a machine. In Austria after 1855 Peter Ritter Von Rittenger constructed working heat pumps that were said to be 80 percent efficient. These devices were used to evaporate salt brine, and similar devices were constructed in Switzerland after 1870.

A theoretical discussion of the heat pump appeared in the *Journal of the Franklin Institute* in 1886. T. G. N. Haldane of Scotland comprehensively pursued the application of heat pumps to the heating of buildings after the mid-1920s. Haldane tested air-to-water heat pump systems in his home and concluded that the vapor-compression refrigeration cycle could, under certain conditions, provide a more economical means to heat buildings and swimming pools than fossil fuels. Haldane further proposed using a reversed cycle so the heat pump could be used to cool buildings or make ice. Haldane's heat pump had a coefficient of performance (COP) between 2 and 3, depending on operating conditions.

During the 1930s and 1940s a number of residential and commercial heat pump installations were made in the United States. They were of all types, and the heating COPs of these systems, where results were known, seem to have ranged from 2 to 5. Hundreds of articles and papers were published discussing the theory, application, and installed examples of heat pumps. Despite this activity, most heating systems installed were conventional fossil-fuel furnaces and boilers due to their lower first cost, broad acceptance, and well-established manufacturing, sales, and service infrastructure.

Comfort cooling was one benefit of a heat pump that was typically cited in the literature of the 1930s and 1940s. However there was not a great consumer demand for general comfort cooling at that time, particularly for residences. True, there were isolated pockets of interest, as in movie theaters and some commercial buildings, but genuine mass consumer demand for comfort cooling did not develop until the mid-twentieth century. Thus, there was no financial incentive for mass production of unitary comfort cooling air conditioners in the 1930s and 1940s, and therefore no incentive to produce unitary heat pumps. There were a few isolated attempts, such as a package heat pump marketed by DeLaVergne in 1933, but the efforts were short-lived.

Electric utilities did have a vested financial interest in heat pumps, since most designs used electricity for heating and cooling. But utilities were not manufacturers or consumers. Despite their attempts to promote heat pumps with articles and showcase system installations, they failed to create the demand for the product.

By the 1960s, central residential air conditioning was becoming increasingly popular. Equipment had developed to the point that a number of manufacturers were producing and marketing unitary air conditioning equipment. Some of these manufacturers did attempt to resurrect the idea of applying the heat pump to residential, store, and small office heating

and cooling. Systems were designed and marketed, but suffered dismally in the market. System components used did not stand up to the demands of summer and winter operation over wide weather conditions. Compressors and reversing valves in particular saw enough failures that most manufacturers withdrew their products from the marketplace. The air-to-air heat pumps of the 1960s used timed defrost cycles, initiating defrosts even when they were not necessary. Thus the system efficiency was unnecessarily reduced.

Spurred on by the energy crisis of the 1970s interest in heat pumps renewed. Rising price of fossil fuels caused manufacturers to take another look at heat pumps. Owners of "all electric homes," stung by the high cost of electric heat, were looking for a way to reduce heating costs with minimal existing heating system redesign. A heat pump could save 30 to 60 percent of the cost of electric resistance heating.

Remembering the problems of the 1960s, manufacturers redesigned system components. The system efficiency was generally higher than in the 1960s because system components were more efficient. Most air-to-air heat pumps of the period used a demand-controlled defrost cycle, further increasing overall energy efficiency. By the mid-1970s most of the major manufacturers of unitary air conditioning equipment were offering heat pumps. They were particularly popular in areas of moderate winters where air-to-air heat pumps could operate at higher COPs.

Conventional heat pumps continued to be available through the 1970s to the 1990s. The trend has been toward progressively higher energy efficiencies. The average efficiency of new heat pumps increased 60 percent (based on cooling performance) between 1976 and 1998. Minimum efficiencies were mandated for residential heat pumps in 1987 and for commercial equipment in 1992, a significant factor contributing to the rise in efficiency over the past twenty years. For heat pumps used in commercial buildings, the efficiency standards were derived from the model standards published by the American Society of Heating, Refrigerating and Air Conditioning Engineers (ASHRAE). The U.S. Department of Energy may review these standards in the next few years and setting more stringent standards (for both residential and commercial heat pumps) if technically and economically feasible.

CURRENT PRACTICE AND THE FUTURE

Heat pump systems are now rated with a Heating Seasonal Performance Factor (HSPF) and a seasonal energy efficiency ratio (SEER). The HSPF is calculated by the annual heating system output in BTU by the heating electricity usage in watt-hours. The SEER is an estimate of annual cooling output in Btu divided by the cooling electricity usage in watt-hours. The HSPF and SEER are calculated at specific rating points, standardized by the Air Conditioning and Refrigeration Institute (ARI), an industry trade organization that performs testing advocacy, and education. ARI conducts a certification program, participated in by almost all manufacturers of heat pumps. The program gives the consumer access to unbiased and uniform comparisons among various systems and manufacturers. Rating various systems at the same operating point allows accurate comparisons between systems. Heat pumps introduced in 1999 have an NSPF of 6.8 or greater and a SEER of 10.0 or greater, with high efficiency units having an HSPF of as much as 9 and a SEER of 13 or greater.

There is more interest than ever in the geothermal heat pumps. The cost of such systems has decreased with use of plastic piping for water or ground loops. Geothermal systems are proving particularly advantageous in colder winter areas, since ground temperature is much higher than outdoor temperatures. In addition, there is no efficiency-robbing defrost cycle that is necessary in air-to-air heat pumps.

New compressor technology employing rotary scroll type compressors is replacing previously-used reciprocating technology. Scroll compressors operate at higher efficiencies over wider operating conditions. Heat pumps employing scroll compressors use less supplemental heat at low outdoor temperatures. Electrical and electronic technology are making variable-capacity compressors cost-effective for the next century. Compressor performance can be optimized for a wider operation range, and matching compressor capacity to the actual demand for heating or cooling increases the system efficiency, saving energy. Application of microprocessors to control systems further permits fine-tuning of system operation. Energy efficiency is also being increased by development of higher efficiency electric motors for compressors and fans.

Increased environmental awareness will no doubt spur increasing interest in solar-assisted heat pumps in the future. Such systems can operate at higher heating COPs than more conventional heat pumps. A solar-assisted heat pump system uses a solar heating system in parallel with a conventional heat pump. The solar heating system reduces the operating time of the heat pump, and also reduces the need for supplemental electric or fossil-fuel heating during colder weather.

Bernard A. Nagengast

See also: Air Conditioning; Heat Transfer.

BIBLIOGRAPHY

Air Conditioning and Refrigeration Institute. (1999) "How a Heat Pump Works." Consumer Information section. <http://www.ari.org>.

Air Conditioning and Refrigeration Institute. (1999) "Heat, Cool, Save Energy with a Heat Pump." Arlington, VA: ARI.

Air Conditioning and Refrigeration Institute. (1999) *ARI Unitary Directory*. <http://www.ari.org>.

American Society of Heating, Refrigerating and Air Conditioning Engineers. (1978). *ASHRAE Composite Index of Technical Articles 1959–1976*. Atlanta: ASHRAE.

American Society of Heating, Refrigerating and Air Conditioning Engineers. (1996). *Absorption/Sorption Heat Pumps and Refrigeration Systems*. Atlanta: ASHRAE.

American Society of Heating, Refrigerating and Air Conditioning Engineers. (1996). *ASHRAE Handbook: Heating, Ventilating, and Air-Conditioning Systems and Equipment*. Atlanta: ASHRAE.

American Society of Heating, Refrigerating and Air Conditioning Engineers. (1999). *ASHRAE Handbook: Heating, Ventilating, and Air-Conditioning Applications*. Atlanta: ASHRAE.

Bose, J.; Parker, J.; and McQuiston, F. (1985). *Design/Data Manual for Closed-Loop Ground-Coupled Heat Pump Systems*. Atlanta: American Society of Heating, Refrigerating and Air Conditioning Engineers.

Evans, O. (1805). *The Young Steam Engineer's Guide*, p. 137. Philadelphia: H. C. Carey and I. Lea.

Haldane, T. G. N. (1930). "The Heat Pump: An Economical Method of Producing Low-Grade Heat from Electricity." *Journal of the Institution of Electrical Engineers* 68: 666–675.

Howell, R.; Sauer, H.; and Coad, W. (1997) *Principles of Heating, Ventilating and Air-Conditioning*. Atlanta: American Society of Heating, Refrigerating and Air Conditioning Engineers.

Kavanaugh, S., and Rafferty, K. (1997). *Ground Source Heat Pumps*. Atlanta: American Society of Heating, Refrigerating and Air Conditioning Engineers.

Lorsch, H. G. (1993). *Air Conditioning Design*. Atlanta: American Society of Heating, Refrigerating and Air Conditioning Engineers.

Southeastern Electric Exchange. (1947). *Heat Pump Bibliography*. Birmingham, AL: Southern Research Institute.

Sporn, P.; Ambrose, E. R.; and Baumeister, T. (1947). *Heat Pumps*. New York: John Wiley & Sons.

Thompson, W. (1852). "On the Economy of the Heating or Cooling of Buildings by Means of Currents of Air." *Proceedings of the Philosophical Society of Glasgow* 3: 269–272.

University of Pennsylvania. (1975). *Proceedings of Workshop on Solar Energy Heat Pump Systems for Heating and Cooling Buildings*. University Park, PA: University Press.

HEAT TRANSFER

Heat transfer is the energy flow that occurs between bodies as a result of a temperature difference. There are three commonly accepted modes of heat transfer: conduction, convection, and radiation. Although it is common to have two or even all three modes of heat transfer present in a given process, we will initiate the discussion as though each mode of heat transfer is distinct.

CONDUCTION

When a temperature difference exists in or across a body, an energy transfer occurs from the high-temperature region to the low-temperature region. This heat transfer, q, which can occur in gases, liquids, and solids, depends on a change in temperature, ΔT, over a distance, Δx (i.e., $\Delta T / \Delta x$) and a positive constant, k, which is called the thermal conductivity of the material. In equation form, the rate of conductive heat transfer per unit area is written as

$$q/A = -k\Delta T/\Delta x$$

where q is heat transfer, A is normal (or perpendicular to flow of heat), k is thermal conductivity, ΔT is the change in temperature, and Δx is the change in the distance in the direction of the flow. The minus sign is needed to ensure that the heat transfer is positive when heat is transferred from the high-temperature to the low-temperature regions of the body.

The thermal conductivity, k, varies considerably for different kinds of matter. On an order-of-magnitude basis, gases will typically have a conductivity range from 0.01 to 0.1 W/(m–K) (0.006 to 0.06 Btu/h-ft-°F), liquids from 0.1 to 10 W/(m–K) (0.06 to 6 Btu/h-ft-°F), nonmetallic solids from 0.1 to 50 W/(m–K) (0.06 to 30 Btu/h-ft-°F), and metallic solids from 10 to 500 W/(m–K) (6 to 300 Btu/h-ft-°F). Obviously, gases are among the lowest conductors of thermal energy, but nonmetallic materials such as foamed plastics and glass wool also have low values of thermal conductivity and are used as insulating materials. Metals are the best conductors of thermal energy. There is also a direct correlation between thermal conductivity and electrical conductivity—that is, the materials that have a high thermal conductivity also have a high electrical conductivity. Conductive heat transfer is an important factor to consider in the design of buildings and in the calculation of building energy loads.

CONVECTION

Convective heat transfer occurs when a fluid (gas or liquid) is in contact with a body at a different temperature. As a simple example, consider that you are swimming in water at 21°C (70°F), you observe that your body feels cooler than it would if you were in still air at 21°C (70°F). Also, you have observed that you feel cooler in your automobile when the air-conditioner vent is blowing directly at you than when the air stream is directed away from you. Both of these observations are directly related to convective heat transfer, and we might hypothesize that the rate of energy loss from our body due to this mode of heat transfer is dependent on not only the temperature difference but also the type of surrounding fluid and the velocity of the fluid. We can thus define the unit heat transfer for convection, q/A, as follows:

$$q/A = h(T_i - T_\infty)$$

where q is the heat transfer per unit surface area, A is the surface area, h is the convective heat transfer coefficient W/(m² – K), T_i is the temperature of the body, and T_∞ is the temperature of the fluid.

Thus convective heat transfer is a function of the temperature difference between the surface and the fluid; the value of the coefficient, h, depends on the type of fluid surrounding the object, and the velocity of the fluid flowing over the surface. Convective heat transfer is often broken down into two distinct modes: free convection and forced convection. Free convection is normally defined as the heat transfer that occurs in the absence of any external source of velocity—for example, in still air or still water. Forced convection has some external source, such as a pump or a fan, which increases the velocity of the fluid flowing over the surface. Convective heat transfer coefficients range widely in magnitude, from $6W/(m^2 - K)$ (1 Btu/h-ft²-°F) for free convection in air, to more than 200,000 $W/(m^2 - K)$ (35,000 Btu/h-ft²-°F) for pumped liquid sodium. Convective heat transfer plays a very important role in such energy applications as power boilers, where water is boiled to produce high-pressure steam for power generation.

RADIATION

Radiative heat transfer is perhaps the most difficult of the heat transfer mechanisms to understand because so many factors influence this heat transfer mode. Radiative heat transfer does not require a medium through which the heat is transferred, unlike both conduction and convection. The most apparent example of radiative heat transfer is the solar energy we receive from the Sun. The sunlight comes to Earth across 150,000,000 km (93,000,000 miles) through the vacuum of space. Heat transfer by radiation is also not a linear function of temperature, as are both conduction and convection. Radiative energy emission is proportional to the fourth power of the absolute temperature of a body, and radiative heat transfer occurs in proportion to the difference between the fourth power of the absolute temperatures of the two surfaces. In equation form, q/A is defined as:

$$q/A = \sigma (T_1^4 - T_2^4)$$

where q is the heat transfer per unit area, A is the surface area, σ is the Stefan-Boltzmann constant 5.67×10^{-8} $W/(m^2 - K^4)$, T_1 is the temperature of surface one, and T_2 is the temperature of surface two.

In this equation we are assuming that all of the energy leaving surface one is received by surface two, and that both surfaces are ideal emitters of radiant energy. The radiative exchange equation has to be modified to account for real situations—that is, where the surfaces are not ideal and for geometrical arrangements in which surfaces do not exchange their energies only with each other. Two factors are usually added to the above radiative exchange equation to account for deviations from ideal conditions. First, if the surfaces are not perfect emitters, a factor called the emissivity is added. The emissivity is a number less than 1, which accounts for the deviation from nonideal emission conditions. Second, a geometrical factor called a shape factor or view factor is needed to account for the fraction of radiation leaving a body that is intercepted by the other body. If we include both of these factors into the radiative exchange equation, it is modified as follows:

$$q/A = F\varepsilon F_G \varepsilon \ (T_1^4 - T_2^4)$$

where $F\varepsilon$ is a factor based on the emissivity of the surface and F_G is a factor based on geometry.

The geometric factor can be illustrated by considering the amount of sunlight (or radiative heat) received by Earth from the Sun. If you draw a huge sphere with a radius of 150 million km (93 million miles) around the sun that passes through Earth, the geometric factor for the Sun to Earth would be the ratio of the area on that sphere's surface blocked by Earth to the surface area of the sphere. Obviously, Earth receives only a tiny fraction of the total energy emitted from the Sun.

Other factors that complicate the radiative heat transfer process involve the characteristics of the surface that is receiving the radiant energy. The surface may reflect, absorb, or transmit the impinging radiant energy. These characteristics are referred to as the reflectivity, absorbtivity, and transmissivity, respectively, and usually are denoted as ρ, α, and τ. Opaque surfaces will not transmit any incoming radiation (they absorb or reflect all of it), but translucent and clear surfaces will transmit some of the incoming radiation. A further complicating factor is that thermal radiation is wavelength-dependent, and one has to know the wavelength (spectral) characteristics of the material to determine how it will behave when thermal radiation is incident on the surface. Glass is typical of a material with wavelength-dependent properties. You have observed that an automobile sitting in hot sunlight reaches a temperature much higher than ambient temperature conditions. The reason is that radiant energy from the Sun strikes the car windows, and the very-short-wavelength radiation readily passes through the glass. The glass, however, has spectrally dependent properties and absorbs almost all the radiant energy emitted by the heated surfaces within the car (at longer wavelengths), effectively trapping it inside the car. Thus the car interior can achieve temperatures exceeding 65°C (150°F). This is the same principle on which a greenhouse works or by which a passively heated house is warmed by solar energy. The glass selectively transmits the radiation from the Sun and traps the longer-wavelength radiation emitted from surfaces inside the home or the greenhouse. Specially made solar glasses and plastics are used to take advantage of the spectral nature of thermal radiation.

R VALUES

Many everyday heat flows, such as those through windows and walls, involve all three heat transfer mechanisms—conduction, convection, and radiation. In these situations, engineers often approximate the calculation of these heat flows using the concept of R values, or resistance to heat flow. The R value combines the effects of all three mechanisms into a single coefficient.

R Value Example: Wall

Consider the simple wall consisting of a single layer of Sheetrock, insulation, and a layer of siding as shown in Figure 1. We assume it is a cool autumn evening when the room air is 21°C (70°F) and the outside air is 0°C (32°F), so heat will flow from inside the wall to the outside. Convective and radiative heat transfer occurs on both the inside and outside surfaces, and conduction occurs through the Sheetrock, insulation, and siding. On the outside, the convective heat transfer is largely due to wind, and on the inside of the wall, the convective heat transfer is a combination of natural and forced convection due to internal fans and blowers from a heating system. The room surfaces are so close to the outside wall temperature that it is natural to expect the radiative heat transfer to be very small, but it actually accounts for more than half of the heat flow at the inside surface of the wall in this situation. When treated using the R value con-

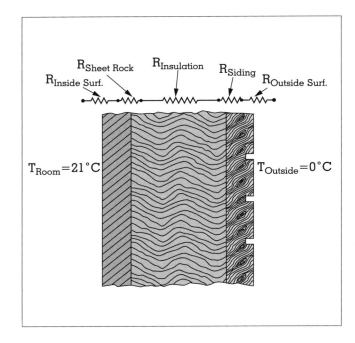

Figure 1.
Schematic diagram of a simple wall.

cept, the total R value of the wall is the sum of individual R values due to the inside surface, the Sheetrock, the insulation, the siding, and the outer surface. Hence the heat flow through this wall may be written as

$$Q = A(T_{room} - T_{outside})/(R_{inside\ surf.} + R_{Sheetrock} + R_{insulation} + R_{siding} + R_{outside\ surf.})$$

The resistances of the Sheetrock, siding, and insulation may be viewed as conductive resistances, while the resistance at the two surfaces combine the effects of convection and radiation between the surface and its surroundings. Typical values for these resistances in units of $(W/m^2 - W/m^2\text{-}°C)^{-1}$ $((Btu/h\text{-}ft^2\text{-}°F)^{-1})$ are as follows:

$R_{inside\ surf}$	= 0.12	(0.68)
$R_{Sheetrock}$	= 0.08	(0.45)
$R_{insulation}$	= 1.94	(11.0)
R_{siding}	= 0.14	(0.80)
$R_{outside\ surf}$	= 0.04	(0.25)
R_{total}	= 2.32	(13.18)

For this case, the heat flow through the wall will be $(21-0)°C/2.32\ (W/m^2\text{-}°C)^{-1} = 9.05\ W/m^2$ (2.87

Btu/h-ft²). While the surfaces, Sheetrock, and siding each impede heat flow, 80 percent of the resistance to heat flow in this wall comes from the insulation. If the insulation is removed, and the cavity is filled with air, the resistance of the gap will be 0.16 (W/m²-°C)⁻¹ (0.9 (Btu/h-ft²-°F)⁻¹) and the total resistance of the wall will drop to 0.54 (W/m²-°C)⁻¹ (3.08 (Btu/h-ft²-°F)⁻¹) resulting in a heat flow of 38.89 W/m² (12.99 Btu/h-ft²). The actual heat flow would probably be somewhat different, because the R-value approach assumes that the specified conditions have persisted long enough that the heat flow is "steady-state," so it is not changing as time goes on. In this example the surface resistance at the outer wall is less than half that at the inner wall, since the resistance value at the outer wall corresponds to a wall exposed to a wind velocity of about 3.6 m/s (8 mph), which substantially lowers the resistance of this surface to heat flow.

If the wall in the example had sunlight shining on it, the heat absorbed on the outer surface of the wall would reduce the flow of heat from inside to outside (or could reverse it, in bright sunshine), even if the temperatures were the same.

R Value Example: Window

A window consisting of a single piece of clear glass can also be treated with R-value analysis. As with the wall, there is convective and radiative heat transfer at the two surfaces and conductive heat transfer through the glass. The resistance of the window is due to the two surface resistances and to the conductive resistance of the glass, R_{glass}. For typical window glass, R_{glass} = 0.003 (W/m²-°C)⁻¹ (0.02 (Btu/h-ft²-°F)⁻¹) so the total resistance of the window is R_{window} = (0.12 + 0.003 + 0.04) (W/m²-°C)⁻¹ = 0.163 (W/m²-°C)⁻¹ (0.95 (Btu/h-ft²-°F)⁻¹). Thus the heat flow will be q = (21 – 0)°C/0.163(W/m²-°C)⁻¹ = 128.8 W/m² (40.8 Btu/h-ft²), or fourteen times as much as that through the insulated wall. It is interesting to note that the heat flow through an ancient window made from a piece of oilskin, or even a "window" made from a piece of computer paper, would not increase by more than 2 percent from that of the glass window, because the resistance of the glass is so small.

When sunlight is shining through a window, the heat transfer becomes more complicated. Consider Figure 2.

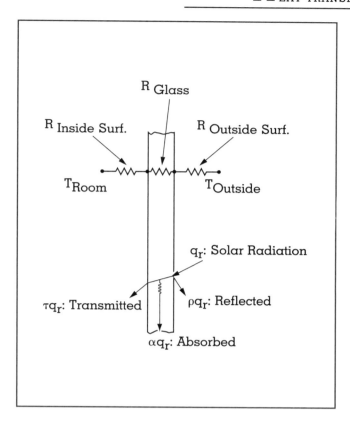

Figure 2.
Schematic diagram of a window.

Suppose the outdoor and indoor temperatures are still 0°C and 21°C, but the window now has 600 W/m² (190 Btu/ft²-h) of sunlight striking it (q_r), typical of a fairly sunny window on an autumn day. About 8 percent (ρq_r) will be reflected back into the atmosphere, about 5 percent (αq_r), or 30 W/m², will be absorbed in the glass, and the remaining 87 percent (τq_r), or 522W/m², will be transmitted into the room, providing light before it strikes the walls, floor, ceiling, and furnishings in the room. Again, some will be reflected, providing indirect lighting, and the remainder will be absorbed and converted to heat. Eventually, all—except any that may be reflected back out the window—will be absorbed and converted to heat in the room. Thus the amount of sunlight coming through the window is about four times as great as the amount of heat flowing outward. The 30 W/m² that is absorbed in the glass heats the glass slightly, reducing the conductive heat flow through the glass to 121 W/m², so the net effect of the sunlight is to result in the window providing 401 W/m² of heating (522–121) to the room instead of losing 128 W/m².

In this example, we have assumed that the outside temperature is lower than inside; therefore the heat flow due to the temperature difference is from inside to outside. In the summer it will be hotter outside, and heat will flow from outside to inside, adding to the heat gained from the sunlight. Because windows represent a large source of heat gain or the heat loss in a building, a number of schemes are used to reduce the heat gains in summer and heat losses in winter. These include the use of double (two glass panes) or triple glazing for windows. The glass is separated by an air space, which serves as an added insulating layer to reduce heat transfer. Reflective films are used to reduce heat gain from sunlight. Some of these films also serve to reduce the heat transfer by longwave (nonsolar) radiation as well.

There are numerous other examples where heat transfer plays an important role in energy-using systems. One is the production of steam in large boilers for power production, where steam is boiled from water. Hot combustion gases transfer heat to the water by radiation, conduction through the pipe walls, and convection. The boiling of the water to produce steam is a special case of convective heat transfer where very high heat transfer coefficients are obtained during the boiling process. Another very practical example of convective cooling is in automobiles. The "radiator" is, in fact, a "convector" where pumped water is forced through cooling tubes and is cooled by air forced over the tubes and fins of the radiator. Forced convective heat transfer occurs at both the air side and the water side of the "radiator." Cooling towers and air-conditioning coils are other practical examples of combined heat transfer, largely combined conduction and convection.

While heat transfer processes are very useful in the energy field, there are many other industries that rely heavily on heat transfer. The production of chemicals, the cooling of electronic equipment, and food preparation (both freezing and cooking) rely heavily on a thorough knowledge of heat transfer.

W. Dan Turner
David E. Claridge

See also: Air Conditioning; Furnaces and Boilers; Heat and Heating; Heat Pumps; Insulation; Refrigerators and Freezers; Solar Energy; Solar Energy, Historical Evolution of the Use of; Thermodynamics; Windows.

BIBLIOGRAPHY

ASHRAE. (1997). *1997 ASHRAE Handbook: Fundamentals*. Atlanta: Author.

Incropera, F. P., and DeWitt, D. P. (1996). *Fundamentals of Heat and Mass Transfer*, 4th ed. New York: John Wiley & Sons.

Turner, W., ed. (1982). *Energy Management Handbook*. New York: John Wiley & Sons.

HEAVISIDE, OLIVER (1850–1925)

Oliver Heaviside was born May 13, 1850, in London, England. He was a physicist and mathematician whose theoretical work played a large part in the understanding of radio transmission and long-distance telephony. Educated at Camden House School in London, Heaviside came fifth in the College of Preceptors examination out of five hundred pupils. He left school in 1866 and continued to study. He learned Morse code and studied electricity and languages. In 1868 he went to Denmark and became a telegraph operator, and in 1870 he was appointed Chief Telegraph Operator.

In 1871 Heaviside returned to England to take up a post with the Great Northern Telegraph Company dealing with overseas traffic. In 1875 he had to leave the job due to increasing deafness. Heaviside was encouraged to continue his electrical research by his uncle, Charles Wheatstone, who with W. F. Cooke patented the electric telegraph in 1837, and who later devised the Wheatstone bridge, an electrical network for measuring resistance.

In 1872 Heaviside's first paper, "Comparing of Electromotive Forces," was published. Heaviside's second paper was published in 1873 and attracted the attention of Scottish physicist James Clerk Maxwell. In 1873 Heaviside was inspired by Maxwell's treatise on electricity and magnetism. It took Heaviside several years to fully understand Maxwell's book, which he then set aside to follow his own course of thinking. Finding Maxwell's conventional mathematics difficult to apply to practical matters, Heaviside introduced a simpler treatment that, in his opinion, did not impair accuracy. It proved to be controversial. His proofs did not conform to the standards of the time and his methods on partial differential equations had varying success.

In 1874 Heaviside established the ordinary symbolic method of analyzing alternating current circuits in common use today. It was a technique developed about fifteen years before AC came into commercial use. He emphasized the role of metallic circuits as guides, rather than conductors of AC currents. He discussed in detail causes of line distortion and suggested ways of alleviating it.

Between 1880 and 1887 Heaviside developed his operational form of calculus, a branch of mathematics that permits the manipulation of varying quantities applicable to practical problems, including such matters as electrical circuits with varying voltages and currents. Heaviside received a great honor in 1891 when he was elected a Fellow of The Royal Society in recognition of his work on electromagnetic waves. He was later the first recipient of the Society's Faraday Medal. The first volume of his great work "Electromagnetic Theory" was published in 1893 and a second volume in 1899. A third volume was published in 1912, but he died before completing the fourth.

In 1902 Heaviside's famous prediction about an ionized layer in the atmosphere that would deflect radio waves was published in an article titled "Telegraphy," in the tenth edition of "Encyclopaedia Britannica." The idea came when he was considering the analogy between the movement of electric waves along a pair of conducting wires and over a conducting earth. Discussing the possibility of radio waves being guided around a curved path he suggested: "There may possibly be a sufficiently conducting layer in the upper air. If so, the waves will, so to speak, catch on to it more or less. Then guidance will be by the sea on one side and the upper layer on the other." The layer was first named the Heaviside Layer and later the Kennelly-Heaviside Layer, as a similar prediction had been made around the same time by Arthur Kennelly at Harvard University. The hypothesis was proved correct in 1924 when radio waves received from the upper atmosphere showed that deflection of upward waves took place at a height of approximately 100 kilometers.

Heaviside made a significant contribution to electrical communications when he advocated the introduction of additional inductance in long-distance telephony cables although there was then no practical means to add it. His idea was eventually patented in 1904 by Michael Campbell of AT&T after Heaviside and George Pupin of Columbia University had shown it was possible to apply inductance in the form of uniformly spaced loading coils. By 1920 engineers had installed such loading on thousands of miles of cable, particularly in the United States.

During his lifetime, Heaviside made extensive contributions to pure mathematics and its practical applications to alternating current, vector analysis and telegraphy. He introduced new concepts that later became commonplace thinking and expressions for every electrical engineer, and invented much of the language now basic to communication engineering, including such words as "capacitance," "inductance," "impedance," and "attenuation."

Heaviside's last years were spent as an embittered recluse at Torquay, Devon where he allowed only a few people to visit him. For much of his life he suffered from recurring jaundice that was to prove fatal. He died on February 3, 1925. Heaviside's work has been an inspiration to countless electrical engineers and mathematicians. Time has enhanced the esteem in which he is held; succeeding generations have spent many hours studying his writings. As a lasting honor, craters on Mars and the Earth's moon were named after him.

Alan S. Heather

See also: Wheatstone, Charles.

BIBLIOGRAPHY

Institution of Electrical Engineers. (1950). *The Heaviside Centenary Volume.* London: Author.

Josephs, H. J. (1963). *Oliver Heaviside: A Biography.* London: Author.

Nahin, P. (1988). *Oliver Heaviside: Sage in Solitude: The Life, Work, and Times of an Electrical Genius of the Victorian Age.* New York: I.E.E.E.

Searle, G. F. C. (1987). *Oliver Heaviside, the Man.* St Albans: CAM Publishing.

Yavetz, I. (1995). *From Obscurity to Enigma: The Work of Oliver Heaviside, 1872–1889.* Basel: Birkhauser Verlag.

HELMHOLTZ, HERMANN VON (1821–1894)

One of the most versatile scientists who ever lived, Hermann von Helmholtz was born in Potsdam, Germany, in 1821, the son of a "Gymnasium" (high school) teacher. From an early age he wanted to be a physicist, but his family could not afford the money required for his education. Instead, his father persuaded him to take up medicine, since his son's education as a physician would be subsidized by the state on the condition that he serve as a doctor in the Prussian army after he received his degree. Helmholtz attended the Institute for Medicine and Surgery in Berlin from 1838 to 1842 and fulfilled his obligation as an army surgeon from 1843 to 1848. His real interest, however, was always research; even in the army barracks he set up a small laboratory for research in physiology and physics.

On July 23, 1847, Helmholtz presented a paper on the conservation of energy at a meeting of the Berlin Physical Society. It was a talk at a reasonably high level of mathematical sophistication intended to convince physicists that energy was always conserved in any physical process. Although it was rejected by J. C. Poggendorf, the editor of the *Annalen der Physik*, as being too long and too mathematical for his readers, it eventually appeared in pamphlet form and was soon recognized as one of the most important papers in nineteenth-century science. This bold and path-breaking paper, written when he was only twenty-six, was Helmholtz's first and most fundamental statement of the conservation-of-energy principle. It came at a critical moment in the history of science when scientists and philosophers were waging a battle over whether conservation of energy was a truly universal principle. In his 1847 paper Helmholtz had shown convincingly that it was.

Helmholtz stated that his only purpose in his 1847 paper was to provide a careful investigation and arrangement of accepted facts about energy conservation for the benefit of both the physicists and the physiologists who attended his lecture. He never claimed priority for himself in this discovery, but later conceded that honor to J. R. Mayer and James Joule. When Helmholtz delivered this paper, however, he knew little about Joule's research, which was going on at about the same time, and nothing at all about Robert Mayer's 1842 paper, which had appeared in a journal not normally read by many scientists.

Some years later, in 1861, when Helmholtz was much involved in physiology, he realized that he had discussed physiology in only two paragraphs of his 1847 paper, and so supplemented it by a fuller consideration of energy transformations in organic systems. This he included in a talk before the Royal Society of London on April 12, 1861. Here he generously referred to the important contributions of Sadi Carnot, Mayer and Joule in establishing the principle of conservation of energy on a firm foundation.

On the basis of this work and other research, in 1849 Helmholtz was appointed professor of physiology at the University of Koönigsberg. There he devoted himself to the physiology of the eye, first explaining the mechanism of lens accommodation. In 1851 his invention of the ophthalmoscope, still the basic instrument used by eye doctors to peer at the retina of the eye, immediately made Helmholtz famous. In 1852 he also became the first experimentalist to measure the speed of nerve impulses in the human body.

Helmholtz did important research on another sense organ, the ear, and explained how it was able to detect differences in pitch. He showed how the quality of a sound depended on the number, nature and relative intensities of the harmonics present in the sound.

On the basis of his studies on the eye and ear, in 1858 Helmholtz was appointed Professor of Anatomy in Heidelberg. His thirteen years in Heidelberg gave him the opportunity to work closely with other gifted scientists including physicist Gustav Kirchhoff and chemist Robert Bunsen, in what was then called "an era of brilliance such as has seldom existed for any university and will not readily be seen again." In 1871 Helmholtz abandoned anatomy and physiology in favor of physics. He accepted the most prestigious chair of physics in Germany at the University of Berlin and spent the rest of his life there.

As he grew older, Helmholtz became more and more interested in the mathematical side of physics and made noteworthy theoretical contributions to classical mechanics, fluid mechanics, thermodynamics and electrodynamics. He devoted the last decade of his life to an attempt to unify all of physics under one fundamental principle, the principle of least action. This attempt, while evidence of Helmholtz's philosphical bent, was no more successful than was Albert Einstein's later quest for a unified field theory. Helmholtz died in 1894 as the result of a fall suffered on board ship while on his way back to Germany from the United States, after representing Germany at the Electrical Congress in Chicago in August, 1893.

It is difficult to exaggerate the influence Helmholtz had on nineteenth-century science, not only in Germany but throughout the world. It was during his lifetime that Germany gained its preeminence in science, which it was not to lose until World War II. His own research contributions, together with the impetus he gave to talented students by his teaching, research guidance, and popular lectures, had much to do with the scientific renaissance that Germany experienced during his lifetime.

Helmholtz was a sensitive and sickly man all his life, plagued by severe migraine headaches and fainting spells. He sought relief from his pain in music and the other arts, and in mountain climbing in the Alps. It is intriguing to imagine what he might have accomplished had he been in good health for all his seventy-three years. On the occasion of Helmholtz's death, Lord Kelvin stated: "In the historical record of science the name of Helmholtz stands unique in grandeur, as a master and leader in mathematics, and in biology, and in physics." Many scientists felt that Kelvin had short-changed Helmholtz by this statement; Medicine, physiology, chemistry, and philosophy certainly deserved to be added to the list of fields Helmholtz had mastered during his long and productive scientific career.

Joseph F. Mulligan

BIBLIOGRAPHY

Cahan, D., ed. (1993). *Hermann von Helmholtz and the Foundations of Nineteenth-Century Science.* Berkeley, California: University of California Press.

Cahan, D., ed. (1995). *Science and Culture. Popular and Philosophical Essays of Hermann von Helmholtz.* Chicago: University of Chicago Press.

Helmholtz, H. v. (1847). "The Conservation of Force: A Physical Memoir." In *Selected Writings of Hermann von Helmholtz* (1971), ed. R. Kahl, pp. 3–55. Middletown, CT: Wesleyan University Press.

Helmholtz, H. v. (1861). "The Application of the Law of Conservation of Force to Organic Nature." In *Selected Writings of Hermann von Helmholtz* (1971), ed. R. Kahl, pp. 109–121. Middletown, CT: Wesleyan University Press.

Kahl, R., ed. (1971). *Selected Writings of Hermann von Helmholtz.* Middletown, CT: Wesleyan University Press.

Koenigsberger, L. (1965). *Hermann von Helmholtz* (One-volume condensation of three-volume 1902 German edition), tr. Frances A. Welby. New York: Dover Publications.

Mulligan, J. F. (1989). "Hermann von Helmholtz and His Students." *American Journal of Physics* 57:68–74.

Turner, R. S. (1972). "Helmholtz, Hermann von." In *Dictionary of Scientific Biography,* ed. Charles Coulston Gillispie, Vol. 6, pp. 241–253. New York: Scribner.

HERTZ, HEINRICH RUDOLF (1857–1894)

Heinrich Hertz was born into a well-to-do German family in Hamburg on February 22, 1857. He was an exceptionally talented student, doing equally well in the humanities and sciences. He read extensively, tried his hand at sculpting, and even built scientific apparatus on a lathe at home. After a year of military service, Hertz studied structural engineering in Munich, but gave it up for physics when he realized he had the ability to contribute something substantial to that field. At the age of 20 he wrote to his parents: "… I would rather be a great scientific investigator than a great engineer, but I would rather be a second-rate engineer than a second-rate investigator."

In 1878 Hertz enrolled at the University of Berlin to study under Hermann von Helmholtz, the leading German physicist of the time. He obtained his degree *magna cum laude* in 1880 with a theoretical dissertation on the electromagnetic induction of currents in

Heinrich Rudolf Hertz. (Library of Congress)

conducting spheres rotating in a magnetic field. He remained at the Berlin Physics Institute as assistant to Helmholtz until 1883; during these years he published fifteen research papers on a great variety of topics in physics.

In 1883 Hertz was appointed *Privatdozent* for mathematical physics at Kiel, and after two years became a full professor at the *Technische Hochschule* in Karlsruhe. In 1889 Hertz left Karlsruhe to assume his last academic post as Professor of Physics at the Friedrich-Wilhelm University in Bonn. Five years later, following a long period of declining health and many painful operations, Heinrich Hertz died in Bonn of blood poisoning on January 1, 1894, a few months before his thirty-seventh birthday.

After Hertz's death Helmholtz paid tribute to his former student as a "consummate physicist," who uniquely combined mathematical ability, theoretical insight, and experimental skill. These qualities enabled Hertz to make many important contributions to physics, of which only those relating more directly to energy are outlined below.

Hertz's most direct involvement with all aspects of the energy question was the research he did to prepare his inaugural lecture to the faculty at the *Technische Hochschule* in Karlsruhe, delivered on April 20, 1885 and entitled: "On the Energy Balance of the Earth." The manuscript for this lecture has only recently been found and published in both German and English. Although written more than a century ago, this impressive document records both Hertz's insightful view of the Earth's energy situation at that time and his remarkably good order-of-magnitude estimates of the energy sources then known to be available to the Earth.

The most important contribution Hertz made in this inaugural lecture was his prediction, based on his estimates of the energy sources available, that ultimately the Earth was completely dependent on the Sun for the light and heat it needed to support life. Of course, this picture would change after Henri Becquerel discovered radioactivity in 1896, and thus introduced the nuclear age of physics.

The research that brought Hertz undying fame as a physicist was that on electromagnetic waves, performed in 1886–1889 in Karlsruhe. By his elegant experiments he confirmed the theoretical prediction of James Clerk Maxwell that electromagnetic waves in what are now called the microwave and radiowave regions of the spectrum travel through a vacuum at the speed of light. He also demonstrated that microwaves of 66-cm wavelength exhibit the same properties of reflection, refraction, interference and polarization as do light waves. Hertz's research also provided conclusive evidence that electromagnetic energy cannot be transmitted from place to place instantaneously, but only at a finite velocity, that of light. Hertz never considered the possibility of using electromagnetic waves for wireless communication over long distances. His sole interest was in understanding the world about him — "the intellectual mastery of nature," in the words of Helmholtz.

In the course of his research on electromagnetic waves Hertz discovered the photoelectric effect. He showed that for the metals he used as targets, incident radiation in the ultraviolet was required to release negative charges from the metal. Research by Philipp Lenard, Wilhelm Hallwachs, J. J. Thomson, and other physicists finally led Albert Einstein to his famous 1905 equation for the photoelectric effect, which includes the idea that electromagnetic energy is "quantized" in units of $h\nu$, where h is Planck's constant and ν is the frequency of the "bundle of energy" or "photon." Einstein's equation is simply a conservation-of-energy equation, stating that the energy of the incident photon (hν) is used partially to provide the energy needed to extract the negative particle (electron) from the metal, with the rest of the photon's energy going into the kinetic energy of the extracted electron. Hertz's discovery of the photoelectric effect in 1887 therefore led eventually to Einstein's energy-conservation equation for submicroscopic systems.

Hertz's last piece of experimental work was done when his health was deteriorating and he was devoting most of his research time to intensive theoretical work on the logical foundations of mechanics. In 1892 in his laboratory in Bonn, he discovered that cathode rays could pass through thin metallic foils. He published a short paper on the subject, but did not pursue the matter further. Instead he handed his apparatus and his ideas over to Philipp Lenard (1862–1947), his assistant in Bonn. Lenard pushed Hertz's suggested research so far that Lenard received the Nobel Prize in Physics in 1905 "for his work on cathode rays."

During his brief life Hertz moved back and forth with extraordinary ease and great dedication between intense theoretical study at his desk and equally demanding experimental work in his laboratory. As Robert S. Cohen wrote of Hertz in 1956, "His like is rare enough within science ... but his fusion of theory and experiment with a creative interest in philosophical and logical foundations [as revealed particularly in his *Principles of Mechanics*] is nearly unique."

Joseph F. Mulligan

BIBLIOGRAPHY

Fölsing, A. (1997). *Heinrich Hertz. Eine Biographie.* Hamburg: Hoffmann und Campe.

Hertz, H. G. and Mulligan, J. F., eds. (1998). "Der Energiehaushalt der Erde," von Heinrich Hertz, *Fridericiana* (Zeitschrift der Universität Karlsruhe), Heft 54:3–15.

McCormmach, R. (1972). "Hertz, Heinrich Rudolf." In *Dictionary of Scientific Biography,* ed. Charles Coulston Gillispie, Vol. 6, 340–350. New York: Scribner.

Mulligan, J. F., ed. (1994). *Heinrich Rudolf Hertz: A Collection of Articles and Addresses*; with an Introductory Biography by the editor. New York: Garland.

Mulligan, J. F., and Hertz, H. Gerhard. (1997). "An Unpublished Lecture by Heinrich Hertz: 'On the Energy Balance of the Earth.'" *American Journal of Physics* 65: 36–45.

Planck, M. (1894). *Heinrich Rudolf Hertz: A Memorial Address.* English translation in *Heinrich Rudolf Hertz: A Collection of Articles and Addresses* (1994), ed. J. F. Mulligan, pp. 383–403. New York: Garland.

Susskind, C. (1995). *Heinrich Hertz. A Short Life.* San Francisco: San Francisco Press, Inc.

HIGH OCCUPANCY VEHICLE LANES

See: Traffic Flow Management

HISTORICAL PERSPECTIVES AND SOCIAL CONSEQUENCES

As with so many other phenomena that define our civilization, we owe the idea of energy—from *en* (in) and *ergon* (work)—to ancient Greeks. In his *Metaphysics,* Aristotle gave the term a primarily kinetic meaning: "The term "actuality" (*energeia*) ... has been extended to other things from motions, where it was mostly used; for *actuality* is thought to be motion most of all" (*Metaphysics*, Theta 3, p. 149).

For Greeks the word and its cognate terms filled a much larger conceptual niche than they do in modern scientific usage. In some of Aristotle's writings *energeia* stands in opposition to mere disposition, *hexis*; in others it carries the vigor of the style. The verb *energein* meant to be in action, implying constant motion, work, production, and change. The classical concept of *energeia* was thus a philosophical generalization, an intuitive expression embracing the totality of transitory processes, the shift from the potential to the actual. Although the perception was clearly holistic, it did not embrace the modern notion of the underlying commonality of diverse energies, the fact that their conversions can perform useful work.

This understanding was clearly formulated only by the middle of the nineteenth century; conceptualization of energy thus made hardly any advances during more than two thousand years following Aristotle's writings. Interestingly, even many founders of modern science held some dubious notions concerning energy. To Galileo Galilei, heat was an illusion of senses, an outcome of mental alchemies; Francis Bacon thought that heat could not generate motion or motion, heat.

But neither the absence of any unified understanding of the phenomenon nor the prevalence of erroneous interpretations of various energy conversions prevented a great deal of empirical progress in harnessing diverse energies and in gradually improving efficiencies of some of their conversions. Seen from a biophysical point of view, all human activities are nothing but a series of energy conversions, and so it is inevitable that different energy sources and changing conversion techniques have imposed obvious limits on the scope of our action—or opened up new possibilities for human development.

PREHISTORIC AND ANCIENT CULTURES

From the perspective of general energetics, the long span of human prehistoric development can be seen as the quest for a more efficient use of somatic energy, the muscular exertions used primarily to secure a basic food supply and then to gradually improve shelters, acquire more material possessions, and evolve a variety of cultural expressions. This quest was always limited by fundamental bioenergetic considerations: Fifty to ninety watts is the limit of useful work that healthy adults can sustain for prolonged periods of time (of course, short bursts of effort could reach hundreds of watts).

Human labor dominated all subsistence foraging activities, as the food acquired by gathering and hunting sufficed merely to maintain the essential metabolic functions and to support very slow population growth. Societies not very different from this ancestral archetype survived in some parts of the world (South Africa, Australia) well into the twentieth century: Because they commanded very little energy beyond their subsistence food needs, they had very few material possessions and no permanent abodes.

Simple wooden, stone, and leather tools—including digging sticks, bows, arrows, spears, knives and scrapers—were used to increase and extend the inherently limited musclepower in collecting, hunting, and processing tasks. The only extrasomatic energy conversion mastered by foraging societies was the use of fire for warmth and cooking. The earliest dates for controlled use of fire for warmth and cooking remain arguable: it may have been nearly 500,000 years ago, but a more certain time is about 250,000 years ago.

Energy returns in more diversified foraging (energy in food/energy spent in collecting and hunting) varied widely: they were barely positive for some types of hunting (particularly for small arboreal animals), high for gathering tubers (up to fortyfold), and very high for coastal collecting and hunting of marine species ranging from shellfish to whales. Some foraging societies able to secure high energy returns built permanent dwellings and also channeled the surplus energies into more elaborate tools and remarkable artistic expressions.

Energy returns actually declined when some foragers began practicing shifting cultivation, and they declined further with sedentary agriculture—but these modes of subsistence were gradually adopted because they made it possible to support much larger population densities. Carrying capacities of foraging societies were as low as 0.01 person per square kilometer in arid regions and almost as high as 1 person per square kilometer in species-rich coastal sites. In contrast, shifting agricultures would commonly support 20 to 30 people per square kilometer, and even the earliest, extensive forms of settled farming could feed 100 to 200 people per square kilometer (or one to two people from a hectare of cultivated land). Shifting agriculturalists extended the use of fire to the removal of forest vegetation and acquired a larger assortment of tools.

Human labor—with varying contributions by slave, corvee, and free work—continued to be the dominant source of mechanical energy throughout antiquity, and combustion of biomass fuels remained the most important conversion of extrasomatic energy in all ancient civilizations. Indeed, these two energy sources remained critically important until the creation of modern industrial civilization. But, in contrast to the prehistoric era, energetic perspectives on human history reveal a fascinating quest for harnessing extrasomatic energies combined with ingenious efforts to increase the efficiency of available energy conversions. Both the collective achievements

Painting of a Phoenician merchant galley, c. 7th century B.C.E., crossing the Red Sea. (Corbis Corporation)

of human societies and individual standards of living have always been highly correlated with success in diversifying energy sources and increasing efficiencies of their use.

With more ingenious tools (based on such fundamental mechanical principles as the lever and the inclined plane) and with thoughtful organization of often very complex tasks, societies commanding just musclepower were able to accomplish such feats as the building of admirably planned megalithic structures (such as Stonehenge and the colossi of Easter Island), pyramids (Egypt, Mesoamerica), and stone temples, walls, and fortresses on four continents. In contrast to this concerted deployment of human labor in construction—amounting to one to five kilowatts of sustained power in work gangs of ten to fifty people—no Old High culture took steps to the really large-scale manufacture of goods.

The use of fire was extended during antiquity to produce bricks and to smelt various metals, starting with copper (before 4000 B.C.E.) and progressing relatively rapidly to iron (the metal was used extensively in parts of the Old World after 1400 B.C.E.). Conversion of wood to charcoal introduced a fuel of higher energy density (about 50% higher than air-dried wood) and a superior quality. Being virtually pure carbon, charcoal was nearly smokeless, and hence much better suited for interior heating, and its combustion could produce high temperatures needed

for smelting metal ores. But the inefficient method of charcoal production meant that a unit of charcoal's energy required commonly five to six units of wood.

ENERGY CONVERSIONS IN TRADITIONAL SOCIETIES

All ancient societies were eventually able to harness a variety of extrasomatic energies other than the combustion of biomass fuels and the use of pack animals for transport. Capturing wind by simple sails made seagoing transport much cheaper than moving goods or people on land, and it enabled many societies (most notably the Phoenicians and the Greeks, and later, of course, the Romans) to extend their cultural reach far beyond their core regions. However, simple square or rectangular sails were rather inefficient airfoils unsuitable for sailing in the open ocean. Transoceanic voyages had to wait for many centuries not only for the adoption of better sails but also for better keels, rudders, and compasses.

Cattle were the first animals used as sources of draft power in agriculture, and the coincidence of the first clearly documented cases of cattle domestication and plow farming are hardly surprising. Cattle were used in many regions also for lifting irrigation water and for processing harvested crops, and everywhere for transportation. Individual small animals could deliver no more than 100 to 300 watts of useful power, and a pair of good oxen (harnessed by head or neck yokes) could work at a sustained rate of 600 to 800 watts. A peasant working with a hoe would need 100 to 200 hours to prepare a hectare of land for planting cereals. Even with a simple wooden plow pulled by a single medium-size ox, that task could be done in just over 30 hours. Hoe-dependent farming could have never attained the scale of cultivation made possible by domesticated draft animals.

Because of their size and anatomy, horses are inherently more powerful draft animals than cattle: A good horse can work at a rate of 500 to 800 watts, an equivalent of eight to ten men. But this superiority was not translated into actual performance until the invention and general adoption of an efficient harness. The oldest preserved images of working horses do not show them laboring in fields, but rather pulling light ceremonial or attack carriages while using a throat-and-girth harness that is not suited for heavy fieldwork. The breastband harness was better, but only the invention of the collar harness (in China, just before the beginning of the common era) turned horses into superior draft animals.

By the ninth century an improved version of the collar harness reached Europe, where its use remained largely unchanged until horses were replaced by machines more than 700 years later. Millions of working horses worldwide use it still. Two other improvements that made horses into superior draft animals were the diffusion of horseshoes, and better feeding provided by cultivation of grain feeds. A pair of good horses could sustain one and a half kilowatts for many hours, and peak exertions of individual large animals surpassed two kilowatts. Horse teams of the late nineteenth century (with four to forty animals) were powerful enough to open up heavy grassland soils for pioneer farming and to pull the first mechanical harvesters and, later, the first modern combines.

Harnessing of wind by simple sails aside, the diversification of inanimate energy sources began about 2,000 years ago, and their slowly expanding contributions were limited by low efficiencies of prevailing conversion techniques as well as by their cost. Roman civilization, Islam, dynastic China, and premodern Europe came up with many ingenious solutions to convert kinetic energies of wind and water into useful work, easing many everyday tasks as well as allowing for previously impossible mechanical feats.

The first inanimate source of energy harnessed by a machine was flowing water. The origin of the earliest waterwheels remains uncertain. Vertical wheels, favored by the Romans, were much more efficient than the horizontal ones, and their first use was turning the millstones by right-angle gears. Use of waterwheels gradually expanded far beyond grain milling, and their power was used for sawing, wood turning, oil pressing, paper making, cloth fulling, tanning, ore crushing, iron making, wire pulling, stamping, cutting, metal grinding, blacksmithing, wood and metal burning, majolica glazing, and polishing.

Most of these tasks were eventually also done by windmills, whose diffusion began about a millennium after the wider use of waterwheels (the first clear European records come from the last decades of the twelfth century). While simple Asian machines had horizontally mounted sails, European mills were vertically mounted rotaries whose driving shafts could be turned into the wind. Useful power of both waterwheels and windmills continued to increase only slowly: Even during the early decades of the eighteenth century, European waterwheels averaged less than four kilowatts, and a typical large eighteenth-century Dutch mill with a thirty-meter span could develop seven and a half kilowatts (an equivalent of ten good horses).

Per capita consumption of biomass fuels in these premodern societies was commonly well below twenty gigajoules a year (an equivalent of less than one metric ton of wood). Radical changes came only with the widespread use of coal (and coke) and with the adoption of steam engines for many stationary and mobile uses. These changes were gradual and decidedly evolutionary: The notion of the eighteenth-century coal-based Industrial Revolution is historically inaccurate. Both in Europe and in North America it was waterpower, rather than coal combustion, that was the prime mover of rapidly expanding textile and other manufacturing industries, and in many countries charcoal-based smelting dominated iron production until the last decades of the nineteenth century.

EMERGENCE OF FOSSIL-FUELED ECONOMIES

In parts of Europe and northern China coal was mined, and used directly as household fuel and in small-scale manufacturing, for centuries preceding the first commercial use of Newcomen's inefficient

Two men stand on the Victoria Express Engine Series 1070, which was designed by Matthew Kirtly, the locomotive superintendent of Midland Railway in England during the early 1870s. (Corbis Corporation)

steam engine in the early decades of the eighteenth century. After 1770 James Watt transformed the existing steam engine into a prime mover of unprecedented power: Although his improved design was still rather inefficient, average capacity of steam engines built by Watt's company was about twenty kilowatts, more than five times higher than the mean for typical contemporary watermills, nearly three times larger than that for windmills, and twenty-five times that of a good horse.

During the nineteenth century the size, and the efficiency, of steam engines rose rapidly. The largest stationary machines designed during the 1890s were about thirty times more powerful (about one megawatt) than those in 1800, and the efficiency of the best units was ten times higher than during Watt's time. Railways and steamships greatly expanded and speeded up long-distance transport and international trade.

As the twentieth century began, two new prime movers were greatly extending the power of fossil-fueled civilization. Internal-combustion engines (Otto and Diesel varieties), developed and perfected by a number of French and German engineers between 1860 and 1900, opened the possibilities of unprecedented personal mobility, first when installed in cars, trucks, and buses, and later when used to propel the first airplanes. The steam turbine, invented by Charles Parsons, patented in 1884 and then rapidly

commercialized by his company, made large-scale generation of electricity affordable. The closing decades of the nineteenth century also saw the emergence of a new fossil fuel industry: exploration, drilling, transportation (by pipelines and tankers), and refining of crude oil.

But at that time it was only North America and most of Europe that were in the midst of a radical shift from the combination of animate energies and low-power inanimate prime movers, to the dominance of fossil fuels converted by new, much more powerful, and much more efficient machines. Average annual per capita energy consumption had more than doubled, compared to the pre–fossil fuel era, with most of the fuel spent on industrial production and transportation of people and goods. Acting via industrialization and urbanization, growing rates of energy consumption were reflected in a higher standard of living.

In the early stages of economic growth, these benefits were limited because the fossil fuels were overwhelmingly channeled into building up an industrial base. Slowly increasing consumption of household goods—better cookware, clothes and furniture—were the first signs of improvement. Afterward, with better food supply and improving health care, a lower infant mortality and longer life expectancy.

Eventually, basic material and health advantages started spilling into the countryside. The educational levels of urban populations began to rise, and there were increasing signs of incipient affluence. The emergence of electricity as the most versatile and most convenient form of energy greatly aided the whole process of modernization.

These socioeconomic transformations were accompanied by a deepening understanding of basic principles underlying energy conversions. Practical advances in new machine design as well as new observations in the chemistry of plants and animals were behind some of the most fundamental theoretical breakthroughs of the late eighteenth and the nineteenth centuries that eventually created a unified understanding of energy.

THEORETICAL UNDERSTANDING OF ENERGY

James Watt's invention of a simple indicator, a recording steam gauge, opened the way for detailed studies of engine cycles that contributed immeasurably to the emergence of thermodynamics during the following century. During the 1820s Sadi Carnot used a purely theoretical approach to investigate the ways of producing kinetic energy from heat to set down the principles applicable to all imaginable heat engines, regardless of their working substance and their method of operation.

Not long afterward Justus von Liebig ascribed the generation of carbon dioxide and water to food oxidation, offering a basically correct view of human and animal metabolism. During the 1840s Julius Robert Mayer, a German physician, was the first researcher to note the equivalence of food intake and oxygen consumption. Mayer saw muscles as heat engines energized by oxidation of the blood and offered calculations proving the sufficiency of food's chemical energy to supply the mechanical energy necessary for work as well as to maintain constant body temperature. All of this led him to estimate that mammals are about 20 percent efficient as machines and to establish the mechanical equivalent of heat, and thus to formulate, in 1851, the law of conservation of energy, commonly known as the first law of thermodynamics.

Independent, and more accurate, quantification of the equivalence of work and heat came from an English physicist, James Prescott Joule, whose first publication in 1850 put the conversion rate at 838 foot-pounds and a later revision at 772 foot-pounds,

a difference of less than 1 percent from the actual value. Soon afterward William Thomson (Lord Kelvin) identified the sun as the principal source of kinetic energy available to man and wrote about nature's universal tendency toward the dissipation of mechanical energy. Rudolf Clausius sharpened this understanding by showing in 1867 that heat energy at low temperature is the outcome of these dissipations; he named the transformational content entropy, a term derived from the Greek *trope*, for transformation. As the energy content of the universe is fixed, but its distribution is uneven, its conversions seek uniform distribution, and the entropy of the universe tends to maximum.

This second law of thermodynamics—the universal tendency toward heat death and disorder—became perhaps the most influential, and frequently misunderstood, cosmic generalization. Only at the absolute zero (-273°C) is the entropy nil. This third law completes the set of thermodynamic fundamentals. Josiah Willard Gibbs applied the thermodynamic concepts to chemistry and introduced the important notion of free energy. This energy actually available for doing work is determined by subtracting the product of temperature and entropy change from the total energy entering a chemical reaction.

The second law exercised a powerful influence on scientists thinking about energetic foundations of civilization during the closing decades of the nineteenth century. Edward Sacher viewed economies as systems for winning the greatest possible amount of energy from nature and tried to correlate stages of cultural progress with per capita access to fuels. The contrast of rising fuel demands and inexorable thermodynamic losses led to anxious calls for energy conservation. Wilhelm Ostwald, the 1909 Nobel laureate in chemistry, formulated his energetic imperative, admonishing to waste no energy but to value it as mankind makes the inevitable transition to a permanent economy based on solar radiation. Another Nobel laureate, Frederick Soddy (in 1921, in chemistry), investigated the magnitude of the earth's natural resources of energy and was the first scientist to make the often-quoted distinction between utilizing natural energy flows (spending the interest on a legacy) and fossil fuels (spending the legacy itself).

The twentieth century brought a fundamental extension of the first law, with Albert Einstein's follow-up of his famous special-relativity paper published in 1905. Soon after his publication Einstein, writing to a friend, realized that the principle of relativity requires that the mass of a body is a direct measure of its energy content, which means that light transfers mass. During the next two years Einstein formalized this "amusing and attractive thought" in a series of papers firmly establishing the equivalence of mass and energy. In the last of these papers, in 1907, he described a system behaving like a material point with mass

$$M_o = \mu + E_o c^2$$

and noted that this result is of extraordinary importance because the inertial mass of a physical system is equivalent with an energy content μc^2.

HIGH-ENERGY CIVILIZATION

The central role played by abundant, affordable, and varied energy conversions in sustaining and improving modern civilization is thus a matter of indisputable scientific understanding. That the public often ignores or overlooks this role is, in a way, a great compliment to the success of modern science and engineering, which have managed to make the conversions of extrasomatic energies so ubiquitous and so inexpensive that most people hardly give much thought to the centrality of fossil fuels and electricity in our civilization.

A variety of coals, crude oils, and natural gases supplies the bulk of the world's primary energy, and these flows of fossil energies are complemented by primary electricity derived from flowing water, nuclear fission, wind, and solar radiation. Increasing shares of fossil fuels have been used indirectly, after being converted to electricity, the most versatile form of energy. While most of our prime movers—internal-combustion engines, steam and gas turbines, and electric motors—reached their growth plateaus in terms of their unit power, their overall capacity, and their efficiency, keep increasing.

Most of the citizens of affluent economies have been enjoying numerous benefits resulting from these large, and more efficient, energy flows for several generations, but the successive waves of the energy-driven revolution are only now diffusing throughout low-income countries. Although the process has been uneven in its spatial and temporal progress, there is no doubt that we are witnessing the emergence of a global civilization marked by mass consumption, with its many physical comforts (and frequent ostentatious displays), high personal mobility, longer periods of schooling, and growing expenditures on leisure and health.

Correlations of this sequence with average per capita energy consumption have been unmistakable. National peculiarities (from climatic to economic singularities) preclude any simple classification, but three basic categories are evident. No country with average annual primary commercial energy consumption of fewer than one hundred kilograms of oil equivalent can guarantee even basic necessities to all of its inhabitants. Bangladesh and Ethiopia of the 1990s were the most populous nations in this category, and China belonged there before 1950.

As the rate of energy consumption approaches 1 metric ton of oil equivalent (or 42 gigajoules), industrialization advances, incomes rise, and quality of life improves noticeably. China of the 1980s, Japan of the 1930s and again of the 1950s, and Western Europe and the United States between 1870 and 1890 are outstanding examples of this group. Widespread affluence requires, even with efficient energy use, at least 2 metric tons of oil equivalent (more than 80 gigajoules) per capita a year. France made it during the 1960s, Japan during the 1970s. During the 1990s mean per capita consumption rates were below 150 gigajoules (fewer than 4 metric tons of crude oil equivalent) in Europe and Japan, but more than 300 gigajoules in North America, reflecting the profligate use of liquid fuels for transport and the wasteful use of household energy.

A typical premodern family—that is, a peasant household of five to eight people—controlled no more than three to four kilowatts, roughly split between the animate energy of the working members of the family and a couple of draft animals, and the energy in wood used for heating and cooking. By flipping switches of their electric appliances and setting thermostats of their heating and cooling units, a modern American household of four people controls power of about fifteen to twenty-five kilowatts—and internal-combustion engines in their cars add ten times as much installed power. The average person in modern North American society thus controls about ten times as much power as his or her preindustrial ancestor—and given the much higher conversion efficiencies of modern processes (e.g., a simple wood

A Qantas Boeing 747–400 commercial airplane flies above the opera house in Sydney, Australia. (Corbis Corporation)

stove converts 10 to 15% of fuel to useful heat compared to 80 to 90% for high-performance natural-gas furnaces), the differences in terms of useful power are easily twenty to fiftyfold.

Maximum differences are even more stunning: Even when holding reins of a dozen large horses, a well-to-do traditional farmer controlled no more than about ten kilowatts of useful power. In contrast, a pilot of a Boeing 747 commands some sixty megawatts in the plane's four engines, a duty engineer of a large electricity-generating plant may be in fingertip control of more than one gigawatt, and a launch order by a U.S. president could unleash exawatts (10^{18} watts) in thermonuclear warheads. This immense concentration of power under individual control has been one of the hallmarks of modern high-energy society.

The grand transformation wrought by this surfeit of easily controllable energies defines modern civilization. Greatly increased personal mobility, a change that began with the application of steam power to land and water transport, has been vastly expanded thanks to the mass production of vehicles and airplanes powered by internal-combustion engines and gas turbines (the only notable twentieth-century addition to common prime movers). Wide-bodied jet airplanes in general, and the Boeing 747 in particular, have been among the leading agents of globalization. The latest stage of this energy-driven globalization trend is vastly enlarged access to information made possible by mass diffusion of personal computers, by electronic storage of data and images, and by high-volume wireless and optical fiber transmissions.

The easing of women's household work in the Western world is another particularly noteworthy social transformation wrought by efficient energy conversions. For generations rising fuel consumption made little difference for everyday household work; indeed, it could make it worse. As the standards of hygiene and social expectations rose with better education, women's work in Western countries often got harder. Electricity was the eventual liberator. Regardless of the availability of other energy forms, it was only its introduction that did away with exhausting and often dangerous labor: Electric irons, vacuum cleaners, electric stoves, refrigeration, washing machines and electric dryers transformed housework, beginning in the early decades of the twentieth century. Hundreds of millions of women throughout the poor world are still waiting for this energy-driven liberation.

Air conditioning has been another revolutionary application of electricity. The technique was first patented by William Carrier in 1902 but widespread adoption came only after 1960, opening up first the American Sunbelt to mass migration from northern states, then increasing the appeal of subtropical and tropical tourist destinations, and since the 1980s becoming also a common possession for richer urbanites throughout the industrializing world.

And yet the energy-civilization link should not be overrated. Basic indicators of physical quality of life show little or no increase with average per capita energy consumption rising above one hundred giga-joules per year—and it would be very unconvincing to argue that North American consumption levels, twice as high as in Western Europe, make Americans and Canadians twice as content, happy, or secure. Indeed, international and historic comparisons show clearly that higher energy use will not assure reliable food supply, does not confer political stability, does not necessarily enhance personal security, does not inevitably lead to a more enlightened governance, and may not bring any widely shared improvements in the standard of living.

CHALLENGES AHEAD

During the past generation, a number of poor countries moved to the intermediate energy consumption category. Still, in terms of total population, the distribution of global energy use remains extremely skewed. In 1950 only about 250 million people (one-tenth of the global population) consumed more than two metric tons of oil equivalent a year per capita, yet they claimed 60 percent of the world's primary energy (excluding biomass). By 1999 such populations were about a fifth of all mankind, and they claimed nearly three-quarters of all fossil fuels and electricity. In contrast, the poorest quarter of humanity used less than 5 percent of all commercial energies.

Stunning as they are, these averages do not capture the real difference in living standards. Poor countries devote a much smaller share of their total energy consumption to private household and transportation uses. The actual difference in typical direct per capita energy use among the richest and the poorest quarters of the mankind is thus closer to being fortyfold rather than "just" twentyfold. This enormous disparity reflects the chronic gap in economic achievement and in the prevailing quality of life and contributes to persistent global political instability.

Narrowing this gap will require higher substantially increased output of fossil fuels and electricity, but an appreciable share of new energy supply should come from more efficient conversions. Energy intensities (i.e., the amount of primary energy per unit of

the GDP) of all affluent economies have been declining during most of the twentieth century, and some modernizing countries have been moving in the same direction even faster; perhaps most notably, China has more than halved its energy intensity since 1980!

Low energy prices provide little incentive for sustaining the efficiency revolution that flourished after OPEC's crude-oil price rises of the 1970s, but higher conversion efficiencies should be pursued regardless; in the absence of (highly unlikely) voluntary frugality, they are the only means of reducing overall energy throughput and hence minimizing the impact of energy use on the biosphere. Energy industries and conversions have many environmental impacts, but it now appears that the threat of relatively rapid global warming, rather than the shrinking resource base, will be the most likely reason for reducing our civilization's high dependence on fossil fuels.

So far we have been successful in preventing the use of the most destructive energy conversion, the explosion of thermonuclear weapons, in combat. Although the risk of armed superpower conflict has decreased with the demise of the Soviet empire, thousands of nuclear weapons remain deployed around the world. Our challenge during the twenty-first century will be fourfold: avoiding nuclear conflict; extending the benefits of high-energy society to billions of people in low-income countries; decoupling the development of rich societies from continuous growth of energy consumption; and preserving the integrity of the biosphere.

Vaclav Smil

See also: Carnot, Nicolas Leonard Sadi; Clausius, Rudolf Julius Emmanuel; Culture and Energy Usage; Ethical and Moral Aspects of Energy Use; Gibbs, Jonah Willard; Industry and Business, History of Energy Use and; Joule, James Prescott; Kinetic Energy, Historical Evolution of the Use of; Mayer, Julius Robert von; Refining, History of; Thomson, William; Watt, James.

BIBLIOGRAPHY

Alexander, R. McN. (1992). *The Human Machine.* New York: Columbia University Press.
Aristotle. (1966). *Metaphysics*, tr. by H. H. Apostle. Bloomington: Indiana University Press.
Cardwell, D. S. L. (1971). *From Watt to Clausius.* Ithaca, NY: Cornell University Press.

Clausius, R. (1867). *Abhandlungen uüber die mechanische Waärmetheorie.* Braunschweig, Ger.: F. Vieweg.

Cook, E. (1976). *Man, Energy, Society.* San Francisco: W. H. Freeman.

Cottrell, F. (1955). *Energy and Society.* New York: McGraw-Hill.

Cowan, C. W., and Watson, P. J., eds. (1992). *The Origins of Agriculture.* Washington DC: Smithsonian Institution Press.

Dickinson, H. W. (1939). *A Short History of the Steam Engine.* Cambridge, Eng.: Cambridge University Press.

Düring, I. (1966). *Aristoteles Darstellung und Interpretation seines Denken.* Heidelberg, W. Ger.: Carl Winter.

Durnin, J. V. G. A., and Passmore, R. (1967). *Energy, Work, and Leisure.* London: Heinemann Educational Books.

Einstein, A. (1905). "Zur Elektrodynamik bewegter Koörper." *Annalen der Physik* 17:891–921.

Einstein, A. (1907). "Relativitaätsprinzip und die aus demselben gezogenen Folgerungen." *Jahrbuch der Radioaktivitaät* 4:411–462.

Finniston, M., et al., eds. (1992). *Oxford Illustrated Encyclopedia of Invention and Technology.* Oxford, Eng.: Oxford University Press.

Forbes, R. J. (1964–1972). *Studies in Ancient Technology.* Leiden, Neth.: E. J. Brill.

Georgescu-Roegen, N. (1971). *The Entropy and the Economic Process.* Cambridge, MA: Harvard University Press.

Goudsblom, J. (1992). *Fire and Civilization.* London: Allen Lane.

Joule, J. P. (1850). *On Mechanical Equivalent of Heat.* London: R. and J. E. Taylor.

Langdon, J. (1986). *Horses, Oxen and Technological Innovation.* Cambridge, Eng.: Cambridge University Press.

Lindsay, R. B., ed. (1975). *Energy: Historical Development of the Concept.* Stroudsburg, PA: Dowden, Hutchinson, & Ross.

Mayer, J. R. (1851). *Bemerkungen über das mechanische Aequivalent der Waärme.* Heilbronn, Ger.: J. V.Landherr.

McLaren, D. J., and Skinner, B. J., eds. (1987). *Resources and World Development.* Chichester, Eng.: John Wiley & Sons.

National Research Council. (1986). *Electricity in Economic Growth.* Washington, DC: National Academy Press.

Nef, J. U. (1932). *The Rise of the British Coal Industry.* London: Routledge.

Odum, H. T. (1971). *Environment, Power, and Society.* New York: Wiley-Interscience.

Ostwald, W. (1909). *Energetische Grundlagen der Kulturwissenschaften.* Leipzig, Ger.: Alfred Kroöner.

Pacey, A. (1990). *Technology in World Civilization.* Cambridge, MA: MIT Press.

Reynolds, J. (1970). *Windmills and Watermills.* London: Hugh Evelyn.

Rose, D. J. (1986). *Learning about Energy.* New York: Plenum Press.

Schipper, L., and Meyers, S. (1992). *Energy Efficiency and Human Activity.* New York: Cambridge University Press.

Singer, Charles, et al., eds. (1954–1984). *A History of Technology.* Oxford, Eng.: Clarendon Press.

Smil, V. (1991). *General Energetics.* New York: John Wiley & Sons.

Smil, V. (1994). *Energy in World History.* Boulder, CO: Westview Press.

Soddy, F. (1926). *Wealth, Virtual Wealth, and Debt: The Solution of the Economic Paradox.* New York: E. P. Dutton.

World Energy Council Commission. (1993). *Energy for Tomorow's World.* London: Kogan Page.

HORSEPOWER

See: Units of Energy

HOUDRY, EUGENE JULES (1892–1962)

EDUCATION AND WAR EXPERIENCE

Eugene Jules Houdry, one of the fathers of petroleum refining, was born in Dumont, France, outside of Paris, on April 18, 1892, and grew up during an era of rapid technological change and innovation. Houdry was blessed with considerable emotional and financial support from his family and encouraged to reach for lofty and risky goals.

With the urging of his father, a wealthy steel manufacturer, Houdry studied mechanical engineering at the École des Arts et Métiers in Chalons-sur-Marne near Paris. He earned the government's gold model for highest scholastic achievement in his class and captained its national champion soccer team. Graduating in 1911, Houdry briefly worked in his father's metalworking business, Houdry et Fils, but joined the army prior to the outbreak of World War I. As a lieutenant in the tank corps, Houdry was seriously wounded in 1917 in the first great tank battle, the Battle of Javincourt. He was awarded the Croix de Guerre, the French military decoration created in 1915 to reward feats of bravery, and membership in the Legion d'Honneur for extraordinary bravery.

FROM TANKS TO AUTOMOBILE RACING TO CATALYTIC CRACKING AND CATALYST DEVELOPMENT

After the war he rejoined Houdry et Fils, but his interest drifted from steel to car-racing and then to improving the fuel performance of car racing engines. Houdry recognized the need for more efficient motor fuels during a visit to Ford and while attending the Memorial Day Indianapolis 500 Race. In 1922 Houdry learned of a superior gasoline being produced from lignite by E. A. Prudhomme, a Nice chemist. Undeterred by unfamiliarity with catalysis or chemical engineering, Houdry enlisted Prudhomme in developing a workable lignite-to-gasoline process in a laboratory privately financed by a group of Houdry's friends. In their scheme, solid lignite would be initially broken down by heat (thermally cracked) to produce a viscous hydrocarbon oil and tars. The oil was then somehow further converted (actually, catalytically cracked) to gasoline by contacting the oil with various clays. Houdry's wartime experiences impressed upon him France's need for indigenous petroleum resources, which provided further impetus to his pursuing the project. With government support, the alliance built a plant that could convert sixty tons of lignite per day to oil and gasoline. The process was not economic, and the unit shut down in 1929. Across the ocean, a similar fate befell Almer McAfee at Gulf Oil in the same year. McAfee is credited with developing the first commercially viable catalytic cracking process using aluminum chloride as the catalyst.

The poor economics for McAfee's catalytic process lay in the substantial improvements being made in the technology of thermally cracking crude oil. Notable among these competing efforts were those of William Burton of Standard Oil (Indiana), whose high-pressure and high-temperature thermal technology more than doubled the potential yield of gasoline from crude oil and thereby substantially reduced the cost of thermal cracking.

Houdry was dejected by the failure of his process to compete against the more mature thermal cracking technology, but his seven-year effort was not in vain. Houdry had been devoting his attention initially to using a naturally occurring aluminosilicate clay, fuller's earth, as the catalyst and later to using more catalytically active hydrosilicates of aluminum prepared by washing clays, such as bentonite with acid. This approach ran contrary to the mainstream of catalytic cracking investigators, who were working unsuccessfully with nickel-containing catalysts, or to McAfee, who was using anhydrous aluminum chloride as the catalyst and making it practical by effecting tenfold reductions in its cost of manufacture.

The catalytic cracking activity of Houdry's catalysts and McAfee's aluminum chloride were puzzling to researchers at the time: It worked, but they could not understand why. As later determined by Hans Tropsch of UOP Research Laboratories, both silica-alumna and aluminum chloride behave as acid catalysts capable of donating protons or accepting electron pairs to form "carbonium ion" reaction intermediates. The concept of such active intermediates explained the propensity of catalytic cracking to favor formation of higher-octane branched chain paraffins and aromatics as compared with thermal cracking's disposition toward producing lower-octane straight chain paraffins. It also explained the greater yields of gasoline resulting from other hydrocarbon reactions occurring in the process, namely isomerization (e.g., conversion of zero octane n-pentane to isopentane), alkylation (e.g., conversion of isobutane and isobutene to "isooctane"), polymerization (e.g., propylene to hexene) and dehydrogenation (e.g., hexene to benzene).

Remarkably, seventy years after Houdry's utilization of the catalytic properties of activated clay and the subsequent development of crystalline aluminosilicate catalysts that are a magnitude more catalytically active, the same fundamental principles remain the basis for the modern manufacture of gasoline, heating oils, and petrochemicals.

COMMERCIALIZATION AND IMPACT

Houdry concentrated his personal efforts on developing a viable processing scheme, solving the engineering problems, scaling the process to commercial size, and developing requisite equipment. In 1930,

H. F. Sheets of Vacuum Oil Company, who learned of Houdry's work and shared his vision for converting vaporized petroleum to gasoline catalytically, invited him to the United States. After a successful trial run, Houdry moved his laboratory and associates from France to Paulsboro, New Jersey, to form a joint venture, Houdry Process Corporation, with Vacuum Oil Company. In that year Vacuum Oil Company merged with Standard Oil of New York to become Socony-Vacuum Company (much later Mobil Oil Corporation).

In 1933, a 200-barrel-per-day Houdry unit was put into operation. The financial strains of the Great Depression coupled with formidable technical problems necessitated inclusion of Sun Oil Company in a fruitful three-way codevelopment partnership. In the next few years the Houdry process underwent considerable changes, especially in developing an innovative method for periodically burning off a buildup of coke within the catalyst to restore catalyst activity lost after only ten minutes of usage and employing a molten salt heat exchanger to replace heat absorbed by the endothermic cracking reactions.

In 1936, Socony-Vacuum built a 2,000-barrel-per-day semicommercial Houdry unit at Paulsboro, closely followed the next year by Sun Oil's 15,000-barrel-per-day commercial plant in its Marcus Hook, Pennsylvania, refinery. With a capability to produce a higher-octane and thus better-performing gasoline, the new development was touted as "the new miracle of gasoline chemistry." The timing was fortuitous in giving the Allies a major advantage in World War II. During the first two years of the war Houdry units produced 90 percent of all catalytically cracked gasoline. Minimal additional refining, notably washing with sulfuric acid to remove low-octane straight-chain olefins and deleterious sulfur components, allowed the manufacture of exceedingly scarce high-test aviation gasoline. Allied planes were 10 to 30 percent superior in engine power, payload, speed and altitude during these critical early years of the war.

The Houdry fixed-bed cyclic units were soon displaced in the 1940s by the superior Fluid Catalytic Cracking process pioneered by Warren K. Lewis of MIT and Eger Murphree and his team of engineers at Standard Oil of New Jersey (now Exxon). Murphree and his team demonstrated that hundreds of tons of fine catalyst could be continuously moved like a fluid between the cracking reactor and a separate vessel for catalyst regeneration. Nevertheless, the fixed bed remains an important reactor type in petrochemicals, and the Houdry reactor remains a classic example of their high state of development.

BEYOND CATALYTIC CRACKING

Houdry's entire career was characterized by foresight, imagination, leadership, boldness, and persistence. He was a demanding individual, single-minded of purpose and a workaholic, yet with considerable charm. Houdry was one of twenty whose opposition to the Vichy government resulted in their French citizenship being revoked. Later during the war, when natural rubber supplies were cut off, Houdry invented a catalytic process for producing butadiene, an essential reactant in manufacturing synthetic rubber. Following World War II, motivated by a concern for the environment, Houdry showed foresight in turning his attention to the health hazards of automobiles and industrial pollution, well before they were widely acknowledged as major problems. He founded Oxy-Catalyst Company and holds one of the early U.S. patents for catalytically reducing the amounts of carbon monoxide and unburned hydrocarbons in automobile exhausts, anticipating the development of the catalyst processes now used with modern automobiles. He died on July 18, 1962, at age seventy.

Barry L. Tarmy

See also: Catalysts; Combustion; Gasoline and Additives; Heat and Heating; Petroleum Consumption; Refineries; Refining, History of.

BIBLIOGRAPHY

American Chemical Society. (1996). *A National Historic Chemical Landmark: The Houdry Process for the Catalytic Conversion of Crude Petroleum to High-Octane Gasoline.* Washington, DC: American Chemical Society.
Buonora, P. T. (1998). "Almer McAfee at Gulf Oil." *Chemical Heritage* 16(2):5–7, 44–46.
Enos, J. L. (1962). *Petroleum Progress and Profits: A History of Process Innovation.* Cambridge, MA: MIT Press.
Mosely, C. G. (1984). "Eugene Houdry, Catalytic Cracking and World War II Aviation Gasoline." *Journal of Chemical Education* 61:655–656.
Spitz, P. H. (1988). *Petrochemicals: The Rise of an Industry.* New York: John Wiley & Sons.
Thomas, C. L.; Anderson, N. K.; Becker, H. R.; and McAfee J. (1943). "Cracking with Catalysts." *Petroleum Refiner* 22(1):95–100, 365–370.

HUB AND SPOKE SYSTEMS

See: Air Travel

HYBRID VEHICLES

INTRODUCTION

By incorporating into the driveline of the vehicle the capability to generate electricity onboard the vehicle from a chemical fuel, a hybrid-electric vehicle has the characteristics of both an electric vehicle and a conventional internal combustion engine (ICE) vehicle and can be operated either on wall-plug electricity stored in a battery or from a liquid fuel (e.g., gasoline) obtained at a service station. This essay discusses technologies for hybrid-electric vehicles that can attain significantly higher fuel economy and lower emissions than conventional ICE vehicles of the same size, performance, and comfort.

HYBRID-ELECTRIC VEHICLE DESIGN OPTIONS

There are a large number of ways an electric motor, engine, generator, transmission, battery, and other energy storage devices can be arranged to make up a hybrid-electric driveline. Most of them fall into one of two configurations—series and parallel. In the series configuration (Figure 1, Top), the battery and engine/generator act in series to provide the electrical energy to power the electric motor, which provides all the torque tot he wheel of the vehicle. In a series hybrid, all the mechanical output of the engine is used by the generator to produce electricity to either power the vehicle or recharge the battery. This is the driveline system used in diesel-electric locomotives. In the parallel configuration (Figure 1, Bottom), the engine and the electric motor act in parallel to provide torque to the wheel of the vehicle. In the parallel hybrid, the mechanical output of the engine can be used to both power the vehicle directly and to

recharge the battery or other storage devices using the motor as a generator. In recent years, a third type of hybrid configuration, the dual mode, is being developed that combines the series and hybrid configurations. As shown in Figure 1b, the engine output can be split to drive the wheel (parallel mode) and to power a generator to produce electricity (series mode). This configuration is the most flexible and efficient, but it is also likely to be the most complex and costly.

A range-extended electric vehicle would most likely use the series configuration if the design is intended to minimize annual urban emissions. It would be designed for full-performance on the electric drive alone. The series hybrid vehicle can be operated on battery power alone up to its all-electric range with no sacrifice in performance (acceleration or top speed) and all the energy to power the vehicle would come from the wall plug. This type of hybrid vehicle is often referred to as a "California hybrid" because it most closely meets the zero emission vehicle (ZEV) requirement. The engine would be used only on those days when the vehicle is driven long distances.

Hybrid vehicles designed to maximize fuel economy in an all-purpose vehicle could use the series, parallel, or dual configurations depending on the characteristics of the engine to be used and acceptable complexity of the driveline and its control. Parallel hybrid configurations will require frequent on-off operation of the engine, mechanical components to combine the engine and motor torque, and complex control algorithms to smoothly blend the engine and motor outputs to power the vehicle efficiently. The parallel hybrid would likely be designed so that its acceleration performance would be less than optimum on either the electric motor or engine alone and require the use of both drive components together to go zero to sixty mph in 10–12 sec. Such a hybrid vehicle would not function as a ZEV in urban/freeway driving unless the driver was willing to accept reduced acceleration performance. The parallel configuration can be designed to get better mileage than the series configuration. A fuel cell hybrid vehicle would necessarily be a series hybrid because the fuel cell produces only electricity and no mechanical output. The dual mode hybrid is intended to maximize fuel economy and thus be designed like a parallel hybrid, but with a relatively small generator that could be powered by the engine. The engine in the dual mode hybrid would operate in

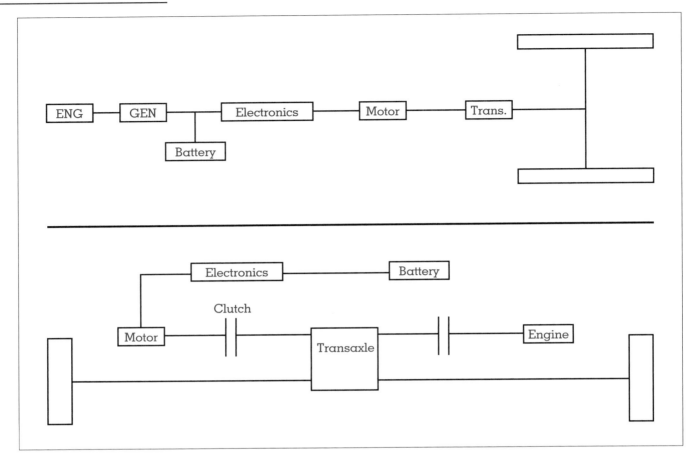

Figure 1.
Hybrid Vehicle Driveline Schematics.
NOTE: Top: Series; Bottom: Parallel

the on/off mode, but be cycled on and off less frequently than in the parallel configuration.

COMPONENT TECHNOLOGIES: STATUS AND PROSPECTS

Motor and Electronics

Recent advances in both motor and inverter electronics technologies have important implications for the design of high performance hybrid vehicles. In 2000 the size and weight of the motor and the electronics combined are significantly smaller (by a factor of two to three—see Table 1) than that of an engine and transmission of the same peak power in 1998. The size advantage of the electrical components will be even greater as the operating voltage of the electrical drive system is increased above the 300–400V that is common today. This means that in packaging the hybrid driveline, the electrical drive components take up only a fraction of the space available and finding room for the mechanical components, such as the engine, transmission or torque coupler, can present a difficult challenge.

Electrical Energy Storage

The key component in the hybrid driveline that permits it to operate more efficiently than the engine/transmission in a conventional car is the electrical energy storage unit. It must store sufficient energy (kWh) to provide the all-electric range of the vehicle or to permit the engine or fuel cell to operate near the average power required by the vehicle (in a load-leveled mode). It also must have sufficient power capability to meet on demand the peak power of the motor/electronics for vehicle acceleration or

Engine Type	kg/kW	l/kW	Maximum Efficiency (%)	Emissions (gm/kWh) (Nominal)			Manufacturer/Developer
				HC	CO	NOX	
Spark Ignition							
Valve Injection	2.0	4.0	32	3.0	20	8.0	Auto Companies
Direct Injection	2.5	4.1	38	5.0	4.0	3.5	Auto Companies
Rotary	.8	1.0	30	3.0	4.0	5.0	Moller International
Two-Stroke	1.0	1.75	30	3.0	3.0	6.0	Orbital
Diesel							
Prechamber/Turbo	2.6	4.2	35	.1	1.0	3.0	Auto Companies
Direct Injection/Turbo	3.0	4.4	42	1.0	1.0	15.0	Auto Companies
Gas Turbine							
Metal							
With Recup. Incl. generator	2.5	5.3	30	.08	.30 (catalytic combustion)	.03	Capstone
Ceramic							
with Recup.	2.0	4.0	40	.08	.30 (catalytic combustion)	.03	Allison
Stirling							
H₂ Working Fluid/ Wobble Plate Drive	2.7-3	2-2.5	30-35	.01	.15	.22	Stirling Thermal Motors

Table 1.
Hybrid Vehicle Engine Characteristics

regenerative braking. In most cases, the energy storage unit in a hybrid vehicle is sized by the peak power requirement. Because the size (weight and volume) of the energy storage unit (often a battery) in the hybrid vehicle is smaller than the battery in a battery-powered electric vehicle (EV), the power density (W/kg and W/liter) requirements of the energy storage unit in the hybrid vehicle are greater than for the battery in an electric vehicle. For example, power densities of 200–300W/kg are satisfactory for use in an EV, but power densities of 500–1,000W/kg are needed for hybrid vehicles. Considerable progress has been made in the development of high power batteries (often referred to as pulse batteries) of various types. In hybrid vehicles, the energy density of the energy storage unit is of secondary importance; compromises in energy density have been made to reach the high power density of pulse batteries. For example, nickel metal hydride batteries designed for EVs have an energy density of 70Wh/kg and a power density of 250 W/kg, while those designed for hybrid vehicles have an energy density of 40–45Wh/kg and a peak power density of 600–700W/kg. Another important consideration for energy storage units for use in hybrid vehicles is the need to minimize the losses associated with transferring energy into and out of the unit, because in the hybrid, a reasonable fraction of the electrical energy produced onboard the vehicle is stored before it is used to power the vehicle. In order to minimize the energy storage loss, the round-trip efficiency (energy out/energy in) should be at least 90 percent. This means that the useable peak power capability of a battery is much less that the power into a match impedance load at which efficiency of the transfer is less than 50 percent. Useable power is only about 20 percent of the peak power capacity. This is another reason that the design of an energy storage units for hybrid vehicles is much more difficult than for battery-powered electric vehicles and energy storage technology is often described as the enabling technology for hybrid vehicles.

A new energy storage technology that is well suited for hybrid vehicles is the electrochemical ultracapacitor, often referred to as the double-layer capacitor. Ultracapacitors for vehicle applications have been under development since about 1990. The construction of an ultracapacitor is much like a battery in that it consists of two electrodes, a separator, and is filled with an electrolyte. The electrodes have a very high surface area (1,000–1,500 m^2/gm) with much of the surface area in micropores 20 angstroms or less in diameter. The energy is stored in the double-layer (charge separation) formed in the micropores of the electrode material. Most of the ultracapacitors presently available use activated carbon as the electrode material. The cell voltage is primarily dependent on the electrolyte used. If the electrolyte is aqueous (sulfuric acid or KOH) the maximum cell voltage is 1V; if an organic electrolyte (propylene carbonate) is used, the maximum cell voltage is 3V. As in the case of batteries, high voltage units (300–400V) can be assembled by placing many ultracapacitor cells in series.

Batteries have much higher energy density and capacitors have much higher power capacity. The technical challenge for developing ultracapacitors for vehicle applications is to increase the energy density (Wh/kg and Wh/liter) to a sufficiently high value that the weight and volume of a pack to store the required energy (500 Wh for a passenger car) is small enough to be packaged in the vehicle. Power density and cycle life are usually not a problem with ultracapacitors. The cost ($/Wh) of ultracapacitors is presently too high—being about $100/Wh. It must be reduced by at least an order of magnitude (a factor of 10) before this new technology will be used in passenger cars. Nevertheless, ultracapacitors are a promising new technology for electrical energy storage in hybrid vehicles.

Engines and Auxiliary Power Units

The characteristics of engines for hybrid vehicles are shown in Table 1. Most of the hybrid vehicles designed and built have used four stroke gasoline or diesel engines. Nearly all gasoline engines are now fuel-injected and both gasoline and diesel engines are computer controlled. Continuing improvements are being made in these engines in terms of size, weight, efficiency, and emissions. These improvements and the common use of computer control make it fairly easy to adapt the conventional engines to hybrid vehicle applications. The major difficulty in this regard is to find an engine of appropriate power rating for hybrid vehicle application. Most automotive engines have power of 60kW (75hp) and greater, which is too large for use in most hybrid vehicle designs. Experience has shown that good sources of engines for hybrid vehicles are the minicars designed for the small car markets on Japan and Europe.

Several advanced engines have been developed especially for hybrid vehicles. These include a high-expansion ration gasoline engine (Atkinson cycle) developed by Toyota for their Prius hybrid vehicle which they started to market in Japan in 1997. Stirling Thermal Motors (STM) under contract to General Motors (GM) designed and fabricated a Stirling engine (30kW) for use on GM's series hybrid vehicle built as part of the Department of Energy (DOE) hybrid vehicle program. Capstone Technology developed a 25kW recuperative gas turbine engine for use in a flywheel hybrid vehicle built by Rosen Motors. The characteristics of these engines are indicated in Table 1. The most successful of these engine development projects was the Prius engine of Toyota. The other two engines were too large and were not efficient enough to warrant further development for hybrid vehicle applications. The new Toyota engine in the Prius is a four cylinder (1.5 liter), four stroke gasoline engine that utilizes variable inlet valve timing to vary the effective compression ratio from 9.3 to 4.8 in an engine having a mechanical compression ratio of 13.5. Varying the effective volume of air during the intake stroke permits operation of the engine at part load with reduced pumping and throttling losses. This results in an increase in engine efficiency. The expansion ratio at all times is set by the high mechanical compression ratio of the engine. This new engine was optimized for operation in the hybrid mode and had a brake specific fuel consumption of about 235 gm/kWh for output powers between 10 and 40kW. This corresponds to an efficiency of 37 percent, which is very high for a four stroke gasoline engine of the same peak power rating.

The engine output in a hybrid vehicle can be utilized to generate electricity on board the vehicle or to provide torque to the driveshaft of the vehicle. In the first case, the engine output torque drives a generator and the combination of the engine and the generator is termed an auxiliary power unit (APU). The generator can be either an ac induction or a brushless dc permanent magnet machine. The size of the generator for a given power rating depends to a large extent on the voltage of the system and the rpm at which the generator rotates. For a 400V system and a maximum of 8,000–10,000, the size and weight of the APU are 0.7kg/kW and 0.8 liter/kW, respectively, including the electronic controls. The efficiency of the generator system will vary between 90 to 95 percent depending on the power output. The losses associated with the production and storage of the electrical energy onboard the vehicle in a series hybrid are significant (10 to 20%) and cannot be neglected in predicting the fuel economy of the vehicle.

Mechanical Components

The transmission, clutch, and other mechanical components needed in a hybrid vehicle to combine the output of the engine with the electric motor and generator and the main driveshaft of the vehicle are critical to the efficient and smooth operation of parallel and dual mode hybrid vehicles. The design of these components is relatively straightforward and not much different than that for similar components for conventional engine-powered vehicles. In a parallel hybrid (Figure 1b), the engine is connected to the drive shaft through a clutch that opens and closes as the engine power is needed. The speed ratio between the engine and the wheels is determined by the gear ratios in the transmission. Mechanical design of the engine clutch so that it has a long life and smooth operation is one of the critical tasks in the development of the parallel hybrid vehicle. Many hybrid vehicles are built with manual transmissions, because automatic transmissions with a torque converter have unacceptably high losses. A recent development is the use of a continuously variable transmission (CVT) in a parallel hybrid driveline. The operation of the CVT is like an automatic transmission from the driver's point-of-view with the advantages of lower losses and a wider range of continuous gear ratios, which result in efficient driveline operation in both the electric and hybrid modes. The disadvantages of the CVT are that control of the system is more difficult than with a manual transmission and the steel belt used in the CVT is much less tolerant of abuse (sudden changes in speed and torque). Nevertheless, it appears that in the future years CVTs will have application in parallel hybrid drivelines.

There are several arrangements of the dual mode hybrid driveline. In the simplest arrangement (Figure 2), the generator can be used as a motor to start the engine and/or supplemental torque of the traction motor to drive the vehicle. In this dual mode configuration, the batteries can be recharged either by the generator or by the traction motor acting as a generator. This simple dual mode system does not require a transmission. A second dual mode system utilizes a planetary gear set to couple the engine, generator, and main driveshaft. In this arrangement, the speed ratio between the engine and the main driveshaft depends on the fraction of the engine power that is applied to the generator. This second arrangement is used by Toyota in the Prius hybrid car. This system is less flexible and less efficient than the first system in which the engine and generator are directly connected on the same drive shaft, but it does not require a clutch, which must be opened/closed smoothly and reliably under computer control. Operation of the Toyota dual mode hybrid driveline has proven to be smooth and reasonably efficient.

Fuel Cells

Fuel cells can be utilized in electric hybrid vehicles as the means of converting chemical fuel to electricity. Rapid progress has been made in the development of fuel cells, especially proton exchange membrane (PEM) fuel cells, for transportation applications. This progress has resulted in a large reduction in the size and weight of the fuel cell stack and as a result, there is now little doubt that the fuel cell of the required power (20–50kW) can be packaged under the hood of a passenger car. The primary question regarding fuel cells in light duty vehicles is how they will be fueled. The simplest approach is to use high pressure hydrogen as has been done in the most successful bus demonstration to date. This approach is satisfactory for small test and demonstration programs, but the development of the infrastructure for using hydrogen as a fuel in transportation will take many years. Considerable work is underway to develop fuel processors (reformers) to generate hydrogen onboard the vehicle from various chemical fuels (e.g., methanol or hydrocarbon distillates). Most of the hydrogen used for industrial and transportation applications is presently generated by reforming natural gas using well-developed technology. A promising approach to fuel processing to hydrogen (H^+ and electrons) onboard the vehicle is direct oxi-

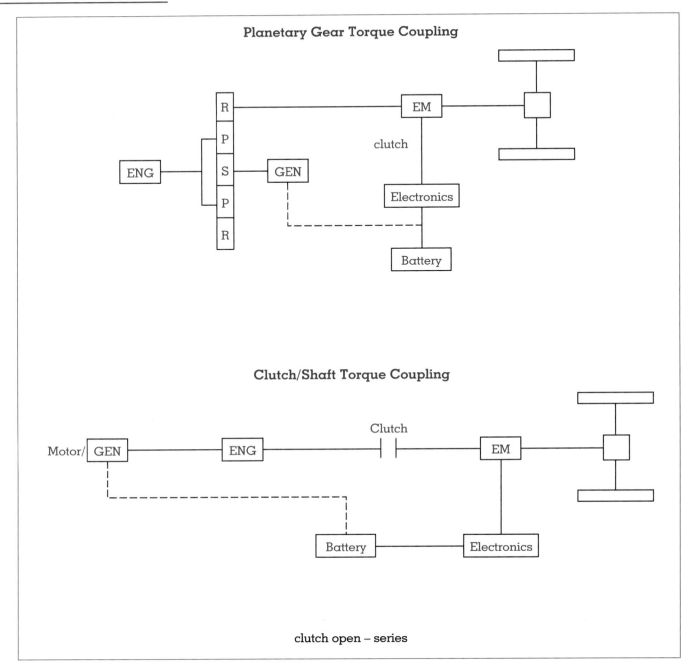

Figure 2
Dual Mode Hybrid Vehicle Driveline Schematics

dation of methanol within the fuel cell stack. When technology for the efficient, direct conversion of a liquid fuel to hydrogen within the PEM fuel cell is developed, the commercialization of fuel cells in light duty vehicles will occur rapidly.

CONTROL STRATEGIES

Control Strategies for Series Hybrid Vehicles

The intent of the control strategy is to maintain the state-of-charge of the energy storage unit within a prescribed range regardless of the driving cycle and the resultant power demand on the driveline. This should be done so that the onboard electrical generator (engine/generator or fuel cell) is operated at high efficiency and low emissions. This is done more easily when the energy storage capacity is reasonably large as with a battery than when it is small as using ultracapacitors. The strategy used for vehicles having a significant all-electric range is to discharge the battery to a prescribed state-of-charge (20 to 30%) and then to turn on the engine to maintain the battery within 10 to 20 percent of that condition. Electrical energy is generated at a rate slightly greater than the average power demand of the vehicle to account for losses associated with storing the energy. In the case of an engine/generator, a minimum power level is set so that the engine is never operated below it. Proper selection of this minimum power can have an important effect on fuel economy. When the battery charge reaches the maximum permitted, the engine is turned off and it remains off until the battery state-of-charge falls to the engine turn-on state-of-charge. When the series hybrid is operated so that the battery is permitted to discharge to a relatively low state-of-charge, it is termed a charge depleting hybrid. If the battery is maintained at a high state-of-charge (60 to 70%), it is termed a charge sustaining hybrid and the battery is seldom, if ever, recharged from the wall-plug. A significant fraction of the energy used by charge depleting hybrid vehicles is from the wall-plug and their average annual emissions and energy consumption are dependent on the use-pattern (miles of travel per day) of the vehicle and how the electricity used to recharge the batteries is generated.

Control Strategies for Parallel Hybrid Vehicles

The control strategies for parallel hybrid vehicles are more complicated than those for series hybrids primarily because they are dependent on both vehicle speed and state-of-charge of the energy storage unit and should include a criteria for splitting the driveline torque between the engine and the electric motor. The intent of the strategy is to permit the electric motor to provide the torque if it can at vehicle speeds below a prescribed value and permit the engine to provide the torque at higher speeds. If the vehicle is operating in the all-electric mode, the motor provides the torque and the engine is not turned on regardless of the torque demand or vehicle speed. Since the all-

electric range of a hybrid vehicle is usually less than 80 km, operation of the vehicle should change automatically to the hybrid mode when the all-electric range is exceeded. The control strategy in the hybrid mode can be either charge sustaining or charge depleting. In the case of charge sustaining, the battery state-of-charge is maintained at a near constant value by a control strategy using electrical energy produced by the engine and the motor acting as a generator and consequently little electrical energy is used from the wall-plug. For the charge depleting case, the control strategy permits the battery state-of-charge to decrease as the vehicle is driven and the battery is then recharged from the wall-plug at night. Parallel hybrids usually have a multi-speed transmission so the control strategy must also include a gear shifting algorithm that depends on whether the motor or engine or both are producing torque. A continuously variable transmission (CVT) would be particularly attractive for use in a parallel hybrid driveline.

In order to achieve high fuel economy with a parallel hybrid, it is necessary to avoid engine operation below some minimum engine torque (or effective throttle setting) where the engine brake specific fuel consumption (gm/kWh) is relatively high and to manage engine turn on and off carefully to minimize emissions and wasted fuel. In urban driving, the control strategies for parallel hybrids often result in the engine being turned on and off frequently because the vehicle speed and power demands vary rapidly in stop-and-go driving. The effects of this on-off engine operation on fuel usage and emissions for the parallel hybrids are neglected in most simulations at the present time, so further analysis and vehicle testing is needed to determine whether the high fuel economy and low emissions projected for parallel hybrids can be attained. The control strategies for parallel hybrids are necessarily more complex than those for series hybrids and the uncertainty in the simulation results for parallel hybrids are greater.

Control Strategies for Dual Mode Hybrid Vehicles

The control strategies for dual mode hybrid vehicles are a combination of those used for series and parallel hybrids. There are so many possible hardware arrangements and associated control strategies, it is not possible to summarize them in a simple manner as was the case for series and parallel hybrids. The objective of the dual mode operation is to use the possibility of battery charging simultaneously with the use of the engine and electric motor to power the vehicle as a means of maintaining engine operation at high efficiency at all times. At highway speeds, the engine can be used directly to power the vehicle with the engine operating at high efficiency. This mode of operation is essentially that of a parallel hybrid. At low vehicle speeds, when the battery does not need charging (state-of-charge greater than a specified value), the driveline would operate in an electric-only mode if the electric motor can provide the power required by the vehicle. If the power demand is greater than that available from the electric motor, the engine is turned on to assist the electric motor. At low speeds when the battery requires charging, the engine output is split between powering the generator and the vehicle. The possibility of splitting the engine output in this way at low vehicle speeds is the distinguishing feature of the dual mode hybrid configuration. This permits the engine to be operated near its maximum efficiency at all times and the battery to be recharged, when needed, regardless of the vehicle speed and power demand. The dual mode arrangement also reduces the need for on-off engine operation as required in the series and parallel control strategies. Dual mode hybrids are operated with the battery state-of-charge maintained in a narrow range (charge sustaining) and thus require no recharging of the battery from the wall-plug. The Toyota Prius hybrid vehicle uses this dual mode operating strategy.

PERFORMANCE OF HYBRID VEHICLES

The added complexity of the various hybrid vehicle designs relative to battery-powered electric vehicles and conventional engine-powered vehicles is evident. In order to justify this added complexity, hybrid vehicles must be more marketable than pure electric vehicles and have higher fuel economy and lower emissions than conventional engine-powered vehicles. All the vehicle types (electric, hybrid, and conventional) can be designed to have the same acceleration performance and top speed by the proper selection of the driveline components. Acceleration times of 0–96 km/hr (60 mph) in less than 8 sec and top speeds in excess of 120 km/hr for passenger cars have been demonstrated for both pure electric and hybrid vehicles. The primary advantage of the hybrid vehicle compared to the electric vehicle is that its range and

refueling time can be the same or better than the conventional vehicle because it is refueled in the same manner—at the fuel pump. The key comparisons of interest are the fuel economy, emissions, and costs of hybrid and conventional vehicles.

Computer simulations have been performed for both midsize and compact, lightweight vehicle designs for driving on the Federal Urban Driving Schedule (FUDS) and the Federal Highway Driving Schedule (FHWDS). These driving cycles (speed vs. time) are intended to simulate vehicle operation in city and highway driving. For each vehicle type, computer simulations were run using gasoline fuel injected engines, diesel engines, and Stirling engines. Electricity was generated on board the hybrid vehicles by coupling the engines to a generator or by utilizing a fuel cell fueled using compressed hydrogen. In all cases, electrical energy was recovered into the energy storage unit during braking using the traction motor as a generator. The control strategies used in the simulations were essentially those previously discussed in the section on control strategies. In all cases, the engines and fuel cell were operated in an on-off mode to maintain the energy storage unit in the state-of-charge range specified by the control strategy. The minimum power setting of the engines and fuel cell when they were "on" were set so their efficiency was not outside the high efficiency portion of their operating maps.

Fuel Economy

Fuel economy simulation results for various engines in series hybrids are compared in Table 2 for the FUDS and FHWDS driving cycles. For both the midsize and compact cars, fuel economy depends significantly on the technology used in the driveline. The use of diesel engines results in the highest fuel economy (miles per gallon of diesel fuel); however, from the energy consumption (kJ/mi) and CO_2 emission (gm CO_2/mi) points-of-view, the advantage of diesel engine relative to gasoline-fueled engines should be discounted to reflect the higher energy and the carbon content per gallon of diesel fuel compared to gasoline. These discount factors are 15 to 20 percent. The simulation results also indicate that for the same type of engine, the fuel economy can be 10 to 20 percent higher using ultracapacitors in place of batteries as the energy storage device. The highest fuel economics are projected for vehicles using fuel cells. The fuel economies (gasoline equivalent) of the fuel cell vehicles using compressed hydrogen are

Vehicle	Engine	Energy Storage	Miles per Gallon	
			FUDS	Highway
Midsize	Honda Gasoline	Ni. Mt. Hy. Bat.	36.1	45.4
	Direct Injection Gasoline	Ni. Mt. Hy. Bat.	47.3	56.0
	Sw. Ch. Diesel	Ni. Mt. Hy. Bat.	49.7	56.8
	Direct Injection Diesel	Ni. Mt. Hy. Bat.	60.5	71.1
	Stirling	Ni. Mt. Hy. Bat.	50.0	57.2
	Honda Gasoline	Capacitor	44.3	47.3
	Sw. Ch. Diesel	Capacitor	62.3	65.7
	Fuel Cell (H_2)	Ni. Mt. Hy. Bat.	89.2	105.0
Lightweight Compact	Honda Gasoline	Ni. Mt. Hy. Bat.	69.6	71.4
	Direct Injection Gasoline	Ni. Mt. Hy. Bat.	82.9	84.9
	Sw. Ch. Diesel	Ni. Mt. Hy. Bat.	98.2	95.6
	Direct Injection Diesel	Ni. Mt. Hy. Bat.	107.3	110.4
	Stirling	Ni. Mt. Hy. Bat.	89.5	92.7
	Honda Gasoline	Capacitor	81.4	75.5
	Sw. Ch. Diesel	Capacitor	109.8	104.0
	Fuel Cell (H_2)	Ni. Mt. Hy. Bat.	16.3	17.9

Table 2.
Summary of Hybrid Vehicle Fuel Economy Results on the FUDS and Highway Driving Cycles using Various Engines and a Fuel Cell
(1) mpg diesel fuel for diesel engine and mpg gasoline equivalent for fuel cell powered vehicles

about twice those of hybrid vehicles with direct injected gasoline engines and about 80 percent higher than vehicles with diesel engines. All the fuel cell vehicle designs utilized a fuel cell load-leveled with a nickel metal hydride battery permitting it to operate at high efficiency at all times.

In comparisons between the fuel economies of conventional passenger cars and those using series hybrid divelines, the hybrid vehicles have the same weight and road load as the conventional cars. Still, the utilization of the hybrid driveline resulted in about a 50 percent improvement in fuel economy for the FUDS cycle and about a 10 percent improvement on the FHWDS (highway cycle). The fuel economy of the conventional cars was taken from the EPA Fuel Economy Guide corrected by 10 percent for the FUDS and 22 percent for the highway cycle. These corrections were made, because the actual dynamometer fuel economy test data had been reduced by those factors so that the published fuel economies would be in better agreement with values experienced in the real world.

The fuel economy of series and parallel hybrid vehicles are compared in Table 3 for both the compact, lightweight, and midsize cars. The series hybrids are assumed to operate only in the charge sustaining mode (no battery recharging from the wall plug), but the parallel hybrids can operate in either the charge sustaining or charge depleting mode. In the case of the parallel hybrid in the charge depleting mode, the fuel economy is given for gasoline alone and at the powerplant (pp) including energy needed to recharge the batteries from the wall plug. For hybrid vehicles using gasoline engines (port injected), the fuel economy of the parallel hybrid vehicles in the charge sustaining mode (batteries charged from the engine—not from the wall plug) is 9 to 12 percent higher than that of the series hybrids. For the powerplant efficiency (33%) assumed in the calculations, the parallel hybrids operating in the charge depleting mode (battery charged only from the wall plug) had only 1 to 4 percent higher equivalent fuel economy than the same vehicle operating in the charge sustaining mode. If the batteries were recharged using electricity from a higher efficiency powerplant, the fuel economy advantage of the parallel hybrid in the charge depleting mode would be lighter.

Vehicle	Fuel Economy (mpg) Gasoline Engine		Fuel Economy (mpg) Swirl Chamber Diesel	
	FUDS	Highway	FUDS	Highway
Small, lightweight				
Series Hybrid				
Charge Sustaining	62.4	71.1	75.7	85.2
Parallel Hybrid				
Charge Sustaining	68.2	79.5	75.9	88.8
Charge Depleting				
gasoline alone	90.1	86.1	95.6	94.0
Including power plant	71.1	80.5	75.8	88.1
Midsize (1995 materials)				
Series Hybrid				
Charge Sustaining	39.2	48.2	47.9	58.5
Parallel Hybrid				
Charge Sustaining	42.8	54.1	48.0	60.6
Charge Depleting				
gasoline alone	55.3	56.6	59.0	62.3
Including power plant	45.4	54.7	49.1	60.6

Table 3.
Comparisons of the Fuel Economy for Series and Parallel Hybrid Vehicles
(1) Fuel economy shown for diesel engines is gasoline equivalent

Full Fuel Cycle Emissions

The full fuel cycle emissions of the hybrid vehicles are the total of all the emissions associated with the operation of the vehicle and the production, distribution, and dispensing of the fuel and electricity to the vehicle. The total emissions can be calculated for all the vehicle designs utilizing as inputs the vehicle simulation results for the electricity consumption, fuel economy, and exhaust emissions in the all-electric and hybrid modes and the upstream refueling, evaporative, and fuel production emissions based on energy usage—bothfuel and electricity. Both regulated emissions (nonmethane organic gases [NMOG], CO, NO$_x$) and CO$_2$ emissions can be calculated.

Regulated Emissions for Hybrid, Electric, and Conventional Cars

Hybrid vehicles operated in the charge depleting mode (battery charged from the wall plug) have total emissions comparable to those of electric vehicles if their all-electric range is 50 mi or greater. Hybrid vehicles operating in the charge sustaining mode have much greater NMOG emissions than electric vehicles when the refueling and evaporative emissions are included. The calculated total emissions of the electric vehicles are close to the equivalent zero emission vehicle (EZEV) emissions when the battery charging is done in the Los Angeles (LA) basin. These comparisons are based on the total NMOG, CO, and NO$_x$ emissions for the FUDS and highway driving cycles for electric vehicles and conventional ICE vehicles as well as hybrid vehicles. A baseline use pattern of 7,500 miles per year random, city travel, and a round trip to work of 15 miles was assumed.

Total CO$_2$ Emissions

The difference in the CO$_2$ emissions between operating a hybrid vehicle in charge depleting and charge sustaining modes, regardless of its all-electric range, is not large using nickel metal hydride batteries. The CO$_2$ emissions of the gasoline and diesel engine powered hybrids vary only about 25%—not as much as might be expected based on the differences in their fuel economies—because of the higher energy content and the higher carbon-to-hydrogen ratio of the diesel fuel. The fuel cell powered, hydrogen fueled vehicles are projected to have the lowest CO_2 emissions by 25 to 30 percent when compared to the most efficient of the engine powered vehicles even when the hydrogen is produced by reforming natural gas. The CO_2 emissions of the conventional ICE vehicles are directly proportional to their fuel economy, which is projected to be significantly less than the hybrid vehicles. ICE powered vehicles will have low CO_2 emissions only when their fuel economy is greatly increased. CO_2 emissions of vehicles are highly dependent on the technologies used to power them and can vary by a factor of at least two for the same size and weight of vehicle.

TESTING OF HYBRID VEHICLES

There have been relatively few tests of hybrid vehicles in recent years. One example of such tests is that of the Toyota Prius by the EPA. Special care was taken in those tests to account for changes in the net state-of-charge of the batteries on the vehicle. The simulation results were obtained using the same hybrid vehicle simulation program used to obtain the fuel economy projections given in Table 2. There is good agreement between the measured and calculated fuel economies for the Prius. The EPA emissions data indicate that the CO and NO_x emissions of the Prius are well below the California ULEV standards and that the NMOG emissions are only slightly higher than the 0.04 gm/mi ULEV standard. The fuel economy of the Prius on the FUDS cycle (in city driving) is 56 percent higher than a 1998 Corolla (equipped with a 4-speed lockup automatic transmission) and 11 percent higher for highway driving. These two improvements in fuel economy for a hybrid/electric car compared to a conventional ICE car of the same size are consistent with those discussed earlier in the section on fuel economy.

PROSPECTS FOR MARKETING HYBRID CARS

The societal advantages of the hybrid-electric vehicles will come to fruition only when a significant fraction of vehicle purchasers decide to buy one of them. This will occur if the purchase of the hybrid vehicles makes economic sense to them and the vehicle meets their needs. Otherwise vehicle buyers will continue to purchase conventional ICE-powered vehicles. The key to any workable marketing strategy is the availability of hybrid driveline technologies that make the transition from engine-based to electric-based drivelines manageable and attractive to the consumer with only modest financial incentives. The state of development of the new driveline technologies at the time of introduction must be such that vehicles meet the needs of the first owners and they find vehicles to be reliable and cost-effective to operate. Otherwise the market for the new technologies will not increase and the introduction of the new technologies at that time will be counterproductive. Even after the technical and economic feasibility of a new technology is shown in prototype vehicles, a large financial commitment is needed to perform the preproduction engineering and testing of the vehicles before the vehicles can be introduced for sale.

Starting in the fall of 1997, Toyota offered for sale in Japan the Prius Hybrid at a price close to that of the comparable conventional car. The initial response of the public was enthusiastic and the production rate quickly rose to more than 1,000 vehicles per month. Toyota is planning to introduce a redesigned Prius in the United States in the fall of 2000. Honda began selling a subcompact hybrid/electric car, the Insight, in the United States in the fall of 1999 at a price of less than $20,000, which is about $5,000 higher than the Honda Civic. According the EPA tests, the corporate average fuel economy (CAFÉ) (combined city and highway cycles) of the Honda Insight is 76 mpg.

Recent advances in exhaust emission technologies have resulted in the certification by several auto manufacturers of conventional gasoline fueled cars that can meet the California ULEV and SULEV standards. It seems important that the driving force for the eventual introduction of the hybrid-electric and fuel cell cars will be improved fuel economy and lower CO_2 emissions and not lower regulated emission standards. It appears likely that a significant increase in the price of energy (either because of scarcity or higher taxes), regulation (for example, the CAFÉ standard), or financial incentives to purchase and license hybrid vehicles will be necessary before advanced technology hybrid vehicles become popular.

Andrew Burke

See also: Capacitors and Ultracapacitors; Drivetrains; Electric Motor Systems; Electric Vehicles; Engines; Fuel Cell Vehicles; Fuel Cells.

BIBLIOGRAPHY

Ahmed, S. (1997). "Partial Oxidation Reformer Development for Fuel Cell Vehicles." Proceedings of the 32nd Intersociety Energy Conversion Engineering Conference. Paper 97081 (August).

Burke, A. F. (1996). "Hybrid/Electric Vehicle Design Options and Evaluations." Paper 920447 (February). Warrendale, PA: Society of Automotive Engineers.

Burke, A. F., and Miller, M. (1997). "Assessment of the Greenhouse Gas Emission Reduction Potential of Ultra-clean Hybrid-Electric Vehicles." Institute of Trans-portation Studies. Report No. UCD-ITS-RR-97-24 (December). Davis: University of California.

Conway, B. E. (1999). Electrochemical Capacitors: Scientific Fundamentals and Technological Applications. New York. Kluwer Academic/Plenum Publishers.

Craig, P. (1997). The "Capstone Turbogenerator as an Alternative Power Source." Paper 970292 (February). Pittsburgh, PA: Society of Automotive Engineers.

U.S. Department of Energy. (1997). "Fuel Economy Guide—Model Year 1998." Publication DOE/EE-0102. Washington, DC: U.S. Government Printing Office.

Gottesfeld, S., and Zawodzinski, T. A. (1998). Polymer Electrolyte Fuel Cells, Advances in Electrochemical Science and Engineering, ed. R. Alkire et al. New York: Wiley.

Helman, K. H.; Peralta, M. R.; and Piotrowski, G. K. (1998). "Evaluation of a Toyota Prius Hybrid System (THS)." U.S. Environmental Protection Agency, Report No. EPA 420-R-98-006(August). Washington, DC: U.S. Government Printing Office.

Hermance, D. (1999). "The Toyota Hybrid System." Pres. At SAE TOPTEC, Hybrid Electric Vehicles: Here and Now (27 May).

Honda Multi-Matic News Release. (1997). "Technical Description of the System and its Operation." Torrance, CA: Author.

Honda Unveils World's Highest Mileage Hybrid Vehicle. (1999). Honda Environmental News (March). Torrance, CA: Author.

Howard, P. F. (1996). "The Ballard Zero-Emission Fuel Cell Engine." Prese. At Commercializing Fuel Cell Vehicles, InterTech Conference (September).

Johansson, L., and Rose, T. (1997). "A Stirling Power Unit for Series Hybrid Vehicles." Stirling Thermal Motors, Poster at the 1997 Customer Co-ordination Meeting (October).

Kluger, M. A., and Fussner, D. R. (1997). "An Overview of Current CVT Mechanisms, Forces, and Efficiencies." Paper 970688. Pittsburgh, PA: Society of Automotive Engineers.

Michel, S., and Frank, A. (1998). "Design of a Control Strategy for the Continuously Variable Transmission for the UC Davis Future Car (Parallel Hybrid Vehicle)." Department of Mechanical Engineering Report (September). Davis, University of California.

Moore, R. M.; Gottesfeld, S.; and Zelenay, P. (1999). "A Comparison Between Direct-Methanol and Direct Hydrogen Fuel Cell Vehicles." SAE Future Transportation Technologies Conference. Paper 99FTT-48 (August).

NECAR 4—DaimlerChrysler's Sixth Fuel Cell Vehicle in Five Years. (1999). Press Release. March 17.

Pepply, B. A., et al. (1997). "Hydrogen Generation for Fuel Cell Power Systems by High Pressure Catalytic Methanol Steam Reforming." Proceedings of the 32nd Intersociety Energy Conversion Engineering Conference. Paper 97093 (August).

Prater, K. B. (1996). "Solid Polymer Fuel Cells for Transport and Stationary Applications." Journal of Power Sources 61:105–109.

Society of Automotive Engineers. (1998). "Recommended Practice for Measuring the Exhaust Emissions and Fuel Economy of Hybrid-Electric Vehicles." Document J1711 (September). Pittsburgh, PA: Author.

Ren, X.; Springer, T. E.; and Gottesfeld, S. (1998). "Direct Methanol Fuel Cell: Transport Properties of the Polymer Electrolyte Membrane and Cell Performance." Vol. 98-27. Proc. 2nd International Symposium on Proton Conducting Membrane Fuel Cells. Pennington, NJ: Electrochemical Society.

Takaoka, T., et al. (1998). "A High-Expansion—Ratio Gasoline Engine for the Toyota Hybrid System." Toyota Review 47(2).

Toyota. (1997). "Toyota's Green Machine." Business Week (December 15):108–110.

Vaughn, M. (1997). "Reinventing the Wheel." Autoweek (March 3).

HYDROELECTRIC ENERGY

Hydroelectric energy—electric power created by the kinetic energy of moving water—plays an important role in supplying the world's electricity. In 1996, nearly 13 trillion kilowatt-hours of electricity were generated worldwide; almost one-fifth of this electricity was produced with hydroelectricity. On average hydropower provides about ten percent of the U.S. electricity supply, although the annual amount of electricity generated by hydroelectric resources varies due to fluctuations in precipitation. In many parts of the world, reliance on hydropower is much higher than in the United States. This is particularly true for countries in South America where abundant hydroelectric resources exist. In Brazil, for example, 92 percent of the 287 billion kilowatt-hours of electricity generated in 1996 were generated by hydroelectricity.

There are many benefits for using hydro resources to produce electricity. First, hydropower is a renewable resource; oil, natural gas, and coal reserves may be depleted over time. Second, hydro resources are indigenous. A country that has developed its hydroelectric resources does not have to depend on other nations for its electricity; hydroelectricity secures a country's access to energy supplies. Third, hydroelectricity is environmentally friendly. It does not emit greenhouse gases, and hydroelectric dams can be used to control floods, divert water for irrigation purposes, and improve navigation on a river.

There are, however, disadvantages to developing hydroelectric power. Hydroelectric dams typically require a great deal of land resources. In conventional hydroelectric projects, a dam typically is built to create a reservoir that will hold the large amounts of water needed to produce power. Further, constructing a hydroelectric dam may harm the ecosystem and affect the population surrounding a hydro project. Environmentalists often are concerned about the adverse impact of disrupting the flow of a river for fish populations and other animal and plant species. People often must be relocated so that a dam's reservoir may be created. Large-scale dams cause the greatest environmental changes and can be very controversial. China's 18.2 gigawatt Three Gorges Dam project—the world's largest hydroelectric—will require the relocation of an estimated 1.2 million people so that a 412-mile reservoir can be built to serve the dam.

Another potential problem for hydroelectricity is the possibility of electricity supply disruptions. A severe drought can mean that there will not be enough water to operate a hydroelectric facility. Communities with very high dependence on the hydroelectric resources may find themselves struggling with electricity shortages in the form of brown-outs and black-outs.

This article begins with a description of how hydroelectricity works, from the beginning of the hydrological cycle to the point at which electricity is transmitted to homes and businesses. The history of the dam is outlined and how dams evolved from structures used for providing a fresh water supply to irrigation and finally to providing electricity. The history of hydropower is considered and the different hydroelectric systems (i.e., conventional, run-of-

river, and pumped storage hydroelectricity) currently in use are discussed.

HOW DOES HYDROELECTRICITY WORK?

Hydroelectricity depends on nature's hydrologic cycle (Figure 1). Water is provided in the form of rain, which fills the reservoirs that fuel the hydroelectric plant. Most of this water comes from oceans, but rivers and lakes and other, smaller bodies of water also contribute. The heat of the sun causes the water from these sources to evaporate (that is, to change the water from its liquid state into a gaseous one). The water remains in the air as an invisible vapor until it condenses and changes first into clouds and eventually into rain. Condensation is the opposite of evaporation. It occurs when the water vapor changes from its gaseous state back into its liquid state.

Condensation occurs when air temperatures cool. The cooling occurs in one of two ways. Either the air vapor cools as it rises and expands or as it comes into contact with a cool object such as a cold landmass or an ice-covered area. Air rises for several reasons. It can be forced up as it encounters a cooler, denser body of air, or when it meets mountains or other raised land masses. It can rise as it meets a very warm surface, like a desert, and become more buoyant than the surrounding air. Air also can be forced to rise by storms—during tornadoes particles of air circling to the center of a cyclone collide and are forced up. When the water vapor collides with a cold object, it can become fog, dew, or frost as it condenses. The vapor cools as it rises into the atmosphere and condenses to form clouds and, sometimes, rain.

In order for rain to form, there must be particles in the air (i.e., dust or salt) around which the raindrop can form and which are at temperatures above freezing. When the particles are cooled to temperatures below the freezing point water condenses around them in layers. The particles grow heavy enough that they eventually fall through the clouds in the form of raindrops or— if the air temperature is below the freezing point all the way to the ground — as snow, sleet, or hail.

Much of the rain that reaches the ground runs off the surface of the land and flows into streams, rivers, ponds and lakes. Small streams lead to bigger ones, then to rivers, and eventually back to the oceans where the evaporation process begins all over again.

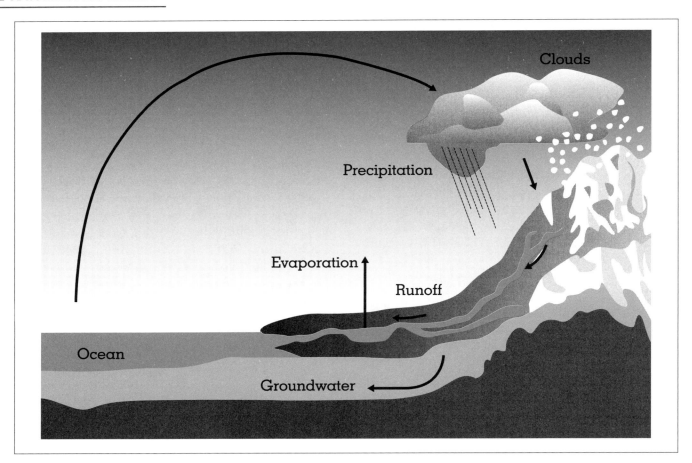

Figure 1.
The Hydrologic Cycle.
SOURCE: U.S. Department of Energy Hydropower Program.

Although water continuously changes from solid to liquid to gas, the amount of water on the earth remains the same; there is as much water today as there was hundreds of millions of years ago.

This water cycle—the process of moving water from oceans to streams and back again— is essential to the generation of hydroelectricity. Moving water can be used to perform work and, in particular, hydroelectric power plants employ water to produce electricity. The combination of abundant rainfall and the right geographical conditions is essential for hydroelectric generation.

Hydroelectric power is generated by flowing water driving a turbine connected to an electric generator. The two basic types of hydroelectric systems are those based on falling water and those based on natural river current, both of which rely on gravitational energy. Gravitational forces pull the water down either from a height or through the natural current of a river. The gravitational energy is converted to kinetic energy. Some of this kinetic energy is converted to mechanical (or rotational) energy by propelling turbine blades that activate a generator and create electricity as they spin.

The amount of energy created by a hydroelectric project depends largely upon two factors: the pressure of the water acting on the turbine and the volume of water available. Water that falls 1,000 feet generates about twice as much electric power as the same volume of water falling only 500 feet. In addition, if the amount of water available doubles, so does the amount of energy.

The falling water hydrosystem is comprised of a dam, a reservoir, and a power generating unit (Figure

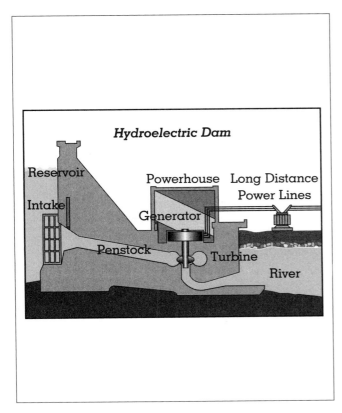

Figure 2.
Conventional, Falling Water Hydroelectric System.
SOURCE: U.S. Department of Energy Hydropower Program.

2). The dam is constructed so that a reservoir is created within which water accumulates and may be stored until it is needed. Water is released as required to meet electricity demands of customers. At the bottom of the dam wall is the water intake. The water intake controls when and how much water is moved into a steel called the "penstock." Gravity causes the water to fall through the penstock. This pipe delivers the running water to a turbine—a propeller-like machine with blades like a large fan. The water pushes against the turbine blades and the blades turn. Because the turbine is connected to an electric generator, as the turbine gains speed, it powers the generator and electricity is produced. The largest falling water facility in the United States is the Grand Coulee hydroelectric project on the Columbia River, in Washington State. Indeed, the largest power production facilities in North America are found at the Grand Coulee Dam project where an average 21 billion kilowatt-hours of electricity are produced each year.

The second type of hydroelectric plant is called a run-of-the-river system. In this case, the force of the river current applies pressure to the turbine blades to produce electricity. Run-of-the-river systems do not usually have reservoirs and cannot store substantial quantities of water. As a result, power production from this type of system depends on the river flow—the electricity supply is highly dependent upon seasonal fluctuations in output. Run-of-river projects are most successful when there are large flows in flat rivers or when a high natural geological drop is present, and when the required electricity output is below the maximum potential of the site.

Hydropower systems, as in all electricity-producing systems, require a generator to create the electricity. An electric generator is a device that converts mechanical energy into electric energy. The process is based on the relationship between magnetism and electricity. When a wire or any other electrically conductive material moves across a magnetic field, an electric current occurs in the wire. In a power plant, a strong electromagnet, called a rotor, is attached to the end of a shaft. The shaft is used to spin the rotor inside a cylindrical iron shell with slots—called a stator. Conducting wires are wound through the slots of the stator. When the rotor spins at a high rate of speed, an electric current flows through the conducting wires.

In a hydroelectric system, flowing water is used to propel a turbine that spins the shaft connected to the generator and creates the electric current. The kinetic energy of the moving water is changed into rotational energy and thereby causes the turbine blades to rotate. The turbine is attached to an electric generator and the rotational energy of the turbine is then converted to electric energy. The electricity then leaves the generator and is carried to the transformers where the electricity can travel through electric power lines and is supplied to residential, commercial, and industrial consumers. After the water has fallen through the turbine, it continues to flow downriver and to the ocean where the water cycle begins all over again.

Pumped storage hydroelectricity is an extended version of the falling water hydroelectric system. In a pumped storage system, two water sources are required—a reservoir located at the top of the dam structure and another water source at the bottom. Water released at one level is turned into kinetic energy by its discharge through high-pressure shafts that direct the downflow through the turbines connected to the generator. The water flows through the hydroelectric generating system and is collected in a lower reservoir. The water is pumped back to the upper reservoir once the initial generation process is complete. Generally this is done using reversible turbines—that is, turbines that can operate when the direction of spinning is reversed. The pump motors are powered by conventional electricity from the national grid. The pumping process usually occurs overnight when electricity demand is at its lowest. Although the pumped storage sites are not net energy producers—pumped storage sites use more energy pumping the water up to the higher reservoir than is recovered when it is released—they are still a valuable addition to electricity supply systems. They offer a valuable reserve of electricity when consumer demand rises unexpectedly or under exceptional weather conditions. Pumped storage systems are normally used as "peaking" units.

HISTORY AND EVOLUTION OF HYDROELECTRIC DAMS

Dams have existed for thousands of years. The oldest known dam, the Sadd el-Kafara (Arabic for "Dam of the Pagans"), was constructed over 4,500 years ago twenty miles south of Cairo, Egypt. The 348-feet wide, 37-feet high dam was constructed to create a reservoir in the Wadi el-Garawi. The dam was built with limestone blocks, set in rows of steps about eleven inches high. It appears that the dam was supposed to be used to create a reservoir that would supply drinking water for people and animals working in a nearby quarry. Scientists and archaeologists believe that the Sadd el-Kafara failed after only a few years of use because there is no evidence of siltation at the remains of the dam. When water flows into the reservoir created by a dam, the silt—sand and other debris carried with the stream—is allowed to settle, rather than be borne further downstream by the force of the flow of a stream or river. In the still water behind a dam this sediment is deposited on the bottom of the reservoir.

The earliest dams were built for utilitarian purposes, to create reservoirs for drinking supplies—as in the Sadd el-Kafara—or to prevent flooding or for irrigation purposes. Early dams were even used to create lakes for recreational purposes. In the middle

of the first century, the Roman emperor, Nero, constructed three dams to create three lakes to add to the aesthetic beauty of his villa.

Today a variety of dam structures are utilized. These can be classified as either embankment dams or concrete dams. Embankment dams are constructed with locally available natural resources. There are several different types of embankment dams, including earth dams—which are constructed primarily of compacted earth; tailings dams—constructed from mine wastes; and rockfill dams—which are constructed with dumped or compacted rock. The shape of these dams is usually dependent on the natural settling angle of the materials used to build them. Concrete, bitumen, or clay is often used to prevent water from seeping through the dam. This can be in the form of a thin layer of concrete or bitumen facing which acts as a seal, or in the form of a central core wall of clay or other fine materials constructed within the dam, allowing water to penetrate the upstream side of the structure, but preventing it from moving beyond the clay core.

Concrete dams are more permanent structures than embankment dams. Concrete dams can be categorized as gravity, arch, or buttress. Because concrete is a fairly expensive material, different construction techniques were developed to reduce the quantity of concrete needed. This is highly dependent upon geological considerations. In particular, the rock foundations must be able to support the forces imposed by the dam, and the seismic effects of potential earthquakes. Gravity dams work by holding water back by way of their own weight. A gravity dam can be described as a long series of heavy vertical, trapezoidal structural elements firmly anchored at the base. It is generally a straight wall of masonry that resists the applied water-pressure by its sheer weight. The strength of a gravity dam ultimately depends on its weight and the strength of its base.

An arch dam, on the other hand, relies on its shape to withstand the pressure of the water behind it. The arch curves back upstream and the force exerted by the water is transferred through the dam into the river valley walls and to the river floor. They are normally constructed in deep gorges where the geological foundations are very sound. The United States's Hoover Dam is an example of a concrete arch dam.

The buttress dam uses much less concrete than the gravity dam, and also relies on its shape to transfer the water load. Buttress dams were developed in areas where materials were scarce or expensive, but labor was available and cheap. They have been built primarily for purposes of irrigation. The dams are particularly suited for wide valleys. They have a thin facing supported at an incline by a series of buttresses. The buttresses themselves come in a variety of shapes, including the multiple arch and the simple slab deck (see Figure 3). The weight of the concrete is transferred to bedrock through the downstream legs or "buttresses" of the structure. The Coolidge Dam near Globe, Arizona is an example of a buttress dam, constructed of three huge domes of reinforced concrete.

Using water as a source of power actually dates back more than 2,000 years when the Greeks used water to turn wheels to grind wheat into flour. The first recorded use of water power was a clock, built around 250 B.C.E. and since that time, falling water has provided power to grind corn, wheat, and sugar cane, as well as to saw mills. In the tenth century, water wheels were used extensively in the Middle East for milling and irrigation. One of these dams, built at Dizful—in what is now Iran—raised the water 190 feet and supplied the residents with water to grind corn and sugar cane.

The discoveries associated with electromagnetism in the early nineteenth century had a major impact on the development of hydropower. The development of electric power generators and the fact that electric power was the only form of energy in a "ready to use" state which can be transmitted over long distances has been particularly significant to hydroelectricity. Dams located away from population centers could be useful generators if there was a way to supply the consumers. Water turbines were also developed during the nineteenth century as a natural successor to the water wheel. The high performance and small size of the turbine relative to the water wheel were important advancements. Combining the technology of electric generators, turbines, and dams resulted in the development of hydroelectric power.

Water was first used to generate electricity in 1880 in Grand Rapids, Michigan when a water turbine was used to provide storefront lighting to the city. In 1882—only two years after Thomas Edison demonstrated the incandescent light bulb—the first hydroelectric station to use Edison's system was installed on the Fox River at Appleton, Wisconsin. In 1881, construction began on the first hydroelectric generat-

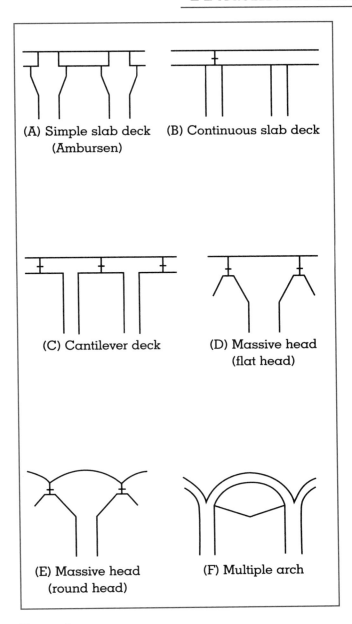

(A) Simple slab deck (Ambursen)

(B) Continuous slab deck

(C) Cantilever deck

(D) Massive head (flat head)

(E) Massive head (round head)

(F) Multiple arch

Figure 3.
Types of buttress dams.
SOURCE: University of Colorado, Department of Civil, Environmental, and Architectural Engineering.

ing station on the Niagara River in Niagara Falls on the New York-Canadian border. For twenty years, this project met the small electricity needs of the city. Water from the upper Niagara River fell 86 feet down a flume (a narrow gorge with a stream running through it) onto spinning water wheels on the lower river to generate electricity. The electricity was used to run the equipment of a paper company, other small factories, and sixteen open arc-lights on village streets. By 1896, the first long-distance transmission of electricity allowed the Niagara Falls project to provide electricity to Buffalo, New York—some 26 miles away.

By the early twentieth century, hydroelectric power was providing more than 40 percent of electricity generation in the United States. In 1940, hydropower supplied about three-fourths of all the electricity consumed in the West and Pacific Northwest, and still supplied about one-third of the total U.S. electricity supply. Although hydroelectricity's share of total electricity generation has since fallen to about 10 percent in the United States, hydroelectricity provides almost one-fifth of the world's total electricity generation today.

Hydroengineering has evolved over the past century so that it has become possible to build larger and larger hydroelectric projects. In 2000, the largest hydroelectric project in the United States was the Grand Coulee power plant on the Columbia river in Washington State. Grand Coulee is also the third largest currently-operating hydroelectric project in the world. The Grand Coulee project began operating in 1941 with 20 megawatts of installed capacity. The project has been expanded so that in 2000 it operated with an installed capacity of 6,180 megawatts.

The world's largest hydroelectric plant is the Itaipú power plant, located on the Paraná river separating Brazil and Paraguay. This 12,600-megawatt hydroelectric project is jointly-owned by Brazil and Paraguay. It is comprised of eighteen generating units, each with an installed capacity of 700 megawatts. The plant produces an estimated 75 billion kilowatt-hours each year. In 1994, Itaipú supplied 28 percent of all the electricity consumed in Brazil's south, southeast, and central-west regions, and 72 percent of Paraguay's total energy. Construction on this mammoth project began in 1975 and was not completed until 1991 at a cost of about $18 billion (U.S.). It required fifteen times more concrete than the Channel Tunnel that connects France and the United Kingdom; and the amount of steel and iron used in its construction would have built 380 Eiffel towers. The main dam of Itaipú is a hollow gravity type dam, but a concrete buttress dam and two embankment dams were also incorporated into the system design. At its highest point, the main dam is 643 feet—more than twice as

high as the Statue of Liberty; it extends nearly five miles across the Paraná River.

The large-scale disruptions to the environment of projects like the Itaipú hydroelectric system are profound and this is one of the major disadvantages to constructing large-scale hydroelectric facilities. As a result of construction on Itaipú, over 270 square miles of forest land has been negatively impacted, mostly on the Paraguayan side of the Paraná River where an estimated 85 percent of the forest was destroyed during the early years of construction. Despite programs to minimize the environmental damage through migration to reserves of the wildlife and plants facing extinction, several plant species became extinct mostly because they did not survive the transplant process.

Itaipú no longer represents the upper limit of large-scale hydroelectric expansion. At the end of 1994, the Chinese government announced the official launching of construction on the Three Gorges Dam hydroelectric project. When completed in 2009, this 18,200-megawatt project will provide the same amount of electricity as thirty 600-megawatt coal-fired plants. Project advocates expect the dam to produce as much as 85 billion kilowatt-hours of electricity per year, which is still only about 10 percent of the 881 billion kilowatt-hours of electricity consumed by China in 1995. It will also be used to control flooding along the Yangtze River and will improve navigational capacity, allowing vessels as large as 10,000 tons to sail upstream on the Yangtze as far as Chongqing, 1,500 miles inland from Shanghai. It is a very controversial project that will require the relocation of an estimated 1.2 million people, and that will submerge thirteen cities, 140 towns, 1,352 villages, and some 650 factories in its 412-mile reservoir that is to be created to support the dam. Disrupting the flow of the river will place several rare plant and animals at risk, including the endangered Yangtze River dolphin. Some environmentalists believe the reservoir created for the Three Gorges Dam project will become a huge pollution problem by slowing the flow of the Yangtze River and allowing silt to build up, possibly clogging the planned harbor at Chongqing within a few decades. The World Bank and the U.S. Export-Import Bank have refused to help finance the project, primarily because of the adverse environmental effects this massive hydroelectric project might have. Opponents to the Three Gorges Dam project have stated that smaller dams built on Yangtze River tributaries could produce the same amount of electricity and control flooding along the river, without adversely impacting the environment and at a substantially lower price than this large-scale project.

Hydroelectric projects can have an enormous impact on international relations between surrounding countries. For instance, the controversy between Hungary and Slovakia over the Gabcíkovo dam began at the dam's opening in 1992 and as of 2000 had not been completely resolved, despite a ruling by the International Court of Justice at The Hague in 1997. In its first major environmental case, the Court ruled that both countries were in breach of the 1977 treaty to construct two hydroelectric dams on the Danube River—the Slovakian Gabcíkovo and, 80 miles to the south, the Hungarian Nagymaros. Hungary suspended work on its portion in 1989 due to protests from the populace and international environmental groups. In 1992, Czechoslovakia decided to complete Gabcíkovo without Hungarian cooperation. Today, the $500 million, 180-megawatt project supplies an estimated 12 percent of the electricity consumed in Slovakia.

The Court stated that Hungary was wrong to withdraw from the treaty, but that Slovakia also acted unlawfully by completing its part of the project on its own. After almost a year of talks to resolve the differences between the two countries, Hungary and Slovakia agreed to construct a dam either at Nagymaros—the original site of the Hungarian portion of the project—or at Pilismarot. Unfortunately, soon after the agreement was signed, environmental protests began anew and little progress has been made to resolve the situation to everyone's satisfaction.

Another illustration of the political ramifications often associated with constructing dam projects concerns Turkey's plans to construct the large-scale Southeast Anatolia (the so-called GAP) hydroelectric and irrigation project. When completed, GAP will include twenty-one dams, nineteen hydroelectric plants (generating 27 billion kilowatthours of electricity), and a network of tunnels and irrigation canals. Neighboring countries, Syria and Iraq, have voiced concerns about the large scale of the hydroelectric scheme. Both countries argue that they consider the flow of the historic rivers that are to be affected by GAP to be sacrosanct. Scarce water resources in many countries of the Middle East make the disputes over the resource and the potential impact one country may have on the water supplies of another country particularly sensitive.

In the United States, in 2000 there were still over 5,600 undeveloped hydropower sites with a potential combined capacity of around 30,000 megawatts, according to estimates by the U.S. Department of Energy. It is, however, unlikely that a substantial amount of this capacity will ever be developed. Indeed, there is presently a stronger movement in this country to dismantle dams and to restore the natural flow of the rivers in the hopes that ecosystems damaged by the dams (such as fish populations that declined when dams obstructed migratory patterns) may be repaired. The U.S. Department of Interior (DOI) has been working to decommission many hydroelectric dams in the country and to restore rivers to their pre-dam states. In 1997, the Federal Energy Regulatory Commission (FERC) ordered the removal of the 160-year old Edwards Dam on the Kennebec River in Augusta, Maine and the demolition of the dam began in July 1999. Although Edwards Dam has only a 3.5-megawatt installed generating capacity and provided less than one-tenth of one percent of Maine's annual energy consumption, the event is significant in that it represented the first time that FERC had used its dam-removal authority and imposed an involuntary removal order.

In 1998, DOI announced an agreement to remove the 12-megawatt Elwha Dam near Port Angeles, Washington, but removal has been delayed indefinitely because Congress withheld the funds needed to finance the project. Some small dams in the United States have been successfully removed—such as the 8-foot Jackson Street Dam used to divert water for irrigation in Medford, Oregon and Roy's Dam on the San Geronimo Creek outside San Francisco, California. However, efforts to dismantle many of the larger dams slated for removal in the United States have been delayed by Congressional action, including four Lower Snake River dams and a partially built Elk Creek Dam in Oregon.

Linda E. Doman

See also: Electric Power, Generation of; Electric Power Transmission and Distribution Systems; Magnetohydrodynamics.

BIBLIOGRAPHY

The American University, School of International Service (1997). <http://www.gurukul.ucc.american.edu/salla/sis-home.htm>.

Baird, S. (1993). *Energy Fact Sheet: Hydro-Electric Power*. Ontario: Energy Educators of Ontario.

Biolchini dos Santos, M., and de Assis, R. (1998). *Information System of the Brazilian Embassy in London (INFOLONDRES)* "Brazil in the School." <http://www.brazil.org.uk>.

Blue Bridge Enterprises, Inc. (1996). "The Three Gorges Dam—II: The Super Dam Along the Yangtze River." New York: Author.

California State Government. (2000). *Energy Education from California Energy Commission*. "The Energy Story — Hydro." <http://www.energy.ca.gov>.

First Hydro Company. (1998). "About Pumped Storage." <http://www.fhc.co.uk>.

Hydro Tasmania. (1999). *Water Power: An Introduction to Hydro-electricity in Tasmania*. <http://www.hydro.com.au>.

Mufson, S. (1997). "The Yangtze Dam: Feat or Folly?" *The Washington Post*. (November 7).

National Hydropower Association. (2000). *Hydro Facts*. <http://www.hydro.org>.

New York Power Authority. (1999). *NYPA: Our History: The Niagara Power Project*. <http://www.nypa.gov>.

Powercor Australia, Ltd. (1997). *At School*. <http://www.powercor.com.au.>.

Smith, Norman (1972). *A History of Dams*. Seacaucus, NJ: Citadel Press.

Tyson, J. L. (1997). "Ardent Foe Takes on China Dam." *The Christian Science Monitor*. (November 12).

U.S. Bureau of Reclamation, Power Program. (1999). *The Hydropower Industry*. <http://www.usbr.gov>.

U.S. Department of Energy, Energy Information Administration. (1999). *International Energy Annual 1997*. Washington, DC: U.S. Government Printing Office.

U.S. Department of Energy, Energy Information Administration. (1998). *Electric Power Annual 1997: Volume I*. Washington, DC: U.S. Government Printing Office.

U.S. Department of Energy, Energy Information Administration. (1997). *Inventory of Power Plants in the United States as of January 1, 1998*. Washington, DC: U.S. Government Printing Office.

U.S. Department of Energy, Energy Information Administration. (1998). *International Energy Outlook 1998*. Washington, DC: U.S. Government Printing Office.

U.S. Department of Energy, Idaho National Engineering and Environment Laboratory (2000) *Hydropower Program* <http://www.inel.gov>.

U.S. Department of Interior. (2000). *Hoover Dam: National Historic Landmark*. <http://www.hooverdam.com>.

U. S. Geological Survey. (1998). *Water Science for Schools*. "Hydroelectric Power: How it Works." <http://ga.water.usgs.gov/edu>.

University of Colorado at Boulder, Department of Civil, Environmental, and Architectural Engineering. (1999). *Intro to Civil and Architectural Engineering*. "Geo Applications: Dams." <http://bechtel.colorado.edu/courseware/mos-cven/ce/3.2/texts/dams.html>.

HYDROGEN

Hydrogen is a high-quality energy carrier that can be employed with high conversion efficiency and essentially zero emissions at the point of use. Hydrogen can be made from a variety of widely available primary energy sources, including natural gas, coal, biomass (agricultural or forestry residues or energy crops), wastes, solar, wind, or nuclear power. Technologies for production, storage, and transmission of hydrogen are well established in the chemical industries. Hydrogen transportation, heating, and power generation systems have been technically demonstrated, and in principle, hydrogen could replace current fuels in all their present uses. If hydrogen is made from renewable or decarbonized fossil sources, it would be possible to produce and use energy on a large scale system with essentially no emissions of air pollutants (nitrogen oxides, carbon monoxide, sulfur oxides, volatile hydrocarbons or particulates) or greenhouse gases during fuel production, transmission, or use. Because of hydrogen's desirable environmental characteristics, its use is being proposed to reduce emissions of air pollutants and greenhouse gases. However, technical and economic challenges remain in implementing a hydrogen energy system.

CHARACTERISTICS OF HYDROGEN AS A FUEL

Table 1 summarizes some physical characteristics of hydrogen relevant to its use as a fuel. Specifically,

- Hydrogen is a low-density gas at ambient conditions. It liquefies at –253°C. For storage hydrogen must be compressed or liquefied.
- Hydrogen has the highest heating value per kilogram of any fuel (which makes it attractive as rocket fuel), although it has the lowest molecular weight. The specific heat of hydrogen is higher than that of other fuels.
- Hydrogen has a wider range of flammability limits and a lower ignition energy than other fuels.

- Hydrogen can be used with very little or no pollution for energy applications. When hydrogen is burned in air, the main combustion product is water, with traces of nitrogen oxides. When hydrogen is used to produce electricity in a fuel cell, the only emission is water vapor.

HISTORICAL PERSPECTIVE

Although it is not considered a commercial fuel today, hydrogen has been used for energy since the 1800s. Hydrogen is a major component (up to 50% by volume) of manufactured fuel gases ("town gas") derived from gasification of coal, wood, or wastes. Town gas was widely used in urban homes for heating and cooking in the United States from the mid-1800s until the 1940s, and is still used in many locations around the world (including parts of Europe, South America, and China) where natural gas is unavailable or costly. Hydrogen-rich synthetic gases also have been used for electric generation. Hydrogen is an important feedstock for oil refining, and indirectly contributes to the energy content of petroleum-derived fuels such as gasoline. Liquid hydrogen also is a rocket fuel, and it has been proposed as a fuel for supersonic aircraft. Despite hydrogen's many applications, primary energy use of hydrogen for energy applications (including oil refining) is perhaps 1 percent of the world total.

The idea of boosting hydrogen's role—a "hydrogen economy" or large-scale hydrogen energy system—has been explored several times, first in the 1950s and 1960s as a complement to a largely nuclear electric energy system (where hydrogen was produced electrolytically from off-peak nuclear power), and later, in the 1970s and 1980s, as a storage mechanism for intermittent renewable electricity such as photovoltaics, hydroelectric and wind power. More recently, the idea of a hydrogen energy system based on production of hydrogen from fossil fuels with separation and sequestration (e.g., secure storage underground in depleted gas wells or deep saline aquifers) of by-product CO_2 has been proposed.

Concerns about global climate change have motivated new interest in low-carbon or noncarbon fuels. Recent rapid progress and industrial interest in low-temperature fuel cells (which prefer hydrogen as a fuel) for transportation and power applications have also led to a reexamination of hydrogen as a fuel.

Molecular weight of H_2 (g/mole)	2.016
Mass density of H_2 gas (kg/Nm³) at standard conditions (P=1 atm=0.101 MPa, T=0°C)	0.09
Higher Heating Value (MJ/kg)	141.9
Lower Heating Value (MJ/kg)	120.0
Gas Constant (kJ/kg/°K)	4.125
Specific heat (c_p) (kJ/kg/°K) at 20°C	14.27
Flammability Limits in air (% volume)	4.0-75.0
Detonability Limits in air (% volume)	18.3-59.0
Diffusion velocity in air (meter/sec)	2.0
Buoyant velocity in air (meter/sec)	1.2-9.0
Ignition energy at stoichiometric mixture (milliJoules)	0.02
Ignition energy at lower flammability limit (milliJoules)	10
Temperature of liquefaction (°C)	-253
Mass density of liquid H_2 at -253°C (kg/Nm³)	70.9
Toxicity	non-toxic

Table 1
Physical Properties of Hydrogen

HYDROGEN ENERGY TECHNOLOGIES

Hydrogen Production

Hydrogen is the most abundant element in the universe and is found in a variety of compounds, including hydrocarbons (e.g., fossil fuels or biomass) and water. Since free hydrogen does not occur naturally on earth in large quantities, it must be produced from hydrogen-containing compounds.

More than 90 percent of hydrogen today is made thermochemically by processing hydrocarbons (such as natural gas, coal, biomass, or wastes) in high-temperature chemical reactors to make a synthetic gas or "syngas," comprised of hydrogen, CO, CO_2, H_2O and CH_4. The syngas is further processed to increase the hydrogen content, and pure hydrogen is separated out of the mixture. An example of making hydrogen thermochemically from natural gas is shown in Figure 1.

Where low-cost electricity is available, water electrolysis is used to produce hydrogen. In water electrolysis

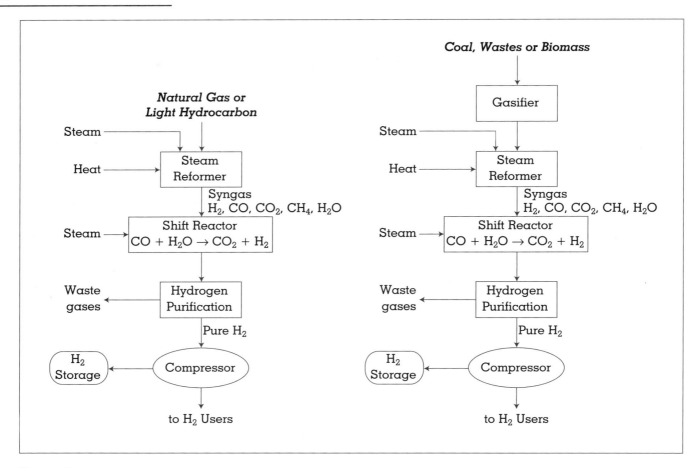

Figure 1.
Thermochemical Hydrogen Production Methods.

electricity is passed through a conducting aqueous electrolyte, breaking down water into its constituent elements, hydrogen and oxygen (see Figure 2). The hydrogen can be compressed and stored for later use. Any source of electricity can be used, including intermittent (time-varying) sources such as off-peak power and solar or wind energies. Fundamental research is being conducted on experimental methods of hydrogen production, including direct conversion of sunlight to hydrogen in electrochemical cells, and hydrogen production by biological systems such as algae or bacteria.

Hydrogen Transmission and Distribution

The technologies for routine handling of large quantities of hydrogen have been developed in the chemical industry. Hydrogen can be liquefied at low temperature ($-253°C$) and delivered by cryogenic tank truck or compressed to high pressure and delivered by truck or gas pipelines.

If hydrogen were widely used as an energy carrier, it would be technically feasible to build a hydrogen pipeline network similar to today's natural-gas pipeline system. It has been suggested that the existing natural-gas pipeline system might be converted to transmit hydrogen. With modifications of seals, meters, and end-use equipment, this could be done if pipeline materials were found to be compatible. However, hydrogen embrittlement (hydrogen-induced crack growth in pipeline steels that are subject to changes in pressure) could be an issue, especially for long-distance gas pipelines. Rather than retrofitting existing pipelines, new hydrogen pipelines might be built utilizing existing rights-of-way.

Hydrogen Storage

Unlike gasoline or alcohol fuels, which are easily handled liquids at ambient conditions, hydrogen is a

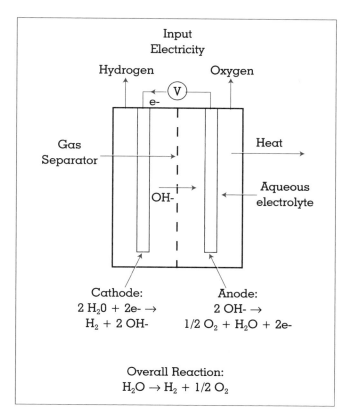

Figure 2.
Hydrogen Production Via Water Electrolysis.

lightweight gas and has the lowest volumetric energy density of any fuel at normal temperature and pressure. Thus hydrogen must be stored as a compressed gas (in high-pressure gas cylinders), as a cryogenic liquid at –253°C (in a special insulated dewar), or in a hydrogen compound where the hydrogen is easily removed by applying heat (such as a metal hydride). All these storage methods for hydrogen are well known in the chemical industry. Innovative storage methods such as hydrogen adsorption in carbon nanostructures are being researched.

Hydrogen onboard storage systems for vehicles are bulkier, heavier, and costlier than those for liquid fuels or compressed natural gas, but are less bulky and less heavy than presently envisaged electric batteries. Even with these constraints, it appears that hydrogen could be stored at acceptable cost, weight, and volume for vehicle applications. This is true because hydrogen can be used so efficiently that relatively little fuel is needed onboard to travel a long distance.

HYDROGEN FOR TRANSPORTATION AND POWER APPLICATIONS

Hydrogen Engines

Hydrogen engines resemble those using more familiar fuels, such as natural gas, gasoline, or diesel fuel. There are several key differences: (1) the emissions from a hydrogen engine are primarily water vapor with traces of nitrogen oxides, (2) the engine can be made more energy-efficient than with other fuels, and (3) there is much less need for postcombustion clean-up systems such as catalytic converters. Use of hydrogen in engines would substantially reduce pollutant emissions and would improve efficiency somewhat.

While hydrogen engines have some advantages over natural-gas engines, the hydrogen fuel cell offers a true "quantum leap" in both emissions and efficiency.

Hydrogen Fuel Cells

A fuel cell is shown in Figure 3. A fuel cell is an electrochemical device that converts the chemical energy in a fuel (hydrogen) and an oxidant (oxygen in air or pure oxygen) directly to electricity, water, and heat. The emissions are water vapor. The electrical conversion efficiency can be quite high. Up to 60 percent of the energy in the hydrogen is converted to electricity, even at small scale. This compares to 25 to 30 percent conversion efficiency for small fossil-fuel-powered engines.

The fuel cell works as follows. Hydrogen and oxygen have "chemical potential," an attraction driving them to combine chemically to produce water. In a fuel cell hydrogen is introduced at one electrode (the anode) and oxygen at the other electrode (cathode). The two reactants (hydrogen and oxygen) are physically separated by an electrolyte, which can conduct hydrogen ions (protons) but not electrons. The electrodes are impregnated with a catalyst, usually platinum, which allows the hydrogen to dissociate into a proton and an electron. To reach the oxygen, the proton travels across the electrolyte, and the electron goes through an external circuit, doing work. The proton, electron and oxygen combine at the cathode to produce water.

Although the principle of fuel cells has been known since 1838, practical applications are fairly recent. The first applications were in the space program, where fuel cells powered the Gemini and Apollo spacecraft. In the 1960s and 1970s, fuel cells

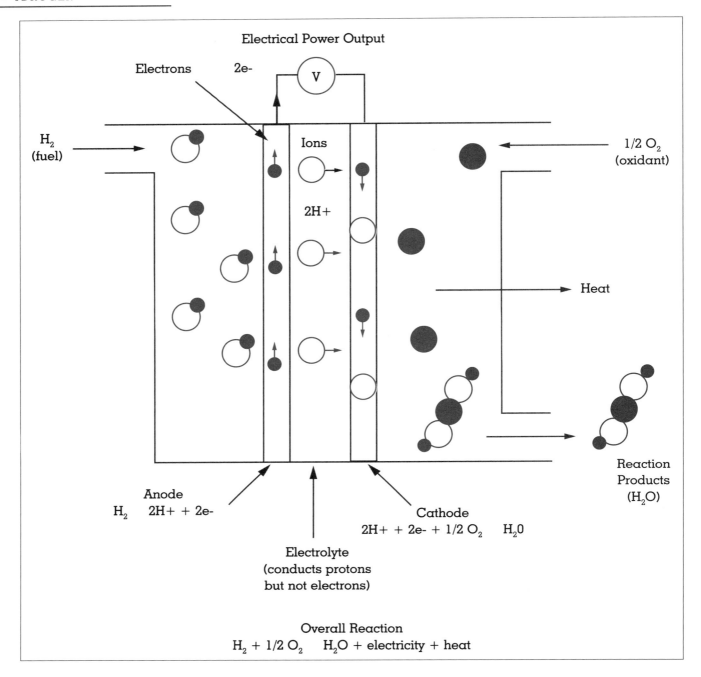

Figure 3.
Diagram of a Hydrogen-Oxygen Fuel Cell.

were used in space and military applications such as submarines. More recently fuel cells have been developed for low-polluting cogeneration of heat and power in buildings. In the past few years there has been a large worldwide effort to commercialize fuel cells for use in zero-emission vehicles.

Hydrogen for Transportation

A number of experimental hydrogen-powered vehicles have been built, dating back to the 1930s. Beginning in the early 1990s, "zero-emission-vehicle" regulations (enacted first in California and later in Massachusetts and New York) and government programs encouraging the development of high-efficiency automobiles (notably the Partnership for a New Generation of Vehicles) led to increased levels of R&D on fuel-cell and electric vehicles. Progress toward a commercial fuel cell vehicle is proceeding at a rapid pace. As of 1999, eight major automobile manufacturers were developing fuel-cell vehicles, and Ford, Daimler Chrysler, GM, and Toyota have announced their intent to commercialize a fuel-cell vehicle by 2005. Several oil companies (Mobil, Shell, Exxon, Texaco, and ARCO) are partners in demonstrations of fuel-cell-vehicle technologies. Shell has recently started a new business unit, Shell Hydrogen.

Hydrogen for Power Generation and Heating

Hydrogen also can be used to heat buildings and for power production. If low-cost fuel cells are commercialized, this could open the way for efficient combined heat and electric power systems for use in commercial and residential buildings. Because of ongoing deregulation, distributed, smaller-scale production of power may play an increasing role in future electric utilities. Several companies are developing fuel cell systems in a range of sizes. The economics are particularly attractive for buildings far from the existing electrical transmission grid, where resources for hydrogen production are present, such as remote sites or islands.

ENVIRONMENTAL AND SAFETY CONSIDERATIONS

Emissions from a Hydrogen Energy System

Hydrogen can be used with zero or near-zero emissions at the point of use. When hydrogen is burned in air, the main combustion product is H_2O, and there are traces of nitrogen oxides, which can be controlled to very low levels. No particulates, carbon monoxide, unburned hydrocarbons, or sulfur oxides are emitted. With hydrogen fuel cells, water vapor is the sole emission. Moreover, the total fuel cycle emissions of pollutants and greenhouse gases (such as CO_2, which could contribute to global climate change) can be much reduced compared to conventional energy systems, which vent all of the CO_2 in the flue gas.

Fuel cycle emissions are all the emissions involved in producing, transmitting, and using an alternative fuel. For example, for hydrogen made from natural gas, there would be emissions of CO_2 and nitrogen oxides at the hydrogen production plant, emissions associated with producing electricity to run hydrogen pipeline compressors (the nature of these emissions would depend on the source of electricity), and zero local emissions if the hydrogen is used in a fuel cell. The more efficient the end-use device (e.g., a fuel-cell vehicle), the lower the fuel cycle emissions per unit of energy service (e.g., emissions per mile traveled).

If hydrogen is made from decarbonized fossil fuels, fuel-cycle emissions can be cut by up to 80 percent. With renewable energy sources such as biomass, solar, or wind, the fuel cycle greenhouse gas emissions are virtually eliminated. It is possible to envision a future energy system based on hydrogen and fuel cells with little or no emissions of pollutants or greenhouse gases in fuel production, distribution, or use.

Resource Issues

In contrast to fossil energy resources such as oil, natural gas, and coal, which are unevenly distributed geographically, primary sources for hydrogen production are available virtually everywhere in the world. The choice of a primary source for hydrogen production can be made based on the best local resource.

Can hydrogen be produced sustainably? Over the next few decades, and probably well into the twenty-first century, fossil sources such as natural gas or coal may offer the lowest costs in many locations, with small contributions from electrolysis powered by low-cost hydropower.

In the longer term (or where locally preferred), renewable resources such as wastes, biomass, solar, or wind might be brought into use. It has been estimated that hydrogen derived from biomass produced on about two-thirds of currently idle cropland in the United States would be sufficient to supply transportation fuel to all the cars in the United States, if they used fuel cells (Ogden and Nitsch, 1993). Municipal solid waste could be gasified to produce transportation fuel for perhaps 25 to 50 percent of the cars in U.S. metropolitan areas (Larson, Worrell, and Chen, 1996). Solar power and wind power are potentially huge resources for electrolytic hydrogen production, which could meet projected global demands for fuels, although the delivered cost is projected to be about two to three times that for hydrogen from natural gas (Ogden and Nitsch, 1993).

Hydrogen Safety

When hydrogen is proposed as a future fuel, the average person may ask about the *Hindenburg*, the *Challenger*, or even the hydrogen bomb. Clearly, consumers will not accept hydrogen or any new fuel unless it is as safe as our current fuels.

Table 2 shows some safety-related physical properties of hydrogen as compared to two commonly accepted fuels, natural gas and gasoline.

In some respects hydrogen is clearly safer than gasoline. For example, hydrogen is very buoyant and disperses quickly from a leak. Experiments have shown that is it difficult to build up a flammable concentration of hydrogen except in an enclosed space, because the hydrogen disperses too rapidly. This contrasts with gasoline, which puddles rather than dispersing, and where fumes can build up and persist. Hydrogen is nontoxic, which also is an advantage.

Other safety concerns are hydrogen's wide flammability limits and low ignition energy. Hydrogen has a wide range of flammability and detonability limits (a wide range of mixtures of hydrogen in air will support a flame or an explosion). In practice, however, it is the lower flammability limit that is of most concern. For example, if the hydrogen concentration builds up in a closed space through a leak, problems might be expected when the lower flammability limit is reached. Here the value is comparable to that for natural gas.

The ignition energy (e.g., energy required in a spark or thermal source to ignite a flammable mixture of fuel in air) is low for all three fuels (hydrogen, gasoline, and natural gas) compared to commonly encountered sources such as electrostatic sparks from people. The ignition energy for hydrogen is about an order of magnitude lower for hydrogen than for methane or gasoline at stoichiometric conditions (at the mixture needed for complete combustion). But at the lower flammability limit, the point where problems are likely to begin, the ignition energy is about the same for methane and hydrogen.

Although safe handling of large quantities of hydrogen is routine in the chemical industries, people question whether the same safety can be achieved for hydrogen vehicle and refueling systems. According to a 1994 hydrogen vehicle safety study by researchers at Sandia National Laboratories, "There is abundant evidence that hydrogen can be handled safely, if its unique properties—sometimes better, sometimes worse and sometimes just different from other fuels—are respected." A 1997 report on hydrogen safety by Ford Motor Company (Ford Motor Company, 1997) concluded that the safety of a hydrogen fuel-cell vehicle would be potentially better than that of a gasoline or propane vehicle, with proper engineering. To assure that safe practices for using hydrogen fuel are employed and standardized, there has been a considerable effort by industry and government groups (within the United States and internationally) in recent years to develop codes and standards for hydrogen and fuel-cell systems.

ECONOMICS OF HYDROGEN PRODUCTION AND USE

If hydrogen technologies are available now and the environmental case is so compelling, why aren't we using hydrogen today? Part of the answer lies in the economics of hydrogen use. There are substantial capital and energy costs involved in hydrogen production. Comparing the delivered cost of hydrogen transportation fuel on an energy cost basis, we find that it is more costly than natural gas or gasoline. However, hydrogen can be used more efficiently than either natural gas or gasoline, so the cost per unit of energy service is comparable. Studies of the projected cost of hydrogen-fueled transportation have shown that if fuel-cell vehicles reach projected costs in mass production, the total life-cycle cost of transportation (accounting for vehicle capital costs, operation and maintenance costs, and fuel) will be similar to that for today's gasoline vehicles.

SCENARIOS FOR DEVELOPING A HYDROGEN ENERGY SYSTEM

The technical building blocks for a future hydrogen energy system already exist. The technologies for producing, storing, and distributing hydrogen are well known and widely used in the chemical industries today. Hydrogen end-use technologies—fuel cells, hydrogen vehicles, and power and heating systems—are undergoing rapid development. Still, the costs and the logistics of changing our current energy system mean that building a large-scale hydrogen energy system probably would take many decades.

Because hydrogen can be made from many different sources, a future hydrogen energy system could evolve in many ways. In industrialized countries,

	Hydrogen	Methane	Gasoline
Flammability Limits (% volume)	4.0-75.0	5.3-15.0	1.0-7.6
Detonability Limits (% volume)	18.3-59.0	6.3-13.5	1.1-3.3
Diffusion velocity in air (meter/sec)	2.0	0.51	0.17
Buoyant velocity in air (meter/sec)	1.2-9.0	0.8-6.0	non-buoyant
Ignition energy at stoichiometric mixture (milliJoules)	0.02	0.29	0.24
Ignition energy at lower flammability limit (milliJoules)	10	20	n.a.
Toxicity	non-toxic	non-toxic	toxic in concentrations > 500 ppm

Table 2
Safety-Related Properties of Hydrogen, Methane, and Gasoline

hydrogen might get started by "piggybacking" on the existing energy infrastructure. Initially, hydrogen could be made where it was needed from more widely available energy carriers, avoiding the need to build an extensive hydrogen pipeline distribution system. For example, in the United States, where low-cost natural gas is widely distributed, hydrogen probably will be made initially from natural gas, in small reformers located near the hydrogen demand (e.g., at refueling stations). As demand increased, centralized production with local pipeline distribution would become more economically attractive. Eventually hydrogen might be produced centrally and distributed in local gas pipelines to users, as natural gas is today. A variety of sources of hydrogen might be brought in at this time. In developing countries, where relatively little energy infrastructure currently exists, hydrogen from local resources such as biomass may be more important from the beginning.

Joan M. Ogden

See also: Fuel Cells; Nuclear Energy; Refineries; Reserves and Resources.

BIBLIOGRAPHY

Ford Motor Company. (1997). *Direct Hydrogen-Fueled Proton Exchange Membrane Fuel Cell System for Transportation Applications: Hydrogen Vehicle Safety Report.* Contract No. DE-AC02-94CE50389. Washington, DC: United States Department of Energy.

Herzog, H.; Drake, E.; and Adams, E. (1997). *CO₂ Capture, Reuse, and Storage Technologies for Mitigating Global Climate Change.* Washington, DC: United States Department of Energy.

Hord, J. (1976). "Is Hydrogen a Safe Fuel?" *International Journal of Hydrogen Energy* 3:157–176.

James, B. D.; Baum, G. N.; Lomax, F. D.; Thomas, C. E.; Kuhn, I. F. (1996). *Comparison of Onboard Hydrogen Storage for Fuel Cell Vehicles.* Washington, DC: United States Department of Energy.

National Research Council. (1998). *Review of the Research Program of the Partnership for a New Generation of Vehicles: Fourth Report.* Washington DC: National Academy Press.

Ogden, J. (1999). "Prospects for Building a Hydrogen Energy Infrastructure," In *Annual Reviews of Energy and the Environment*, ed. R. Socolow. Palo Alto, CA: Annual Reviews.

Ogden J., and Nitsch, J. (1993). "Solar Hydrogen." In *Renewable Energy Sources of Electricity and Fuels*, ed. T. Johannsson, H. Kelly, A. K. N. Reddy, and R. H. Williams. Washington DC: Island Press.

Ogden, J.; Steinbugler, M.; and Kreutz, T. (1999). "A Comparison of Hydrogen, Methanol, and Gasoline as Fuels for Fuel Cell Vehicles." *Journal of Power Sources* 79:143–168.

Socolow, R., ed. (1997). *Fuels Decarbonization and Carbon Sequestration: Report on a Workshop.* Princeton, NJ: Princeton University Center for Energy and Environmental Studies.

Taylor, J. B.; Alderson, J. E. A.; Kalyanam, K. M.; Lyle, A. B.; and Phillips, L. A. (1986). "Technical and Economic Assessment of Methods for the Storage of Large

Quantities of Hydrogen." *International Journal of Hydrogen Energy* 11:5.

Thomas, C. E.; Kuhn, I. F.; James, B. D.; Lomax, F. D.; and Baum, G. N. (1998). "Affordable Hydrogen Supply Pathways for Fuel Cell Vehicles." *International Journal of Hydrogen Energy* 23(6).

Williams, R. H. (1998). "Fuel Decarbonization for Fuel Cell Applications and Sequestration of the Separated CO_2." In *Eco-restructuring: Implications for Sustainable Development*, ed. W. Ayres. Tokyo: UN University Press.

Winter, C.-J., and Nitsch, J. (1998). *Hydrogen as an Energy Carrier*. New York: Springer-Verlag.

HYDROGEN BOMB

See: Nuclear Fusion

I

IMPORT/EXPORT MARKET FOR ENERGY

The import and export of energy on the international market helps create world economic growth. Countries that need energy because they do not have very large domestic reserves must import energy to sustain their economies. Countries that have large reserves export energy to create growth in their energy industries. The export and import of energy also contributes to specialization and comparative advantage in the world economy, lowering costs of production, increasing productivity, and expanding world economic output.

Comparative advantage in energy trade creates more robust economic growth, similar to any economic trade. For example, an oil worker has a comparative advantage over the farmer in producing oil as a farmer has a comparative advantage over the oil worker in producing food. An oil worker who specializes in pumping oil out of the ground is better off trading the oil for food from a farmer. The oil worker could grow his own food, but he could better utilize his time by simply pumping oil and trading for food. He specializes in oil production because he can obtain more food that way, or more of other goods and services. The farmer likewise could try to produce his own oil, but would spend an enormous amount of time and resources exploring for and developing an oil field. Both the farmer and the oil producer gain from trade. The medium of exchange for this trading is money, and both the farmer and the oil worker make more money by specializing and trading than by being totally self-sufficient.

The dynamics of specialization and comparative advantage that apply to the individuals work similarly for nations as well. On the world market, energy exporters specialize in producing energy, and trade their energy to other countries for other goods and services. Energy importers specialize in producing other goods and services, and sell those goods and services for energy. Saudi Arabia, for example, obtains high-tech computers from the United States, automobiles from Japan, and manufactured goods from Europe in exchange for its oil. If Saudi Arabia had to produce all its own computers, their computers would cost a lot more and probably not be nearly as good. If Europe, Japan and the United States had to produce all their own oil by, for example, converting coal into oil, oil would be much more costly, and that cost would inhibit economic growth. The trade in energy therefore benefits both sides.

The largest international energy trading on a Btu basis is with oil, partly because oil is the world's most valuable and versatile existing energy resource. The international natural gas trade is also increasing but because of the portability problems with natural gas (far less energy content given volume making it more expensive and more difficult to transport) its greater potential is limited. But within geographical regions around the world. Electricity is traded internationally more frequently now that many countries have deregulated their utility markets. The Middle East is the most prolific energy exporter, exporting about five and a half billion barrels of oil a year in 1998, which was over 50 percent of all world oil exports, and one-third of all world energy exports. That share looks to increase in the future as many older oil-producing regions, including the United States and Russia, have declining production.

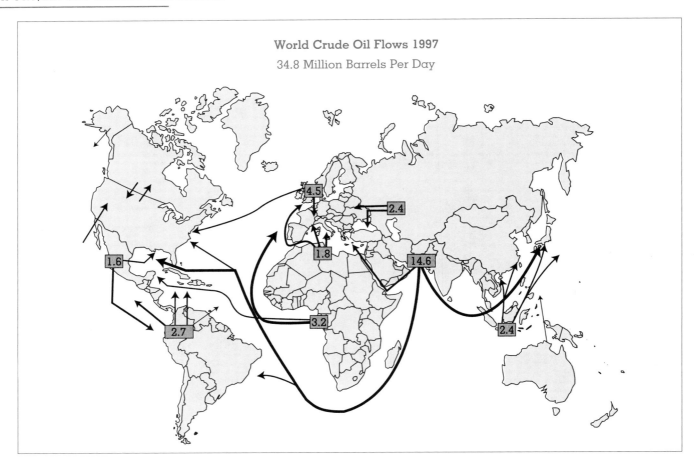

Figure 1.

SOURCE: Energy Information Agency, <http://www.eia.doe.gov/emeu/security/oilflow2.gif>.

MARKET POWER IN ENERGY TRADE

Energy exports not only depend on geological reserves of energy and internal consumption requirements of a country, but also on the market power of that country in energy markets. For example, in geological terms, 65 percent of world oil reserves reside in only five countries in the Middle East—Saudi Arabia, Iran, Iraq, Kuwait, and the United Arab Emirates—also called the big five Middle Eastern oil producers. The big five had proven oil reserves of more than 660 billion barrels of oil in 1998, with a theoretical potential to produce and export as much as 180 million barrels of oil a day (mbd), yet they produced and exported only 18 mbd on average in 1998. By restricting their oil exploration, production, and exports, the big five cause oil prices to rise on the world market. Therefore, it is not enough to look at world oil reserves and the theoretical potential oil

production to determine how long oil will last or when an oil shortage will occur; market power must also be taken into consideration. World oil production will undoubtedly last for hundreds of years, but at low levels. Production will be restricted by the use of market power so that the price of oil on the world market will be exceptionally high, decreasing the demand for oil.

The percent of world oil production from the Middle East has changed over the years. In 1950, the Middle East produced about 16 percent of world production, but by 1975 it had a 35 percent market share of all production. That share declined to 25 percent by 1990 due to a decline in world demand, and stayed at that level through 1998. Eventually that share will rise. The big five Middle Eastern oil producers have the largest oil reserves in the world and therefore have the greatest potential to supply increasing world demand, giving them an increasing market share of world production, which in turn will give them enough market power to substantially increase prices by limiting their production. It is the share of total world production rather than the share of the export market itself that determines market power. Middle Eastern oil market power may be a good thing, however, since by raising oil prices and limiting production the world's oil reserves will last longer, albeit at a higher price.

THE HISTORY OF U.S. ENERGY IMPORTS

The United States became the world's first producer of deep crude oil from an oil well when in 1859 Colonel Edwin Drake successfully used a pipe drilled into the ground to obtain oil. From then until about 1970, the United States was virtually energy-independent with only some oil and gas imports from Mexico and Canada. While U.S. reserves of coal, natural gas and uranium continue to be large enough to supply internal demand with enough left over to export, the supply of oil took a sharp turn downward. After 1970, even while U.S. demand continued to increase at a steep 6.5 percent per year, the supply of U.S. oil began to decline, necessitating sharp increases in U.S. oil imports.

Suddenly there was a significant change in the import/export dynamics of the world. From 1970 to 1973 the United States increased its proportion of imported oil from 10 percent to 23 percent of domestic demand, a substantial 2 mbd increase in imports. At the same time, Saudi Arabia increased its exports by 120 percent, or about 4 mbd, to become the world's most important oil exporter. This change in the balance of exports on the international market gave Saudi Arabia and the rest of the Organization of Petroleum Exporting Countries (OPEC) tremendous market power. It was only a matter of time before they would use it.

The 1973 Yom Kippur war caused Arab oil producers to boycott oil exports to the United States and some allies. While the boycott may look like the culprit in the 1973 oil price shock, the cuts in exports actually did not last long, with the oil price increases continuing long after the Arab/Israeli war was over. The real reason why oil prices surged up and stayed there was because world oil demand was high, while Saudi Arabia and OPEC kept supplies low. OPEC used its market power to curtail its own oil exports in the face of surging world demand. The result was an incredible quadrupling of the international price for oil in only one year. The sharp increase in oil prices created an immediate call for energy independence.

After the first oil shock, world demand for oil began to subside somewhat, but then increased again. Exports by OPEC members increased simultaneously as imports by the United States and other countries also increased. In 1979 Iran had a revolution that cut exports of Iranian oil to virtually nothing. The Iranian supply cut pushed oil prices into another shock. Again oil prices jumped high, but this time the extremely high price pushed oil demand down. After 1980, world oil demand declined for awhile and then stayed low due to the structural changes in the world's economies. Firms began using less oil and more coal, natural gas and nuclear power. The resulting softening of demand cut oil exports by OPEC members, particularly Saudi Arabia.

By 1986, world demand for oil was so low, and non-OPEC oil producers were exporting oil so copiously, that the price of oil took a tremendous dive. The combination of low demand and more abundant supplies could not sustain a high world oil price. U.S. demand, for example, had reached 19 mbd in 1979, but fell to 15 mbd by 1984. At the same time, imports of oil fell from 8 mbd to 4.25 mbd (i.e., from 42 percent to 28 percent of demand). This same trend occurred in all the developed countries resulting in the price plunge. However, since the crash in oil prices, oil demand has turned up again. After 1984, U.S. demand increased at a rate of 1.5 percent per year to 22 mbd by 1998. Yet, U.S. oil supplies continued to fall, even taking into account the great Alaska North Slope oil fields. By 1996, the United States was importing 50 percent of its oil demand, or 8.5 mbd, which has created calls for energy independence for America.

ENERGY INDEPENDENCE

The idea behind energy independence is that if all energy production occurs within a country's borders, then that country's economy will be insulated from any energy supply disruptions. The country would then have less unemployment and less economic decline if and when the world's energy exporters, especially OPEC members, cut their supplies. France, has developed a strong nuclear power industry so that it would not have to import as much oil from the Middle East. A close policy alternative for energy independence is to have all imported energy come from friendly countries. In 1998, the United States imported much of its oil from Mexico and Venezuela rather than from Iraq, Libya, or Iran.

One of the problems with attaining energy independence has been the cost of alternative energies. In the late 1970s and early 1980s, the United States tried to produce oil from oil shale in order to become energy-independent. However oil from oil shale cost twice as much as oil from OPEC. The problem was that if the United States were to rely on shale oil, then the economy would have been worse off, with even more unemployment than if it imported relatively cheaper OPEC oil. The expansion of the oil shale industry itself would have caused an increase in employment in that particular industry, but at a drastically higher cost of energy to the economy as a whole. The higher cost of domestically produced energy would have caused the rest of the U.S. economy to go into more of a recession than it already was in. This is why a policy of energy independence has not been successful.

Another U.S. policy to attain energy independence was to force all Alaskan North Slope crude oil to be consumed inside the United States and not be allowed to be exported. The problem was that North Slope crude oil is relatively heavy and not suitable for west coast fuel needs. The mismatch of supply and demand caused California refineries to sell heavy distillate fuels abroad and import lighter fuel additives. Furthermore, the forced selling of Alaska crude oil on a very saturated west coast market caused Alaska crude prices to be $1 to $5 per barrel less than the international price, resulting in less oil exploration and development in Alaska. The upshot of all this was lower tax revenue, a loss of jobs in the oil fields, and less oil exploration and development on the North Slope. The United States actually exported heavy bunker fuel oil at a loss, as opposed to the profit that could have been attained by simply exporting crude oil directly.

FRIENDLY COUNTRY IMPORTS

According to some experts, the next best solution for energy independence is to make sure that all imported energy comes from friendly countries. The idea is to ensure a guaranteed source of energy in case a major energy exporter were to suddenly cut off its

exports. The United States, for example, can buy all its oil from Mexico or Venezuela and be independent of crises in the Middle East. The problem with this strategy is that markets have a tendency to be independent of political expediency.

In a world market, oil is often bought by and sold to the highest bidder. Since all oil is fungible, the oil exporter usually does not care where the oil is sold as long as he receives top dollar for it. Oil import competitors such as Japan will always be ready to bid higher prices for oil no matter where it comes from. For example, the United States may import oil from Mexico. If a supply disruption occurred in Iran so that Japan could not get enough oil, Japan would immediately offer a higher price to Mexico for its oil. Even though Mexico is friendly to the United States it is not responsible for the U.S. economy. Mexico wants to get as much money as it can for its oil, and so will sell its oil to the highest bidder regardless of politics. In this case, Mexico would sell its oil to Japan if Japan were the highest bidder. This is how events in the Middle East can affect U.S. oil imports even if the U.S. imports its oil from somewhere other than the Middle East.

If Mexico did sell its oil to Japan rather than the United States, then the United States would immediately offer a high price for oil from another country, for example, Libya. Even though Libya has been a political foe of Americas, Libya would be willing to sell its oil to the United States if the United States paid more than European countries. In fact, events outside the United States have more to do with the price of oil than where the United States gets its oil. Whether U.S. oil is obtained from Alaska, Mexico, or Libya, it is the supply and demand for oil on the world market that determines the price of oil, not the location of supplies. Oil demand in East Asia, political tension in the Middle East, and lower oil production in the former Soviet republics, are much bigger determinants of U.S. energy prices than the source of imported oil. The world price of oil is independent of secure supplies from friendly countries, and as such it is the world oil market that affects the price of oil and brings about the economic instability caused by changing oil prices. The only way to ensure a stable low price for oil is to make sure that the international oil market stays competitive and free from supply disruptions caused by war or terrorist acts. This is why the United States,

Europe, and the rest of the world defended Kuwait and Saudi Arabia in the Gulf War in 1991. However, as long as the United States or any other country is willing to pay the world price for oil, then it can purchase as much oil as it wants.

THE AMAZING INCREASE IN U.S. COKE IMPORTS

Coal is a very useful energy resource in the production of steel. However, before coal can be used in steel mills, it must be converted into another form called coke, which is the source of the carbon monoxide used in steel-making. The coal is heated up in a low-oxygen chamber and burned slightly to produce coke, which is similar to wood charcoal. The United States has a 200-year supply of coal, and therefore vast potential supplies of coke. Nevertheless, the United States imported more coke than it exported in 1998, mostly from Japan and China. Even though imported coke represents only eight percent of U.S. consumption, it is still surprising that the United States, which has the largest reserves of coal in the world, would import coke at all.

That imports of coke have increased in the United States is mostly due to the high cost of transportation within the United States and the lower cost of shipping by foreign firms. It is cheaper to extract coal in China, coke it and ship it to California than to buy coke from U.S. producers and transport it within the United States. The trade in coke doesn't hurt the U.S. economy any more than the trade in jet aircrafts to China hurts China. Mutual trade is always beneficial to trading partners.

HYDRO-QUEBEC CANADIAN ELECTRICITY

Quebec, Canada, has great hydroelectric generating potential. Hydroelectricity is a very cheap source of electric power and relatively environmentally benign. As the U.S. continues to deregulate its power industry, Quebec hydroelectricity can be marketed for export to the U.S. northeast where electricity is very costly. It is even possible for Quebec to obtain market power in the U.S. electric power market. However, electric power was much less regional in 1999 than it was in the 1980s and before. In 1999 the technology existed to transport electric power over high voltage power lines as far as two thousand miles. This means the lowest priced electric producer could compete in the power market from very far away. If Quebec begins exporting electric power to the U.S. northeast, it will compete with electric producers as far away as Texas and Oklahoma that can produce cheap electricity from natural gas.

The overriding concern for energy imports and exports should not be where energy comes from or whether an industrialized country should be completely self reliant for its energy needs. The overriding concern is simply the price of energy. Whether energy is produced inside a country or not, world supply and demand for energy, energy market power, and the costs of energy production are the biggest factors in determining energy prices. The price of energy then affects how well the world's economy grows. This is the reason the United States and the rest of the world are concerned about all economic and energy events around the world, including energy supply cutoffs, recessions, and a generally increasing world demand for energy.

Douglas B. Reynolds

See also: Hydroelectric Energy.

BIBLIOGRAPHY

Adelman, M. A. (1993). *The Economics of Petroleum Supply*. Cambridge MA: MIT Press.

Brantly, J. E. (1960). *History of Oil Drilling*. Houston, TX: Gulf Publishing Co.

BP Amoco. (1996). *Statistical Review of World Energy 1996*. London, England: Author.

Cuff, D. J., and Young, W. J. (1986). *The United States Energy Atlas,* 2nd ed. New York: Macmillan.

Energy Information Administration. (1997). *International Energy Annual 1997*. Washington DC: Author.

Energy Information Administration. (1999). *International Energy Data*. Washington, DC: Author.

Exxon. (1984). *Middle East Oil and Gas*. Exxon Background Series, December 1984. New York: Author.

Federal Energy Regulatory Commission. (1996). *Promoting Wholesale Competition Through Open Access Non-Discriminatory Transmission Services by Public Utilities*. Final Order No. 888, 75-FERC 61, 080, Docket No. RM95-8-000, April, 1996. Washington DC: Author.

IMPULSE TURBINE

See: Turbines, Steam

INCANDESCENT LAMP

See: Lighting

INCINERATION

See: Waste-to-Energy Technology; Cogeneration;
Materials

INDUCTION MOTOR

See: Electric Motor Systems

INDUSTRY AND BUSINESS, ENERGY AS A FACTOR OF PRODUCTION IN

The production of goods and services requires energy as an input, which is called a factor of production. Energy sources vary in their effectiveness as a factor of production, depending on their energy characteristics. The energy characteristics are measured in energy grades, which indicate the levels of usefulness of any given energy input. Low-grade energy resources are less useful to an economy than high-grade resources, because any given process will be able to produce more economic value from a high-grade energy resource than from a low-grade energy resource. For example, it is easier to fly a passenger jet aircraft using jet fuel rather than coal because jet fuel has more suitable energy characteristics.

ENERGY GRADES

There are four energy grades used to measure energy characteristics: weight, volume, area, and state. The weight grade is British thermal units (Btus) per pound of the energy resource. For example, coal has 10,000 Btus per pound, while oil has 20,000 Btus per pound, making oil the higher-grade resource. The volume grade is Btus per cubic foot of the energy resource. Oil has 1 million Btus per cubic foot while compressed natural gas, at 3,000 pounds per square inch, has 177,000 Btus per cubic foot, which makes oil the higher-grade resource. The weight and volume grades are important determinants for how easy energy is to transport. Light, compact energy sources are much easier to store and use than heavy, voluminous energy sources. The area grade is Btus per acre where the energy resource is found in its original state. Wood has 1 billion to 5 billion Btus per acre as a forest, whereas coal has 10 billion to 1 trillion Btus per acre in a mine. The area grade generally determines how costly it is to extract or produce energy. Energy that is diffuse over an area, such as trees in a forest, tends to require more capital and labor to extract each Btu of energy than concentrated sources. The state grade is the original physical state of the energy resource, such as a liquid, gas, or solid as measured at standard atmospheric temperature and pressure. The highest state grade is the liquid state, followed by the gas state, the solid state, and the field state. The liquid state is the highest state grade because liquids are easier to use than gases and solids. The field state is the lowest state grade, since energy from energy fields such as solar energy is difficult to store. The field state is any kind of energy field such as a magnetic field, an electric field, or a radiation field. Nuclear energy is a field state grade, since it derives its energy from a radiation field. The state grades are fundamental in determining how well various energy resources can produce economically valuable outputs.

High-grade energy resources can create higher-valued, lower-cost outputs. For example, oil is one of the highest-grade energy resources there is. It is a liquid that is a very high state grade. It has a high weight and volume grade, better than any other energy resource except nuclear fuels, making it cost-effective to carry with a mobile machine. For example, chain saws and automobiles work better with a light-weight fuel than with a heavy fuel, because there is less fuel weight to carry. Oil also has a high area grade—100

The solar-powered car Mad Dog III is put through its paces before it is packed up on August 26, 1999, to travel to Australia to compete in the World Solar Challenge. Its 760 solar cells generate 1,200 watts of power and give it a cruising speed of 75km/hr. (Corbis Corporation)

billion to 1 trillion Btus per acre in oil fields—which makes it easier and cheaper to produce. Because oil is a liquid, it is easy to extract with no mining, and it is easy to convert into a refined liquid fuel. Finally, liquid fuels are the easiest of all energy resources to store and transport. Liquid fuels can be used in internal-combustion engines, which are lighter in weight and have more power per pound than external-combustion engines (steam engines) that operate on solid fuel (coal). And although internal-combustion engines can operate more cleanly on natural gas, the infrastructure needed to compress and store gas is much more complex and expensive.

THE ENERGY UTILIZATION CHAIN (EUC)

For any energy to produce goods and services, it goes through a sequence of usage called an energy utilization chain (EUC). The EUC determines how energy will be used to produce goods and services. Link 1 of the EUC is simply obtaining the energy source. This includes exploration and extraction of the energy or in some way producing it. Link 2 is energy conversion. More often than not, energy must be refined or converted into a more useful form of energy for consumption to occur. Link 3 is energy transportation and storage. All energy must be brought to the consumer or firm for use and if necessary be stored for later use. Link 4 is energy consumption. This is where energy is burned or used up. Conservation of energy resources is also a part of link 4—that is, consuming less energy resources in link 4 is conservation of energy. Link 5 is the energy service. The ultimate end of using and consuming any energy resource is to provide some sort of service to society. The service can be used directly by consumers or be an input into the production of other goods and services.

The use of oil for transportation, such as in automobiles and in aircraft, for example, is generally called the oil EUC. The chain of EUC links are the following: the exploration and production of oil; the transportation of the oil by pipeline or tanker; the refining of the oil in a refinery; the transportation of the gasoline to filling stations; the consumption of the gasoline in automobiles, which is sometimes conserved by the use of high-mileage cars; and finally the transportation service from driving the automobile. The EUC then explains the system for using energy. If an alternative energy resource is used to replace oil, then some or all of the EUC links must change. For example, replacing

the oil EUC with coal converted into oil, requires only changing the first two links of the oil EUC. Using solar energy to replace the oil EUC may require all of the links to change. For example, gasoline-driven automobiles would be replaced with electric vehicles, which give different services than normal automobiles. Electric vehicles based on solar energy may have less range of operation, carry less cargo, and take longer to refuel than ordinary automobiles.

The oil EUC is one of the highest-value EUCs in the economy. Alternative EUCs for transportation are the natural gas EUC, the solar/electric EUC, the synthetic fuels EUC, and so on. The reason it is so costly to use these alternative EUCs to replace the oil EUC is because of the low energy grades of the energy sources for the alternative EUCs. For example, to replace the oil EUC with an ethanol alcohol fuel, the alcohol EUC would have to use industrial distilleries to convert grain into alcohol. However, grain has a low area grade of about 40 million Btus per acre, as compared to oil's 1 trillion Btus per acre. Oil has about five magnitudes greater energy per acre in its original state, making it much cheaper to produce. It takes about 100,000 acres of farmland planted in grain to equal one acre of an oil field. In terms of supply, it takes about three times as much capital and 10 times as much labor to extract a Btu of ethanol from farmland as to extract a Btu of gasoline from an oil field. If all U.S. farmland were used to make ethanol, it could replace only 35 percent of U.S. oil needs. In addition, the grain needs to be converted from a solid-state grade energy resource into a liquid. Because oil is already a liquid, this transition is much easier and cheaper.

ALTERNATIVE EUCS

In the future, as oil and other energy resources begin to deplete, the economy will need to use alternative EUCs. One alternative EUC is the solar EUC. Solar energy is a field grade—that is, it is and energy radiated in light waves similar to an electric field. Because it is a field, it cannot be stored easily. An acre of solar collectors can catch about 65 million Btus per hour. One hour of an oil field operation on one acre of land can produce 2 billion Btus. The oil requires much less capital to extract its energy than does solar energy. This is why it is cheaper to obtain energy from oil than from solar energy; in addition, it is easier to convert, store, and transport the oil energy, allowing it to produce cheaper and more useful energy services than solar energy can. Once oil supplies decline substantially, the economy may be forced to use the solar EUC, but at a substantially higher cost than for the oil EUC.

Alternative EUCs provide the economy with energy services such as transportation at a certain cost. Usually, low-cost, high-value EUCs are used wherever possible and are used before higher-cost, lower-value EUCs. However, as a high-grade energy resource declines, lower-grade energy resources must be used, meaning the economy must begin to use higher-cost, lower-valued EUCs—that is, the cost of energy as a factor of production may rise, at least in the short run. In the longer run, breakthroughs in advanced energy sources, such as from hydrogen, may actually lower energy costs. The cost structure of alternative EUCs, though, depends on the cost of inputs. Ironically, one of the inputs that goes into every EUC is energy itself. It takes energy to produce energy. For example in the case of the synthetic fuels EUC, oil shale is used as an energy source to replace crude oil. To produce oil from oil shale, the oil shale must be converted from a solid-state energy grade into a liquid energy grade, which requires much more capital and labor inputs than does converting crude oil into fuels. However, the capital and labor require energy to produce oil from oil shale. When the price of oil goes up, the costs of the labor and capital inputs also rise causing the price of the shale oil to go up. In 1970, before the first oil shock, oil from oil shale cost about $3 per barrel to produce, while oil cost $1.50 per barrel. However, by 1982, when oil was $30 per barrel, shale oil cost $60 per barrel to produce. The high cost of oil made other inputs into the synthetic fuels EUC cost more, creating a higher price for the synthetic fuel, which in turn resulted in an inflation cost spiral. The nature of energy grades and the inflation cost spirals of inputs into the EUCs tend to make it difficult to pin down the cost of energy.

In general and in the short run, alternative EUCs have higher costs and lower-valued services than currently used EUCs. As the economy begins to use alternative EUCs, they tend to cost more, possibly creating an inflationary cost spiral. Technology can help to make alternative EUCs cost less and create more value, but is not likely to change the physical characteristics of alternative energy resources in the short run, making alternative EUCs overall less valuable than currently used high-grade-energy EUCs. One way to deal with higher-cost, lower-value alternative EUCs is for society to change its lifestyle. The

value of alternative lifestyles must be evaluated in comparison to the cost of alternative EUCs. Higher-cost EUCs may force society into changing lifestyles to be able to afford energy services.

In the longer run, perhaps new alternative EUCs, even superior to the current high-grade-energy EUCs, can be developed. This possibility offers the opportunity of defeating any energy-induced inflationary cost spiral that might have developed, and opening up new possibilities of economic expansion driven by inexpensive energy as a factor of production. Given the critical role of energy to the production process, our economic output in the future is dependent on further advances in energy technology.

Douglas B. Reynolds

See also: Auditing of Energy Use; Capital Investment Decisions; Economically Efficient Energy Choices; Energy Economics; Industry and Business, Productivity and Energy Efficiency in.

BIBLIOGRAPHY

Cuff, D. J., and Young, W. J. (1986). *The United States Energy Atlas*, 2nd ed. New York: Macmillian.

Graham, S. (1983). "U.S. Pumps $2 Billion into States Oil Shale." *Rocky Mountain News*, July 31, pp. 1, 22, 25.

Katell, S., and Wellman, P. (1971). *Mining and Conversion of Oil Shale in a Gas Combustion Retort*, Bureau of Mines Oil Shale Program Technical Progress Report 44. Washington, DC: U.S. Department of the Interior.

Reynolds, D. B. (1994). "Energy Grades and Economic Growth." *Journal of Energy and Development* 19(2):245–264

Reynolds, D. B. (1998). "Entropy Subsidies." *Energy Policy* 26(2):113–118.

Ricci, L. (1982). *Synfuels Engineering*. New York: McGraw-Hill.

INDUSTRY AND BUSINESS, PRODUCTIVITY AND ENERGY EFFICIENCY IN

Improving productivity is one of the central concerns of businesses. In buildings, energy-efficient tech-

nologies and design strategies can improve labor productivity (output of goods and services per hour
worked) far in excess of the improvement in energy
productivity (output per unit energy consumed).
Similarly, in manufacturing, energy efficiency can
improve total factor productivity (product output as a
function of all labor, capital, energy, and materials
consumed in its production) far in excess of the
improvement in energy productivity.

OFFICE PRODUCTIVITY AND ENERGY EFFICIENCY

Offices and buildings are not typically designed to
minimize either energy use or labor costs (by maximizing worker productivity). Almost everyone
involved in building construction—such as the
developer, architects, and engineers—is rewarded by
the ability to minimize the initial cost of a building,
as opposed to its life-cycle cost. Moreover, the
designers are rarely the ones who will be paying the
energy bill or the salaries of the people working in the
building. The missed opportunity is revealed by the
total life-cycle costs of a building (Table 1).

A systematic approach to energy-efficient design
can cost-effectively cut energy costs by 25 percent to
50 percent, as has been documented in both new construction and retrofit. The evidence on how energy-
efficient design can increase worker productivity has
been limited, so the issue has been a controversial one.

A growing body of international research suggests
that specific design approaches can simultaneous save
energy and increase productivity. The fundamental
goal in productivity-enhancing design is to focus on
the end users—workers—giving them the lighting,
heating, and cooling they need for the job. This end-
use approach maximizes productivity by ensuring,
for instance, that workers don't have too much light
or inadequate light quality for the job. It maximizes
energy savings because, in most cases, this end-use
approach eliminates excess and/or inefficient lighting, heating, and cooling, and because new technologies that provide higher quality services (such as
better light quality) typically use far less energy than
the technologies they replace.

Daylighting—use of natural light—is a key strategy because it is both the highest quality lighting and
the most energy-saving, when it is systematically
integrated into a design. In Costa Mesa, California,
VeriFone achieved both large energy savings and pro-

Initial cost (including land and construction)	2 percent
Operation & Maintenance (including energy)	6 percent
People Costs	92 percent

Table 1.
30-Year Life-Cycle Costs of a Building

ductivity gains when it renovated a 76,000-square-
foot building containing offices, a warehouse, and
light manufacturing. The upgrade included energy-
efficient air handlers, high-performance windows, 60
percent more insulation than is required by code, a
natural-gas-fired cooling system, occupancy sensors,
and a comprehensive daylighting strategy, including a
series of skylights.

On sunny days, workers in the remanufacturing
area construct circuit boards with only natural light
and small task lighting. In the office area, on the other
hand, the design minimizes direct solar glare on
computers, while providing enough daylight to allow
workers there to see changes associated with the sun's
daily and seasonal variation.

The building beat California's strict building code
by 60 percent, with a 7.5-year payback on energy-
efficient technologies based on energy savings alone.
Workers in the building experienced an increase in
productivity of more than 5 percent and a drop in
absenteeism of 45 percent, which brought the payback to under a year—a return on investment of
more than 100 percent.

INDIVIDUAL CONTROL OVER THE WORKPLACE ENVIRONMENT

An increasingly popular strategy is to give individuals control over their workplace conditions. The
benefits of this approach were documented at West
Bend Mutual Insurance Company's 150,000-square-
foot building headquarters in West Bend, Wisconsin.
The design used a host of energy-saving design features, including efficient lighting, windows, shell
insulation, and HVAC (heating, ventilation, and air
conditioning).

In the new building, all enclosed offices have individual temperature control. A key feature is the

Environmentally Responsive Work-stations (ERWs). Workers in open-office areas have direct, individual control over both the temperature and air-flow. Radiant heaters and vents are built directly into their furniture and are controlled by a panel on their desks, which also provides direct control of task lighting and of white noise levels (to mask out nearby noises). A motion sensor in each ERW turns it off when the worker leaves the space, and brings it back on when he or she returns.

The ERWs give workers direct control over their environment, so that individuals working near each other can and often do have very different temperatures in their spaces. No longer is the entire HVAC system driven by a manager, or by a few vocal employees who want it hotter or colder than everyone else. The motion sensors save even more energy. The lighting in the old building had been provided by overhead fluorescent lamps, not task lamps. The workers in the new building all have task lights and they can adjust them with controls according to their preference for brightness. The annual electricity costs of $2.16 per square foot for the old building dropped to $1.32 per square foot for the new building, a 40 percent reduction.

The Center for Architectural Research and the Center for Services Research and Education at the Rensselaer Polytechnic Institute (RPI) in Troy, New York conducted a detailed study of productivity in the old building in the 26 weeks before the move, and in the new building for 24 weeks after the move. To learn just how much of the productivity gain was due to the ERWs, the units were turned off randomly during a two-week period for a fraction of the workers. The researchers concluded, "Our best estimate is that ERWs were responsible for an increase in productivity of about 2.8 percent relative to productivity levels in the old building." The company's annual salary base is $13 million, so a 2.8 percent gain in productivity is worth about $364,000—three times the energy savings.

FACTORY PRODUCTIVITY AND ENERGY EFFICIENCY

Just as in offices, energy efficiency improvements in manufacturing can generate increases in overall productivity far in excess of the gains in output per unit energy. This occurs in two principal ways: improved process control and systemic process redesign.

Process Control

Many energy-efficient technologies bring with them advanced controls that provide unique and largely untapped opportunities for productivity gains. In factories, probably the biggest opportunity is available in the area of variable-speed drives for motors. Variable- or adjustable-speed drives are electronic controls that let motors run more efficiently at partial loads. These drives not only save a great deal of energy, but also improve control over the entire production process. Microprocessors keep these drives at precise flow rates. Moreover, when the production process needs to be redesigned, adjustable drives run the motor at any required new speed without losing significant energy efficiency. Two examples are illustrative.

In Long Beach, Toyota Auto Body of California (TABC) manufactures and paints the rear deck of Toyota pickup trucks. In 1994, the company installed variable-speed motor drives for controlling the air flow in the paint booths. Applying paint properly to truck beds requires control over the temperature, air flow/balance, and humidity in the paint booths. Before the upgrade, manually-positioned dampers regulated airflow into the booths. Since the upgrade, the dampers are left wide open, while the fan motor speed changes automatically and precisely with touch screen controls, which also provide continuous monitoring of the airflow.

The improvements to the motor systems reduced the energy consumed in painting truck beds by 50 percent. In addition, before the upgrade, TABC had a production defect ratio of 3 out of every 100 units. After the upgrade, the ratio dropped to 0.1 per hundred. In 1997, the plant received a special award for achieving zero defects. The value of the improvement in quality is hard to put a price tag on, but TABC's senior electrical engineer Petar Reskusic says, "In terms of customer satisfaction, it's worth even more than the energy savings."

The Department of Energy documented another typical case at the Greenville Tube plant in Clarksville, Arkansas, which produces one million feet of customized stainless steel tubing per month for automotive, aerospace, and other high-tech businesses. Greenville's production process involves pulling, or "drawing," stainless steel tubing through dies to reduce their diameter and/or wall thickness. The power distribution system and motor drive were inefficient and antiquated, leading to overheating, overloading, and poor control of the motor at low speed. A larger, but more efficient, motor (200 hp) was installed along with a computerized control system for $37,000. Electricity consumption dropped 34 percent, saving $7,000 a year, which would have meant slightly more than a five-year payback.

The greater horsepower meant that many of the tubes needed fewer draws: On average, one draw was eliminated from half the tubes processed. Each draw has a number of costly ancillary operations, including degreasing, cutting off the piece that the motor system latches on to, and annealing. Reducing the number of draws provided total labor cost savings of $24,000 a year, savings in stainless steel scrap of $41,000, and additional direct savings of $5,000. Thus, total annual savings from this single motor system upgrade was $77,000, yielding a simple payback of just over five months, or a return on investment in excess of 200 percent.

Process Redesign

The second principal way that energy efficiency can bring about an increase in industrial productivity is by achieving efficiency through process redesign. That process redesign could simultaneously increase productivity, eliminate wasted resources, and save energy has been understood as far back as Henry Ford. Perhaps the most successful modern realization and explication of the connection between redesigning processes to eliminate waste and boost productivity was achieved in the post-war development of the "lean production" system.

To understand lean production it is important to distinguish between improving processes and improving operations:

- An automated warehouse is an *operations* improvement: It speeds up and makes the operation of storing items more efficient. Eliminating all or part of the need for the warehouse by tuning production better to the market is a *process* improvement.
- Conveyor belts, cranes, and forklift trucks are *operations* improvements: They speed and aid the act of transporting goods. Elimination of the need for transport in the first place is a *process* improvement.
- Finding faster and easier ways to remove glue, paint, oil, burrs and other undesirables from products are *operations* improvements; finding ways not to put them there in the first place is a *process* improvement.

These examples show that when one improves the process, one does not merely cut out unnecessary operations, critical though that is to increasing productivity; one invariably reduces energy consumption as well as environmental impact. Reducing warehouse space reduces the need for energy to heat, cool, and light it. Reducing transportation reduces fuel use and exhaust fumes. Eliminating "undesirables" means no glue, paint, oil or scrap; it also avoids the resulting clean-up and disposal.

Thus, there is an intimate connection between redesigning processes to increase productivity (by eliminating wasted time, which is the essence of lean production) and eliminating wasted energy and resources. The most-time consuming steps in any process also tend to produce the most pollution and use the most energy. An integrated redesign can minimize everything.

For instance, in 1996, 3M company announced a breakthrough in the process for making medical adhesive tapes, a process that reduces energy consumption by 77 percent. The new process also cuts solvent use by 2.4 million pounds, lowers manufacturing costs, and cuts manufacturing cycle time by 25 percent. The proprietary process took researchers nine years from conception through final implementation.

The Sealtest ice cream plant in Framingham, Massachusetts, modified its refrigeration and air-handling system, cutting energy use by one third. The new system blew more air and colder air, and it defrosted the air handler faster. As a result, the time required to harden the ice cream was cut in half. The overall result was a 10 percent across-the-board increase in productivity, which is worth more to the company than the energy savings.

Process redesign drives a company toward a number of practices associated with productivity gain, including cross-functional teams, prevention-oriented design, and continuous improvement. One of the best programs for continuously capturing both energy and productivity gains was developed by Dow Chemical's Louisiana Division.

In 1982, the Division began a contest in which workers were invited to propose energy-saving projects with a high return on investment. The first year's result—27 winners requiring a capital investment of $1.7 million providing an average return on investment of 173 percent—was somewhat surprising. What was astonishing to everyone was that, year after year, Dow's workers kept finding such projects. Even

Reducing warehouse space reduces the need for energy to heat, cool, and light it. (Corbis-Bettmann)

as fuel prices declined, the savings kept growing. Contest winners increasingly achieved their economic gains through process redesign to improve production yield and capacity. By 1988, these productivity gains exceeded the energy and environmental gains.

Even after 10 years and nearly 700 projects, the two thousand employees continued to identify high-return projects. The contests in 1991, 1992, and 1993 each had in excess of 100 winners, with an average return on investment of 300 percent. Total energy savings to Dow from the projects of those three years exceeded $10 million, while productivity gains came to about $50 million.

CONCLUSION

Not all energy-efficiency improvements have a significant impact on productivity in offices or factories. Nonetheless, productivity gains are then possible from a systematic approach to design. The key features that simultaneously save energy and enhance productivity in office and building design are (1) a focus on the end user, (2) improved workplace environment, especially daylighting, and (3) individual control over the workplace environment. The key features that save energy and enhance productivity in factories are (1) improved process control, and (2) systematic process redesign.

Joseph Romm

See also: Building Design, Commercial; Building Design, Energy Codes and.

BIBLIOGRAPHY

Brill, M.; Margulis, S.; and Konar, E. (1984). *Using Office Design to Increase Productivity*, Volume 1. Buffalo, NY: Workplace Design and Productivity, Inc.

Ford, H. (1988). *Today and Tomorrow*, reprint. Cambridge, MA: Productivity Press.

Kroner, W.; Stark-Martin, J. A.; and Willemain, T. (1982). *Using Advanced Office Technology to Increase Productivity.* Troy, NY: The Center for Architectural Research.

Loftness, Vivian, et al. (1998). *The Intelligent Workplace Advantage,* CD-ROM. Pittsburgh: Center for Building Performance and Diagnostics, School of Architecture, Carnegie Mellon University.

Romm, J. (1999). *Cool Companies: How the Best Businesses Boost Profits and Productivity By Cutting Greenhouse Gas Emissions.* Washington, DC: Island Press.

Shingo, S. (1990). *Modern Approaches to Manufacturing Improvement: The Shingo System,* ed. Alan Robinson. Cambridge, MA: Productivity Press.

U.S. Department of Energy, Office of Energy Efficiency and Renewable Energy. (1997). "The Challenge: Improving the Efficiency of a Tube Drawing Bench." Washington, DC: Author.

Womack, J., and Jones, D. (1996). *Lean Thinking: Banish Waste and Create Wealth in Your Corporation.* New York: Simon & Schuster.

INSULATION

The term "thermal insulation" refers to a material or combination of materials that slows the transfer of heat from high temperature (hot) regions to low temperature (cold) regions. Thermal insulation is placed between regions or surfaces having different temperatures to reduce heat flux—heat flow rate per unit area. In general, the heat flow increases as the temperature difference increases. The heat flux also depends on the type of material between the hot and cold surfaces. If the material between the surfaces is a thermal insulation—a material with relatively low values of a property called apparent thermal conductivity—then the heat flux will be small. As an added bonus, most building thermal insulation also function as effective sound or acoustical insulation.

The development of better thermal insulation is important since space heating and cooling account for the majority of energy consumption in the residential sector, and are second only to lighting in the commercial sector. Because of advances in insulation and more efficient heating systems, the U.S. Department of Energy projects that the energy used for space heating will drop at least 25 percent per household by 2020 relative to 1997 usage.

HISTORICAL DEVELOPMENT

The use of thermal insulation dates back to ancient times, when primitive man used animal skins for clothing and built structures for protection from the elements. Primitive insulation included fibrous materials such as animal fur or wool, feathers, straw, or woven goods. Bricks and stone, while not highly efficient thermal insulation, provided protection from the elements, reduced the loss of heat from fires, and provided large masses that moderate temperature changes and store heat.

Origins of the science associated with thermal insulations coincide with the development of thermodynamics and the physics associated with heat transfer. These technical subjects date to the eighteenth century. Early observations that a particular material was useful as thermal insulation were not likely guided by formal theory but rather by trial and error. Sawdust was used, for example, in the nineteenth century to insulate ice storage buildings.

MODERN DEVELOPMENT

Both building and industrial insulation experienced widespread use in the twentieth century. Because the use of thermal insulation in buildings significantly reduces the energy required to heat and cool living or working space, the use of thermal insulation has grown as expectations for comfort and the cost of energy have increased. The amount of insulation used in a particular application is economically justified by balancing the cost of the insulation against the value of energy saved during the lifetime of the insulation. For buildings, it depends on geographic location; for industry, it depends on the process temperature and safety considerations. The outside surface of insulation that is installed around a steam pipe, for example, will be cooler than the metal surface of the pipe.

Thermal insulation is available over a wide range of temperatures, from near absolute zero (–273°C) (–459.4°F) to perhaps 3,000°C (5,432°F). Applications include residential and commercial buildings, high- or low-temperature industrial processes, ground and air vehicles, and shipping containers. The materials and systems in use can be broadly characterized as air-filled fibrous or porous, cellular solids, closed-cell polymer foams containing a gas other than air, evacuated powder-filled panels, or reflective foil systems.

Thermal insulation in use today generally affects the flow of heat by conduction, convection, or radiation. The extent to which a given type of insulation affects each mechanism varies. In many cases an insulation provides resistance to heat flow because it contains air, a relatively low thermal conductivity gas. In general, solids conduct heat the best, liquids are less conductive, and gases are relatively poor heat conductors. Heat can move across an evacuated space by radiation but not by convection or conduction.

The characterization of insulation is simplified by the use of the term "apparent thermal conductivity" (k_a). Apparent thermal conductivity is used to signify that all three modes of heat transfer have been included in the evaluation of k_a. The k_a of an insulation can be determined by a standard laboratory test without specifying the heat transfer mechanism. The thermal resistivity (R^*) of an insulation is the reciprocal of k_a. R^* is the R-value per unit thickness of insulation, which is one inch in the inch-pound (IP) system. The ratio of insulation thickness over k_a is called the thermal resistance or R-value of the insulation. The property k_a is often referred to as the k-factor. In some cases the Greek letter λ is used. These terms are appropriate for fibrous or porous insulation containing air or some other low-thermal-conductivity gas or vacuum. Insulation such as cellulose, cork, fiberglass, foamed rubber, rock wool, perlite or vermiculite powder, and cellular plastics or glasses fall into this category of insulation sometimes called mass insulation. The thermal resistivity is useful for discussing insulations since the R-value is the product of thickness (in appropriate units) and k_a. Insulation in the building industry is commonly marketed on the basis of R-value while industrial users often prefer k_a, or k-factor. The greater the R-value or the smaller the k_a, the greater the insulation's effectiveness. The physical units commonly associated with the terms k_a, R^*, and R are listed in Table 1. The R-values of insulation of the type mentioned above are directly proportional to thickness as long as the thickness exceeds a minimum value, which is a few inches in most conditions.

In the case of thermal insulation that primarily reduces thermal radiation across air spaces, the term k_a is not used. This type of insulation is called reflective insulation, and R is not always directly proportional to thickness. The R-value of a reflective system is the temperature difference across the system divided by the heat flux.

Term	Inch-Pound System (IP)	Scientific-International (SI)
k_a	Btu·in./ft²·h·°F	W/m·k
R^*	ft²·h·°F/Btu in.	m·K/W
R	ft²·h·°F/Btu	m²·K/W
Thickness	in.	m
Heat flow rate	Btu/h	W
Heat flux	Btu/h·ft²	W/m²

Table 1.
Units Associated with Thermal Insulation

Fibrous, open-cell, and particulate insulations reduce heat flow because of the low thermal conductivity of the air in the insulation, which occupies a large fraction of the material's volume. The conduction across the material is limited because the air in the void fraction has low thermal conductivity. The structure provided by the solid restricts the movement of the air (convection) and limits radiation across the insulated space by intercepting and scattering thermal radiation. The type of "mass" insulation eliminates internal convection, reduces radiation, and increases conduction slightly because a small fraction of air volume is replaced by solid volume. The R-values of this type of insulation generally decrease with increasing temperature because the thermal conductivity of air increases with temperature. The R-value of mass-type insulation also depends on density, which is directly related to the amount of solid present. In fiberglass insulation at low density and 37.7°C (100°F), for example, R^* increases as density increases primarily because of reduced radiation. At high densities, R^* decreases because of increased solid conduction.

Closed-cell plastic foams such as polystyrene, polyurethane, and polyisocyanurate initially containing a gas other than air represent an important class of insulations. The gases that are used in the manufacturing process (blowing agents) usually have thermal conductivites less than that of air. Gases such as propane, carbon dioxide, and fluorinated hydrocarbons are examples of gases used as blowing agents. Closed-cell foams containing these or similar low-thermal-conductivity gases have R^* values up to about two times those of air-filled products. These products retain their relatively high R^* values as long as the blowing agent is not lost or unduly diluted by air. The lifetime average R-values for these products are less than the R-values of fresh products because air diffuses into the cells with time and the blowing agent diffuses out, thus changing the chemical composition and thermal conductivity of the gas mixture in the cells. Insulation board stock made of closed-cell polymeric foam are often faced with a gas-impermeable barrier material such as metal foil to reduce or prevent gas diffusion from taking place and thus maintain their R-values with time.

Fiberglass and rock wool insulations are produced by spinning and cooling to produce fine fibers. Fiberglass is produced from molten glass, binder or adhesive is added to produce batt-type insulation, and dye can be added for color. Rock wool insulation is produced from molten slag from iron or copper production. Cellulosic insulation is manufactured from recycled newsprint or cardboard, using equipment that produces small particles or flakes. Fire-retardant chemicals such as boric acid, ammonium sulfate, borax, and ammoniated polyphosphoric acid are added to reduce the smoldering potential and surface burning. Perlite and vermiculite particulate insulations are produced from ores passed through high-temperature furnaces. The high-temperature processing results in a lightweight particle with low bulk density. These particulate insulations are not flammable.

Reflective insulation and reflective insulation systems reduce the heat flux across enclosed air spaces using surfaces that are poor thermal radiators. These insulations have little effect on conduction across the air space, usually decrease the natural convection, and significantly reduce radiation. The R-values for reflective insulations like mass insulations depend on temperature. The terms k_a and $R*$, however, are not appropriate. The R-values for reflective systems depend on the number and positioning of the reflective surfaces, the heat flow direction, and the temperature difference across the system. The unique feature of reflective insulations is that they include low-emittance, high-reflectance surfaces. The low-emittance material commonly used is aluminum foil, with emittance in the range of 0.03 to 0.05. The emittance scale goes from 0 to 1, with 0 representing no radiation being emitted. The aluminum foil is commonly bonded to paper, cardboard, plastic film, or polyethylene bubble pack. The backing materials provide mechanical strength

and support needed for handling and attachment. R-values for reflective insulations are generally measured for heat flow directions up, horizontal, and down.

Thermal insulations used in industrial applications include high-density fiberglass, high-density rock wool, particle insulations such as perlite and vermiculite, solids formed from perlite or calcium silicate, and occasionally reflective systems. The temperature range for industrial applications is much greater than that for buildings, so there is a wider variety of materials available. Fine perlite powder, for example, is used in low-temperature applications, while formed perlite or calcium silicate is used for high-temperature applications. Calcium silicate, perlite, high-density fiberglass, high-density rock wool, and foam glass are used for applications such as pipe insulation and furnace insulation. Each material has a range of temperatures in which it can be used, so it is important to match the material with the application. Reflective insulation produced from polished aluminum or stainless steel is also used for high-temperature applications.

Selected references that include thermal insulation data are listed in the bibliography at the end of this article. Table 2 contains nominal thermal resistivities of ten commonly used insulations. The thermal resistivities in Table 2 are at 23.9°C (75°F) and include the effects of aging and settling.

To obtain thermal resistance from $R*$ multiply $R*$ by the thickness of the insulation in the appropriate units, inches or meters. In the case of $R*$ in IP units (ft²·h·°F/Btu·in.), the thickness is in inches to give R_{IP} in ft²·h·°F/Btu, the units commonly used in the United States. For R_{SI}, the thickness is in meters to obtain units m²·K/W. Conversely, the thickness required for a specific R-value is the ratio $R_{IP}/R*_{IP}$ or $R_{SI}/R*_{SI}$. To obtain $R_{IP} = 30$, for example, a thickness of 10.3 inches of a fiberglass batt insulation with $R*_{IP} = 2.9$ is required.

The $R*$s of a fibrous or cellular insulation like those in Table 2 generally decrease as the temperature increases. In the case of closed-cell polymeric foams like polyurethane or polyisocyanurate board, the $R*$ may decrease if the insulation temperature drops below the condensation temperature of the blowing agent in the cells. This is because of changes in the gas-phase composition and therefore the gas-phase thermal conductivity. The $R*$ of insulations also depends on density when all other factors are constant. The relationship between $R*$ and density

Insulation Type	R*(ft²·h·°F/Btu·in.)[a]	R*(m·K/W)[b]
Fiberglass batts (standard)	2.9 - 3.8	20.1 - 26.3
Fiberglass batts (high performance)	3.7 - 4.3	25.7 - 29.8
Loose-fill fiberglass	2.3 - 2.7	15.9 - 18.7
Loose-fill rock wool	2.7 - 3.0	18.7 - 20.8
Loose-fill cellulose	3.4 - 3.8	23.6 - 26.3
Perlite or vermiculite	2.4 - 3.7	16.6 - 25.7
Expanded polystyrene board	3.6 - 4.0	25.5 - 27.7
Extruded polystyrene board	4.5 - 5.0	31.2 - 34.7
Polyisocyanurate board, unfaced	5.6 - 6.3	38.8 - 43.7
Plyurethane foam	5.6 - 6.3	38.8 - 43.7

Table 2.
Nominal Values for R* at 23.9°C (75°F)
Note: (a) The values listed are for one inch of thickness.
(b) The values listed are for one meter of thickness.

depends on the type of insulation in question. In general, the compression or settling of an insulation results in a decrease in the R value.

Reflective insulations consist of enclosed air spaces bounded by low-emittance surfaces. The distance across the air spaces is set by the insulation designer. The spacing is generally in the range of 0.5 to 2.0 in, with one to seven air spaces provided by different products. The low-emittance (high-reflectance) surfaces serve as radiation barriers and, as a result, thermal radiation across the enclosed air spaces is dramatically reduced. The open air spaces, however, can have natural convection (air movement) occurring so the R-values depend on the heat-flow direction and the magnitude of the temperature difference across each air space in the system. The R-value of a reflective system is greatest when the heat-flow direction is downward, and least when the heat-flow direction is upward.

TRENDS

The design and manufacturing of cellulose, fiberglass, and rock wool are constantly being improved.

The resulting increases in R*, however, are relatively small. The fiberglass and rock wool producers seek optimum fiber diameters and improved radiation attenuation. The cellulosic insulation producers have changed from hammer mills producing relatively high-density products to fiberizers that produce relatively low-density products. The result in most cases is improved R* with less material required. In some cases adhesives are added to building insulations to improve mechanical stability so that long-term settling of the insulation is reduced. Adhesives also have been added to loose-fill fiberglass insulation to produce a blow-in-blanket insulation (BIBS) widely used as insulation in cavity walls.

More advanced insulations are also under development. These insulations, sometimes called superinsulations, have R* that exceed 20 ft²·h·°F/Btu·in. This can be accomplished with encapsulated fine powders in an evacuated space. Superinsulations have been used commercially in the walls of refrigerators and freezers. The encapsulating film, which is usually plastic film, metallized film, or a combination, provides a barrier to the inward diffusion of air and water that would result in loss of the vacuum. The effective life of such insulations depends on the effectiveness of the encapsulating material. A number of powders, including silica, milled perlite, and calcium silicate powder, have been used as filler in evacuated superinsulations. In general, the smaller the particle size, the more effective and durable the insulation packet. Evacuated multilayer reflective insulations have been used in space applications in past years.

Silica aerogels, a newly developing type of material, also have been produced as thermal insulations with superinsulation characteristics. The nanometer-size cells limit the gas phase conduction that can take place. The aerogels are transparent to visible light, so they have potential as window insulation. The use of superinsulations at present is limited by cost and the need to have a design that protects the evacuated packets or aerogels from mechanical damage.

David W. Yarbrough

See also: Building Design, Commercial; Building Design, Energy Codes and; Building Design, Residential; Domestic Energy Use; Economically Efficient Energy Choices; Efficiency of Energy Use.

BIBLIOGRAPHY

Goss, W. P., and Miller, R. G. (1989). "Literature Review of Measurement and Predictions of Reflective Building Insulation System Performance." *ASHRAE Transactions* 95(2):651–664.

Govan, F. A.; Greason, D. M.; and McAllister, J. D., eds. (1981). *Thermal Insulation, Materials, and Systems for energy Conservation in the 80s* (STP 789). West Conshohocken, PA: American Society for Testing and Materials.

Graves, R. S., and Wysocki, D. C., eds. (1991). *Insulation Materials: Testing and Applications*, vol. 2 (STP 1116). West Conshohocken, PA: American Society for Testing and Materials.

Graves, R. S., and Zarr, R. P., eds. (1997). *Insulation Materials: Testing and Applications*, vol. 3 (STP 1320). West Conshohocken, PA: American Society for Testing and Materials.

Industrial Insulation (ORNL/M-4678). (1995). Oak Ridge, TN: Oak Ridge National Laboratory.

Kreith, F., ed. (1999). *The CRC Handbook of Thermal Engineering*. Boca Raton, FL: CRC Press.

Kreith, F., and Bohn, M. S. (1997). *Principles of Heat Transfer*. Boston: PWS.

McElroy, D. L., and Kimpflen, J. F., eds. (1990). *Insulation Materials: Testing and Applications*, vol. 1 (STP 1030). West Conshocken, PA: American Society for Testing and Materials.

McElroy, D. L., and Tye, R. P., eds. (1978). *Thermal Insulation Performance* (STP 718). West Conshohocken, PA: American Society for Testing and Materials.

Powell, F. J., and Matthews, S. L., eds. (1984). *Thermal Insulation: Materials and Systems* (STP 922). West Conshohocken, PA: American Society for Testing and Materials.

Pratt, A. W. (1981). *Heat Transmission in Buildings*. New York: John Wiley and Sons.

Rohsenow, W. M.; Hartnett, J. P.; and Cho, Y. I. (1998). *Handbook of Heat Transfer*, 3rd ed. New York: McGraw-Hill.

Turner, W. C., and Malloy, J. F. (1998). *Thermal Insulation Handbook*. New York: McGraw-Hill.

Tye, R. P., ed. (1977). *Thermal Transmission Measurements of Insualtion* (STP 660). West Conchohocken, PA: American Society for Testing and Materials.

INSULATOR, ELECTRIC

See: Electricity

INTELLIGENT TRANSPORTATION SYSTEMS

See: Traffic Flow Management

INTERNAL COMBUSTION ENGINES

See: Engines

IPATIEFF, VLADIMIR NIKOLAEVITCH (1967–1952)

Vladimir Nikolaevitch Ipatieff, one of the founding fathers of high-pressure catalysis, was the innovating force behind some of this country's most important petroleum processing technologies in the years leading up to World War II.

THE EARLY YEARS

Ipatieff was born in Russia in 1867. As a member of the privileged (i.e., noble) class, Ipatieff prepared for a military career. Early on in his education, Ipatieff gravitated toward the sciences, and in particular chemistry.

Ipatieff's formal training in chemistry began in earnest when, at the age of twenty-two, he entered the Mikhail Artillery Academy in St. Petersburg. The Academy was founded to give technical training to officers who were to serve as engineers and in other positions within the Russian military.

Chemical training at the Academy was particularly strong. The school taught Ipatieff how to apply textbook concepts to actual plant conditions. As a student, and then later as an instructor at the Academy, he regularly visited local metallurgical and chemical plants to examine first-hand industrial production.

While still a student at the Academy, Ipatieff began to make a name for himself in the Russian chemical community as he began to publish some of his laboratory findings. His first professional milestone as a chemist came in 1890 when he joined Russia's Physical-Chemical Society. Here he came into close contact with Russia's most famous chemists, including Dimitri Mendeleev, discoverer of the periodic table and one of the founders of the Society. In 1891, upon graduating from the school, he was appointed lecturer in chemistry at the Academy where he also continued to undertake original chemical research for his doctoral dissertation. In 1895, he was made assistant professor and, upon completion and acceptance of his dissertation in 1899, he became a full professor of chemistry.

Right up to World War I, Ipatieff, as part of his responsibilities at the Academy, traveled extensively to conferences and seminars throughout Europe and the United States. As a result, he made the acquaintance of some of the major chemists of the day.

During these years he gained an international reputation in the area of experimental chemistry in general and high pressure catalysis in particular. At the Academy he designed, built, and directed one of the first permanent high-pressure laboratories in Europe. Ipatieff investigated the composition and synthesis of numerous aliphatic compounds, including isoprene and the complex alcohols. He demonstrated for the first time the catalytic effect of the metal reactor walls on reactions, a result which would have important commercial implications. Ipatieff also undertook the first sophisticated experiments in catalytic polymerization, isomerization, and dehydrogenation of organic materials, including alcohols and petroleum.

He became intimately familiar with a wide range of catalytic materials—including aluminum oxide, silica, and clay, as well as nickel, platinum, zinc, and copper—and their role individually and as mixtures in effecting chemical transformation. One of Ipatieff's most important lines of research was his breakthrough work on the nature and mechanisms of catalytic promoters on organic reactions.

Although Ipatieff was dedicated to research, his practical training at the Academy prepared him well for applying his research to commercial problems. Ipatieff consulted to industry and government on a regular basis. He was often called upon to make commercial grade catalysts for a variety of applications. One of his most important early assignments, obtaining gasoline from Caspian sea petroleum, introduced Ipatieff to industrial research in fuels. He also used his expertise to undertake research for the government and helped expand Russia's explosives capacity during World War I.

THE POSTREVOLUTIONARY YEARS

Ipatieff's life changed dramatically with the Russian Revolution in 1917. The Bolsheviks were hesitant to retain in official positions those who were too closely allied with the old regime.

Ipatieff, however, remained loyal to Russia, even in its new form. He was also a pragmatist who understood that he must find a way to accommodate the new way of life in Russia, and gain the confidence of those currently holding power if he was to continue his chemical research.

Over the next few years, Ipatieff called upon the wide contacts he had made within Russian scientific and administrative circles before the revolution, and his growing international reputation, to negotiate with the Bolsheviks a favorable place for his research activity. An important element in his success was his ability to convince the government of the strategic importance of his work, which (he pointed out) had just been demonstrated during the war. Under Lenin, who knew of Ipatieff and his work, the scientist received government support. He remained at the Academy, which was retained to train future Bolshevik military officers. Most importantly, Ipatieff was given an official and prestigious research position as Director of the newly created Institute for High Pressures.

In this favorable environment, Ipatieff undertook some of his most important work in high pressure reactions. He studied the effect of catalysts and high pressures on reactions taking place in hydrogen atmospheres, and the effect of such metals as platinum and nickel on forming cyclic compounds from olefin-based materials. This was to become very important later in the 1930s and 1940s in the United States with the development of the platforming process.

When Stalin came to power, the political climate

in Russia turned against the creative individuals—scientists, musicians, and writers—especially those who had strong roots in Czarist Russia. During the 1920s Ipatieff saw his colleagues and coworkers arrested, sent to labor camps, or executed.

THE LATER YEARS: AMERICAN PETROLEUM TECHNOLOGY

The opportunity for Ipatieff to emigrate to the United States came through a job offer by Dr. Gustav Egloff, Director of Research at the Universal Oil Products Co. (UOP). UOP, a Chicago-based firm, was a petroleum process development firm known in particular for its commercialization of the first continuous thermal cracking technology (i.e., the Dubbs Process). In the late 1920s, UOP wanted to go into catalytic cracking research. Having met Ipatieff in Germany while attending an international chemical conference, Egloff, who was familiar with Ipatieff's reputation, suggested that Ipatieff think of relocating to Chicago to direct UOP in these efforts.

Reluctantly, in 1929 Ipatieff left his homeland for the United States. He remained at UOP for the remainder of his life. Northwestern University, which had close ties to UOP, appointed Ipatieff to a professorship and directorship of the university's high-pressure research laboratory.

At UOP, Ipatieff had the opportunity to apply his former research in catalytic promoters and high-pressure technique to develop important catalytic petroleum processing technologies. In contrast to the way he conducted science, Ipatieff's technical efforts were conducted in teams comprised of a wide assortment of specialists.

Through the 1930s, Ipatieff led UOP in its effort to develop two catalytic processes for the production of high-octane fuel: alkylation and polymerization—the first, a reaction of a hydrocarbon with an olefin (double-bonded compound); the second, the formation of long molecules from smaller ones. Both processes produce high-octane blending compounds that increase the quality of cracked gasoline.

Ipatieff's work on promoted phosphoric acid and hydrogen fluoride catalysts, which extended his earlier research in Russia, provided the key to commercializing alkylation and polymerization technologies. Shell and other refiners quickly established industrial scale operations. These technologies, generally operating together and in tandem with cracking oper-

ations, created vast amounts of high quality aviation fuel. They were a vital part of the Allied war effort during World War II.

Ipatieff's final contribution to catalytic technology was more indirect but essential. His guidance and suggestions to Vladimir Haensel, a fellow Russian émigré who worked on catalytic reforming at UOP and studied under Ipatieff at Northwestern, were a significant contribution to Haensel's development of the high-pressure reforming technology known as platforming. An extension of Ipatieff's previous work involving platinum catalysts, high pressure, and hydrogen environments, platforming represents the first continuous catalytic reforming process. It produces large tonnages of ultra-high octane gasoline materials and critical organic intermediates (benzene, toluene, and xylene), previous obtained only in limited quantities from coal tar. Platforming, along with fluid catalytic cracking, is generally considered one of the great petrochemical innovations of the century.

Ipatieff never received the honor he coveted the most, the Nobel Prize. However, he continued to publish voluminously and, ever the practical scientist, he obtained numerous patents. He continued to receive honors in the United States and internationally. He became a member of the National Academy of Sciences and received the prestigious Gibbs medal for his many achievements. Ipatieff lived long enough to see the petroleum industry transformed with process technologies that he created and that were rooted in his early scientific research. The first platforming plant, the culmination of his life's work in catalytic research, came on line just shortly before his death in 1952.

Sanford L. Moskowitz

See also: Catalysts; Refining, History of.

BIBLIOGRAPHY

Enos, J. (1962). *Petroleum, Progress and Profits: A History of Process Innovation.* Cambridge, MA: MIT Press.
Frankenburg, W. G., et. al., eds. (1954). *Advances in Catalysis and Related Subjects.* New York: Academic Press.
Haensel, V. (1935). "Polymerization of Butylene." M.S., MIT.
Haensel, V. (1983). "The Development of the Platforming Process—Some Personal and Catalytic Recollections." In *Heterogeneous Catalysis: Selected American Histories*, eds. B. H. Davis and W. P. Hettinger. American Chemical Society Symposium Series No. 222. Washington, DC: American Chemical Society.

Ipatieff, V. N. (1936). *Catalytic Reactions at High Pressures.* New York: Macmillan.

Ipatieff, V. N. (1946). *The Life of a Chemist: Memoirs of V. N. Ipatieff,* trans. V. Haensel and Mrs. R. H. Lusher, eds. X. J. Eudin, H. D. Fisher, and H. H. Fisher. Stanford, CA: Stanford University Press.

Remsberg, C., and Higden, M. (1994). *Ideas for Rent: The UOP Story.* Chicago, IL: Universal Oil Products Co.

Rideal, E. K., and Taylor, H. S. (1926). *Catalysis in Theory and Practice.* London: Macmillan.

Spitz, P. (1988). *Petrochemicals: The Rise of an Industry.* New York: John Wiley and Sons.

Williamson, H. F. (1963). *The American Petroleum Industry, Vol. II: The Age of Energy, 1899-1959.* Evanston, IL: Northwestern University Press.

JET ENGINE

See: Aircraft

JET FUEL

See: Aviation Fuel

JOULE

See: Units of Energy

JOULE, JAMES PRESCOTT (1818–1889)

James Joule was born in Salford, near Manchester, England, on December 24, 1818. He was the second son of a wealthy brewery owner and was educated at home by private tutors. For three years he was fortunate enough to have the eminent British chemist, John Dalton as his chemistry teacher. He never attended a university; as a consequence, while he was bright enough to learn a great deal of physics on his own, he remained, like Michael Faraday, unskilled in advanced mathematics.

Joule had the means to devote his time to what became the passion of his life — obtaining highly accurate experimental results in physics, for which he displayed a precocious aptitude. His genius showed itself in his ability to devise new methods, whenever needed, to improve on the accuracy of his quantitative results.

Joule had no real profession except as an amateur scientist, and no job except for some involvement in running the family brewery. Since his father was ill and forced to retire in 1833, his son had to become more involved in the affairs of the brewery from 1833 to 1854, when the brewery was sold by his family. While Joule was working at the brewery, he carried out his experiments before 9:00 A.M., when the factory opened, and after 6:00 P.M., when it closed. Because his father built a laboratory for him in his home, in 1854 he had the time and means to devote himself completely to physics research. Later in life, he suffered severe financial misfortune, but the Royal Society and Queen Victoria in 1878 each provided a £200 subsidy for Joule to continue his important researches.

In 1847 Joule married Amelia Grimes, and they had two children who survived them. Another son was born on June 8, 1854, but died later that month. This was followed by an even greater tragedy—within a few months Joule's wife also passed away. Joule never remarried, but spent the rest of his life with his two children in a variety of residences near Manchester.

Joule died in Sale, Cheshire, England, on October 11, 1889. He always remained a modest, unassuming man, and a sincerely religious one (even though he

James Prescott Joule. (Library of Congress)

was in the habit of falling asleep during sermons). Two years before his death he said to his brother, "I have done two or three little things, but nothing to make a fuss about." Those "two or three little things" were so important for the advancement of science that Joule was elected in 1850 as a fellow of the Royal Society of London, received the Copley Medal (its highest award) in 1866, and was elected president of the British Association for the Advancement of Science in 1872 and again in 1887. Joule is memorialized by a tablet in Westminster Abbey, and constantly comes to the attention of physicists whenever they use the unit of energy now officially called the joule (J).

JOULE'S CONTRIBUTIONS TO THE PHYSICS OF ENERGY

Joule's interest in the conservation of energy developed as a consequence of some work he did in his teens on electric motors. In 1841 he proposed, on the basis of his experiments, that the rate at which heat Q is generated by a constant electric current i passing through a wire of electrical resistance R is:

$$DQ/Dt = i^2R$$

now called Joule's Law.

From 1841 to 1847 Joule worked steadily on measuring the heat produced by electrical processes (Joule's Law), mechanical processes (rotating paddles churning water or mercury), and frictional processes (the rubbing of materials together, as Count Rumford had done in 1798). In each case he compared the amount of energy entering the system with the heat produced. He proved his mettle as a physicist by spending endless days ferreting out the causes of errors in his experiments and then modifying his experimental set-up to eliminate them. In this way he produced a remarkably precise and accurate value for the constant that relates the energy entering the system (in joules) with the heat produced (in calories). This constant is now called Joule's Equivalent, or the mechanical equivalent of heat.

In 1847 Joule published a paper that contained an overwhelming amount of experimental data. All his results averaged out to a value of 4.15 J/cal (in modern units), with a spread about this mean of only five percent. The best modern value of Joule's Equivalent is 4.184 J/cal, and so his results were accurate to better than one percent. This was truly amazing, for the heat measurements Joule performed were the most difficult in all of physics at that time.

At the British Association meeting at Oxford in June 1847, at which Joule presented his results, his audience's reaction was much more subdued and uninterested than he had expected. Joule fully believed that his paper would have passed unnoticed had not the 23-year-old William Thomson (later Lord Kelvin) asked a number of penetrating questions. These awakened his colleagues to the significance of Joule's work as a proof of the conservation-of-energy principle (now commonly called the first law of thermodynamics) under a variety of experimental conditions and involving many different types of energy.

This event marked the turning point in Joule's career. From 1847 on, when Joule spoke, scientists listened. His research results were one of the two major contributions to the establishment of the first law of thermodynamics, the other being that of the

German physician Julius Robert Mayer. Mayer's work, although historically important for its insights into the conservation-of-energy principle, was however tainted by errors in physics and an unacceptable reliance on philosophical arguments.

In addition to his work on the conservation of energy, Joule made a number of other important contributions to physics. In 1846 he discovered the phenomenon of magnetostriction, in which an iron rod was found to change its length slightly when magnetized. In 1852, together with William Thomson, he showed that when a gas is allowed to expand into a vacuum, its temperature drops slightly. This "Joule-Thomson effect" is still very useful in the production of low temperatures.

Joule believed that nature was ultimately simple, and strove to find the simple relationships (like Joule's law in electricity), which he was convinced must exist between important physical quantities. His phenomenal success in finding such relationships in the laboratory made a crucial contribution to the understanding of energy and its conservation in all physical, chemical and biological processes.

Joseph F. Mulligan

BIBLIOGRAPHY

Cardwell, D. S. L. (1971). *From Watt to Clausius.* Ithaca, New York: Cornell University Press.

Cardwell, D. S. L. (1989). *James Joule: A Biography.* Manchester, England: Manchester University Press.

Crowther, J. G. (1936). "James Prescott Joule." In *Men of Science.* New York: W.W. Norton.

Joule, J. P. (1963). *The Scientific Papers of James Prescott Joule,* 2 vols. London: Dawson's.

Rosenfeld, L. (1973). "Joule, James Prescott." In *Dictionary of Scientific Biography,* ed. Charles Coulston Gillispie, Vol. 7, pp. 180–182. New York: Scribner.

Steffens, H. J. (1979). *James Prescott Joule and the Development of the Concept of Energy.* New York: Science History Publications.

Wood, A. (1925). *Joule and the Study of Energy.* London: G. Bell and Sons, Ltd.

KAMERLINGH ONNES, HEIKE (1853–1926)

Heike Kamerlingh Onnes was born on September 21, 1853, in Groningen, the Netherlands, and died on February 21, 1926, in Leiden. His father owned a roof-tile factory. Heike Kamerlingh Onnes entered the University of Groningen to study physics. When the government threatened to permanently shut down the university for monetary reasons, he led a delegation to the seat of the government in the Hague as president of the student government. During his studies he won several prizes in physics; his Ph.D. thesis demonstrates superior mathematical abilities. Despite delicate health through much of his life, he showed an enormous capacity for work.

Kamerlingh Onnes does not fit the description of a loner; on the contrary, he created one of the first laboratories to be set up as if it were a factory. Per Dahl in his comparison between Kamerlingh Onnes and his British counterpart, James Dewar, states that Onnes, to be sure was paternalistic, opinionated, and a man of strong principles—traits not uncommon among the moguls of late nineteenth-century science—but that he proved to be a benevolent leader, kind and scrupulously fair in his relations with friends and pupils alike—behavior that was certainly within the norms of his time.

After Kamerlingh Onnes was appointed professor in experimental physics at the University of Leiden in 1882, he stated in his inaugural lecture that physics is capable of improving the well-being of society and proclaimed that this should be accomplished primari-

ly through quantitative measurements. He laid out what he was going to do and, to the surprise of onlookers, that is what he did. He was an excellent organizer, and set out to equip his laboratory on a grand scale. The city of Leiden did not yet provide electricity, so he acquired a gas motor and generator and made his own electricity. Pumps and compressors were barely available, so he had them made in the machine shops of the laboratory. There was a need for measuring instruments, so he created an instrument makers' school. There was an even greater demand for glassware (Dewars, McLeod gauges and connecting tubes, etc.), so he created a glassblowers' school that became famous in its own right.

All currents that had to be measured were sent to a central "measurement room" in which many mirror galvanometers were situated on top of vibration-free columns that were separated from the foundations of the building. One should realize that the many announcements in the early literature of the liquefaction of specific gases pertained to not much more than a mist or a few drops; Kamerlingh Onnes planned to make liquid gases by the gallon. A separate hydrogen liquefaction plant was located in a special room with a roof that could be blown off easily.

The availability of large quantities of liquid helium as well as an excellent support staff led to the undertaking of many experiments at 8 K (the boiling temperature of helium) as well as the lower temperatures obtained by pumping. One subset was the measurement of the resistivity (conductivity) of metals, since this property was useful as a secondary thermometer. Although a linear decline was observed, various speculations were made as to what the result would be when zero absolute temperature was reached. In April 1911 came the surprising discovery that the resistivity in mercury disappeared. At first the sur-

Heike Kamerlingh Onnes. (Library of Congress)

prise was a sharp jump to what was thought to be a small value. The first reaction to this baffling result was to suspect a measurement error. All electrical machines in the laboratory were shut down to be sure that there were no unforeseen current leaks and the experiment was repeated several times. Very careful verifications finally showed that the effect was real and that the resistance was indeed unmeasurably small ("sinks below 10^{-4} of the resistance at 0°C"), and moreover it was found that the effect existed also in other metals, even those that could not be purified as well as mercury.

Initially there was speculation about building an "iron-less" magnet, but this hope was dashed when it was discovered that a small field destroyed the superconductivity. Not until 1960 when materials where found that could sustain high fields, did superconductivity show promise for building strong magnets. The other obstacle was the need for a low-temperature environment. Raising the critical temperature had been a goal for many years, and a spectacular breakthrough was made in 1986.

The result was called by Dutch physicist H. A. Lorentz "perhaps the most beautiful pearl of all [of Kamerlingh Onnes's discoveries]." However, as H. B. G. Casimir describes in his memoirs, he refused to give any credit to the graduate student who observed the phenomenon and who realized its importance.

Although the discovery can be called accidental, one may ask what led up to it. The decline in resistivity of metals when the temperature was lowered clearly invited further study. This program (J. van den Handel calls it Kamerlingh Onnes's second major field of research; liquefaction being the first) was both intrinsically interesting as well as relevant to the construction of a good secondary thermometer. It was known that the resistivity was a linear function of the temperature but was noticed to level off at lower temperatures. The height of this plateau was found to depend on the amount of impurities, using a series of experiments with gold, since the amount of admixture in this metal can be easily controlled. To lower this plateau, a metal of very high purity was needed. Since zone melting, in the modern sense, did not exist, the choice fell on mercury, because this metal could be purified by distillation, and the purified liquid was then placed in glass capillaries. When the liquid in these capillaries was frozen, it formed a "wire." Moreover, it became clear that these wires were free of dislocations, to use a modern term, because it was found that pulling of the wires resulted in increased values of the residual resistance. Hence mercury was the best bet to see how far the linear part of the resistivity curve could be extended to lower temperatures. The hope to have a linear resistor at very low temperatures was certainly a driving factor for the research that led to the discovery of superconductivity.

Heike Kamerlingh Onnes was awarded the Nobel Prize in physics in 1913.

Paul H. E. Meijer

See also: Electricity; Energy Intensity Trends; Heat and Heating; Heat Transfer; Magnetism and Magnets; Molecular Energy; Refrigerators and Freezers; Thermal Energy.

BIBLIOGRAPHY

Bruyn Ouboter, R. de. (1997). "Heike Kamerlingh Onnes's Discovery of Superconductivity." *Scientific American* 276:98–103.

Casimir, H. B. G. (1983). *Haphazard Reality: Half a Century of Science.* New York: Harper and Row.

Dahl, P. F. (1992). *Superconductivity: Its Historical Roots and Development from Mercury to the Ceramic Oxides.* New York: American Institute of Physics.

Handel, J. van den. (1973). "Kamerlingh Onnes, Heike." In *Dictionary of Scientific Biography*, Vol. 7, ed. C.C. Gillispie. New York: Charles Scribner's Sons.

Meijer, P. H. E. (1994). "Kamerlingh Onnes and the Discovery of Superconductivity." *American Journal of Physics* 62:1105–1108.

Nobel, J. de. (1996). "The Discovery of Superconductivity." *Physics Today* 49(9):40–42.

KEROSENE

Kerosene is that fraction of crude petroleum that boils between about 330° to 550° F (183° to 305° C) (i.e. the distillate cut between gasoline and heavy fuel oil). What is usually thought of as kerosene, the domestic fuel or "range oil" commonly used for space heaters, is only a small part of this fraction. Diesel fuel, aviation jet fuel, and No. 1 and No. 2 heating oils also come from this general boiling range.

HISTORY

Petroleum was discovered by Drake in Pennsylvania in 1859. It is a mixture of hydrocarbon compounds ranging from highly flammable volatile materials like propane and gasoline to heavy fuel oil, waxes, and asphalt. It was soon discovered that crude oil could be separated by distillatoin in to various factions according to their boiling point (molecular weight). Kerosene, a middle distillate, was the first product made from petroleum that had a substantial commercial market. The Pennsylvania crudes happened to be low in sulfurous and aromatic components, so the straight run (unrefined) distillate was relatively odorless and clean-burning. The distillate rapidly replaced whale oil in lamps, and coal in home room heaters, and by 1880 kerosene accounted for 75 percent by volume of the crude oil produced. However, after the turn of the century, the advent of electric lights soon made kerosene lamps obsolete. Later, central home heating based on heating oil and natural gas largely replaced space heating based on kerosene, wood, and coal.

By 1980 production of "range oil" type kerosene was less than one percent of the crude oil refined. By the end of the twentiteth century, kerosene's use for space heating, water heating, cooking, and lighting was largely limited to camps, cabins, and other facilities remote from centralized energy sources. Kerosene is a relatively safe fuel, but to prevent fires care must be taken not to tip over operating kerosene appliances. While burner designs have been improved to minimize accidental upset, carbon monoxide formation is still a potential hazard. This deadly gas can result from use of dirty, poorly maintained burners or if the heater is used in a space with insufficient ventilation.

Kerosene is also used as a solvent for herbicides and insect sprays. However, most of the kerosene fraction in crude oils is used to make Diesel engine fuel and aviation jet fuel.

PROPERTIES

Kerosene is an oily liquid with a characteristic odor and taste. It is insoluble in water, but is miscible with most organic solvents. Structurally, it is composed mostly of saturated hydrocarbon molecules containing twelve to fifteen carbon atoms. When a sweet (low sulfur content) paraffinic crude oil is cut to the proper boiling range in a refinery's atmospheric pipe still, the resulting kerosene may only require a drying step before use. However, if the crude oil contains aromatic ring compounds like xylenes, they must be eliminated by solvent extraction because they burn with a smoky flame. Olefinic (unsaturated) molecules in this boiling range must also be removed by refining, usually by hydrogenation. These compounds tend to form color bodies and polymerize, thus imparting poor storage stability. Sulfur compounds are undesirable because of their foul odor and formation of air polluting compounds. They can also be removed by hydrogenation.

Other important properties include flash point, volatility, viscosity, specific gravity, cloud point, pour point, and smoke point. Most of these properties are related directly to the boiling range of the kerosene and are not independently variable. The flash point, an index of fire hazard, measures the readiness of a fuel to ignite when exposed to a flame. It is usually mandated by law or government regulation to be 120° or 130° F (48° or 72° C). Volatility, as measured

	Kerosene	Diesel Fuel	#2 Heating Oil
Gravity, A.P.I.	40	37	34
Boiling Range, °F.	325 - 500	350 - 650	325 - 645
Viscosity, SSU @ 100° F.	33	35	35
Flash Point, °F.	130	140	150
Sulfer, Weight Percent	0.05 - 0.12	0.30	0.40

Table 1.
Comparison of Heating Fuels

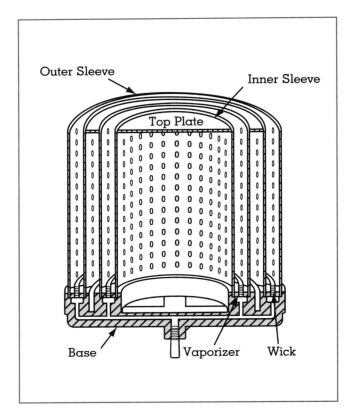

Figure 1.
Cross-section of a range burner.

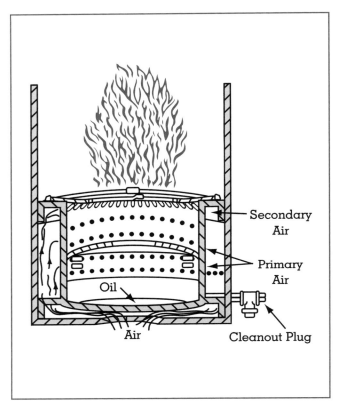

Figure 2.
Cross-section of a vaporizing pot burner.

by the boiling range, determines the ease with which the fuel can evaporate in a burner or lamp. Viscosity measures the fuel's resistance to flow, and thus determines the ease with which it can be pumped or atomized in a burner nozzle. Specific gravity is the ratio of the weight of a given volume of fuel to the same volume of water. Denser, higher gravity fuels have higher heating values. Typical kerosenes have heating values of about 139,000 BTU per gallon. Cloud and pour points indicate the temperature at which the precipitation of waxy constituents affect performance in cold climates. As wax begins to form (cloud point) there is danger of plugging fuel nozzles and filters. At an even lower temperature (the pour point), enough wax has formed so that the fuel sets up and can no longer flow as a liquid. The smoke point is the flame height in millimeters at which a kerosene lamp begins to smoke.

Other properties of interest are carbon residue, sediment, and acidity or neutralization number. These measure respectively the tendency of a fuel to foul combustors with soot deposits, to foul filters with dirt and rust, and to corrode metal equipment. Cetane number measures the ability of a fuel to ignite spontaneously under high temperature and pressure, and it only applies to fuel used in Diesel engines. Typical properties of fuels in the kerosene boiling range are given in Table 1.

Because of its clean burning characteristics, kerosene commands a higher price than other fuels in its boiling range.

EQUIPMENT

The technology of kerosene burners is quite mature. The most popular kerosene heater is the perforated sleeve vaporizing burner or range burner (Figure 1). It consists of a pressed steel base with concentric, interconnected grooves and perforated metal sleeves, between which combustion takes place. Kerosene is maintained at a depth of about 1/4 inch in the grooves. As the base heats up, oil vaporizes from the surface, and the flame lights from asbestos wicks. Combustion air is induced by natural draft. The flame is blue, and the burner is essentially silent, odorless, and smokeless.

For larger capacity space heaters vaporizing pot burners are used (Figure 2). They consist of a metal pot perforated with holes for combustion air. Oil flows into the bottom of the pot by gravity, and is vaporized from the hot surface. Fuel vapors mix with primary air from the lower holes, then with additional secondary air in the upper section, and burn at the top of the vessel. Between periods of high demand, pot burners idle at a low fire mode. The flames burn in the bottom section of the pot, at a fraction of the high fire fuel feed rate, using only primary air. Thus no extraneous pilot light or ignition source is required for automatic operation.

Kerosene lamps have a flat cloth wick. Flame height is determined by the height of the wick, which is controlled by a ratchet knob. A glass chimney ensures both safety and a stable, draft-free flame.

Herman Bieber

BIBLIOGRAPHY

Gary, J. H., and Handwerk, G. E. (1993). *Petroleum Refining Technology and Economics*, 3rd ed. New York: M. Dekker.

Ramose, D. (1983). *Kerosene Heaters*. Blue Ridge Summit, PA: Tab Books.

U.S. Consumer Product Safety Commission. (1988). *What You Should Know About Kerosene Heaters*. Washington, DC: U.S. Government Printing Office.

Wood, Q. E. (1986). *Quaker State Roots Go Deep in World's First Oil Field*. New York: Newcomen Society of the United States.

Yergin, D. (1991). *The Prize: the Epic Quest For Oil, Money and Power*. New York: Simon & Schuster.

KILOWATT

See: Units of Energy

KINETIC ENERGY

Energy is "capacity (or ability) to do work," and work is "the result of a force acting through some distance." A car running into the rear of a stalled car exerts a force on it, pushing it some distance, doing work in the process. The capacity of the moving car to do work is termed its kinetic energy. The greater a car's speed and/or mass, the greater its capacity to do work—that is, the greater its kinetic energy.

Formally, the kinetic energy (K) of a mass (m) moving with speed (v) is defined as $K = 1/2 \ mv^2$. Kinetic energy is measured in joules (J) when m and v are expressed in kilograms (kg) and meters per second (m/s). A 1,000-kg car traveling 15 m/s (about 30 miles per hour) has 112,500 J of kinetic energy. Kinetic energy depends much more on speed than on mass. That is because doubling the mass of an object doubles the kinetic energy, but doubling the speed quadruples the kinetic energy. A 4,000-kg tractor trailer traveling at 30 m/s has the same kinetic energy as a 1,000-kg car traveling at 60 m/s.

Invariably, energy of use to a society is kinetic. A car is useful when it is in motion. Water in motion is useful for driving turbines in a hydroelectric plant. Electricity, the most versatile of all forms of energy, involves electric charges (electrons) in motion. And thermal energy, which provides energy for a steam turbine, is associated with the kinetic energy of molecules.

Most energy converters convert sources of potential energy to forms of kinetic energy that are useful. Gasoline in the tank of an automobile has potential energy. When burned, potential energy is converted to heat, a form of kinetic energy. Uranium in the core of a nuclear reactor has nuclear potential energy. Conversion of nuclear potential energy through nuclear fission reactions produces nuclei and neutrons with kinetic energy. This kinetic energy is the source for the electric energy produced by the nuclear power plant. Water atop a dam has gravitational potential energy. Flowing toward the bottom of the dam, the water continually loses potential energy but gains kinetic energy (and speed). The potential energy the water had at the top of the dam is converted entirely to kinetic energy at the bottom of the dam.

Every molecule in the air around you has kinetic energy because each has mass and is in incessant motion. According to the kinetic theory of gases, each of the molecules has average kinetic energy given by $E = 3/2 \, kT$ where k is the Boltzmann constant, having a value of 1.38×10^{-23} joules per Kelvin, and T is the Kelvin temperature. Interestingly, the average kinetic energy is independent of the mass of the molecule. Thus, nitrogen molecules and oxygen molecules in the air we breathe have the same average kinetic energy. On the other hand, their average speeds differ because their masses differ. When considering the fundamental meaning of temperature, it is appropriate to think of temperature as measuring the average kinetic energy of a molecule in a gas.

Flywheels of reasonable size and speed can store energy comparable to that of batteries and have promise for storing energy for electric vehicles. When a car is brought to rest by braking, the kinetic energy of the car is converted into heat that dissipates into the environment. Conceivably, the car could be brought to rest by transferring the linear kinetic energy of the car to rotational energy in a flywheel, with very little lost as heat during the conversion. The rotational energy then could be recovered by allowing the flywheel to operate an electric generator.

Joseph Priest

See also: Conservation of Energy; Flywheels; Gravitational Energy; Nuclear Energy; Potential Energy.

BIBLIOGRAPHY

Hobson, A. (1995). *Physics: Concepts and Connections*. Englewood Cliffs, NJ: Prentice-Hall.

Priest, J. (2000). *Energy: Principles, Problems, Alternatives*, 5th ed. Dubuque, IA: Kendall/Hunt Publishing Company.

Serway, R. A. (1998). *Principles of Physics*, 2nd ed. Fort Worth: Saunders College Publishing.

KINETIC ENERGY, HISTORICAL EVOLUTION OF THE USE OF

Historically humans have used three natural sources of kinetic energy: wind, water, and tides.

THE ANCIENT WORLD (TO 500 C.E.)

Although early humans often inadvertently tapped into the kinetic energy of moving air or water to do things such as separate grain from chaff or float downstream, the deliberate use of kinetic energy to power machinery came only in the historical era.

Only one machine in classical antiquity made deliberate use of the kinetic energy of wind: the sailing vessel. As early as 3000 B.C.E., paintings illustrated Egyptian vessels using sails. By the first millenium B.C.E., the use of sails was common for long-distance, water-borne trade. However, ancient sails worked poorly. The standard sail was square, mounted on a mast at right angles to the ship's long axis. It was effective only if the wind was dead astern. It was barely adequate if the wind was abeam, and totally inadequate in head winds. As a result, ancient mariners usually timed sailings to correspond with favorable wind direction and often averaged only 1 to 1.5 knots.

Very late in the ancient period the square sail was challenged by a more effective design: the triangular lateen sail, aligned with the vessel's long axis (i.e., fore-and-aft). The origin of this rigging is uncertain; unlike square sails, lateen sails operated as fabric aerofoils and permitted vessels to sail more closely into headwinds.

The only evidence for the use of wind in antiquity to power other machinery occurs in the *Pneumatica*

The sails on these Phoenician ships exhibit an early use of naturally occuring kinetic energy. (Corbis Corporation)

of Hero of Alexandria dating from the first century C.E. Hero described a toy-like device with four small sails, or blades, attached perpendicularly to one end of a horizontal axle. Wind struck the blades and turned the device. Pegs mounted on the opposite end of the axle provided reciprocating motion to a small air pump feeding an organ. There is no evidence that the idea was expanded to a larger scale, and some experts are suspicious of the authenticity of this portion of the *Pneumatica*.

Water, the other source of kinetic energy used in antiquity, saw wider application to machinery than did wind. The first evidence of the use of waterpower comes from the first century B.C.E., simultaneously in both China and the Mediterranean region. In China the preferred method of tapping the power of falling water was the horizontal water wheel, named after the plane of rotation of the wheel. Around the Mediterranean, the preferred form was the vertical

water wheel. The vertical wheel came in two major forms: undershot and overshot. An undershot wheel had flat blades. Water struck the blades beneath the wheel and turned it by impact. The overshot wheel emerged later—the first evidence dates from the third century C.E. The overshot wheel's periphery consisted of containers, called buckets. Water, led over the top of the wheel by a trough, was deposited in the buckets; weight rather than impact turned the wheel.

The diffusion of waterpower was initially slow—perhaps due to its relatively high capital costs, its geographical inflexibility, and the abundance of manual labor in both the classical Mediterranean world and in China. Only in the declining days of the Roman Empire, for example, did watermills become the standard means of grinding grain in some areas, displacing animal- and human-powered mills.

THE MEDIEVAL WORLD (500–1500 C.E.)

Knowledge of how to tap the energy of wind and water was passed by both the Romans and Han Chinese to their successors. In the conservative eastern realm of the old Roman Empire (the Byzantine Empire), little was done to develop either wind power or water power, although Byzantine vessels did make increased use of lateen sails after the eighth century. On the other hand, the western part of the old Roman Empire, perhaps due to labor shortages, saw a very significant increase in the use of both wind and water power, especially between 900 and 1300.

The expansion of medieval Europe's use of natural sources of kinetic energy was most significant in waterpower. Although many Roman watermills were destroyed during the collapse of centralized authority in the fifth through seventh centuries, recovery and expansion beyond Roman levels was relatively rapid. By the eleventh century, sparsely populated England had over 5,600 watermills, a level of dependence on nonhuman energy then unparalleled in human history. Nor was England unusual. Similar concentrations could be found elsewhere. The Paris basin, for example, had around 1,000 watermills in the same era, and canal networks around medieval European cities were often designed to benefit mills.

In coastal estuaries, European craftsman by the twelfth century or earlier had also begun to use water wheels to make use of the movement of the tides. The usual arrangement required impounding incoming tides with a barrier to create a reservoir. During low tide, this water was released through a gate against the blades of a watermill.

With few exceptions, waterpower had been used in classical antiquity for only one purpose—grinding grain. By the tenth, and definitely by the eleventh century, European technicians had begun significantly to expand the applications of waterpower. By 1500 waterpower was used to grind not only wheat, but mustard seed, gunpowder, and flint for use as a glaze. It was used to crush ore, olives, and rags (to make paper). It was used to bore pipes, saw wood, draw wire, full (shrink and thicken) wool, pump water, lift rocks, shape metal, and pump bellows.

Nor was Europe the only civilization to make increased use of waterpower in this era. Although not blessed with the same abundance of stable, easy-to-tap streams, the Islamic world also increased its use of waterpower. Medieval travelers mention numerous mills and water-lifting wheels (norias) along rivers near Islamic cities such as Baghdad, Damascus, Antioch, and Nishapur. China, too, saw the expanded use of waterpower, for powering metallurgical bellows, grinding grain, spinning hemp, driving fans, crushing minerals, and winnowing rice.

The wood water wheels used in the medieval period were, by modern standards, inefficient. Medieval undershot and horizontal wheels probably had an efficiency of about 15 percent to 25 percent, medieval overshot wheels about 50 percent to 60 percent. Commonly, their power output was only about 2–5 hp. But relative to the alternatives available at the time—human or animal power—they offered a very substantial gain in power.

Water provided a more reliable source of energy than the winds, but labor shortages in medieval Europe also encouraged the further development of wind power. The sailing vessel continued to be the most important wind-powered machine, and it saw significant improvement. Between 1000 and 1500, Chinese and European shipbuilders began to use multiple masts, stern post rudders, and deeper keels to tap larger volumes of air more effectively, and Europeans by 1500 were making effective use of combined square and lateen riggings.

The first solid evidence of the use of air to provide mechanical power to something other than sailing vessels comes from tenth-century Islamic travel accounts that describe wind-powered mills for grinding grain in eastern Persia. These were horizontal windmills, so named because the plane of rotation of their sails, or blades, was horizontal. The rotors were two-story devices, surrounded by walls with openings facing the prevalent wind direction. Wind passed through these orifices, struck the exposed blades, and exited through other orifices in the rear. In heavy winds they might produce as much as 15 hp, but more typically their output was about 2 to 4 hp.

Diffusion was very slow. The earliest clear Chinese reference to a windmill occurs only in 1219. The Chinese adopted the horizontal rotor used in Central Asia, but equipped it with pivoted blades, like venetian blinds. These windmills did not require shielding walls like Central Asian mills, and could tap wind from any quarter.

European technicians departed much more radically from Central Asian designs. Drawing from close experience with the vertical watermill, by 1100–1150

An early wind mill (c. 1430), with an automatic elevator for lifting flour bags. The post was designed to turn in accordance with the direction of the wind. (Corbis Corporation)

they had developed a vertical windmill called a post mill. The blades, gearing, and millstones/machinery of the post mill were all placed on or in a structure that pivoted on a large post so operators could keep the mill's blades pointed into the wind. European windmills typically had four blades, or sails, consisting of cloth-covered wooden frames. They were mounted on a horizontal axle and set at a small angle with respect to their vertical plane of rotation.

The European windmill diffused rapidly, especially along the Baltic and North Sea coasts. By the fourteenth century they had become a major source of power. Eventually, England had as many as 10,000 windmills, with comparable numbers in Holland, France, Germany, and Finland. In some areas of Holland one could find several hundred windmills in a few square miles.

Around 1300, Europeans improved on the post mill by devising the tower mill. In the tower mill, only a cap (containing the rotor axle and a brake

wheel), mounted atop a large, stationary tower, had to be turned into the wind. The tower design allowed the construction of larger windmills containing multiple pairs of millstones, living quarters for the miller and his family, and sometimes machinery for sawing wood or crushing materials.

By the end of the medieval period, growing European reliance on wind and waterpower had created the world's first society with a substantial dependence on inanimate power sources.

THE EARLY MODERN ERA (1500–1880)

Dependence on wind or waterpower does not seem to have grown substantially in other civilizations after the medieval era, but this was not the case in Europe. By 1700, most feasible sites along streams convenient to some mercantile centers were occupied by water-powered mills, making it difficult for newer industries, such as cotton spinning, to find good locations. Moreover, European colonists carried waterpower technology with them to North and South America. By 1840 the United States, for example, had nearly 40,000 water-powered mills.

Pressure on available, easily tapped streams pushed the development of waterpower in several different directions simultaneously between 1700 and 1850. One direction involved the application of quantitative techniques to analyze wheel performance. In 1704 Antoine Parent carried out the first sophisticated theoretical analysis of water wheels. Erroneously believing that all wheels operated like undershot wheels, he calculated the maximum possible efficiency of water wheels at an astoundingly low 4/27 (15%). When John Smeaton in the 1750s systematically tested model undershot and overshot wheels, he discovered Parent's error. Smeaton found the optimum efficiency of an undershot vertical wheel was 50 percent (his model wheels actually achieved around 33 percent), and that of a weight-driven overshot wheel was much higher, approaching 100 percent (his model wheels achieved about 67–70%). Smeaton thereafter made it his practice to install overshot wheels where possible. Where impossible, he installed an intermediate type of vertical wheel called the breast wheel. With breast wheels, water was led onto the wheel at or near axle level. A close-fitting masonry or wooden casing held the water on the wheel's blades or buckets so it acted by weight (as in an overshot wheel) rather than by impact.

The mechanization of various elements of textile production, especially carding, spinning, and weaving, after 1770 created important new applications for waterpower. As textile factories grew larger, engineers modified the traditional wooden water wheel to enable it to better tap the kinetic energy of falling water. In addition to following Smeaton's example and using weight-driven wheels as much as possible, they replaced wooden buckets, with thinner sheet-iron buckets and used wrought iron tie-rods and cast iron axles to replace the massive wooden timbers used to support traditional water wheels. By 1850 iron industrial water wheels, with efficiencies of between 60 percent and 80 percent, had an average output of perhaps 15–20 hp, three to five times higher than traditional wooden wheels. Moreover, iron industrial wheels developing over 100 hp were not uncommon, and a rare one even exceeded 200 hp. The iron-wood hybrid wheel erected in 1851 for the Burden Iron works near Troy, New York, was 62 feet (18.9 m) in diameter, by 22 feet (6.7 m) wide, and generated around 280 hp.

Despite these improvements, in the more industrialized parts of the world the steam engine began to displace the waterwheel as the leading industrial prime mover beginning around 1810–1830. Steam was not able, however, to completely displace waterpower, in part because waterpower technology continued to evolve. Increased use of mathematical tools provided one means of improvement. In the 1760s the French engineer Jean Charles Borda demonstrated, theoretically, that for a water wheel to tap all of the kinetic energy of flowing water, it was necessary for the water to enter the wheel without impact and leave it without velocity.

In the 1820s, another French engineer, Jean Victor Poncelet, working from Borda's theory, designed an undershot vertical wheel with curved blades. Water entered the wheel from below without impact by gently flowing up the curved blades. It then reversed itself, flowed back down the curved blades, and departed the wheel with no velocity relative to the wheel itself. Theoretically the wheel had an efficiency of 100 percent; practically, it developed 60 percent to 80 percent, far higher than a traditional undershot wheel.

In the late 1820s another French engineer, Benoit Fourneyron, applied the ideas of Borda and Poncelet to horizontal water wheels. Fourneyron led water into a stationary inner wheel equipped with fixed,

The "Iron Turbine" windmill is a more modern example of windpower. (Corbis Corporation)

curved guide vanes. These vanes directed the water against a mobile outer wheel, also equipped with curved blades. These curved vanes and blades ensured that the water entered the wheel with minimum impact and left with no velocity relative to the wheel itself. Moreover, because water was applied to the entire periphery of the outer wheel at once, instead of to only a portion of the blades, Fourneyron's wheel developed much more power for its size than a comparable vertical wheel could have. This new type of water wheel was called a water turbine. Its high velocity and the central importance of the motion of the water on the wheel (water pressure) to its operation distinguished it from the traditional water wheel.

By 1837 Fourneyron had water turbines operating successfully on both small falls and large ones. At St. Blasien in Germany, a Fourneyron turbine fed by a pipe, or penstock, used a fall of 354 feet (107.9 m), far more than any conventional water wheel could hope to. It developed 60 hp with a wheel only 1.5 foot (0.46 m) in diameter that weighed less than 40 pounds (18.2 kg).

Engineers quickly recognized that turbines could be arranged in a variety of ways. For example, water could be fed to the wheel internally (as Fourneyron's machine did), externally, axially, or by a combination. Between 1830 and 1850 a host of European and American engineers experimented with almost every conceivable arrangement. The turbines that resulted—the most popular being the mixed-flow 'Francis' turbine—quickly demonstrated their superiority to traditional vertical wheels in most respects. Turbines were much smaller per unit of energy produced, and cheaper. They could operate submerged when traditional wheels could not. They turned much faster, and they did all of this while operating at high efficiencies (typically 75% to 85%). By 1860 most new water-powered wheels were turbines rather than vertical wheels.

The continued growth and concentration of industry in urban centers, however, most of which had very limited waterpower resources, meant that steam power continued to displace water power in importance, even if the development of the water turbine delayed the process.

European engineers and inventors also made significant improvements in windmills between 1500 and 1850, but windmills fared worse in competition with steam than water wheels and water turbines.

The sails of the windmills that emerged from Europe's medieval period were inclined to the plane of rotation at a uniform angle of about 20°. By the seventeenth century, millwrights in Holland had improved the efficiency of some mills by moving away from fixed angles and giving the windmill blades a twist from root to tip (i.e., varying the inclination continuously along the blade's length) and by putting a camber on the leading edge, features found on modern propeller blades. Another advance was the fantail. Developed by Edmund Lee in 1745, the fantail was a small vertical windmill set perpendicular to the plane of rotation of the main sail assembly and geared so that a change in wind direction would power the fantail and rotate the main sails back into the wind. In 1772 Andrew Meikle developed the spring sail, which used hinged wooden shutters mounted on the blades, operating like venetian blinds, to secure better speed control. Other inventors linked governors to spring sails to automatically control wind speeds.

As with water wheels, engineers also began to carry out systematic, quantified experiments on windmills to test designs. In 1759 John Smeaton published a set of well-designed experiments on a model windmill. Since he had no means of developing a steady flow of air, he mounted 21-inch (53-cm) long windmill blades on a 5.5-feet (1.54-m) long arm that was pulled in a circle in still air to simulate the flow of air against a windmill. His experiments refuted prevalent theory, which recommended that the blades of windmills be inclined at a 35° angle. They confirmed that the 18–20° angle generally used in practice was far better.

Smeaton's work also confirmed the even greater effectiveness of giving the blades a twist, that is, inclining sails at a variable angle like a propeller. By 1800 a large, well-designed Dutch windmill might develop as much as 50 hp at its axle in exceptional winds (normally 10–12 hp), although gearing inefficiencies probably reduced the net output at the machinery during such winds to around 10–12 hp.

Despite the empirical improvements in windmill design and attempts to apply quantitative methods to its design, the energy that could be tapped from winds was too erratic and had too low a density to compete with newly developed thermal energy sources, notably the steam engine. Following the development of an effective rotary steam engine by James Watt, and the introduction of mobile, high-pressure steam engines (locomotives) in the early-nineteenth century, that brought the products of steam-powered factories cheaply into the countryside, windmills began a rapid decline.

Only in limited arenas did the windmill survive. It continued to produce flour for local markets in unindustrialized portions of the world. It also found a niche—in a much-modified form—as a small power producer in isolated agricultural settings such as the Great Plains of North America. The American farm windmill of the nineteenth century was a vertical windmill, like the traditional European windmill, but it had a much smaller output (0.2-1 hp in normal breezes). Consisting of a small annular rotor usually no more than 6 feet (2 m) in diameter, with a very large number of blades to produce good starting torque, American windmills were mounted atop a high tower and usually used to pump water. By the early twentieth century several million were in use in America, and the technology had been exported to other flat, arid regions around the world, from Australia to Argentina to India.

THE MODERN ERA (1880–2000)

By the late nineteenth century, both wind and water were in decline as power producers almost everywhere. Steam-powered vessels had, in the late nineteenth century, begun to rapidly replace sailing vessels for hauling bulk merchandise. Only in isolated regions did sailing vessels continue to have a mercantile role. Increasingly, sailing vessels became objects of recreation, not commerce. In areas such as Holland, traditional windmills continued to pump away here and there, but they were on their way to becoming historical relics more than anything else. In the twentieth century even the American windmill went into decline as electric power lines belatedly reached rural areas and electric pumps replaced wind-powered pumps. Meanwhile, growing dependence on steam engines steadily reduced the importance of waterpower in all but a few favorable locations.

The emergence of electricity as a means of transmitting power late in the nineteenth century offered new life to devices that tapped the kinetic power of wind and water, since it offset one of the greatest advantages of steam power: locational flexibility. With electricity, power could be generated far from the point at which it was used.

Around 1890 Charles Brush erected a wind-powered wheel over 56 feet (17 m) in diameter to test the possibilities of wind as an economical generator of dc electricity. Coal- and waterpower-poor Denmark provided much of the early leadership in attempts to link wind machinery with the newly emerging electrical technology. The Danish engineer Poul LaCour, for example, combined the new engineering science of aerodynamics with wind tunnel experiments and new materials such as steel to try to develop a wind engine that would be a reliable and efficient generator of electricity. By 1910 he had several hundred small wind engines operating in Denmark generating dc electricity for charging batteries, but he had few imitators.

The development of the airplane stimulated intensive research on propeller design. In the 1920s and 1930s, a handful of engineers began to experiment with fast-moving, propeller-shaped windmill rotors (wind turbines), abandoning traditional rotor designs. For example, Soviet engineers in 1931 erected a 100 kW (about 130 hp) wind turbine on the Black Sea. In 1942 the American engineer Coslett Palmer Putnam erected the first megawatt-scale wind turbine designed to feed ac into a commercial electrical grid—a very large, two-blade rotor—at Grandpa's Knob in Vermont. It failed after 1,100 hours of operation, and cheap fuel prices after World War II limited further interest in wind-generated electricity.

The fuel crises and environmental concerns of the 1970s revived interest in the use of wind to produce electricity. Governmental subsidies, especially in the United States, supported both the construction of wind turbines and research, and led to the erection of large arrays of wind generators (wind farms) in a few

Modern wind turbine on a hill. (Corbis Corporation)

favorable locations. By the 1990s, over 15,000 wind turbines were in operation in California in three mountain passes, generating a total of about 2.4 million hp (1,800 mW), but wind's contribution to total electrical energy supplies remains miniscule (<1%).

Waterpower fared better. When electrical engineers demonstrated the possibility of long-distance transmission of electrical power between 1875 and 1885, waterpower suddenly assumed new importance. Many of the best waterpower sites had been untapped because they were in remote regions far from manufacturing centers. As long as power had to be directly transmitted by mechanical shafts and belts, the locational flexibility of steam power made it the overwhelming first choice of industry. Electric power transmission made low-cost hydropower more competitive.

The first experiments with commercial water-generated electricity took place in the 1880s, but applications were limited since the predominant form of electricity generated, low voltage dc suffered large power losses in transmission. The introduction of high voltage ac around 1890, with its sharply reduced transmission losses, led to the first hydroelectric plants of significant scale. The spectacular success of the first hydroelectric plant at Niagara Falls in 1895, which used multiple turbine generator units, each developing 5,000 hp (3,750 kW) under a 136-foot head, and transmitting that power via high voltage ac some 22 miles to Buffalo, attracted enormous capital into the field. Between 1895 and 1930 large hydroelectric plants sprang up all across the United States and Europe. In some regions such as Norway and California, hydroelectricity became the dominant form of energy used.

The move toward gigantic hydroelectric developments accelerated in the 1930s, especially in the United States and the Soviet Union, as government-sponsored projects replaced privately funded projects. The Dnieprostroy hydroelectric plant in the Soviet Union, completed in 1932, was equipped with nine 85,000 hp turbine-generator units. The seventeen turbine-generators installed at Hoover Dam in the mid-1930s were even more powerful: rated at 115,000 hp (~86 MW each). More recent hydroelectric plants have grown even larger. For example, Itaipú (1984–1991) on the Paraná River in South America has eighteen turbine-generator units, each with a capacity of around 940,000 hp (700 MW), for a total capacity of around 17 million hp (12,600 MW).

Although the continued growth of energy consumption and the fixed amount of energy available from waterpower has prevented it from returning to its position as humanity's primary source of inanimate power, waterpower, unlike wind power, continues to be an important producer of energy. Today, waterpower is responsible for nearly 20 percent of the world's electrical energy. In some regions where hydrological and terrain conditions are appropriate, for example in Canada, Norway, and New Zealand it supplies a much higher proportion.

The technologies that enabled waterpower to remain important in the twentieth century—turbines, high voltage ac, and large dams—were also applied to harness the tides. In a few select estuaries where tidal variations were significant and construction conditions appropriate, large dams with embedded turbine-generators impounded high tides and used the outflow during low tides to generate electricity. The largest of these was placed in operation in 1966 on the Rance River on the coast of Brittany in France, where an average tidal range of around 28 feet was available. At Rance, a 0.4 mile long dam equipped with twenty-four turbines, generates around 320 MW of electric power. The high capital cost of tidal plants, the very limited number of sites with sufficient tidal variation and suitable construction conditions, and environmental concerns sharply limit the use of this form of kinetic energy. Thus, tides do not produce a significant portion of the world's energy (much less than 1%).

Terry S. Reynolds

See also: Hydroelectric Energy; Smeaton, John; Turbines, Wind; Watt, James.

BIBLIOGRAPHY

Baker, T. L. (1985). *A Field Guide to American Windmills.* Norman, OK: University of Oklahoma Press.

Hay, D. (1991). *Hydroelectric Development in the United States, 1880–1940.* Washington, DC: Edison Electric Institute.

Heymann, M. (1998). "Signs of Hubris: The Shaping of Wind Technology Styles in Germany, Denmark, and the United States, 1940–1990." *Technology and Culture* 39: 641–70.41

Hills, R. L. (1994). *Power from Wind: A History of Windmill Technology.* Cambridge: Cambridge University Press.

Hunter, L. C. (1979). *A History of Industrial Power in the United States, Vol. 1: Waterpower in the Century of the Steam Engine.* Charlottesville, VA: University Press of Virginia.

Hunter, L. C., and Bryant, L. (1991). *A History of Industrial Power in the United States, Vol. 3: The Transmission of Power.* Cambridge, MA: The MIT Press.

Reynolds, T. S. (1983). *Stronger Than a Hundred Men: A History of the Vertical Water Wheel.* Baltimore, MD: Johns Hopkins University Press.

Smil, V. (1994). *Energy in World History.* Boulder, CO: Westview Press.

Smith, N. (1976). *Man and Water: A History of Hydro-Technology.* London: P. Davies.

LAMPS

See: Lighting

LANDFILL GAS

See: Methane

LAND RIGHTS

See: Property Rights

LAPLACE, PIERRE SIMON (1749–1827)

Pierre Simon Laplace, the most influential of the French mathematician-scientists of his time, made many important contributions to celestial mechanics, the theory of heat, the mathematical theory of probability, and other branches of pure and applied mathematics. He was born into a Normandy family of well-to-do peasant farmers and merchants. Intended for an ecclesiastical career, he matriculated at the University of Caen in theology but, discovering his aptitude for mathematics, departed for Paris at age nineteen bearing a letter from his instructors to the mathematician d'Alembert, a leading intellectual figure in prerevolutionary France.

D'Alembert, impressed by Laplace, took him on as his protégé and obtained an appointment for him as a professor of mathematics at the École Militaire, where he taught mathematics to teenage cadets. In this position, Laplace was also expected to present his work to the Académie Royal des Sciences, which was under royal patronage. He presented thirteen papers to the Academy in his first three years in Paris. In 1773, his fifth year in Paris, he was elected an adjunct member of the Academy. By the 1780s he was regarded as one of its leading figures and was appointed to important royal committees, largely based on his published papers on probability, astronomy, and geophysics. In 1785 he was promoted to the rank of senior pensioner in the Academy. However, Laplace was far from a popular figure. He rightly considered himself to be the leading mathematician in France and presumed upon it, pronouncing judgment on matters in other sciences. Even when he was correct, his abrasive manner created enemies. Nevertheless, he played a preeminent role in the Academy committee to reform the system of weights and measures. During this period (1782–1793) he also collaborated with Lavoisier in a series of investigations on the nature of heat and the phenomena of combustion and respiration. The collaboration terminated during the Reign of Terror, when the Academy was suppressed. Toward the end of 1793, probably in fear for his safety, Laplace left Paris until after the fall of Robespierre in the following year.

Pierre Simon Laplace. (Library of Congress)

The functions of the Académie Royal des Sciences were assumed in 1795 by a branch of the newly formed National Institute. Laplace was elected vice president of this reincarnated Academy and then elected president a few months later, in 1796. The duties of this position put him in contact with Napoleon Bonaparte. Three weeks after Napoleon seized power in 1799, Laplace presented him with copies of his work on celestial mechanics. Bonaparte quipped that he would read it "in the first six weeks I have free" and invited Laplace and his wife to dinner. Three weeks later, Napoleon named Laplace his minister of the interior. After six weeks, however, he was replaced; Napoleon thought him a complete failure as an administrator. However, Napoleon continued to heap honors and rewards upon him, regarding him as a decoration of the state. He made Laplace a chancellor of the Senate with a salary that made him wealthy, named him to the Legion of Honor, and raised him to the rank of count of the empire. Laplace's wife was appointed a lady-in-waiting to the Italian court of Napoleon's sister. Laplace responded with adulatory dedications of his works to Napoleon.

When Napoleon fell from power, however, Laplace carefully dissociated himself from the emperor. In the Senate, Laplace voted for the return of the Bourbon monarchy and absented himself from Paris in 1815 during Napoleon's brief hundred-day return from Elba. In 1817 Louis XVIII raised Laplace to the rank of marquis. Laplace remained loyal to the Bourbons for the rest of his life, and his 1826 refusal to sign a petition supporting freedom of the press condemned him as far as the liberals in the Academy were concerned.

Laplace's lifelong work was the successful application of Newtonian gravitation to the entire solar system, accounting for all the observed deviations of the planets and satellites from their theoretical orbits. He began this work in 1773. His five-volume work on celestial mechanics (1798–1827) provided a complete mechanical description of the solar system. During the course of his investigations, Laplace discovered that the force on a body in a gravitational field could always be derived as the gradient of a potential function, a discovery that had profound implications in other branches of physics.

In mathematics, Laplace's name is most often associated with the "Laplace transform," a technique for solving differential equations. Laplace transforms are an often-used mathematical tool of engineers and scientists. In probability theory he invented many techniques for calculating the probabilities of events, and he applied them not only to the usual problems of games but also to problems of civic interest such as population statistics, mortality, and annuities, as well as testimony and verdicts.

In the first and second decades of the nineteenth century, Laplace exercised a powerful influence on physics in France, not only through his publications but also through the power to direct the research of others by virtue of his prestige and his position in the Academy. The genius of Laplace lay in his skill in surmounting mathematical difficulties. However, unlike other great scientists, he never attempted to go beyond the view of the world that existed when his career began. Laplace was an apostle of Newtonian mechanics. He believed, for example, that phenomena such as the behavior of light in material media could be understood on the basis of Newton's corpuscular theory of light and short-range molecular forces acting on the corpuscles of light; he directed research to strengthen this approach. As the weight of evidence in favor of the wave nature of light grew larger, Laplace's influence and power waned. In his last years he developed an elaborate caloric theory of heat, again based on a theory of short-range mechanical forces, but the theory was greeted with indifference and generated no further research.

Leonard S. Taylor

See also: Gravitational Energy; Heat and Heating.

BIBLIOGRAPHY

Arago, F. (1972). *Biographies of Distinguished Scientific Men*, tr. W. H. Smyth, B. Powell, and R. Grant. Freeport, NY: Books for Libraries Press.

Gillespie, C. C. (1997). *Pierre-Simon Laplace.* Princeton NJ: Princeton University Press.

LASERS

It is said that necessity is the mother of invention. This adage says volumes about the early development of the laser. During World War II, U.S. military and civilian scientists searched frantically for improved radar. While these researchers met with only mixed success, their efforts spurred basic research. After the war, using knowledge gained from this line of inquiry, the first successful laser was developed in 1960.

Applications of the laser were found in numerous fields, including science, medicine, industry, and entertainment. By the 1980s, with the widespread use of lasers in industry and in commercial devices such as compact disk players and retail store price scanners, lasers were affecting the daily lives of nearly everyone in the developed world, whether they realized it or not. In a few decades, the laser has gone from a new cutting-edge technology to one that is so pervasive it is difficult to imagine many fields even existing without it.

WHAT IS A LASER?

The word "laser" is an acronym for "light amplification by the stimulated emission of radiation." Lasers of all kinds consist of several basic components: an active medium, an outside energy source, and an optical cavity with carefully designed mirrors on both ends. One of the mirrors is 100 percent reflective

while the other is somewhat less reflective, so a beam can be emitted. The active medium is inside the optical cavity and is excited with an external energy source (typically electricity). For the cavity to emit laser radiation, the active medium has to achieve an unusual energy state called "population inversion."

All atoms consist of electrons orbiting around a nucleus. Normally these electrons prefer to remain in their lowest energy state, or orbit, which is known as ground level. When excited, however, electrons will jump to a higher orbit and then drop back down by radiating energy in the form of light at a discrete wavelength. This release of energy is called spontaneous emission. The energy or wavelength of the emitted light determines its color. Spontaneous emission occurs all by itself and is the source of virtually all light we see from natural sources (stars) and man-made ones (television sets, fluorescent lamps.)

Unlike these light sources, laser technology relies on a concept known as stimulated emission. When an excited atom is stimulated by a photon, light is emitted at precisely the same wavelength and precisely in phase with the light wave that stimulated it.

As energy is added and photons collide with other atoms, more and more electrons gain energy and jump orbits. The excited atoms all emit photons at the exact same wavelength. At some point there are more atoms with electrons in excited states than at ground level. This is population inversion.

As the emitted radiation bounces back and forth between the two mirrors, it becomes coherent. Some of the energy traveling back and forth through the optical cavity is transmitted though the less reflective mirror and becomes a laser beam.

The coherence of the beam makes a laser extremely powerful. The beam is light if a single frequency or color in which all the components are in step with one another. Normal radiation (such as the light from the sun) consists of many different wavelengths of light traveling in different directions. A laser's energy is concentrated, traveling in one direction, and traveling at the same wavelength, so its effect on other matter is extraordinary. A 100-watt lightbulb, for example, can barely light a living room, but a pulsed, 100-watt laser can be used to cut or drill holes in metal.

Another feature of many (though not all) lasers is a cooling system to dissipate energy wasted in the form of heat. Although some are much more efficient than others, all lasers waste significant amounts of energy. For example, a CO_2 laser (the most popular laser for industrial uses) operates at approximately 20 percent efficiency. To produce 100 watts of coherent light, these lasers require 500 watts of input power. A ruby laser is even less efficient, at about 0.1 percent; it would require 100,000 watts of electricity to produce that same 100 watts of laser light. Conversely, one company introduced a semiconductor laser in 1998 that operates at 56 percent, the most efficient up to that time.

DEVELOPMENT OF THE LASER

The first known laser was made by Theodore Maiman at Hughes Research Laboratories in Malibu, California, in 1960, but the seeds of this breakthrough were planted years before. In 1917 Albert Einstein, through his work on the quantum theory of light, theorized that stimulated emission of light radiation could occur. The idea was forgotten, though, until the middle of the century.

In the 1940s, researchers were working on better military radar. Radar is used to locate and track objects using pulses of long-wavelength microwave radiation created according to similar principles as laser radiation. Scientists knew that if they could achieve shorter wavelengths, more accurate images can be obtained using more compact equipment. The race was on to reach the infrared and visible portions of the spectrum. Although technical stumbling blocks prevented researchers from achieving this goal during the war, the pursuit continued during peacetime. The war had taught U.S. military and government officials the importance of technological superiority. Worldwide political tensions kept research grants flowing into America's scientific laboratories.

In 1948 Charles Townes was a professor of physics at Columbia University, working with microwaves. During the early 1950s he speculated that stimulated emission could generate microwaves, but he also knew that a population inversion was necessary. By combining the work done by researchers at Harvard University on this concept as well as his own radar-related work on amplifying a signal, Townes came up with the idea for a closed cavity in which radiation would bounce back and forth, exciting more and more molecules.

By 1954 Townes, with the help of graduate students Herbert Zeiger and James Gordon, developed the maser, an acronym for microwave amplification by stimulated emission of radiation. The maser had

Theodore H. Maiman with a ruby that was used in early laser studies in the 1960s. (Corbis Corporation)

all the components of a laser—the resonator cavity and an active medium (Townes used ammonia)—but the wavelengths produced were much longer. In 1957 Townes developed the equation showing that much smaller wavelengths (in the infrared and visible-light range) were possible. Townes collaborated with Arthur Schawlow, a physicist at AT&T Bell Laboratories, and the two worked together to achieve a laser.

Others had similar ideas. R. Gordon Gould, a graduate student at Columbia, had come to the same conclusions. In November 1957 he wrote up his notes on a possible laser. In 1958 Townes and Schawlow patented a similar idea. Theodore Maiman learned of this research in September 1959 at a conference on quantum electronics that Townes had organized. Afterward he began his own research on a laser, using a pink ruby as the active medium. Ironically, Schawlow had rejected ruby because he felt it would require too much energy. By pumping the ruby with the light from a photographer's flash lamp, however, Maiman created the world's first laser, in June 1960.

The floodgates were now open. Lasers using calcium fluoride, helium-neon, glass, and cesium vapor as active media followed within two years. Others substances followed throughout the early 1960s. By 1962 about 400 companies had some kind of laser research. U.S. government spending on laser development soared, from $1.5 million in 1960 to $20 million in 1963. Scientific journals became inundated with papers, and meeting rooms at conferences overflowed with inquisitive attendees.

Lasers were still tools of elite scientists, though. Moving the technology from the lab and into the marketplace took years. The first devices produced in significant quantities were military systems in the early 1970s, followed by a small but growing number of industrial and medical lasers. The laser's time had finally come.

LASERS AND THEIR APPLICATIONS

Different lasers use different materials as the active medium. The medium can be either solid, liquid, or gas, and there are advantages for each in the amount of energy that can be stored, ease of handling and storage, secondary safety hazards, cooling properties, and physical characteristics of the laser output.

Solid-State Lasers

The term "solid-state laser" refers to lasers that use solids as their active medium. However, two kinds of materials are required: a "host" crystal and an impurity "dopant." The dopant is selected for its ability to form a population inversion. The Nd:YAG laser, for example, uses a small number of neodymium ions as a dopant in the solid YAG (yttrium-aluminum-garnet) crystal. Solid-state lasers are pumped with an outside source such as a flash lamp, arc lamp, or another laser. This energy is then absorbed by the dopant, raising the atoms to an excited state. Solid-state lasers are sought after because the active medium is relatively easy to handle and store. Also, because the wavelength they produce is within the transmission range of glass, they can be used with fiber optics.

Diode (Semiconductor) Lasers

Also using solids but considered separately because of their unique characteristics, diode lasers are the most common lasers in use. Compact size and relia-

bility are the chief benefits of this kind of laser. The two common families of diode lasers contain active mediums composed of GaAlAs (gallium-aluminum-arsenite) or InGaAsP (indium/phosphorus). These media emit radiation in the infrared range. Much like those working with radar in the 1940s and '50s, researchers in the 1980s and '90s found ways to shorten wavelengths of lasers produced by diodes to reach into the range of blue visible light.

Liquid (Dye) Lasers

The common liquid lasers utilize a flowing dye as the active medium and are pumped by a flash lamp or another laser. These are typically more complex systems requiring more maintenance. They can be operated as either CW (continuous wave) or pulsed. One advantage liquid lasers have is they can be tuned for different wavelengths over a 100-nm range.

Gas Lasers

Gas lasers are not unlike fluorescent light bulbs and neon signs. Gas is confined to a hollow tube, and electricity passing through it excites the atoms. The most common gas lasers use carbon dioxide, argon, and helium-neon. Gas lasers are relatively inexpensive and can produce very high-powered beams.

Excimer Lasers

Excimer lasers use gases, but because of their special properties are usually considered as a class of their own. Excimer is short for "excited dimer," which consists of two elements, such as argon and fluorine, that can be chemically combined in an excited state only. These lasers typically emit radiation with very small wavelengths, in the ultraviolet region of the electromagnetic spectrum. This shorter wavelength is an enormous advantage for many applications.

There are literally thousands of uses for lasers. One of the largest applications is telecommunications—sending a signal through fiber optic cables, for example. This application grew rapidly in the 1990s with the phenomenal increase in traffic on the Internet. Optical data storage, such as on compact disks, CD-ROMs, and DVDs, is another important use for lasers. The information age was obviously a boon to this application, and as researchers obtained smaller wavelengths with diode lasers, they were able to fit more information on smaller storage devices.

Another group of applications is collectively known as materials processing. This includes the processes used in manufacturing. Production facilities use lasers to cut, weld, drill, mark, and heat-treat numerous materials such as metals, plastics, wood, ceramics, and even diamonds. Lasers are much more precise than other mechanical means used to process materials, and lasers make it possible to build devices with tiny, even microscopic, dimensions. A subgroup of this category, medical device manufacturing, relies on lasers to machine stents and other devices for implantation into the human body.

Lasers are used for medical procedures as well. The laser has been called a bloodless scalpel because it can be used to simultaneously cut and cauterize tissue. Photodynamic therapy is a cancer treatment that relies on lasers, and laser eye surgery can correct vision, eliminating the need for glasses and contact lenses. Plastic surgeons use lasers for hair removal, skin resurfacing, tattoo removal, and many other applications. New medical imaging and diagnostic methods have been devised thanks to the laser.

Military uses for lasers are abundant, from range-finding to guided munitions to laser aiming devices on firearms. Warfare has been revolutionized by the laser. Law enforcement uses lasers to lift hard-to-recover fingerprints and in laser radar speed guns.

Laser printers, bar code readers, unmanned freeway tollbooths, laser pointers—none of these very common devices would be possible without laser technology. This is just a minor sampling; the list of laser applications goes on and on.

The laser has revolutionized many aspects of science and other disciplines, as well as the daily lives of millions of people. When it was first invented, the laser was referred to by some as a solution looking for a problem because it came about mostly from basic research rather than the active solution to a particular concern. At the time, no one could have predicted the far-reaching effects it would have in the second half of the twentieth century, or that it would come to be considered by many as one of the most influential technological achievements of that time.

Karl J. Hejlik

See also: Industry and Business, Productivity and Energy Efficiency in; Lighting; Matter and Energy; Townes, Charles Hard.

BIBLIOGRAPHY

Bromberg, J. L. (1991). *The Laser in America, 1950–1970.* Cambridge, MA: MIT Press.

Bromberg, J. L. (1992). "Amazing Light." *Invention and Technology.* Spring:18–26.

Charschan, S.(1993). *Guide to Laser Materials Processing.* Orlando, FL: Laser Institute of America.

Hecht, J. (1992). *Laser Pioneers.* Boston: Academic Press.

Muncheryan, H. (1983). *Principles and Practices of Laser Technology.* Blue Ridge Summit, PA: TAB Books.

National Academy of Sciences. (1987). *Lasers: Invention to Application.* Washington, DC: National Academy Press.

National Academy of Sciences. (1998). *Harnessing Light: Optical Science and Engineering for the 21st Century.* Washington, DC: National Academy Press.

Svelto, O., and Hanna, D. (1998). *Principles of Lasers,* 4th ed. New York: Plenum Press.

LAUNCH VEHICLES

See: Spacecraft Energy Systems

LAWRENCE BERKELEY LABORATORY

See: National Energy Laboratories

LAWRENCE LIVERMORE LABORATORY

See: National Energy Laboratories

LEVER

See: Mechanical Transmission of Energy

LEWIS, WARREN K. (1882–1975)

INTRODUCTION

As a central figure in twentieth century petrochemical technology, Warren K. Lewis is widely viewed as the father of American chemical engineering. His work opened up an entirely new and powerful engineering discipline applicable across a broad range of manufacturing industries, including chemical synthesis, steel production, and power generation.

Lewis had an enormous influence on the energy industries. Along with the Russian, Vladimir Ipatieff, Lewis was pivotal in advancing petroleum refining technology. As an instructor and mentor, Lewis left his mark on three generations of chemical engineering students, many of whom entered and influenced the U.S. petroleum and petrochemical industry. As a consultant and technological innovator, Lewis directly applied his principles of chemical engineering, an essentially American development, to revamping and extending the production capability of U.S. catalytic cracking facilities. Lewis' greatest single engineering achievement was his central role during World War II in the design of the fluid catalytic cracking process, an innovation of strategic and economic importance.

THE EARLY YEARS

Warren K. Lewis was born in 1882 on a small farm in southern Delaware. Farm life exerted a strong early influence on the future engineer. Believing that after college he would return to manage the family farm, his academic goal was to learn the fundamentals of agricultural science in order to apply them in a rational manner to improve the farm's productivity. To Lewis, this meant employing a thorough understanding of the mechanical operations of advanced (i.e., efficient) equipment and tools. In 1901 Lewis entered MIT as an undergraduate with a major in mechanical engineering.

THE MIT YEARS

Soon after entering MIT, Lewis came under the influence of Professor William H. Walker, who was instituting a new and innovative program in chemical engineering. In his sophomore year, Lewis transferred to the chemical engineering program, which at that time was part of the chemistry department.

After graduation, and at the urging of Walker, Lewis won a two-year fellowship for study in Germany. Studying under the great physical chemist Abegg at the University of Breslau, Lewis obtained his doctorate in 1908 while developing a thorough grounding in the theories of and mathematical structures associated with phenomena underlying numerous industrial engineering processes.

Spending only a brief period in industry, Walker urged Lewis back to MIT in 1910 to become an assistant professor of chemical engineering. Both Walker and Lewis sought to form closer ties between the chemical engineering program and industry. To this end, and with the cooperation of the great industrial chemist A. D. Little, they established in 1916, MIT's School of Chemical Engineering Practice. The Practice School, a successor to Walker's Research Laboratory of Applied Chemistry (established at MIT in 1908), centered around practical chemical engineering instruction within operating plants in and around the New England area.

In addition to being (by all accounts) a brilliant and riveting teacher, Lewis was an innovative administrator in the cause of furthering chemical engineering at MIT. In 1920, largely due to the efforts of Walker and Lewis, MIT's chemical engineering program became a separate department, with the Practice School becoming part of the newly-formed department. In that year, Lewis became the first head of the chemical engineering department. During his nine years as department head, Lewis fought successfully for a strong, independent chemical engineering program. In their path breaking text, The Principles of Chemical Engineering (1923), Lewis and Walker for the first time systematized the engineering study of unit operations and provided a model of instruction for burgeoning chemical engineering departments. Eventually, Lewis secured for the department approval by MIT to develop a curriculum leading to a B.S. in chemical engineering, one of the first such programs in the country.

ENGINEERING, CONSULTING AND PETROLEUM PROCESSING

Lewis' influence on the U.S. energy sector evolved from his career as a consultant to the petroleum

707

refining industry, and in particular Standard Oil of New Jersey (currently Exxon). Lewis' success as a consultant rested in large part on his ability to apply to practical problems his knowledge of and experience in industrial chemical engineering. Between 1919 and 1927, Lewis virtually restructured Jersey Standard's manufacturing operations. His work spanned both oil production and refining.

One of Lewis' first assignments was to make the distillation process more precise and continuous. By the early 1920s, Lewis introduced to Jersey Standard the use of vacuum stills. These were able to operate at lower temperatures that limited coking and fouling of equipment. Thus production engineers did not have to periodically clean out and repair equipment, which in turn facilitated the transformation of distillation from batch to continuous operations.

Lewis directed subterranean reservoir studies to improve the efficiency of extracting oil from the ground. He designed the first bubble tower to effect more precise and efficient fractionation operations and provided important assistance in developing one of the first continuous thermal cracking process.

In 1927, Lewis and Frank A. Howard, head of the Development Department at Jersey Standard, established a research center at Jersey's Baton Rouge plant to research and commercialize advanced petrochemical technology. This development marks the beginning of the Southwest as a center of petroleum-based technology. Lewis populated the facility with colleagues and graduates from MIT's Chemical Engineering Department and the Practice School. These included Robert Haslam, the head of the Practice School, who became director of Baton Rouge R&D and Eger Murphree, a protégé of Lewis, who was made Manager of Development and Research for Jersey Standard's Development Department.

In the years leading to World War II, Lewis and Haslam hired fifteen additional MIT-affiliated chemical engineers to work at the Baton Rouge plant. These men felt great loyalty to Lewis who they considered their mentor. They were instrumental in championing Lewis and his style of chemical engineering at Jersey Standard. Through the 1930s and during World War II, Lewis and Jersey's R&D people at Bayway and Baton rouge worked closely together to develop some of the most important innovations in the history of petrochemical technology. The most significant of these was fluid catalytic cracking.

WORLD WAR II, FLUID CRACKING, AND THE LATER YEARS

Fluid cracking, which began operations in 1942 and was responsible for providing the Allies with sufficient supplies of aviation fuel and synthetic rubber, was Lewis' last and greatest engineering achievement. Its successful development depended on Lewis' engineering genius, on the close relationship established between Jersey Standard and MIT by Lewis, and the ability of his protégés both within MIT and Jersey Standard to transform his core ideas and designs into commercial reality.

In addition to his work on fluid cracking, Lewis served as Vice Chairman of the Chemistry Division of the National Defense Research Committee during World War II. In this capacity, he directed chemical engineering research on problems associated with the development of the atomic bomb. Lewis officially retired from teaching in 1948, although he continued to meet with students at MIT well into his eighties. Lewis died at the age of 92 in 1975. The American Institute of Chemical Engineers, in honor of his achievements in the academic arena, established an annual award in his name for those who have made significant contributions to education in the field.

Sanford L. Moskowitz

BIBLIOGRAPHY

Enos, J. (1962). *Petroleum, Progress and Profits: A History of Process Innovation.* Cambridge, MA: MIT Press.

Frankenburg, W. G., et al., eds. (1954). *Advances in Catalysis and Related Subjects, Vol. 6.* New York: Academic Press.

Grace, J. R., and Matsen, J. M., eds. (1980). *Fluidization.* New York: Plenum Publishing.

Landau, R., and Rosenberg, N. (1990). "America's High-Tech Triumph." *Invention & Technology.* Fall:58–63.

Lewis, H. C. (1980). "W. K. Lewis, Teacher." In *History of Chemical Engineering*, edited by W. F. Furter. Washington, DC: American Chemical Society.

Larson, H. M., et al. (1971). *History of Standard Oil Company (New Jersey): New Horizons, 1927–1950.* New York: Harper and Row.

Mattill, J. (1991). *The Flagship: The MIT School of Chemical Engineering Practice, 1916–1991.* Cambridge, MA: David H. Koch School of Chemical Engineering Practice, MIT.

Moskowitz, S. L. (1999). "Science, Engineering and the American Technological Climate: The Extent and Limits of Technological Momentum in the Development of the U.S. Vapor-Phase Catalytic Reactor, 1916–1950." Diss. Columbia University.

Murphree, E. V., et al. (1945). "Improved Fluid Process for Catalytic Cracking." *Transactions of the American Institute of Chemical Engineers* 41:19–20.

Othmer, D. F. (1956). *Fluidization*. New York: Reinhold Publishing Corp.

Pigford, R. L. (1976). "Chemical Technology: The Past 100 Years." *Chemical and Engineering News* (April 6):190–195.

Popple, C. S. (1952). *Standard Oil Company (New Jersey) in World War II*. New York: Standard Oil Company (New Jersey).

"Reminiscences of William H. Walker." (1952). *Chemical Engineering* (July):158–178.

Russell, R. P. (1944). "The Genesis of a Giant." *Petroleum Refiner* 23:92–93.

Spitz, P. (1988). *Petrochemicals: The Rise of an Industry*. New York: John Wiley and Sons.

Squires, A. M. (1986). "The Story of Fluid Catalytic Cracking: The First Circulating Fluid Bed." *Proceedings of the First International Conference on Circulating Fluid Beds, November 18–20, 1985*. New York: Pergamon Press.

Weber, H. C. (1980). "The Improbable Achievement: Chemical Engineering at MIT." In *History of Chemical Engineering*, edited by W. F. Furter. Washington, DC: American Chemical Society.

Williams, G. C., and Vivian, J. E. (1980). "Pioneers in Chemical Engineering at MIT." In *History of Chemical Engineering*, edited by W. F. Furter. Washington, DC: American Chemical Society.

Williamson, H. F., et al. (1963). *The American Petroleum Industry, Vol. II: The Age of Energy, 1899–1959*. Evanston, IL: Northwestern University Press.

LIGHTING

Light is essential for human life. Modern societies have created homes, schools, and workplaces that rely on electric light sources. Some of the electricity used to generate the light in these spaces is wasted, largely owing to ignorance. The efficient application of electric, as well as natural, light sources to the human condition is a sophisticated effort, but one that is essential to a sustainable and enjoyable future.

WHAT IS LIGHT?

Humans are a diurnal species, which means that we are active in the day and asleep at night. Indeed, day-light is the primary stimulus to the photobiological system that regulates our sleep-awake cycle. Of course, while we are awake, we see, and we depend a great deal on seeing. Approximately 80 percent of the human brain devoted to sensing the environment is devoted to vision. It is not surprising, then, that from the beginning of human history we have strived to produce and control light.

Until very recent human history, the Sun had been our primary source of light for both seeing and waking. Over the past two millennia, and particularly over the past two centuries, our direct reliance on the Sun for light has diminished. Today it can be argued that large segments of affluent human societies are exposed to light from the Sun only rarely. Many people spend virtually all of their active lives under manufactured light sources of various types.

These manufactured light sources are, perhaps ironically, largely dependent on the Sun. The radiant energy from the Sun has been stored in the fossilized remains of billions of creatures over millions of years and is used to power the electric light sources created by modern humans. The power generated by hydroelectric sources also is a result of solar evaporation and subsequent rainfall. Only nuclear reactors provide power independent of the Sun, which is, of course, the largest nuclear reactor in the solar system.

Ironically, too, without humans there cannot be "light." In other words, we formally define light in human terms. Radiation from the Sun and from other sources varies in frequency. These different frequencies have been categorized for convenience into different bands. The highest frequencies are in a band known as cosmic rays, and the lowest are in the radio frequency band. Between these two extremes is a very small band of radiant energy known as visible radiation, or light. Only radiation in the narrow region of the electromagnetic spectrum visible to humans can be called light. All other radiation, *even that seen by other species* outside this band, cannot technically be referred to as light. We measure the frequency of radiation within the visible band in terms of wavelength. By convention, the visible band ranges from 380 to 780 nanometers; one nanometer is one billionth (10^{-9}) of a meter.

Photoreceptors in the eyes convert radiation in the visible band into neural signals that reach the brain. Photoreceptors are located throughout the retina, a sensory membrane that covers the entire back of the

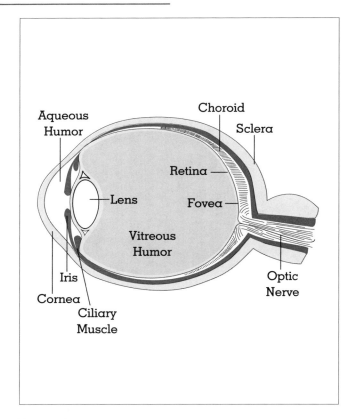

Figure 1.
Cross section of the human eye.

human eye, as shown in Figure 1. One eye contains approximately 130 million photoreceptors. We have, however, two distinct classes of photoreceptors, rods and cones. Rods are more sensitive to light and are used primarily for nighttime vision, whereas cones, of which there are three types, are used for daytime vision. These two classes of photoreceptors enable human vision to comfortably span light levels from a sunny, snow-capped mountain to faint starlight, a billion-to-one range. The three cone types provide us with the ability to convert light into color. Indeed, color is not an inherent property of light. Rather, the human brain "calculates" color from the neural signals generated by the three cone types.

MEASUREMENT OF LIGHT AND COLOR

Light is measured in several ways. Until about 1960, most light measurements were obtained by visual comparison. A standard light source of known, but adjustable, brightness was compared by a trained technician to another light source of unknown, but fixed, brightness. When the two lamps were seen to be equally bright, they were said to produce the same amount of light. Naturally, this technique was fraught with problems of inconsistency and inconvenience. Today photosensitive electronic detectors are used to measure light accurately and reliably. They convert radiant energy into a measurable electrical signal that can be used to quantify the amount of light generated by a source.

The measurement of light is known as photometry. (Color measurement is known as colorimetry; see below.) Photometry can be performed in several different ways, depending on the geometric relationship between a light source and a detector. Most light measurements are based on the flux (photon) density on a detector. This quantity is known as illuminance and represents the rate of photon absorption by a detector of known area. From an illuminance measurement other photometric quantities can be derived. Flux, measured in lumens, is simply the total amount of light generated by a source. Intensity, measured in candelas (cd), is the amount of light projected in a given direction and within a two-dimensional (solid) angle. Luminance, measured in candelas per square meter, is comparable to the human perception of brightness because it accounts for both the amount of light reaching a surface and the amount of light reflected from that surface back to the eye. For many years luminance was known as photometric brightness, but this term is no longer used.

Photometric measurements do not weight all wavelengths in the visible band equally. Rather, a specific weighting function for the electromagnetic spectrum is employed to define "light" (see Figure 2). This function, known as the photopic luminous efficiency function, is not based on the responses of all of the photoreceptors in the eyes. Rather, it is based on the spectral sensitivity of only two cone types found only in the small region of the retina known as the fovea. Responses by the photoreceptors in the fovea enable us to read and see fine detail, but they represent only about 4 percent of all the photoreceptors in the retina. The photopic luminous efficiency function was established by international agreement in 1924 and has been used as the standard weighting function for light ever since.

Light also can be defined in other ways, even though these definitions are rarely used. For exam-

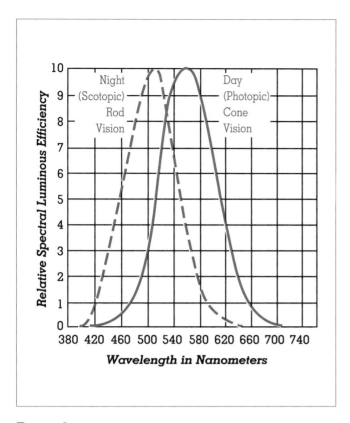

Figure 2.
Photopic (right) and scotopic (left) luminous efficiency functions.

ple, the scotopic luminous efficiency function was established in 1951 by international agreement to represent the spectral sensitivity of rods (see Figure 2). This function is rarely if ever used in photometry because the presence of almost any light source, including moonlight, will raise light levels to a point where some cones also function for vision.

Light sources vary in their ability to produce light—that is, to produce radiation within the photopic luminous efficiency function. In general the efficiency of extracting energy from fossil fuels or other energy sources and converting it into light is very low. For example, the efficiency of light extraction by an open gas flame is only about 0.04 percent. Not until very recently, with the advent of electric light sources, was it possible to substantially increase extraction efficiency. The conversion efficiency of the most efficacious electric light source presently manufactured (low-pressure sodium) is, however, only about 30 percent.

All electric light source efficacies are measured in terms of their ability to generate light per unit electric power, measured in lumens per watt (lm/W). However, efficacy is only one consideration in selecting a light source. For example, the most efficacious light source, low-pressure sodium, essentially produces a single wavelength of light (589 nanometers), making every object color appear as different shades of yellow. For some human activities the absence of color information is not important, but for others it is not only unpleasant but also dangerous (e.g., in surgery). It is beyond the scope of this article to compare light source cost, maintenance, safety, flexibility, and operating conditions, but all of these factors, not just efficacy, are extremely important in selecting the right light source for a particular application.

However, two points about energy efficiency need to be stressed. First, power and energy are two different quantities, and second, high light source efficacy (lm/W) does not always indicate good energy efficiency.

Power is the rate at which a source generates energy, measured in ergs per second. Obviously, then, because energy is the product of power and time, energy conservation techniques can either reduce the power required to generate light or the total time that the power is being supplied to the light source. Many electric light sources can be used with associated electrical dimming circuits to reduce light output by reducing the power supplied to that source. Photosensors, for example, can be used to dim electric lighting levels when daylight enters a room through a window or a skylight. Time clocks, occupant sensors, and even manually operated switches are effective techniques for turning lights off when a room or a building is unoccupied. Both strategies reduce the energy used for lighting; dimming matches the light source intensity with the needs of the people, and switching extinguishes light when no one needs it. Dimming and switching should be used to reduce only wasted lighting energy. It should be reemphasized that light is necessary for human activity. Our societal goal should not be simply to reduce energy consumption but rather to reduce energy wasted by too much light and by lighting that meets no human need.

Light sources are only rarely used without a fixture to house the light source. A fixture makes handling and operating the light source safer and helps provide

Common Lamp Types	Typical CCT	Typical CRI
Incandescent		
household A-lamp	2800	100
tungsten-halogen	3190	100
photoflood	3400	100
Fluorescent		
warm-white	3000	51
cool-white	4150	64
delux cool-white	4160	89
lite-white	4200	40+
daylight	6380	76
RE70 (rare earth phosphor, 70+ CRI)	2700, 3500, 4100, or 5000	70+
RE80 (rare earth phosphor, 80+ CRI)	2700, 3500, 4100, or 5000	80+
C75	7500	90+
full-spectrum	5500	90+
Low pressure		
sodium	1740	-44
HID		
metal halide	4220	67
clear mercury	6410	18
high pressure sodium	2100	24
xenon	5920	94

Table 1.
Typical values of correlated color temperature (CCT) and color rendering index (CRI) for some common electric light sources.

light where it is needed by controlling the direction of light emitted by the source. As already noted, electric lamp efficacy is measured in lumens per watt. However, the total flux generated by a source is measured without regard to direction. Because light should be directed toward an object or an area to achieve a purpose, flux emitted from a fixture is not necessarily the most useful photometric quantity to assess efficacy in a given application. An automobile headlamp, for example, should direct as much of the light generated by the lamp as possible onto the road in front of the automobile. The light directed into the night sky cannot be used, and thereby this lost light reduces the fixture (headlamp) efficacy. Therefore, efficacy measured in terms of lumens per watt does not provide a true measure of fixture efficacy because all the flux from the fixture are not always useful for a given application. The photometric quantity, intensity, is the amount of light flowing within an imaginary cone connecting the illuminated object and the light source. Intensity, measured in candelas, is a measure of how much light generated by the light source actually arrives where it is needed. The amount of light arriving at a specific location is critical for headlamp design, for example. In general, intensity per watt is a valuable measure of efficacy for any lighting system because, ultimately, light must arrive at a specified location to be useful. Other light is wasted, and this wasted light should be reflected in a measure of low fixture efficacy. Different fixtures are produced by manufacturers for different applications using a variety of light sources. The most efficacious light fixture for a given application should be used, but the most efficacious light fixture (cd_{max}/W)

does not always employ the most efficacious light source (lm/W).

The term "color" can be used in two ways to describe the light generated by a source or reflected from an object. As already noted, color appearance is a perceptual phenomenon that, despite much scientific investigation, defies precise quantification. For example, the same physical light may look brown or orange depending on its brightness. Further, this same light may look more or less red, yellow, or brown depending on the apparent color adjacent to it. At present we have only a qualitative understanding of these color appearance phenomena. Color can, however, be quantified very precisely in terms of color matching. This system of quantifying color is known as colorimetry. Any light can be matched in appearance with the right combination of idealized, but quantifiable, red, green, and blue lights. These three lights are known as primary colors or, simply, primaries. Color televisions are practical examples of color matching. They can produce a wide gamut of colors through adjustments of three color pixels: red, green, and blue. Thus, although we can predict precisely what colors will match in terms of these three primaries, we have no precise way to predict whether the pixels will be seen as, for example, brown or orange. Through manual adjustments of the color television pixels, however, each person can reach an acceptable appearance of a televised image.

The color of light generated by a source can be precisely described by colorimetry. From a colorimetric description of the light, other color measures can be derived. Two derived measures—correlated color temperature (CCT) and color rendering index (CRI)—are commonly used to characterize light generated by manufactured sources. It is very important to emphasize that both CCT and CRI are derived from the science of color matching but are used, incorrectly, to describe different aspects of color appearance. Despite the technical error, both CCT and CRI have been found to be practically useful by the lighting industry for color appearance.

CCT refers to the appearance of the light generated by a very hot (i.e., incandescent) object, the temperature of which is measured in kelvins (K). As a body is heated, it begins to produce a reddish-yellow, and then a yellow-white light. As temperature increases, the apparent color of the light changes to blue-white. In astronomy, for example, older, cooler stars appear yellow or red, whereas younger, hotter stars look blue. Paradoxically, electric light sources with CCTs between 2,700 K and 3,200 K generate a yellowish-white light and are termed "warm"; those with CCTs between 4,000 K and 7,500 K produce a bluish-white light and are termed "cool." This paradox seems to have originated from the association between yellow light and a hot fire. The association between apparent color and tactile temperature seems to have been reinforced by the cold feel of glass admitting blue light from the sky on a clear day.

CRI is a measure of how "true" or "natural" colors will appear when illuminated by a light source. Light sources that generate light evenly throughout the visible spectrum, such as daylight, have high values of CRI (maximum CRI = 100); those that have gaps in the visible spectrum (e.g., clear mercury) have low CRI. Electric light sources, particularly fluorescent lamps, have undergone a great deal of development in recent years and now have much higher CRI values than were available in 1970. Although real improvements have been made to the color-rendering properties of these lamps, to some extent these lamp developments are only a game of numbers. Given the technical flaws inherent in CRI for describing color appearance, this measure should not be expected to precisely characterize the color-rendering properties of these lamps. For example, a ten-point difference in CRI values is probably unimportant.

TYPES OF LIGHTING

Much of the history of electric light sources (see Figure 3), fixtures (see Figure 4), and control technologies (dimming and switching) centers around improving energy efficiency. It is often assumed that incandescent lamps were the first electric light source. Actually, carbon arc lamps were the first practical electric light source, preceding the incandescent lamp by almost half a century. Carbon arc lamps employ two carbon electrodes separated by a gap. When current is supplied to the electrodes, the lamp produces a very bright, blue-white arc. These lamps literally burn the electrodes while the lamp is operating, so a clock device is required to continuously feed carbon into the arc to keep the gap width constant.

Carbon arc lamps were first developed in the 1840s, and sometimes elaborate towers were created to provide illumination to streets in a few European

713

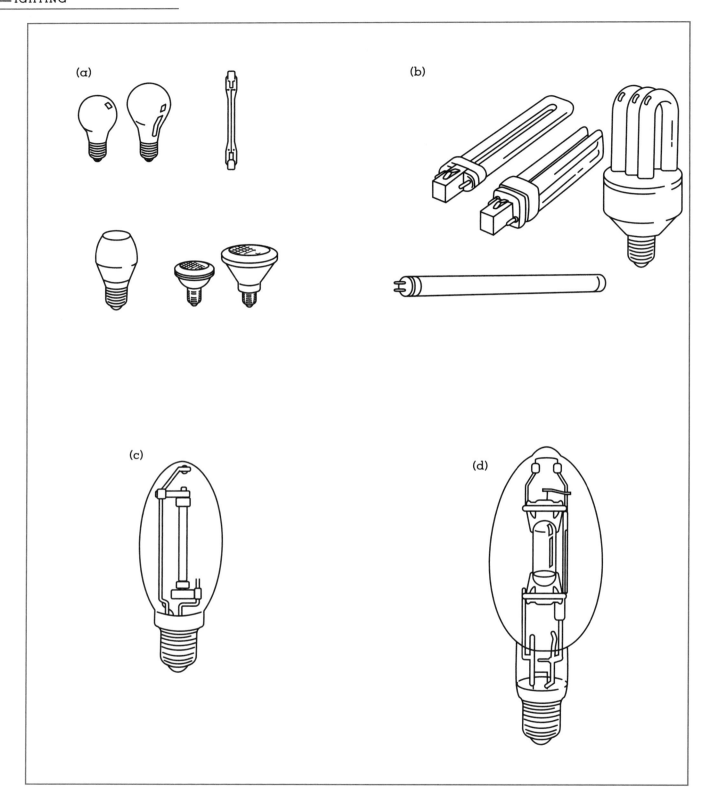

Figure 3.
Lamp types: (a) incandescent and tungsten-halogen lamp shapes; (b) fluorescent and compact fluorescent lamp shapes; (c) typical high-pressure sodium lamp; (d) typical metal halide lamp.

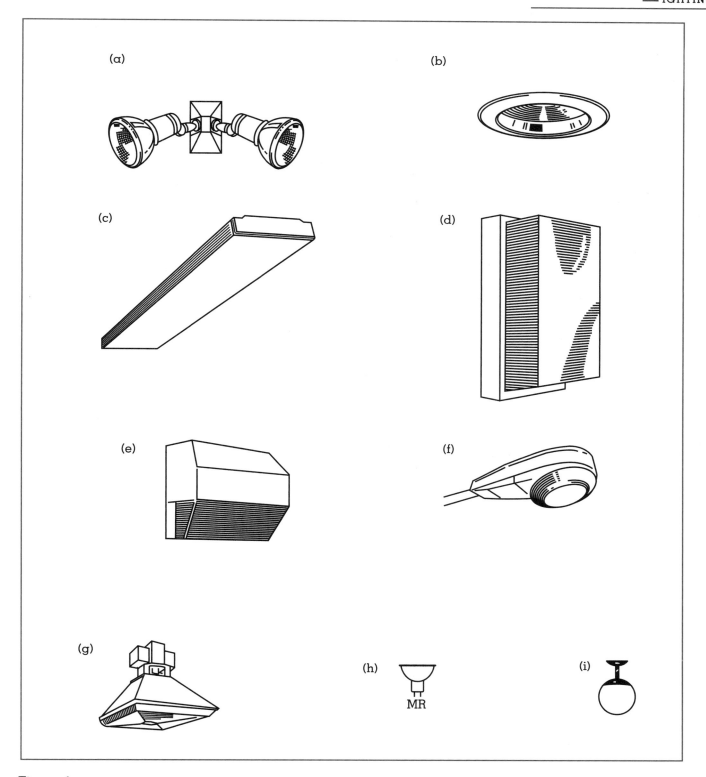

Figure 4.
Luminaire types: (a) parabolic aluminized reflector (PAR) lamp holder; (b) incandescent, compact fluorescent or high-intensity discharge lamp downlight; (c) lensed fluorescent lamp luminaire; (d) wall-mounted compact fluorescent lamp luminaire; (e) wall-mounted high-intensity discharge lamp luminaire; (f) roadway luminaire; (g) low bay industrial luminaire; (h) multi-faceted reflector (MR) lamp; (i) pendant sphere.

IGHTING

cities. Carbon arc lamps are still used today for searchlights because they produce the very bright, concentrated point of light needed for a high-intensity beam. These lamps are impractical for other purposes, however, because it is difficult to segregate and distribute this concentrated light into many small packages useful for human activities indoors. Also, cost and the pollution generated by these lamps make them unacceptable by modern standards.

Incandescence means to heat an object to the point of producing light. The incandescent lamp, then, is a lamp with a filament heated to the point of glowing. The trick accomplished by Thomas Edison and his team at Menlo Park, New Jersey, in 1879 was to produce an inexpensive lamp with a carbon filament that would glow for several hundred hours. After more than a year of experimenting, Edison was able to demonstrate an incandescent lamp producing approximately 2 lumens per watt and lasting several hundred hours. Incremental improvements in that basic design, especially the use of the metal tungsten in filaments, have increased the efficacy of commercially available incandescent lamps to between 10 and 15 lumens per watt. More significantly, perhaps, these lamps are easy to install and cost as little as fifty cents each.

It has been argued that Edison's greatest vision was not the invention of the incandescent lamp but rather his insight two years earlier into the possibilities of providing small "packages" of light where people needed them. A high-voltage power supply with a high resistance light source was required to meet his vision; the incandescent lamp was, therefore, a logical outcome of that insight. The incandescent lamp and the associated electrical distribution system changed human history forever. An inexpensive, manageable light source not only illuminated the night but also made it possible to build the large windowless buildings so prevalent today.

One of the major innovations in incandescent lamps has been the tungsten-halogen lamp, developed in 1959. As the tungsten filament of an incandescent lamp glows, some of it evaporates and is deposited on the inside surface of the bulb. This evaporation not only shortens lamp life, but also the deposited tungsten reduces the light output from the lamp. Iodine, bromine, and chlorine are within the family of elements called halogens. These molecules will combine with the evaporating tungsten and, rather than be deposited on the bulb wall, will

deposit the combined molecules back onto the filament. At this point the halogen molecule disassociates from the tungsten molecule, and the cycle begins again. The halogen cycle prolongs incandescent lamp life and keeps the lamp burning brightly. Special bulb shapes and materials are needed for tungsten-halogen lamps because very high temperatures are required for their operation. The high temperatures increase efficacy so that most tungsten-halogen lamps available to the market produce approximately 15 to 30 lumens per watt.

Probably every reader has heard of the mythical incandescent lightbulb that has burned for more than fifty years. Because we are all susceptible to believing conspiracy theories, we are suspicious that manufacturers could make all lightbulbs last that long, if only they would. Most inexpensive incandescent lamps available in stores are rated for 750 hours of operation. In reality, an incandescent lamp can be made to last a very long time, certainly longer than 750 hours, but there is no "free lunch." Incandescent lamp life can be prolonged considerably if the lamp is operated at low temperatures. Lamps operated at low temperatures, however, produce relatively more heat than light, so the efficacy drops below the rated 10 to 12 lumens per watt. Lamp life could also be improved if the lamp were more expensive. Incandescent lamps used in traffic signals are rated at about 8,000 hours of operation but are approximately five to ten times as expensive as the conventional household incandescent lamp, because of their more durable filament construction. An incandescent lamp can have long life, high efficacy, or low cost, but not all three at the same time. So although the mythical lamp may exist, and indeed it can be made, physics demands that it must either produce light expensively or inefficiently, and probably both.

Fluorescent lamps were the next major innovation in lamp technology. Introduced commercially in 1938, they were a radical improvement in lighting energy efficiency. At nearly 50 lumens per watt, the first fluorescent lighting systems immediately replaced those using incandescent lamps in large industrial and commercial applications. Even today, fluorescent lamp technologies are arguably the most cost-effective, energy-efficient, and reliable source of illumination for interior applications.

Fluorescent lamps are termed a low-pressure discharge lamp. An electric current passes through mercury contained at normal vapor pressure within the bulb. The current vaporizes the mercury and liberates electrons from the molecules. The liberation of electrons from mercury molecules produces radiation, much of which cannot be seen by humans. Phosphors that coat the inside of the bulb absorb the nonvisible radiation. The irradiated phosphor molecules are themselves excited to liberate electrons that *do* emit radiation within the visible band of the electromagnetic spectrum. This multistage process sounds inefficient, but fluorescent lamps can be made to operate at nearly 100 lumens per watt, with lamp life greater than 20,000 hours. New lamp designs as well as improvements to the electrical device needed to start and operate the fluorescent lamp, known as a ballast, have improved system efficacy by more than 40 percent since 1980. Not only has efficacy been improved, but also the color characteristics are better, and the audible noise generated by the ballast has been reduced. Moreover, smaller, compact fluorescent lamps have been introduced and are beginning to replace some of the less efficacious incandescent lamps used most commonly in homes.

Many consumers continue to complain about fluorescent lamps, however, arguing that they "buzz," distort the color of natural objects, and cause headaches. These attitudes are barriers to societal goals for energy efficiency, particularly in the home. As stated above, however, most of the technical issues leading to these complaints have been resolved. Nevertheless, these negative attitudes toward fluorescent lamps persist and are, in fact, reinforced by exposure to the older technologies still in operation and by low-cost, inefficient products being introduced by manufacturers from developing countries. A major barrier to widespread introduction of fluorescent lamps is initial price. Fluorescent lighting systems are much more expensive to purchase than incandescent systems. In the long run, however, the energy savings associated with the fluorescent lighting system would more than pay for its higher initial cost.

The flip side of consumer bias is that some people argue that only fluorescent lamps should be used in buildings. This attitude, while well intentioned, is based on an unsophisticated knowledge of the performance of lighting systems. As previously discussed, the efficacy of a lighting system cannot be characterized by lamp efficacy alone. For example, although a fluorescent lamp and ballast may produce

80 lumens per watt compared to a tungsten-halogen lamp at 20 lumens per watt, the system efficacy of a recessed open downlight ceiling fixture may, in fact, be better with a tungsten-halogen source than with a fluorescent source. This perhaps surprising result is due to the fact that the tungsten-halogen filament is very compact. The light from the small tungsten-halogen filament can be optically controlled much easier than it can from the relatively large fluorescent lamp. Where little or no optical control is necessary, as with general illumination from a standing floor-lamp fixture, fluorescent lighting systems are much more efficacious than tungsten-halogen lighting systems in producing the same visual effect. Again, the application is important for selecting the most efficacious lighting system.

Low-pressure sodium lamps, another low-pressure discharge lamp, produce very bright, monochromatic light at relatively high wattage. These lamps are used exclusively for outdoor applications where a high-light-output lamp can distribute light over a relatively large area. Low pressure sodium lamps have the highest efficacy (180 lumens per watt), but as discussed above, provide people with no color perception. Despite its high efficacy, this monochromatic source has made little inroads into the outdoor lighting market, except in the United Kingdom, which consumes approximately half of all low-pressure sodium lamps manufactured.

High pressure can be used to expand the spectral emission of the sodium gas. Although expanding the sodium spectrum reduces the lamp efficacy to between 60 and 120 lumens per watt, it significantly improves the color characteristics of the lamp. High-pressure sodium lamps were introduced in 1965, and the yellow-orange light produced by these lamps is now found throughout modern societies in roadway, security, and parking lot applications. Newer, color-corrected high-pressure sodium lamps can be found in some indoor applications but have still lower efficacies.

High pressure can also be used to expand the spectral emission of mercury gas. High-pressure, clear mercury lamps are no longer widely used because, even under high pressure, they are relatively inefficient and do not provide good color perception. By mixing halides with the mercury under high pressure, however, good color can be produced at relatively high efficacies ranging from 60 to 110 lumens per watt. These metal halide lamps have been gaining steadily in popularity since they were introduced in 1964, particularly as an outdoor light source associated with retail spaces (e.g., shopping centers, gas stations, and facade lighting). These lamps also are being used in many indoor applications, such as warehouses and shopping malls. They are even being used in modern automobile headlamps as a replacement for tungsten-halogen headlamps.

Perhaps the most radical new development in light sources has come from the electronics industry rather than the traditional lighting industry, established in the nineteenth century. Light-emitting diodes (LEDs) were invented in the 1960s, but only since 1997 have these sources become an important light source for widespread applications. Before then, LEDs were used as indicator lights on electronic equipment and were largely restricted to a few colors, the first being the red LED. Today many colors, including white, can be produced economically with efficacies ranging from 5 to 25 lumens per watt. It has been projected that efficacies of some colored LEDs may approach 100 lumens per watt in the near future.

LEDs are a highly directional light source, ideally suited to applications such as automobile taillights and traffic signals. In architectural applications, LEDs have already proven successful in exit sign applications. The high visibility necessary for transportation applications is also a positive attribute in emergency egress fixtures. In addition, the low power requirements of LEDs in exit signs (fewer than 5 or 6 watts per face) compared to other technologies provide substantial energy savings when multiplied by hours of operation typical of exit signs (usually twenty-four hours per day every day). The long life of LEDs also provides the reliability needed for transportation and egress applications.

In the future, LED fixtures will be produced for architectural applications by clustering the individual LEDs. Coupled with electronic controls, the LED fixtures will enable the color and intensity distribution of light to be customized and easily changed. These systems will provide cost-effective and energy-efficient lighting solutions to future lighting artists, designers, and engineers. Indeed, this technology may change architecture in ways not seen since the time of Edison.

The efficient application of lighting technologies to a residential, commercial, or industrial space is much more than simply picking the lamp with the highest lumens per watt. As described above, it is important to have light where you need it and when

you need it. Color, cost, ease of maintenance, heat, and durability—as well as how people will use, operate, and maintain the space—are among the other factors to consider in selecting the right lighting equipment for an application.

FUTURE TRENDS

The world population continues to expand. Fuel reserves continue to be depleted. Pollution associated with power generation continues to increase. Technical advances in light sources, fixtures, and controls as well as those in power generation provide a modestly optimistic picture of the future. In 1850 the extraction efficiency of light from carbon-based fuels such as open gas flames was about 0.04 percent. Today we have increased that efficiency more than a hundredfold. Estimates from the U.S. Department of Energy in 1999 suggest that efficiency improvements in lighting technologies will reduce the required energy for lighting in commercial buildings, although the amount of commercial floor space will continue to grow in the United States. Even if this projection is correct for the United States, the need for light by people around the world will outpace the increased extraction efficiency offered by new lighting technologies in the decade 2000 to 2010.

Our societal ambition to reduce energy consumption is, without any doubt, both correct and urgent. It must always be remembered, however, that humans always will need light. Our goal, then, should be to reduce wasted lighting energy, energy expended on lighting that meets no real purpose. The technological advances in light source efficacy must also be coupled with technological advances in controls—both optical control to deliver light where it is needed, and power control to deliver the right amount of light when it is needed. Lighting control systems are slowly becoming more sophisticated, utilizing both automatic and manual controls, to tune lights to occupant needs. Estimates made in the late 1990s suggest that lighting energy used in offices can be reduced by as much as 80 percent using controls relative to static lighting systems (Maniccia 1999). We must also strive to better understand exactly what humans need. Recommended illumination levels in North America, for example, have been reduced by roughly two-thirds since the oil embargo of 1972, with no noticeable loss in human productivity or satisfaction.

Arguably, a reason why lighting energy is still

being wasted is ignorance of policymakers, building owners, and developers. Well-intentioned policymakers legislate power but not energy, failing to consider the importance of lighting controls in meeting our societal goals. They also regulate lamp efficacy rather than lighting system efficacy, implicitly failing to recognize that light should be directed to a location where it can be used. Another reason lighting energy is being wasted is the emphasis on the purchase price of lighting equipment. Often the cheapest lighting products are the most energy-wasteful. A lighting system will be operated for many years, and the cost of energy, even at current low prices, far exceeds the initial cost of even the most expensive lighting equipment.

Mark S. Rea

See also: Conservation of Energy; Consumption, Culture and Energy Usage; Economically Efficient Energy; Edison, Thomas Alva; Elctricity; Electricity, History of; Electric Power, Generation of; Power.

BIBLIOGRAPHY

Bierman, A. (1999). "LEDs: From Indicators to Illumination." *Lighting Futures* 3(4):1–5.

Bowers, B. (1998). *Lengthening the Day*. New York: Oxford University Press.

Bright, A. A. (1949). *The Electric-Lamp Industry*. New York: Macmillan.

Cox, J. A. (1979). *A Century of Light*. New York: Benjamin.

Drucker, H. (1997). "1972–1997: Twenty-five Years of Energy and Environmental History: Lessons Learned." In *Twenty-Five Years of Energy and Environmental Policy: Proceedings of the 25th Annual Illinois Energy Conference*. Chicago: University of Illinois at Chicago, Energy Resources Center.

Howell, J. W., and Schroeder, H. (1927). *History of the Incandescent Lamp*. Schenectady, NY: Maqua.

Illuminating Engineering Society. (1947). *IES Lighting Handbook*, 1st ed. New York: Author.

Illuminating Engineering Society. (1952). *IES Lighting Handbook*, 2nd ed. New York: Author.

Illuminating Engineering Society. (1993). *IES Lighting Handbook*, 8th ed. New York: Author.

International Lighting Review. (1979). "1879–1979 [Edison Lamp Centenary Issue]" 30(1):1–13.

Leslie, R. P., and Conway, K. M. (1996). *The Lighting Pattern Book for Homes*, 2nd ed. New York: McGraw-Hill.

Leslie, R. P., and Rogers, P. (1996). *Outdoor Lighting Pattern Book*. New York: McGraw-Hill.

Lighting Research Center. (1994). *A&P Food Market, Old Lyme, Connecticut*, DELTA Portfolio 1(1). Troy, NY: Rensselaer Polytechnic Institute.

Lighting Research Center. (1994). *Linens 'n Things, Patchogue, New York*, DELTA Portfolio 1(2). Troy, NY: Rensselaer Polytechnic Institute.

Lighting Research Center. (1995). *450 South Salina Street, Syracuse, New York*, DELTA Portfolio. 1(3). Troy, NY: Rensselaer Polytechnic Institute.

Lighting Research Center. (1996). *DeGraff Street Industrial Center, Amsterdam, New York*, DELTA Portfolio 1(4). Troy, NY: Rensselaer Polytechnic Institute.

Lighting Research Center. (1996). *Prudential HealthCare, Albany, New York*, DELTA Portfolio 1(5). Troy, NY: Rensselaer Polytechnic Institute.

Lighting Research Center. (1997). *Sacramento Municipal Utility District Customer Service Center, Sacramento, California*. DELTA Portfolio, 2(2). Troy, NY: Rensselaer Polytechnic Institute.

Lighting Research Center. (1997). *Sony Disc Manufacturing, Springfield, Oregon*. DELTA Portfolio 1(6), 2(1). Troy, NY: Rensselaer Polytechnic Institute.

Lighting Research Center. (1998). *Mary McLeod Bethune Elementary School, Rochester City School District, Rochester, New York*. DELTA Portfolio, 2(3). Troy, NY: Rensselaer Polytechnic Institute.

Lighting Research Center. (1999). *South Mall Towers Apartments, Albany, New York*. DELTA Portfolio, 2(4). Troy, NY: Rensselaer Polytechnic Institute.

Lighting Research Center. (1999). *Staples Distribution Center, Killingly, CT*, DELTA Portfolio 2(5). Troy, NY: Rensselaer Polytechnic Institute.

Lighting Research Center. <www.lrc.rpi.edu>.

Lightolier. (1994). *Journey: Lightolier and Lighting in the Twentieth Century*. New York: Lightolier.

Luckiesh, M. (1920). *Artificial Light*. New York: Century.

Maniccia, D.; Ratledge, B.; Rea, M.; and Morrow, W. (1999). "Occupant Use of Manual Lighting Controls in Private Offices." *Journal of the Illuminating Engineering Society* 28(2):42–56.

O'Dea, W. T. (1958). *A Short History of Lighting*. London: HMSO, Ministry of Education, Science Museum.

Rea, M. S., ed. (1991). *Selected Papers on Architectural Lighting*. Bellingham, WA: SPIE—The International Society for Optical Engineering.

LIQUIFIED PETROLEUM GAS

Propane is the most widely used commercial term to describe a family of liquefied petroleum gases (LP-gas or LPG) that also includes ethane and butane.

Name and Formula	Ethane C_2H_6	Propane C_3H_8	N-butane C_4H_{10}
Vapor pressure @ 100 d.F			
lbs/sq. in. absolute	780.0	190.0	51.6
Boiling point of liquid at			
atmospheric pressure-F	-127.5	-43.7	31.1
Weight of liquid @ 60 d.F			
Pounds per gallon	3.11	4.23	4.86
Specific gravity	0.374	0.508	0.584
Gross heat of combustion			
Btu per pound	22,329	21,670	21,315
Btu per cu. ft. @ 60 d.F	1,783	2,558	3,368
Btu per gallon @ 60 d.F	69,433	91,044	103,047
Flammability limits			
Lower % in air	3.0	2.2	1.9
Upper % in air	12.5	9.5	8.5
Freezing point of liquid at			
atmospheric pressure F	-297.8	-305.9	-216.9
Gallons per pound mol @ 60 d.F	9.64	10.41	11.94
Molecular weight	30.07	44.09	58.12

Table 1.
Physical Constants of Selected Hydrocarbons Found in LP Gas.
SOURCE: Handbook/Butane-Propane Gases

Propane is a nontoxic, colorless, odorless hydrocarbon that occurs naturally in natural gas streams and crude oil. At normal atmospheric pressure and temperature, it is a gas; under moderate pressure propane becomes liquid. The ratio of liquid to gas is 270—one unit of liquid expands to 270 units of vapor.

HISTORY

The process of separating "liquefied gases" from natural gasoline was developed in 1912 by Walter O. Snelling, a chemist for the U.S. Federal Bureau of Mines. He wanted to develop better reading light for rural people than the then commonly used candles and kerosene. The first houses were piped for use of this "gasol" in Pennsylvania in 1912. Later, propane was used for cooking and refrigeration. Space heating use soon followed.

Propane became popular as a cutting fuel replacing more expensive acetylene. Another of its first uses was motor fuel. City buses began using it in the 1930s, new farm uses were added in the 1940s including flame weeding, crop drying, hog farrowing, and chicken brooding, and even a locomotive was powered with propane. Recreational uses for barbecues and campers became popular. In the 1950s LP-gas farm tractors and trucks began production. Chicago Transit began converting their city buses to propane.

In the 1960s, use extended to hot air balloons and industrial burners. Residential and commercial use expanded because of new construction without access to natural gas pipelines and the growing popularity of camping. In the 1970s, propane became a popular supplement for natural gas utility systems that were experiencing shortages of natural gas. Refrigerated import terminals were built. When government price controls were imposed in response to the oil embargo, massive vehicle conversions took place to take advantage of the lower-priced and price-controlled propane. Gasoline prices were approaching 100 percent higher and were expected to keep rising.

PHYSICAL PROPERTIES

Propane (C_3H_8) is one of the saturated open-chain hydrocarbons that form the paraffin or alkane series

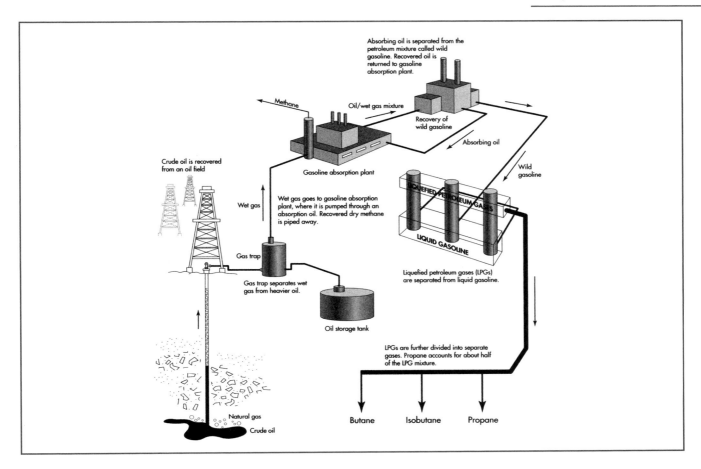

The process of making propane. (Gale Group)

(C_nH_{2n+2}). The parent compound of this family is methane. The propane molecule, with three carbon and eight hydrogen atoms, is third in the series after ethane (C_2H_6). The specific gravity is 0.508–0.510 at 60°F (15.6°C). The melting point is –309.8°F (–189.9°C). (See Table 1 for other selected properties of LP-gases.)

When liquefied, propane has a Btu content of 91,044 per gallon (higher heating value). At ordinary temperatures, propane is relatively unreactive with other chemicals such as acids, alkalis, or oxidizers.

COMMERCIAL PROPERTIES

Upon release from a pressurized container, propane vaporizes immediately. This property makes liquid propane economical to store and to transport by pipeline, rail, barge, or truck. Because they are relatively light, propane tanks and canisters can be carried by hikers or wheeled to outdoor sites for barbecuing, lighting, and campfires. Propane has been nicknamed "the portable gas."

Although propane is nontoxic, exposure to liquefied propane can cause skin burns. In gas form, propane is a simple asphyxiant that can displace oxygen. The gas is highly flammable but has a relatively narrow flammability range of 2.2 to 9.5 percent concentration in air. Numerous safety features are incorporated into propane equipment. Storage tanks are constructed to withstand pressures a minimum of four times actual normal operating pressures. Relief valves are integrated into the tanks to relieve extreme pressures that might occur when the tank is subjected directly to fire. To aid in detection of a leak an odorant such as mercaptan is added to the fuel. Although heavier than air, a propane leak readily dissipates into the atmosphere as it reverts from the liquid phase inside the tank to its normal vapor state.

Propane is desirable as a fuel because of its clean-burning properties, its portability, closed-to-atmosphere storage, and availability even in remote locations. It is an approved alternative fuel listed in the 1990 Clean Air Act Amendments and the National Energy Policy Act of 1992. Current factory-produced propane vehicles meet the U.S. Environmental Protection Agency (EPA) ultralow emission vehicle (ULEV) standards and prototype vehicles meet the even more stringent super ultralow emission (SULEV) standards. Because propane does not contaminate soil or water supplies it is exempt from EPA's leaking underground storage regulations.

Compared with other alternative motor fuel options (reformulated gasoline, compressed or liquefied natural gas, ethanol from corn or coal, methanol and electricity), propane has the lowest greenhouse gas emissions except for natural gas. According to a 1998 study by the Institute of Transportation Studies, greenhouse emissions from propane vehicles are 21.8 percent less than from gasoline or diesel.

PRODUCTION AND DELIVERY

There are two primary sources of commercial propane: natural gas processing (55 percent) and crude oil refining (45 percent). A typical natural gas raw stream is comprised of about 90 percent methane and 3 percent propane, and a remainder of other gases such as ethane, butane and pentane. Ninety percent of the propane used in the United States is produced domestically. Ten percent is imported with 50 percent of imports originating in Canada and Mexico. Domestic production comes from several hundred refineries and gas processing plants. From the production source, propane is transported to bulk storage plants in pipelines, railroad tank cars, trucks and barges. Combinations of several types of transportation may be used depending on distance and economics of each transportation type. From the bulk plant, deliveries are made to end users in delivery trucks of 2,500-3,000 gallon capacity. Imports are by pipeline and refrigerated ships.

CONSUMPTION

Worldwide consumption of propane in 1998 was about 96 billion gallons, of which 15 billion gallons were used in the United States by some 60 million consumers. This represents about 4 percent of the total United States energy market. (Each year an additional 8-22 billion gallons of propane are used by the petrochemical industry where it is reformulated into basic building blocks for the production of a variety of polymeric products and chemicals.)

Homes and commercial establishments are the largest users of propane as a fuel where it is used for space heating, water heating, cooking and clothes drying. Fuel storage tanks can be located underground or above ground. Residential, commercial and industrial appliances and equipment can generally operate on either propane or natural gas with only slight modifications to accommodate the different heating values and fuel/air ratios of the fuels.

Industries use propane for space heating, concrete drying, steel cutting, process heat, and asphalt laying, and to power forklifts. Farms use it for crop drying, irrigation engines, animal barn heating and flame cultivation.

Roy W. Willis
Robert Myers

See also: Import/Export Market for Energy.

BIBLIOGRAPHY

Cannon, R. E. (1998). *The Gas Processing Industry: Origins and Evolution,* 2nd ed. Tulsa: Gas Processors Association.

Delucchi, M. A. (1998). *LPG for Motor Vehicles: A Total Fuel Cycle Analysis of Emissions of Urban Air Pollutants and Greenhouse Gases,* prepared for the Propane Vehicle Council and the Propane Education & Research Council, Washington, DC.

Hayduk, W., ed. (1996). *Propane, Butane and 2-Methylpropane.* New York: Pergamon Press.

National Propane Gas Association. *Propane Gas Facts.* <http://www.propanegas.com/consumer>.

Propane Research & Education Council. <http://www.propanecouncil.org>.

Propane Research & Education Council. *The NEED Project. What Is Propane?* <http://www.propanecouncil.org/need/secondary/htm>.

U.S. Geological Survey. (1995). *The Future of Energy Gases.* Reston, VA: Author.

World LP Gas Association. *LP Gas Uses.* <http://worldlp-gas.com/uses/d_uses.htm>.

LOAD MANAGEMENT

See: Electric Power, Generation of

LOCOMOTIVE TECHNOLOGY

A machine designed to convert the potential energy stored in fuel to tractive effort needed for pulling unpowered wheeled vehicles (usually in a train on railway tracks) is called a "locomotive." The complexity of this definition reflects the fact that "railroads" are a composite technology made up of tracks, locomotives, and trains. There were "trains" of wheeled vehicles pulled by animals or gravity on "tracks" laid to reduce rolling resistance before there were locomotives, but there were no "railroads" until all three components were brought together. The final link was the locomotive—one of the most useful inventions ever made. Within decades of its development locomotives had multiplied overland travel speeds and freight tonnages more than tenfold—in the process revolutionizing land transportation for all time.

HISTORY

Historians usually credit the invention of the locomotive (and hence the innovation of the railroad) to Richard Trevithick, a Cornish mining engineer. In 1804 he assembled a steam-powered locomotive employing used cylinder steam to increase stack draft for a hotter boiler fire, with the resultant puffing smoke and sound characteristic of steam locomotives. Trevithick then used his effective little engine to pull some ten tons of iron and about seventy persons in railed wagons for a distance of nine miles on the Penydarren Iron Works tramroad in South Wales. The Frenchman Nicholas Cugnot had devised a steam-powered, self-propelled street wagon as early as 1769, but before Trevithick's demonstrations there was much disbelief that a locomotive could pull heavy loads without its wheels spinning uselessly on the iron rails.

The word "locomotive" means, literally, "power for moving from one place to another." Early locomotives were given the nickname "iron horse" because of their material and function. Most European languages use similar nomenclature for "railroads," as in the French *chemin de fer*, Italian *ferrovie*, Spanish *ferrocarril* and German *eisenbahn*, all of which mean "iron way" in literal English.

Work is movement of mass through distance, and *power* is work accomplished or energy transferred at a given rate or in a given time. Horsepower is a measure of power—first used, so it is said, by one of the pioneers of steam engine development, James Watt. One horsepower is, approximately, the work a good draft horse could accomplish in a minute—now given as 33,000 foot pounds/minute or 746 watts. A locomotive's power is often rated by the horsepower generated by its prime mover (the engine for converting fuel to power and motion), and the pulling force it can give to the first car in its train (drawbar horsepower). The difference between engine horsepower and drawbar horsepower is power used or lost in the locomotive's transmission, wheel slippage, and auxiliary functions such as producing air pressure for braking or electric current for lighting and other onboard train devices, plus the power needed to move the locomotive itself. A locomotive's tractive effort (work performed by wheel torque at the rails) is estimated in pounds. Actual tractive effort can be directly measured with the use of a wheel dynamometer. What is most important to a railroad, however, is drawbar pull, which equals tractive effort minus the force necessary to move the locomotive.

POWER GENERATION

Locomotives are often classified into three basic types derived from the kind of prime mover and the fuel or energy conversion system employed. These three main types are steam, diesel-electric, and electric (in which the prime mover is actually an off-board power generating station). By historical usage, steam locomotives are often called "engines" and electric locomotives or powered passenger cars are often referred to as "motors" or "traction." Locomotives with internal combustion or turbine prime movers are usually identified further by the type of transmission apparatus, thus "diesel-electric," "LNG diesel," "gas-electric," "turbine-electric," "diesel-hydraulic," or "diesel-mechanical," etc. Innovations in each generic type of locomotive gradually improved its energy efficiency, but inherent features of each type limited technology improvements and ultimately determined (along with other capital and operating cost factors) its success in the marketplace. *Dual mode* or *hybrid* locomotives refer to designs enabling operation with more than one prime mover, such as diesel-electric plus third-rail electrification (see Table 1).

	Steam	Electric	Diesel - DC	Diesel - AC
OUTPUT SIDE:				
Power Generation-Availability per Unit	High	Very High*	Limited	High
Starting Ability	Poor	Excellent	Excellent	Outstanding
Acceleration	Good	Outstanding	Good	Excellent
Operational Control	Poor	Excellent	Good	Excellent
INPUT SIDE:				
Capital Cost - Equipment	Low	Very High	Average	High
- Fuel (and water) Infrastructure	High	Very High	Average	Average
Operating Cost - Labor	High	Low	Average	Average
- Fuel Cost (North America)	High / Varies	High / Varies	Low / Varies	Low / Varies
Maintenance Cost - Equipment	Very High	Average/High	Average	Average/Low
- Shops and Line-side Infrastructure	Very High	High	Low	Low

Table 1.
Technical and Economic Characteristics of Locomotive Types.

Wood was the easiest fuel to use in early steam locomotives, but it was soon realized that the logistics of wood fuel were limiting. Steam engines were developed that could burn coal, peat, or (later) oil where those fuels were more abundant. For intercity railroads (especially in the Americas, Asia, Australia, and Africa), coal remained the fuel of choice for one hundred years. Despite impressive technology development, steam locomotives never could achieve thermal efficiencies greater than about 6 to 8 percent.

As electric and internal combustion motors were developed and improved in the second half of the nineteenth century, railroads began to look at alternatives to steam locomotion. In most urban transit applications, electric traction overtook horse-drawn and cable-drawn cars, as well as any remaining coal burners. Electrification offered smoke-free locomotive operation, the ability to tap hydroelectric energy supplies (important in Europe and Japan), and the benefit of drawing peak power levels when needed for climbing steep mountain grades or rapid acceleration of passenger trains. With regard to intercity rail applications as opposed to urban transit, however, more often than not initial construction costs for electric power distribution systems were prohibitive. The most important extension of intercity rail electrification in many decades has been completed between New Haven, Connecticut and Boston.

Dual mode locomotives offer the ability to operate beyond electrified territory without changing locomotives, but at a cost of deploying additional on-board equipment and incurring consequent suboptimal performance in both modes. Hybrid locomotives have been used for many years in the approaches to New York City commuter terminals.

In America and elsewhere, coal powered steam locomotives gave way to diesel electric traction over the period 1925 to 1960. This three-decade-plus process of "dieselization"—from first innovation to universal application—became a textbook case in the study of the diffusion of new industrial technology. Professor Edwin Mansfield has demonstrated that the economics of dieselization were akin to those of stimulus and response in psychology; railroads that could benefit the most from diesel power implemented the new technology most expeditiously. Consequently (and like jet engines later replacing piston aircraft), dieselization followed a typical logistics curve—slow initial acceptance in the face of skepticism and uncertainty, then rapid deployment as benefits were understood, and finally tapering of demand as opportunities became saturated.

A diesel-electric locomotive uses as its prime mover a large, self-igniting, internal combustion engine of the type invented by Rudolf Diesel and first successfully demonstrated in 1897. Thermal efficiency of these engines exceeded 30 percent, compared

Year	Person/Entity	Development	Year	Person/Entity	Development
1712	**Thomas Newcomen**	**First recorded Newcomen atmospheric (vacuum) steam engine.**	1836	W.G. Whistler	First American steam whistle, on locomotive "Susquehanna."
1776	James Watt	Separately condensing steam engine.	1837	Thomas Rogers	Cast iron wheels with hollow spokes and rims.
1802	Trevithick & Vivian	High-pressure steam engine with feed water heater.	1838	Rogers & Wakely	First successful oil-burning headlight, allowing nighttime operation.
1804	**Richard Trevithick**	**Tramroad stack-blast steam locomotive and cars (first railroad).**	1839	**R. & W. Hawthorn Co.**	**Superheater designed to make more efficient use of steam energy.**
1812	William Chapman	Pivoting locomotive bogie.	1842	**Egide Walschaerts**	**Radial valve gear, enabled better control of steam locomotive operation.**
1814 -15	Hedley, et al., Wylam Colliery RR	"Puffing Billy" with 4 and then 8 geared wheels.			
1814 -16	G. Stephenson, Dodd & Losh	Locomotives with connecting rods for transmission to wheels.	1849	Eugene Bourdon	First practical steam pressure gauge, enabling safe use of higher pressures.
1821	**Michael Faraday**	**Demonstration of electro-magnetic rotation (invention of the electric motor).**	1850	James Samuel	Compound cylinder locomotive (high pressure and low pressure steam cycles).
1825	**George Stephenson**	**First public railway, Stockton and Darlington; engine No. 1, "Locomotion."**	1857	George Griggs	Brick arch firebox for increased steam locomotive energy efficiency.
1827	William Chapman	Equalizing lever for axle bearings and locomotive weight distribution.	1862	Pacific Rail Act (USA)	Authorized transcontinental railroad and promoted standard national track gauge.
1827	Sequin (France); Booth (England)	Multi-tubular boiler with tubes surrounded by water.	1869	Union Pacific and Central Pacific RRs	First transcontinental railroad; May 10 driving of Golden Spike at Promontory, UT.
1829	**Robert Stephenson**	**"The Rocket," drawing three times its weight, reached speed of 12.5 mph.**	1869	George Westinghouse	Compressed air brake enabled heavier and faster locomotives, longer trains.
1829	Delaware & Hudson Canal Co. RR	First locomotives in North America, imported from England, burned coal.	1872	William Robinson	Closed-loop track circuits and wayside signals enabled faster, safer train speeds.
1830	George Stephenson	Locomotive with horizontal inside cylinders.	1872	Robert Davidson	First electric (battery powered) locomotive, pulled 6 tons at 4 mph.
1830	**Peter Cooper**	**"Tom Thumb" runs on B&O from Baltimore to Ellicott Mills, MD.**	1873	Eli H. Janney	Automatic coupler, for faster and safer train make-up.
1831	Ross Winans	Improvements in the construction of locomotive axles with outside journals.	1879	Werner von Siemens	Successful demonstration of the use of electric traction on railways.
1831	Mohawk & Hudson River RR	Locomotive crew cab to protect operators from the elements.	1885	J. van Depoele	Single overhead wire system for electric railways.
1833	Leicester & Swannington Railway	Steam whistle, after train hit horse and cart at crossing.	1886	**Frank J. Sprague**	**Electric-powered operations from axle-mounted traction motors.**
1836	Henry Campbell	Locomotive of the 4-4-0 "American" type with front swivel truck.			

Table 2.
Technological Development of the Locomotive.

1889	S.M. Vauclain	Compound low- and high-pressure cylinders.
1893	George Daniels, New York Central RR	First 100 mph record run (Engine #999)
1895	Frank Sprague	Multiple-unit control system.
1895	GE, Baltimore & Ohio RR	First major railroad electrification in U.S.
1897	**Rudolf Diesel**	**First compression-ignition internal combustion engine.**
1897		Steam-powered locomotive with electrical transmission.
1898	**Wilhelm Schmidt**	**Steam superheater for locomotives.**
1901		Mechanical coal stokers.
1906	Baldwin Locomotive Works	2-6-6-2 articulated compound steam locomotive.
1906	General Electric Co.	Gasoline engine equipped with electric drive.
1914	**Sulzers, Krupp, and the Prussian & Saxon State Railways**	**First large diesel-powered locomotive.**
1923-5	Herman Lemp and GE	Separately excited DC generator for locomotive. Self-regulating loads in diesel engine, generator, and traction motor.
1924	**Ingersoll-Rand and GE**	**First successful diesel-electric locomotive in USA.**
1920s	(end of decade)	Mechanically-driven pressure charger applied to diesel locomotive engine.
1930	Timken Roller Bearing Co.	Steam locomotive using roller bearings on the driving axles.
1934	Union Pacific RR	First distillate-electric streamliner in regular service, the M-10000.
1934	C, B & Q (Burlington) RR	First diesel-electric streamliner run, Denver to Chicago, *The Pioneer Zephyr*.
~1935	**Bucci (Italy)**	**Development of exhaust gas turbocharger for diesel engines.**
1935	Pennsylvania RR	Electrified passenger-train service between New York and Washington.
1937	American Locomotive Co.	First diesel engine turbocharger applied to locomotive in America.
1938	Union Switch & Signal Co.	Communication system for trains.
1939	**GM Electro-Motive Co.**	**First successful diesel-electric locomotive for freight service, and first use of dynamic brakes in road freight locomotive: Demonstrator freight locomotive FT-103; 5400 hp in four units (A-B-B-A).**
1941	Union Pacific RR and ALCO	"Big Boy" 4-8-8-4 steam locomotives, largest ever built (540 long tons).
1941	EMD; A, T & SF (Santa Fe) RR	Diesel-electric road freight locomotives in regular service.
1948	GE & ALCO	First gas-turbine locomotive.
1952	Norfolk & Western Ry Roanoke Shops	Last steam locomotive delivered for use by Class I railroad.
1960	Canadian National RR, Norfolk & Western RR	Last steam locomotive runs in revenue service by US and Canadian trunk lines. Steam use continued longer in Mexico.
1960 -70	**GM-EMD and GE**	**SD40 and U30C class locomotives of 3000 hp and 6 powered axles. Over 5000 were purchased in U.S. alone.**
Late 1970s	ASEA - Brown Bovari (ABB)	Electric locomotive rectifier technology arrived from Europe for Amtrak's Northeast Corridor passenger trains.
1979	Federal Railroad Administration	Proposals for railroad electrification related to second OPEC energy crisis.
~1981	Canadian National	North American "Wide Cab" design to improve comfort and safety of crew.
1984	GE	First micro-processor controlled diesel-electric locomotive (Dash 8).
1984	AAR and Railway Association of Canada	Beginning of industry cooperative effort to develop open-standard Advanced Train Control System (ATCS).

Table 2 (continued).
Technological Development of the Locomotive.

1984 -86	GM-EMD	Testing of 4-axle radial trucks on BN and ATSF.	1989	GM-EMD and Siemens	First successful three phase AC traction high horsepower freight / passenger locomotive (SD60MAC). Compact "brushless" induction rotor ("squirrel cage") traction motor is self-regulating and saves maintenance.
Mid '80s	Canadian Pacific RR, Bombardier	Experimentation with AC traction locomotives. (Conversion of M-640 to AC.)			
1988	GM-EMD and ABB	F69PHAC alternating current locomotive for Amtrak.	1993	Burlington Northern RR & EMD	Largest single locomotive order in history, 460 SD70MAC 4000 hp locomotives; (order later increased to 680 units).
1989	**GM-EMD and Siemens**	**Solid state Voltage Source Inverter (VSI) and Gate Turn-Off (GTO) thyristor technology applied to locomotive. Enabled changing DC to AC power of infinitely variable frequency and amplitude, facilitating high adhesion traction motor control.**	1994	Association of American Railroads	Publication of Locomotive Systems Integration (LSI) standards to improve inter-operability of on-board devices and components.
			1996	GE	**First 6000 hp diesel-electric locomotives using a single diesel engine.**

Table 2 (continued).
Technological Development of the Locomotive.

with about 17 percent for contemporary low pressure oil engines and about 14 percent for a steam turbine. Locomotives use "medium speed" (typically 16 cylinders and 800 to 1200 rpm) diesel engines—in contrast to the "high speed" engines in trucks or "low speed" diesels in ships. Older locomotives powered an electric direct current (DC) generator, which in turn fed electric energy to DC traction motors mounted one to an axle, usually four or six in total, under the frame of the locomotive. Newer locomotives almost always generate alternating current (AC) in an alternator, and then rectify it to DC. If AC traction motors are to be used, the power is then inverted to AC of precisely variable frequency (see below).

Diesel engine output is controlled by throttle settings for the diesel engine ("notch eight" usually means full power), and is automatically balanced to the electrical load on the generator or alternator. Most large diesel-electric locomotives employ a turbo-supercharger, which uses exhaust from the engine to compress air for injection, along with additional fuel, to increase the combustion rate and thus the horsepower of the engine. This action is also automatic, in that the harder the engine runs, the more exhaust, the more power for compression of air, and hence the more oxygen injected for combustion of more fuel. Turbochargers are precision machines and require special maintenance, but they approximately double the engine's horsepower compared with naturally-aspirated diesels. Design improvements in diesel-electric locomotives continue to result in impressive gains in fuel efficiency. Revenue ton-miles per gallon of fuel have nearly doubled since 1975 (see Figure 2).

Top speeds achievable by locomotives depend on their mass, horsepower, gearing, and the quality of track on which they run. Steam and diesel-electric locomotives have been operated at more than 100

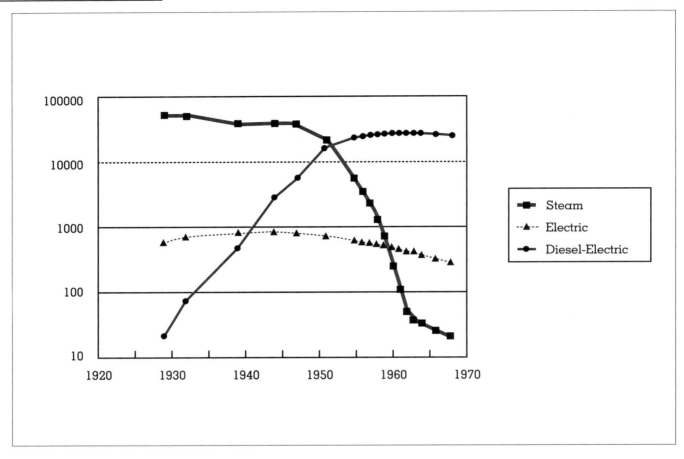

Figure 1.
Logarithmic Scale of Locomotives in Service—Class 1 Railroads.
SOURCE: Association of American Railroads. (1970). *Railroad Facts*, p. 68. Washington, DC: AAR.

mph, and electric locomotives in France first exceeded 200 mph in 1955. The Federal Railroad Administration is currently sponsoring development of a turbine-electric locomotive designed to operate at up to 150 mph.

TRACTIVE EFFORT

After the power rating of the prime mover, a second limitation to any locomotive is the tractive effort it can deliver to the unit's driving wheels without slipping on the rail. The proverbial "inherent advantage" of railroads due to low rolling resistance of steel wheels on steel rails now becomes a limitation, in that a positive coefficient of friction is needed to maintain pulling force at the face of the rail. In the old days, steam locomotive engineers developed great skill at adjusting steam admission to the cylinders and

sanding of the rails to minimize slipping. To start a long train, the engineer would back the engine into the train in order to bunch the train's slack, then accelerate—causing each car in succession to jerk to a roll. This is unnecessary with an appropriately powered diesel locomotive consist.

In 1939 General Motor's Electro-Motive Division sent its famed demonstrator FT-103 diesel-electric units on a triumphant tour of America. One purpose was to show skeptical railroaders that this 5400 hp, four-unit diesel locomotive developed more low speed tractive effort than competitor steam engines, which meant smooth starts and excellent performance on long mountain grades. The effect was much like Trevithick's demonstrations 130 years earlier, and a new generation of railroad locomotive power was assured.

Modern locomotives have sensors that detect

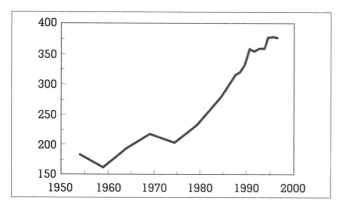

Figure 2.

Revenue Ton-Miles per Gallon of Diesel Fuel—USA Class 1 Railroads.

SOURCE: Association of American Railroads. (1998). *Railroad Facts*, p. 40. Washington, DC: AAR.

when the wheels of any powered axle are about to spin; the microprocessor-based slip control then reduces energy to that traction motor to enable smooth application of maximum pulling power from all wheels. Traditionally, locomotives could achieve (for dry, sanded rail) an adhesion factor of about 25 percent of their weight on driving wheels, but the refinement of wheel slip controls allowed increases in the adhesion factor to about 32 percent. Also to increase torque at the rail, DC locomotive traction motors are wired in series at low speeds, spreading generator voltage and maximizing tractive effort. When speed increases to a certain point, the motors are switched to a parallel connection permitting increased traction motor voltage; this allows the motors to run faster and cooler.

The newest AC locomotives use microprocessor controls and instrumentation that compares ground speed with traction motor rotation to control tractive effort and wheel slip with great precision. These locomotives have equipment that can change the phase angle of the three phase power supplied to the traction motors, and a device called a thyristor (a high speed solid state switch that can handle large ratios of controlled to controlling amperage) to adjust the frequency and amplitude of alternating current cycles. AC locomotives with single axle control will actually reallocate power from a slipping wheelset to those with greater adhesion.

The arrangement of diesel engine, on-board electric power generation, and axle-mounted traction motors

also allows dynamic braking—the ability to reduce speed by converting train momentum back through torque developed between rail and wheel to the locomotive axles and traction motors. The motors, thus converted to electric power generators, resist the kinetic energy of the train and produce electric current, which must be carried away to prevent overheating. In electric locomotives (where it is known as regenerative braking) the current is returned to the catenary or third rail, while in diesel-electric engines the electric power is wasted as heat energy through ventilated grid resistors on the top of the locomotive.

OPERATIONAL CONTROL

The third limitation on performance of any locomotive is the ability, whether through the skill of the engineer or by good design of the technology, to control its overall operation effectively. Steam engines were tricky to operate, to the point of taking on almost humanoid characteristics in popular lore. Getting the boiler fire to burn just right, for example, required hard work and experience. Until the mechanical stoker was invented, the fireman's simple physical limitations constrained performance. An advantage of the diesel-electric power configuration is that loads balance nearly automatically between the engine, generator, and traction motors—making operation by the locomotive engineer or driver much simpler than on a steam engine. The "marriage" and self-correcting nature of the diesel engine, its generator (or alternator), and the traction motors was invented by Herman Lemp of General Electric in 1923–1925, and continues to be used today in all diesel-electric locomotives.

With use of electric and diesel-electric locomotives also came the ability to control power and tractive effort, not just for the unit in which a crew rides, but for multiple units (MU) coupled together in a locomotive consist—and this without need of additional crews in the following units. (A "consist" is the make-up of the train, in this case the full set of locomotives used.) With MU operation, locomotive managers had flexibility to size horsepower of the consist to operating requirements of a particular train and its territory or specific assignment.

In time, other aspects of operational control of the locomotive were made possible by electric and diesel-electric power. One important technology advance

	GM-EMD SD40-2 (1972)	GE U30C (1972)	GM-EMD SD90MAC (1997)	GE AC6000 CW (1997)
Traction Motors	Direct current	Direct current	Alternating current	Alternating current
Diesel Engine HP Rating	3000	3000	6000	6000
Continuous Tractive Effort	83,000 lbs.	90,600 lbs.	170,000 lbs.	166,000 lbs.
Adhesion Factor*	20%	20%	38%	40%
Maximum Braking Effort	61,000 lbs.	61,000 lbs.	115,000 lbs.	117,000 lbs.
Maximum Speed	70 mph	70 mph	75 mph	75 mph

Table 3.
Locomotive Performance Comparison: 1972 to 1997.

represents a logical extension of the principle of MU operation—remote control of additional locomotives placed deep in the train or at its end. This distributed power has proven highly effective in reducing drawbar forces in long, heavy trains, or indeed, in allowing safe handling of longer trains by the locomotive engineer. Experiments were made with distributed power in the 1960s and 1970s, but it was only with the development of reliable microprocessor controls and data radio communications in the 1990s that the concept proved widely effective.

Train control in a different sense of the term is a powerful potential breakthrough technology for railroads. In this concept, accurate knowledge of train position and dispatcher authorities interpreted by an on-board locomotive computer is used to prevent train collisions, overspeed accidents, and incursions into track territory reserved for maintenance crews. The Global Positioning System (GPS), digital data radio communications to and from locomotives, and computer display screens for crews are among the enabling technologies to be used in advanced train control systems, increasingly called positive train control (PTC). The on-board locomotive electronic networks and radio communications systems developed for applications such as power and train control will increasingly be used for locomotive health monitoring, event recording, train dispatching efficiency, on-time (scheduled) train movement, and more responsive customer service. Some of these other business functions are already being deployed on locomotives by railroads, and the Association of American Railroads has developed locomotive systems integration (LSI) standards to promote compatibility of devices from different suppliers.

EMISSIONS

Locomotive emissions have become an important aspect of performance in recent years. While overall fuel efficiency of railroad freight service has been shown by several detailed studies to be approximately three times better than highway motor carrier service, and while rail passenger advocates point out the energy use, congestion mitigation, and superior emissions performance of rail compared with other modes of transportation, environmental protection authorities are interested in reducing major sources of manmade pollution. The 1990 Clean Air Act Amendments specifically mandated extension of emission regulations to locomotives, which the U.S. Environmental Protection Agency (EPA) currently estimates to be responsible for nearly five percent of nationwide emissions of oxides of nitrogen.

In April of 1998, the EPA published a final rule for emission of oxides of nitrogen (NOx), hydrocarbons (HC), carbon monoxide (CO), particulate matter (PM), and smoke opacity for newly manufactured and remanufactured locomotives. The rulemaking took effect in 2000 and is estimated by the EPA to cost the railroads $80 million per year—about $163 per ton of NOx reduced, according to EPA figures. The emissions standards for the several pollutants will be implemented in three tiers—for locomotives manufactured or remanufactured from 1973 to 2001, from 2002 through 2004, and in or after 2005, respectively. The emission levels are based on achievability with reasonable technology and cost outlays. EPA has adopted fleet averaging and banking and trading provisions to provide flexibility in meeting goals more efficiently. The two largest railroads serving the Los Angeles Basin have entered into a separate agreement with California authorities to address the special air quality issues in that area.

LOOKING BACK AND AHEAD

After decades of economic decline and company bankruptcies from the end of World War II through the 1970s, American railroads were largely deregulated in the Staggers Act of 1980. Since then the industry has achieved an unprecedented renaissance in productivity, financial performance, and economic value to customers, due in part to significant technology advances facilitated by deregulation. Because of their labor and energy efficiency, favorable environmental characteristics relative to competing transportation modes, and remarkable ability to absorb and spawn new technologies, railroads have a bright future. Expected increases in traffic volumes in coming decades will require large investments in railroad capacity, including new locomotives. Continuing improvements in locomotive performance, together with the contributions of "intelligent systems" riding these powerful machines, will lead the procession into railroading's third century, as they did its first.

Robert E. Gallamore

See also: Diesel Fuel; Mass Transit; Railway Passenger Service; Steam Engines; Transportation, Freight Delivery and; Trevithick, Richard.

BIBLIOGRAPHY

Armstrong, J. H. (1993). *The Railroad: What It Is, What It Does*, 3rd ed. Omaha: Simmons-Boardman Books, Inc.

Bilz, F., and Holger, S. (1998). "A Decade of Three-phase Traction Technoogy for Diesel-Electric Locomotives in North America." Translated from *ZEV+DET Glassers Annalen* 121(9)/1997. Siemens AG Transportation Systems Group.

Bruce, A. W. (1952). *The Steam Locomotive in America: Its Development in the Twentieth Century*. New York: Bonanza Books.

Doherty, J. M. (1962). *Diesel Locomotive Practice*. London: Odhams Press, Ltd.

Draney, J. (1954). *Diesel Locomotives: Mechanical and Electrical Fundamentals*. Chicago: American Technical Society, Publishers.

Farrington, S. K. (1943). *Railroading from the Head End*. Garden City, NY: Doubleday, Doran & Co.

Grosser, M. (1978). *Diesel: The Man and The Engine*. New York: Atheneum.

Haine, E. A. (1990). *The Steam Locomotive*. New York: Cornwall Books.

Haut, F. J. G. (1970). *The Pictorial History of Electric Locomotives*. New York: A. S. Barnes & Co.

Illustrated Science and Invention Encyclopedia. (1983). Westport, CT: H. S. Stuttman.

Jackson, R., and Triest, M. (1999). "TTCI: Tough Tests for High Tech Power." *Railway Age*, January, pp. 55–58.

Kratville, W. W., ed. (1997). *The Car and Locomotive Cyclopedia*, 6th ed., Section 4:430–619. Omaha: Simmons-Boardman.

Marre, L. A. (1995). *The Contemporary Diesel Spotter's Guide*, 2nd ed. Waukesha, WI: Kalmbach Publishing Co.

Marshall, C. F. D. (1953). *A History of Railway Locomotives*. London: Locomotive Publishing Co.

Marshall, J. (1974). *Rail Facts and Feats*. New York: Two Continents Publishing Group.

Pinkepank, J. A. (1973). *The Second Diesel Spotter's Guide*. Milwaukee: Kalmbach Publishing Co.

Ransome-Wallis, P., ed. (1959). *The Concise Encyclopedia of World Railway Locomotives*. New York: Hawthorn Books.

Rolt, L. T. C. (1962). *The Railway Revolution: George and Robert Stephenson*. New York: St. Martin's Press.

Stretton, C. E. (1989). *The Development of the Locomotive*. London: Bracken Books.

Thomas, D. E., Jr. (1987). *Diesel: Technology and Society in Industrial Germany*. Tuscaloosa, AL: University of Alabama Press.

U.S. Environmental Protection Agency. <http://www.epa.gov/orcdizux/regs/nonroad/locomotv/frm/42097048.htm>.

White, J. H., Jr. (1997). *American Locomotives: An Engineering History, 1830–1880*, revised and expanded ed. Baltimore: Johns Hopkins University Press.

LOS ALAMOS NATIONAL LABORATORY

See: National Energy Laboratories

LUBRICANTS

See: Tribology

LYELL, CHARLES (1797–1875)

Charles Lyell was a founder of modern British geology. One of his most important contributions to science concerned the rates at which the earth's internal energy was released to affect the shape and form of its crust and thus to create the landscape we know today. Geological changes such as creation of valleys, mountain formation, deposition of sediments, and the like were not in his view caused by occasional "catastrophes" but rather the results of ordinary geological processes operating over an immense period of time. In other words the release of energy to produce geological change has occurred at a rate similar to that of the present time, and is largely uniform over the earth's long history. This doctrine of uniformitarianism has been rightly attributed to Lyell, though others before him (e.g., James Hutton) had expressed similar views.

Lyell was born on November 17, 1797, at Kinnordy near Forfar in Scotland. His father was a wealthy landowner with passions for both Italian literature and natural history. His mother came from Yorkshire and while he was still very young the family moved to England, taking a lease of a large house on the fringe of the New Forest in Hampshire. The boy was sent to schools in the locality, though he showed little evidence of scholastic promise. In 1816 he went to Exeter College, Oxford, to read classics, but he also attended lectures by William Buckland, a professor of geology. The effects of these lectures were reinforced by books from his father's ample library and by an encounter with Gideon Mantell, a doctor in the neighboring county of Sussex, renowned for his study of the fossils then being discovered in the chalkbeds and on the coast of southern England.

Sir Charles Lyell. (Library of Congress)

Lyell's 1819 Oxford degree was in classics, and he then began to study law at Lincoln's Inn, London. However, weak eyesight precluded much reading at that time, and this gave him a reason for further geological study in lectures, field studies, tours, and so on. He became a fellow of the Geological Society in 1819, and four years later was made one of its joint secretaries. He undertook geological tours in France, the west country of England, and the Scottish Highlands. A visit with Buckland to Glen Roy revealed to them the now famous Parallel Roads, three raised seabeaches now believed to be of glacial origin but even then posing great problems for conventional geology.

The central problems that Lyell was to address in the next few years involved detailed mechanisms of geological change, the age of the earth; and rates of energy release. The most popular view of geological history had proposed a series of great convulsions or "catastrophes" involving immense amounts of energy, alternating with longer periods of relative quiescence. Most famously, the biblical Flood of Noah was invoked to explain the present crust of the earth, including the regular strata of fossil beds. People who favored water as the chief agent of change ("diluvualists") had to contend with others who thought that fire was chiefly responsible ("vulcanists"). But each believed in a relatively short period of time for it all to happen, and sometimes drew confirmation from certain biblical data that *could* be interpreted to suggest a date of creation a mere few thousand years ago. Much support for this view was given by Buckland and by his Cambridge counterpart Adam Sedgwick, though both were later to change their minds. The opposite views of James Hutton were less popular, though Lyell was increasingly inclined toward them. One of several events that proved critical in his experience was a visit to Sicily in 1828. Before crossing the volcano Etna he had observed strata at the base of the mountain that seemed comparatively recent. Discovering similar deposits on the opposite side he concluded that the vast bulk of Etna rested on what were, in geological terms, recently recent rocks. Hence a vast age for the whole earth was indicated, and he embarked on a crusade for his uniformitarian ideas, publishing his monumental *Principles of Geology* in the early 1830s (and many subsequent editions thereafter). This work brought him lasting fame.

His legal practice largely forgotten, Lyell now devoted the rest of his life to promoting his doctrines and to traveling extensively. In 1832, at the new King's College, London, he became the first professor of geology, a post he occupied for only two years. He received from the Royal Society its Royal Medal in 1834, and a year later he became president of the Geological Society. He became much involved with the Great Exhibition in London in 1851 and in the British Association. He received a knighthood in 1848 and a baronetcy in 1864.

The influence of Charles Lyell on science was profound. Among the recipients of his *Principles* was a young naturalist embarking on a long sea voyage. This was Charles Darwin, who came to develop a strong interest in geology. He also accepted much of Lyell's arguments and, while his own theory of evolution was being formed, relied extensively on Lyellian arguments for an immensely old earth. He became a good friend as well as a disciple, though even in the 1860s Lyell was reluctant to give Darwin his public support, as he saw how thin the fossil evidence really was for transmutation of species. Not all scientists became enthusiastic Lyellians, however. William Thomson (Lord Kelvin) opposed him over the age of the earth, arguing on largely thermodynamic grounds for a shorter time span than Lyell or Darwin wanted. Only with the discovery of subterranean radioactivity were Kelvin's estimates shown to be erroneous and a Lyellian time scale rendered more credible. However, that of all the uniformitarians in the nineteenth century Lyell was the most extreme, and no one identified completely with a literal interpretation of uniformitarian change. There was too much evidence of catastrophic releases of energy in volcanic eruptions, flash flooding, and earthquakes for most people to deny their massive influence of earth history. Today a modified, and reduced, uniformitarianism seems more likely to fit the facts. Lyell's extreme views were probably related to a religious inclination to Unitarianism, which denies God's intervening activity in history through Christ, just as uniformitarianism cannot allow catastrophic interventions in geology. However, Lyell outwardly remained a member of the Church of England.

In 1842 he married Mary Horner, daughter of Leonard Horner, warden of the University of London. Lyell died in London on February 22, 1875.

Colin A. Russell

See also: Geography and Energy Use; Geothermal Energy; Thomson, William.

BIBLIOGRAPHY

Bailey, E. (1962). *Charles Lyell*. London: Thomas Nelson & Sons.
Lyell Centenary Issue. (1976). *British Journal for the History of Science* 9:9–242.
Rudwick, M. J. S. (1985). *The Great Devonian Controversy*. Chicago: University of Chicago Press.
Wilson, L. (1970). "Lyell, Charles." In *Dictionary of Scientific Biography*, ed. C. C. Gillispie. New York: Charles

MACHINES, SIMPLE

See: Mechanical Transmission of Energy

MAGNETIC LEVITATION

The term "maglev" was coined by Howard Coffey in the 1970s as a shortened form of "magnetic levitation" for transportation. One dictionary defines maglev as "having to do with a railroad system using magnets to float a swiftly moving train above its tracks." This is a poor definition since there are no tracks, no need for a train of vehicles, and maglev can work at any speed. A better definition is "a transportation system in which the vehicles are suspended, guided and propelled by magnetic forces without any contact with the guideway."

Maglev is a recent name for an old idea: support a moving vehicle with magnetic fields so there is no contact between the vehicle and a guideway. This apparently simple idea has long inspired inventors, but in spite of sustained efforts by hundreds of people, publication of thousands of technical papers and the expenditure of billions of dollars, there is no commercial maglev system in operation at the start of the Twenty-First Century. This is due to the formidable technical and nontechnical problems that must be solved, the inability to find a design and application that are well matched, and the complexity of the maglev puzzle. This article discusses the past, present, and probable future for this fascinating technology.

EARLY HISTORY: SIMPLE IDEA, A HARD REALITY

The dream of maglev may be as old as the discovery of permanent magnets, such as ones used in magnetic compasses by the Chinese in about 1250 C.E. Permanent magnets can attract or repel each other and it would seem that one might arrange magnets so that a body floats freely in space. The earliest history of maglev is not well known, but in 1839 Earnshaw proved an important theorem which, in simplified form, states that any arrangement of bodies that have attractive or repulsive forces that obey an inverse power law is unstable. For maglev this means that no configuration of permanent magnets, electromagnets with constant current excitation, and ferromagnetic material can be stable, with or without the presence of gravitational forces. If the magnetic field can change, such as via a feedback control mechanism, stability can be achieved. It is also possible to achieve stability by virtue of motion between interacting bodies or by using superconductors arranged so that their persistent currents change appropriately in response to changing position.

A major problem is the design of a propulsion system for a vehicle that has no contact with a guideway. The only reasonable choice is a linear motor that uses magnetic fields to propel the vehicle. The development of a linear motor and its control system is at least as formidable a challenge as the development of a maglev suspension system.

Magnetic scaling laws favor large systems. It takes strong and heavy magnets to suspend a vehicle with any significant spacing between a vehicle and a guideway, and this has made it difficult to build small maglev systems. We have many toy cars, planes, and trains, but no comparable maglev toys that can be used to explain the key ideas.

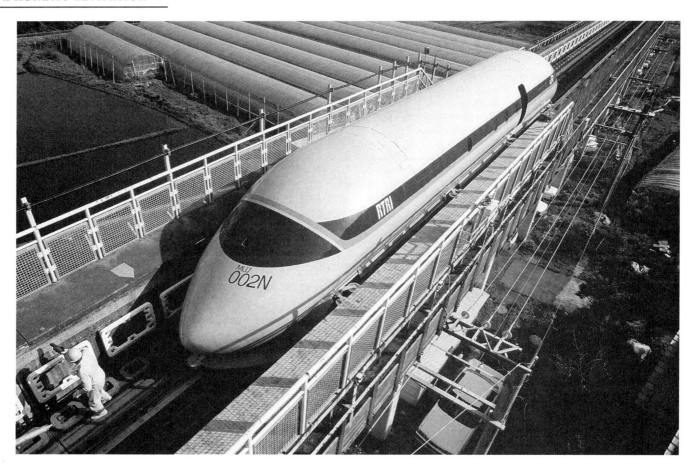

A train that operates by magnetic levitation, at a railroad research center in Japan. (Corbis-Bettmann)

TECHNOLOGICAL ALTERNATIVES FOR SUSPENSION AND PROPULSION

There are four competing maglev technologies: electromagnetic suspension, electrodynamic suspension, linear synchronous motor, and linear induction motor.

Electromagnetic Suspension (EMS)

Figure 1a shows an EMS system in which an array of magnets is attracted upwards to a steel rail. It is possible to design the magnets so that there is an upward force produced by magnetic attraction that cancels the downward gravitational force: the magnets are suspended in space! If steel beams were mounted on either side of a "guideway," then a vehicle with magnets on both sides could move along the guideway and be supported and guided by the steel rails. If permanent magnets, or electromagnets with constant current excitation, were used in the design, the system would be unstable: any disturbance from equilibrium would cause the magnets to move in the direction of the disturbance. The difference between stable and unstable equilibrium is like the difference between balancing a cone on its base and balancing it on its point. EMS was pioneered by B. Graeminger in 1911.

Earnshaw's theorem proves that some instability is inherent. Ideally there is neutral stability in the fore and aft direction, so that we have a choice: the vehicle can be stable laterally and unstable vertically, or vice versa. The usual approach is to use a design similar to Figure 1a and use electromagnets with power controllable current. Position and velocity sensors are used in conjunction with controllers to regulate the magnet currents and thereby achieve vertical stability, and lateral stability can be achieved passively or, possibly, with help from controlled magnetic forces. This approach is called "Electromagnetic Suspension" (EMS) and has been used in many designs.

EMS vehicles have operated at speeds from 0 to 440 km/h (270 mph).

EMS requires about 1 to 2 kW of magnet power for every ton of vehicle mass. At modest and high speeds this power loss is small compared with the power loss due to aerodynamic drag. EMS is the favored approach for urban maglev and is suitable for high speeds if the use of a minimal 10 mm air gap is acceptable.

Electrodynamic Suspension

Figure 1b shows an array of magnets moving over a conducting sheet and being pushed upwards by forces due to induced currents. This electrodynamic suspension that was pioneered by Gordon Danby and James Powell in the 1960s, can not work at zero speed, but at higher speeds it can be inherently stable.

The virtue of EDS designs is their ability to operate with larger air gaps than is feasible with EMS designs. Vehicles suspended in this way have operated at speeds from about 50 to 552 km/h (31 to 343 mph).

All EDS designs are highly underdamped and can even be negatively damped, the equivalent of the instability found in EMS designs. Many maglev designers have underestimated the importance of this damping problem, the equivalent of building a car with solid tires and no shock absorber.

A well designed EDS suspension will create 5 to 10 kW of power dissipation in the guideway for every ton of vehicle mass. This is almost an order of magnitude higher than the power required for EMS. Since the suspension power comes from the linear motor, at low speeds the motor must provide high thrust, and even at moderately high speeds, the suspension power is comparable to aerodynamic drag losses. The high losses can cause serious overheating of the guideway at low speeds. These factors, taken together, make EDS a questionable technology for urban maglev but quite appropriate for high speeds.

Linear Synchronous Motor

One propulsion alternative is to have a powered guideway in which a magnetic field is made to appear to move via electronic means, and it pushes on a magnetic field created by the vehicle. This is called a long stator linear synchronous motor, generally referred to as an LSM. It is possible to have magnets on the guideway and supply the moving field from the vehicle, but this is not a favorable design for maglev. The LSM creates forces by attractive and

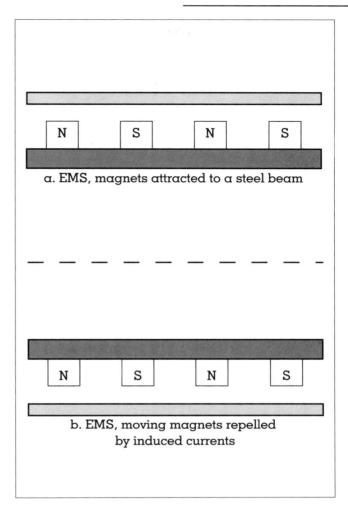

Figure 1.
Electromagnetic suspension: two views.

repulsive magnetic forces, much as an EMS design creates suspension via these forces. The rotary synchronous motor was invented more than 100 years ago and the linear version was pioneered by Henry Kolm and Richard Thornton at MIT in the 1970s.

The advantage of this design is the simplified nature of the vehicle's portion of the propulsion system. In the simplest case the vehicle has a linear array of permanent magnets and is totally passive. This passive vehicle approach has been used successfully for material handling equipment and is probably the best design for small and moderate size vehicles. For larger vehicles the permanent magnets are usually replaced with electromagnets, that also provide the suspension force. This dual use of vehicle magnets for both suspension and propulsion has much merit for high speed transportation systems. The long stator design also has the major advantage that there is no need to transfer large amounts of power to the vehicle and that there is no bulky and heavy power conditioning equipment on the vehicle. Also, the guideway-based power conditioning can be sized according to terrain, with lower power controllers used in regions where high forces and velocity are not required.

A disadvantage of the long stator LSM is the higher cost of the guideway winding as compared with a simple conducting strip that can be used with a linear induction motor. If there are many closely spaced vehicles, each fairly small, then this cost is well justified, but it becomes a problem for some maglev designs.

A second disadvantage is the need to maintain synchronism. Unlike rotary synchronous machines, linear versions are difficult to operate "open loop" with the vehicle automatically maintaining synchronism with a moving field. In practice it is necessary to have a separate controller for each vehicle, and therefore not more than one vehicle can be in one block of guideway. For operation up to a few meters per second, it is possible to use a "stepping mode" with the vehicle following the moving field without feedback, but this method will not work at higher speeds because of inherent "mid frequency resonance" phenomenon. The solution is to use position sensors and electronically synchronize the vehicle with the moving magnetic field.

Linear Induction Motor

A propulsion alternative is to transfer power to the vehicle and then have a magnetic structure on the vehicle that can create a moving field that, in turn, interacts with a conducting guideway to provide propulsion. This is called a short stator linear induction motor (or linear asynchronous motor), generally referred to as a LIM. It is possible to put the powered winding on the guideway and induce currents in a conductor on the vehicle, in which case it would be a long stator design, but this turns out to be an expensive and inefficient approach to maglev propulsion. The LIM creates forces via induced current much as an EDS design creates suspension via induced currents. The rotary induction motor was invented by Nikola Tesla in 1886 and the linear version for transportation was pioneered by Eric Laithwaite in the 1950s.

The advantage of the LIM is the reduced cost of the guideway and the ability to operate closely spaced vehicles without the block limitation of the LSM.

The LIM has been used for propelling both wheel based vehicles and magnetically suspended vehicles over a very wide range of speeds.

The principal disadvantage of the short stator design is the need to transfer power to the vehicle. If this is done with sliding contacts, as with conventional electric trains, then some of the advantages of maglev are lost. If it is done via inductive transfer, then the guideway cost is increased.

A second disadvantage is the need to have substantial amounts of power conditioning on the vehicle. This equipment must be sized to match the highest force and speed for the entire system, even if it rarely needs such high peak power. If the air gap is large this power conditioning equipment can be heavy and expensive.

Note that the popular press often misuses terms like "induction" and "synchronous" when applied to linear motors.

RECENT SYSTEMS

England

A low speed EMS maglev system was constructed in Birmingham in 1984. This design had an air gap of 10 mm and used short stator LIM propulsion with power delivered to the vehicle via sliding contacts and a third rail. It provided transportation between an airport and a train station and worked nearly flawlessly for more than twelve years. When problems did develop there was no one with interest and ability to repair it, so it was removed.

Japan

The Japanese constructed a high-speed maglev system in Yamanashi Prefecture using superconducting technology. This was the successor to several smaller and lower speed vehicles that had been tested in Miyazaki Prefecture starting in 1980. The Yamanashi system uses the null-flux version of EDS and a long stator LSM for propulsion. The latest test facility consists of a two-way, 42.8 km (27 mi.) long guideway, most of it in tunnels. In April 1999 a manned five-car train, MLX01, achieved a speed of 552 km/h (343 mph). In November 1999 two trains passed each other at a relative speed of 1,003 km/h (623 mph). This system shows the potential of EDS, but is relatively expensive to construct and operate and it is questionable whether it will ever become a commercially viable design.

Germany

The Germans constructed a 31.5-km (19.6-mile) test loop in Emsland, starting in 1974. This was the culmination of a research effort started in 1970 and involving experiments with both EDS and EMS and with various types of linear motors. The latest design, called Transrapid 08, has an EMS with 9 to 10 mm (3/8 in.) air gap and LSM propulsion using the same magnets for both suspension and propulsion. It achieved a top speed of 450 km/h (280 mph) in June 1993. Although the air gap is relatively small, the success of the Emsland tests makes it clear that this technology is usable for higher speeds than are ever likely to be commercially feasible with conventional railroad technology. A comparison of Transrapid with the Germany ICE high speed train indicates that maglev consumes 30 percent less energy and the differences are much larger when comparing maglev with automobiles and commercial aviation. There are plans to construct a 292-km (181-mi.) operational system between Hamburg and Berlin. The design calls for five stations and a travel time of less than one hour. The construction cost is estimated to be about the same as for a new ICE high speed rail line, but there is still some doubt whether this project will ever be completed. There are active proposals to construct Transrapid systems in the United States but it is too early to tell whether such a system will ever be built.

A second German effort, designated HSST, has been underway since 1974. This is a lower speed EMS design and uses a LIM with power transfer via sliding contacts. This system has been demonstrated on several occasions and there are pending plans for implementation, but past plans have never been carried out and the future is uncertain.

United States

In 1966 Danby and Powell at Brookhaven National Laboratory conceived of using vehicle-mounted superconducting magnets that induced current into their novel "null-flux" winding on the guideway. This concept helped precipitate maglev research efforts in the early 1970s, but these were soon terminated by political action. In 1987 Senator Patrick Moynihan created the U.S. National Maglev Initiative that was managed by the Federal Railway Administration. It led to four preliminary designs, but the effort was terminated by political action. In the late 1990s a maglev effort was reinitiated, but the combination of political and technical dispute makes the future uncertain.

There is an effort by the U.S. Federal Transit Administration to develop an Urban Maglev system that can be an alternative to urban rail-based transit systems. The objective is to achieve speeds up to 161 km/h (100 mph) with lower noise, lower energy consumption and lower operating cost that any other fixed guideway system. Based on the success of the Birmingham, England system it is likely that a viable urban system can be built. If the lower speed design was well conceived it could evolve into a design suitable for operation at higher speeds, just as the railroad system evolved.

Web Sites for More Information

The best and most up-to-date references are Internet websites. Most of these can be reached via links from the U.S. Department of Transportation site: <http://www.dot.gov>. This site has links to the Federal Railroad Administration and the Federal Transit Administration, each of which has links to other national and international maglev sites. Additional maglev sites can be reached from the *Innovative Transportation Technologies* site: <http://faculty.washington.edu/~jbs/itrans/maglevq. htm>. More details on EMS, LSM, and LIM can be found on the German *Transrapid* and Japanese *HSST* sites. More details on EDS and LSM can be found on the Japanese *Railroad Technical Research Institute* (RTRI) site.

A word of caution: Any publication that claims major advantages for a particular breed of suspension or propulsion should be viewed with skepticism.

PROS AND CONS OF MAGLEV

Advantages

Following are advantages that have been proven by actual tests on working system.

- *More efficient.* Maglev vehicles can operate at a given speed with less energy consumption than for almost any other transportation mode, particularly for short to medium distances. This is because the vehicles can be lighter and more streamlined than is possible for any wheel-based design and commercial aviation is very inefficient for short trips. It has frequently been proposed that the vehicles can operate at very high speeds in a partially evacuated tunnel with dramatic savings in energy and time, and several studies have shown that this is not as far-fetched as it might sound.

- *Less noisy.* Maglev vehicles are quiet. Transrapid test data shows 5 to 10 dB less noise than for a train at the same speed or the same noise for speeds that are 100 km/h (62 mph) higher. Some newer maglev designs with more streamlining have even lower noise.

- *Safer.* The system can be as safe or safer than any other system. If it is well designed with dedicated rights of way, a high level of automation and no physical contact, then the most common cause of accidents will have been eliminated. There have been no significant accidents in all of the maglev tests that have been run. If maglev attracts people away from other modes it could save many lives.

- *Environmentally friendly.* The reduced noise and lack of air pollution is environmentally desirable, and so also is the ability to climb steep grades and operate in smaller diameter tunnels. This combination is unbeatable by any other mode.

- *Faster.* The system can be designed to operate safely at very high speeds. But speed is relative so that 161 km/h (100 mph) operation in an urban area will be dramatically faster than alternatives. Some people believe that maglev designers have put too much emphasis on speed and this has made the designs expensive.

- *Reduced operating cost.* The elimination of bearings and wheel contact will reduce the maintenance time and cost. This is borne out by the high reliability of the Birmingham, England maglev shuttle and the reliability of the Transrapid test system in Emsland, Germany. The lower energy consumption and potential for complete automation creates additional savings.

Disadvantages and Problems

There are several disadvantages and problems that must be solved:

- There is a perception that maglev is too expensive. Part of the problem is a focus on very high speeds where any system would be expensive, if not impossible. For many new construction projects a well designed maglev system need be no more expensive than wheel-based designs, and lower maintenance and energy cost will provide additional savings. But this potential cost advantage must be proven before it is can be used to sell maglev.

- Transportation planners and managers can point to financial disasters in the early application of new technology. They feel, with justification, that no commitment for a commercial system should be made until there is a good demonstration system that proves the technology.

- There are many barriers to innovation in the transportation sector. Large corporations dominate the design and construction of guided systems and they have little incentive to change. They could lose by either failing to deliver a viable system or because another company is stronger in the new technology. There are many government regulations that pertain to any transportation system, and political involvement in transportation decision making tends to favor the status quo.

- There are too many inventors and little cooperation in finding the best combination of suspension, guidance, propulsion, and control. A national competition would be very helpful, such as the locomotive competition in 1829 when George Stephenson's *Rocket* beat several competing designs by hauling a coach of passengers at 39 km/h (24 mph).

PROGNOSIS FOR THE FUTURE

Future success will require a significant group to make a long-term commitment to develop an integrated maglev system. There should be a reduced emphasis on new inventions and increased attention to the myriad problems of integrating suspension, propulsion, guidance, and control. Some group must be willing to fund the development to the point that a potential buyer of a maglev system can clearly see that the advantages outweigh the disadvantages. The probability of this happening is good; the major question is "When, where, and how will maglev reach commercial fruition?"

Richard D. Thornton

See also: Railway Passenger Service.

MAGNETIC FLIGHT

See: Magnetic Levitation

MAGNETISM AND MAGNETS

Magnetism is the phenomenon in which iron is attracted to a natural material called lodestone, the properties of which are similar to a magnet.

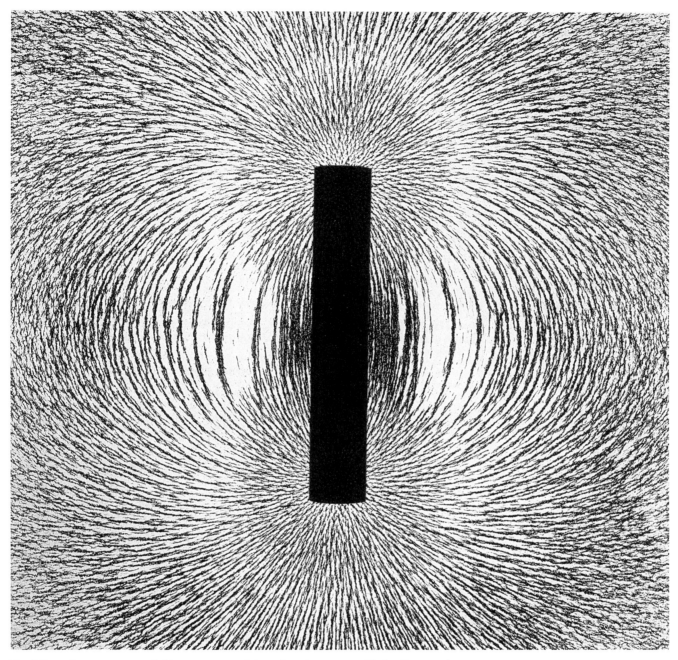

Metal filings form a magnetic field pattern around the poles of a magnet. (Corbis-Bettmann)

LODESTONE AND MAGNETS

Lodestone is a crystalline oxide of iron called magnetite. Until the 1800s, lodestones and the earth were the only sources of magnetism. Iron-based materials are attracted to a lodestone as well as to any other magnet. It is the attraction between a magnet and the iron in a refrigerator door that pins a photograph to the door. Bringing an iron-based material in contact with a magnet or lodestone will make a magnet of the material. Unfold a paperclip and rub it with a magnet. The clip becomes magnetized, with the magnetic properties concentrated near the ends of the clip. These end regions are called poles. Magnetize two unfolded paperclips and, in one parallel orientation, the ends or

741

poles attract each other. Change the orientation of one clip and the poles repel each other. Attach a thread to the middle of one of the clips, suspend it, and one pole will point in the general direction of geographic north. The one pointing north is called an N-pole; the opposite pole is an S-pole. We observe, and say, "unlike poles attract, like poles repel." It is an instructive experiment to cut the magnetized clip into two pieces in an effort to isolate a pole. Interestingly, each of the two pieces has an N-pole and an S-pole. Regardless of how many times a clip is cut in two, magnetic poles always occur in pairs of N-poles and S-poles.

MAGNETIC FIELD

A magnet does not materially change the space around it. Yet, if the magnet were not there, another magnet would not experience a force when brought into the space. Magnetically, the space around the magnet is altered and the modification is thought of as producing a magnetic field. When another magnet is brought into the magnetic field and experiences a force, the magnetic field is the mechanism for exerting the force. The concept of a field applies to gravitational and electric forces as well, and is an extremely important aspect of many energy applications.

Magnetic fields are measured in units of teslas, symbol T. A strong permanent magnet, such as might be found in a physics laboratory, produces a magnetic field of about 0.3 T. Such a magnet is capable of lifting several kilograms of iron. For comparison, the magnetic fields produced by other systems include

10^8 T	neutron stars
10^3 T	short bursts of electric current
10^1 T	strong laboratory superconducting magne
10^{-5} T	Earth's magnetic field
10^{-9} T	interplanetary magnetic fields
10^{-12} T	magnetic field associated with the human body

In 1819, Hans Christian Oersted, professor of physics at Copenhagen University, discovered that a magnet experiences a force when in the vicinity of a wire carrying an electric current. The fact that the magnet experiences a force is evidence that the electric current produces a magnetic field, which eventually led to the development of innumerable devices—electric motors, electric generators, speakers for hi-fidelity amplifiers, and electromagnets, to name a few—based on this principle.

ENERGY APPLICATIONS

Electric motors

Electric motors are found in nearly every room of a typical house. They power everything from the washing machine, refrigerator, and vacuum cleaner, to the hair dryer, fan, garage door opener, and disk drive in a computer. In an automobile, motors adjust seats, raise an antenna, operate windshield wipers, adjust mirrors, run fans, start the engine, and someday may replace the internal combustion engine under the hood. There is probably no device more useful for doing work than electric motors, and their pervasiveness will surely grow.

Fundamentally, an electric motor converts electric energy to rotational energy. Rotation results from magnetic forces between a rotating part (the rotor) and a stationary part (the stator). There are many designs. In the simplest, the rotor is an electromagnet that rotates between the poles of a permanent magnet (Figure 1). The N-poles and S-poles of the electromagnet are determined by the direction of current flow. In the illustration, attraction and repulsion between poles on the rotor and stator cause the electromagnet to rotate clockwise. When the unlike poles approach each other, the direction of current flow is reversed, causing the poles on the electromagnet to change. The alternating magnetic attraction and repulsion between poles keeps the electromagnet rotating in the same direction.

Electric Generators

An electric generator for operating lights is a common sight on many bicycles. The generator has a coil of wire that rotates between the poles of a magnet and looks very much like a motor. Whereas a motor converts electric energy into rotational energy, a generator converts mechanical energy to electric energy. On a bicycle, the tire rubs against a wheel attached to the rotating coil. Some agent, in this case the cyclist pedaling, does the work to turn the coil with the reward being electric energy. In a large electric power plant an electric generator working on the same physical principle is driven by a large steam turbine.

Electric generators are based on the principle that an electric charge experiences a force when it moves in a magnetic field. Electrons in the metallic wires of

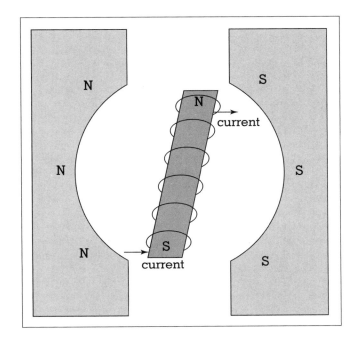

Figure 1.

An elementary motor consists of a rotating electromagnet and a fixed permanent magnet.

the rotor of a generator experience a force when the coil is rotated in a magnetic field. An electric current is produced in a light bulb, for example, when it is connected to the open ends of the wires making up the coils of the rotor.

Magnetic Levitation of Vehicles

Whereas electric motors utilize both attraction and repulsion of magnetic fields, there are energy applications that rely only on attraction or repulsion. Magnetic levitation of vehicles is a good example. A vehicle riding on a track or roadway experiences frictional forces that oppose the movement. Any scheme that can reduce the frictional forces offers improved energy economy. Magnetic levitation involves magnetic forces that hold a vehicle above a roadbed so that the vehicle appears to float on a cushion of air. It does not. It floats on a "magnet" cushion. One method of suspension capitalizes on the idea that like poles repel. The poles are produced by electromagnets rather than using permanent magnets. Another scheme is based on the principle that a metal experiences a force when in the magnetic field of an electromagnet that is energized by an electric current that changes rapidly with time.

Once a vehicle is levitated, it is not in material contact with a track, so cannot be propelled by wheels. The propulsion system, like the levitating system, is based on magnetic principles that are identical to the principles in an ordinary electric motor. The road bed is appropriately configured so that the magnetic field exerts a force on a current-carrying coil secured to the vehicle. In a real sense, the vehicle and magnetic road bed constitute a motor, albeit a linear motor (as opposed to the rotational motor shown in Figure 1). This is called a linear indication motor (LIM). The National Aviation and Space Administration (NASA) believes that this scheme could also be used to launch spacecraft into orbit. A magnetically levitated space vehicle accelerated to a speed of about 600 miles/hour would be catapulted from the ground. Once aloft, a rocket engine would take over and propel the spacecraft into orbit.

Magnetic controls

A magnetic field due to an electric current can be turned on and off simply by turning the current on and off. A piece of iron attached to the end of a spring having the other end fixed can be moved with a magnetic field and returned to its initial position by the spring. The iron piece can then be used to actuate a switch or move a lever on a valve. Applications of this principle include electrically controlled valves in a washing machine and an electrically controlled switch for the starter in an automobile.

Superconducting Magnets

The magnetic field produced by an electric current is proportional to the current; doubling the current doubles the magnetic field. It would seem that an experimenter could achieve any desired magnetic field by creating the necessary electric current in a coil of wire. But wires like those in an electric toaster offer resistance to electric currents, resulting in the production of heat. If the heat is not removed, the wires will melt. Solving this problem is the driving force behind the effort to develop superconducting materials that offer zero electrical resistance to electric current.

Until 1987, achieving zero resistance required cooling electrical conductors to around the temperature of liquid helium (4.2 K or -268.8°C). Nevertheless, practical electromagnets using superconducting wires cooled to around 4 K have been used in the laboratory for several decades. A new class of superconducting materials requiring cooling to around 77 K (-196°C) was discovered in 1987. These materials are difficult to fabricate, but are very attractive because of their substantially higher superconducting temperature. Superconducting materials could be used for wires for transmission lines bringing electricity from an electric power plant to a city. Electromagnets made from superconducting wires can carry much more electric current, and so generate much stronger magnetic fields.

Magnetic Materials

Because magnetic poles occur in N and S pairs, a magnet is referred to as a magnetic dipole. The net magnetic field produced by two magnetic dipoles depends on their orientations. If the N-poles of each point in opposite direction, their contributions to the net magnetic field tend to cancel. If the N-poles of each point in the same direction, their contributions to the net magnetic field tend to add, and the net magnetic field is larger than that of one alone. The more magnets there are with their N-poles pointing in the same direction, the greater is the net magnetic field. In a real sense, iron atoms are like tiny magnetic dipoles. In an unmagnetized piece of iron, the tiny atomic magnetic dipoles are randomly oriented and there is no net magnetic field produced by the iron. But if the dipoles are given a preferred orientation, a net magnetic field around the iron results. The preferred orientation can be achieved by putting the iron in a magnetic field caused by a magnet or an electric current. This is a way of producing a permanent magnet.

Magnetic Recording

The binary system of numbers requires only two numbers, usually designated 0 and 1. Any device having two distinct states can be used to represent the two numbers. A finger pointed up could represent 1; pointing down it could represent 0. Zero and one can be represented with the N-poles of a magnetic dipole pointing in opposite directions. Information is recorded on a floppy disk or magnetic tape in binary fashion. The disk or tape is coated with a magnetizable material. Tiny local areas are magnetized with either the N-pole up or down to represent 0 or 1 . The device that reads the information detects the orientations and translates the information.

Magnetic Resonance Imaging

Magnetic Resonance Imaging (MRI) is a revolutionary diagnostic tool for producing images of the

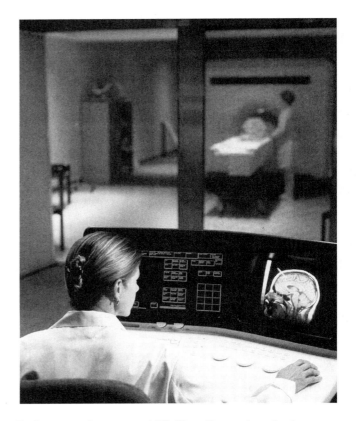

Technicians administer an MRI. (Photo Researchers, Inc.)

interior of a human body. It works because a proton, the sole constituent of a hydrogen atom, behaves like a tiny magnetic dipole. There is a copious source of hydrogen atoms in the body because many components contain water, and every water molecule has two hydrogen atoms. Normally, the atomic dipoles are oriented randomly, but they align when in a strong magnetic field. When stimulated by a short burst of radio frequency electromagnetic radiation, the atomic dipoles are deflected away from the direction of the magnetic field. Following the short burst of radiation, the dipoles rotate (precess) in ever-decreasing circles around the direction of the magnetic field, and in doing so emit a detectable signal before returning to a state of equilibrium. The detected signal can be converted into an image using computer technology. The technique has revolutionized the diagnosis of problems associated with muscles and joints in the human body.

Joseph Priest

See also: Oersted, Hans Christian.

BIBLIOGRAPHY

Elster, A. D. (1986). *Magnetic Resonance Imaging: A Reference Guide and Atlas.* Philadelphia: Lippincott.

Feynman, R. P. (1985). *QED: The Strange Theory of Light and Matter.* Princeton, NJ: Princeton University Press.

Livington, J. D. (1996). *Driving Force: The Natural Magic of Magnets.* Cambridge, MA: Harvard University Press.

Macaulay, D. (1988). *The Way Things Work.* Boston: Houghton Mifflin.

Vranich, J. (1991). *Supertrains: Solutions to America's Transportation Gridlock.* New York: St. Martin's Press.

MAGNETOHYDRO-DYNAMICS

Magnetohydrodynamics (MHD) is a promising technology for electric bulk power generation. MHD is accomplished by forcing an electrically conducting fluid or a plasma through a channel with a magnetic field applied across it and electrodes placed at right angles to flow and field (Figure 1). An MHD plant can be directly fired with coal and there are no moving parts. To achieve extra high efficiencies, MHD is combined in a binary thermodynamic cycle with a conventional steam plant to add an extra 40 percent to the total power output and to boost the overall combined efficiency into the 60 percent range. The high temperature MHD process extracts part of the heat energy in the plasma at the high temperature end. The gas leaving the MHD generator, still at relatively high temperatures, is then used in a conventional bottoming steam plant.

A schematic coalfired MHD generator in Figure 2 is shown using a combustion plasma at up to 3,000 K, seeded with alkali salts for high conductivity, fed directly into the generator with a superconducting magnet providing up to 7 tesla. The combustion products, still at about 2,000 K, then pass through an air preheater, where the hot gas exhausted from the MHD generator preheats the air for the combustor leading to the high temperatures required to create the plasma.

To use lower gas temperatures (under 1,500 K), noble gases like argon and helium offer very high electrical conductivities. Recycling the noble gas leads to the "closed cycle" MHD process.

Nuclear reactors have been used for small generators providing the high temperature and pressure plasma to drive an MHD process, and a nuclear plant could offer a considerable reduction in pollution because of high overall efficiency without CO_2 emission.

EXPERIMENTAL DEVELOPMENTS

In the early part of the nineteenth century Michael Faraday (1832) conducted MHD experiments using

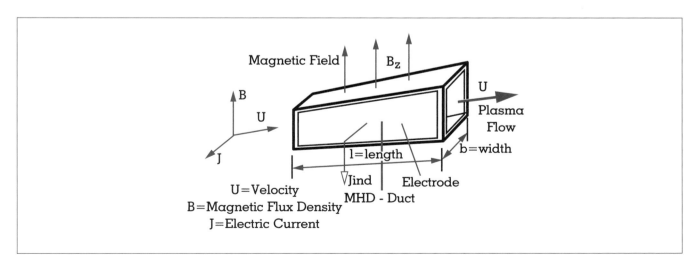

Figure 1.
Linear MHD Generator Channel.

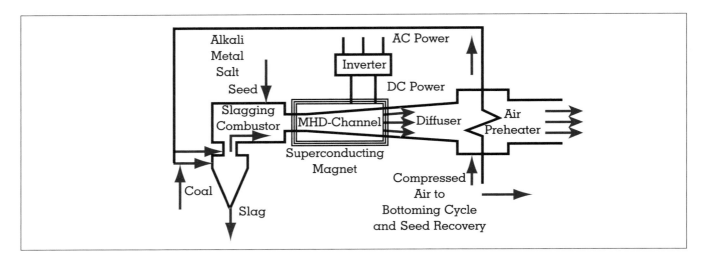

Figure 2.
MHD Linear Generator Flow Chart.

the brackish water of the river Thames flowing through the Earth's magnetic field. The first successful power generation experiment, developed by Richard Rosa in 1959, generated 10 kW with a timber walled channel on the AVCO "Mark 1" facility in Boston, Massachusetts. This success and the possibility of cheap MHD power led in the 1960s to national programs in Britain, the Soviet Union, The Netherlands, France, Germany, Poland, Italy, India, Australia and Israel. In 1965 the AVCO "Mark 5" generator successfully generated 32 MW over a one minute run using alcohol at 45 kg/sec fired with oxygen. AVCO later developed a sophisticated coal fired MHD channel for a 2,000 hour test program and demonstrated technical feasibility under the most stringent conditions.

In 1972 in Moscow, a large experimental facility, the "U-25," used a 250 MW natural gas combustor and generated 20 MW. The Soviets have been using very successfully mobile, pulsed MHD generators throughout the Soviet Union, for seismic studies.

MHD programs in the United States are concentrated in two major facilities. A "Component Development and Integration Facility" is located in Butte, Montana, and a "Coal Fired Flow Facility" at the University of Tennessee to studies coal fired MHD, slag processing, seed handling and downstream systems.

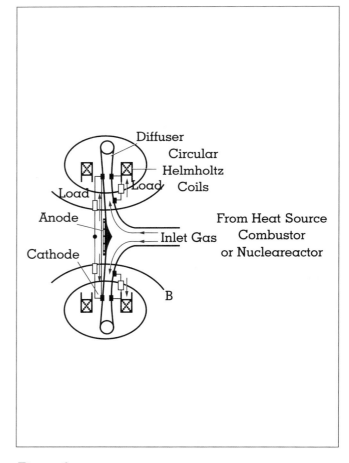

Figure 3.
MHD Disc Generator.

A circular disk MHD generator geometry requires a lower cost circular ring magnet structure as shown in Figure 3. Work in the United States using shock tunnels with large discs proved the feasibility of very high enthalpy extraction. Continuous operation on a fossil fuel fired disk facility was demonstrated at the University of Sydney, Australia. The potential advantages offered by closed cycle disc systems was made evident in the 1980s and 1990s by the Japanese MHD research effort and by the Dutch team at the University of Eindhoven.

A great deal of work has been done in the United States, Russia, Israel, and France showing that low temperature operation is possible using liquid metal as the MHD driver. A liquid metal MHD-generator can be very much smaller because of the much higher conductivity of liquid metal.

POWER SYSTEM REQUIREMENTS

Studies carried out in the United States, Russia, and Japan indicate that a combined cycle MHD-steam plant should be able to achieve an overall power station efficiency of at least 60 percent which is about 20 percent more than offered by a conventional steam plant. This should be possible at capital costs comparable with existing steam plants.

As shown in Figure 4, the operating temperature range for MHD is beyond that of any other generating technology. MHD could still add up to 15 percent at the top end of other combined cycles. MHD is potentially a natural choice for conversion of high temperature energy output from future nuclear power plants whether fission or fusion driven. A combined Nuclear-MHD system design with high efficiency does offer two advantages: (1) a reduction in thermal pollution and (2) no CO_2 emission as with fossil fuel-driven plants.

MHD efforts in many countries, including the United States, have declined substantially, in Russia because of lack of funds and in general because the high costs envisioned in setting up a full-scale power station. If funds become available to set up full-scale power stations and with the advent of high temperature nuclear heat sources, the many advantages provided by MHD may be realised.

Hugh Karl Messerle

See also: Combustion; Electric Power, Generation of Faraday, Michael; Hydroelectric Energy; Mag-

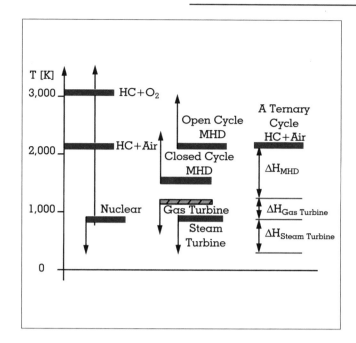

Figure 4.
Temperature Ranges for MHD, Gas Turbine, and Steam Turbine Plants.

netism and Magnets; Nuclear Energy; Thermodynamics.

BIBLIOGRAPHY

Faraday, M. (1832). "Experimental Researches in Electricity." *First Series, Philosophical Transactions of the Royal Society*, pp. 125-162.

Messerle, H. K. (1995). *Magneto-Hydro-Dynamic Electrical Power Generation.* Chichester, England: John Wiley & Sons Ltd.

Petrick, M., and Shumyatsky, B. Y. (1978). *Open-Cycle Magnetohydrodynamic Electrical Power Generation.* Argonne, IL: Argonne National Laboratory.

Rosa, R. J. (1960). "Experimental Magnetohydrodynamic Power Generation." *Applied Physics* 31:735-36.

Rosa, R. J. (1987). *Magnetohydrodynamic Energy Conversion.* Washington, DC: Hemisphere Publishing Corp.

Shioda, S. (1991). "Results of Feasibility Studies on Closed Cycle MHD Power Plants." *Proceedings Plasma Technology Conference*, Sydney, Australia, pp. 189-200.

Simpson, S. W.; Marty, S. M.; and Messerle, H. K. (1989). "Open-Cycle Disk Generators: Laboratory Experiments and Predictions for Base-Load Operation." *MHD An International Journal* 2(1):57-63.

Veilkov, E. P.; Zhukov, B. P.; Scheindlin, A. E.; Vengerskii, V. V.; Shelkov, E. M.; Babakov, Y. P.; Volkov, Y. M.; Zeigarnik, V. A.; Mtveenko, O. G.; and Kolyadin, N. M.

(1983). "Status and Prospects of Geophysical MHD Power." *Proceedings of the 8th International Conference on MHD Electric Power Generation, Moscow.*

Most of the work carried out in MHD power generation is published in the Proceedings of the eleven "International Conferences on MHD Electrical Power Generation" organized by the International Liaison Group on MHD Electrical Power Generation (ILG-MHD) sponsored by UNESCO and originally by IAEA and ENEA, in the annual series of the "Symposia on Engineering Aspects of Magnetohydrodynamics" (SEAM) in the United States and in the Japanese series of CIS MHD Symposia.

MANUFACTURING

Industrial production is the backbone for economic output in almost all countries. Over the past decades, manufacturing industrial production has been growing in most economies. The industrial sector is dominated by the production of a few major energy-intensive commodities, such as steel, paper, cement, and chemicals. In any given country or region, production of these basic commodities follows the general development of the overall economy. Rapidly industrializing countries will have higher demands for infrastructure materials, and more mature markets will have declining or stable consumption levels. The regional differences in consumption patterns (expressed as consumption per capita) will fuel a further growth of consumption in developing countries. In addition to labor costs and costs for raw materials in these "heavy" industries, energy is a very important production cost factor in the effort to achieve higher energy efficiency (see Table 1). Markets in the industrialized countries show a shift toward more service-oriented activities and hence non energy-intensive industries. Because of the great difference in energy intensity between energy intensive industries and all others, changes in output shares of these industries can have a major impact on total industrial energy use. Many commodities (e.g., food and steel) are traded globally, and regional differences in supply and demand will influence total industrial energy use. Production trends also depend on regional availability of resources (e.g., scrap) and capital. Manufacturing energy use also will depend on the energy efficiency with which the economic activities are done. In this article we will assess trends in industrial energy use and energy intensities, followed by a discussion of energy services, uses, and industrial technologies.

GLOBAL MANUFACTURING ENERGY USE

In 1990, manufacturing industry accounted for 42 percent (129 exajoules, EJ) of global energy use. Between 1971 and 1990, industrial energy use grew at a rate of 2.1 percent per year, slightly less than the world total energy demand growth of 2.5 percent per year. This growth rate has slowed in recent years and was virtually flat between 1990 and 1995, primarily because of declines in industrial output in the transitional economies of Eastern Europe and the former Soviet Union. Energy use in the industrial sector is dominated by the industrialized countries, which accounted for 42 percent of world industrial energy use in 1990. Industrial energy consumption in these countries increased at an average rate of 0.6 per year between 1971 and 1990, from 49 EJ to 54 EJ. The share of industrial sector energy consumption within the industrialized countries declined from 40 percent in 1971 to 33 percent in 1995. This decline partly reflects the transition toward a less energy-intensive manufacturing base, as well as continued growth in transportation demand, resulting in large part from the rising importance of personal mobility in passenger transport use.

The industrial sector dominates in the economies in transition, accounting for more than 50 percent of total primary energy demand, the result of the emphasis on materials production, a long-term policy promoted under years of central planning. Average annual growth in industrial energy use in this region was 2.0 percent between 1971 and 1990 (from 26 EJ to 38 EJ), but dropped by an average of -7.3 percent per year between 1990 and 1995.

In the Asian developing countries, industrial energy use grew rapidly between 1971 and 1995, with an annual average growth rate of 5.9, jumping from 9 EJ. It also accounted for the greatest share of primary energy consumption, between 57 percent and 60 percent. The fastest growth in this sector was in China and in other rapidly developing Asian countries. Growth in other developing countries was slightly lower.

The nature and evolution of the industrial sector vary considerably among developing countries. Some economies that are experiencing continued expan-

Sector	1973		1985		1994	
	Energy Intensity (primary energy)	Energy Costs (share of production costs)	Energy Intensity (primary energy)	Energy Costs (share of production costs)	Energy Intensity (primary energy)	Energy Costs (share of production costs)
Iron & Steel	30.5 GJ/t	7%	27.8 GJ/t	11%	25.4 GJ/t	8%
Pulp & Paper	43.1 GJ/t	6%	42.7 GJ/t	6%	32.8 GJ/t	6%
Cement	7.3 GJ/t	40%	5.2 GJ/t	36%	5.4 GJ/t	33%
Primary Aluminum	N/A	14%	17.6 MWh/t	19%	16.2 MWh/t	13%
Petroleum Refining	6.2 GJ/t	4%	4.3 GJ/t	3%	4.5 GJ/t	3%

Table 1.
Energy Intensities and Energy Costs in Selected U.S. Industries.
NOTE: Energy intensity is expressed in primary energy, where the efficiency of electricity generation is assumed to be 33 percent. Energy intensity of primary aluminum production is given in MWh (1000 kWh).
SOURCE: Lawrence Berkeley National Laboratory, Berkeley, California.

sion in energy-intensive industry, such as China and India, show relatively unchanging shares of industrial energy use. In other countries, such as Thailand and Mexico, the share and/or growth of the transportation sector dominate. Many smaller countries have remained primarily agrarian societies with modest manufacturing infrastructure.

ENERGY INTENSITY TRENDS

In aggregate terms, studies have shown that technical efficiency improvements of 1 to 2 percent per year has been observed in the industrial sector in the past, Between 1975 and 1983 during and after the years of major oil price increases, U.S. industrial energy intensity declined by 3.5 percent per year. Between 1984 and 1994 industrial energy intensity declined by less than 1 percent on average (Brown et al., 1998). Figure 1 gives an overview of energy of economic intensity trends in the industrial sector in industrialized countries.

The trends demonstrate the capability of industry to improve energy efficiency when it has the incentive to do so. Energy requirements can be cut by new process development. In addition, the amount of raw materials demanded by a society tends to decline as countries reach certain stages of industrial development, which leads to a decrease in industrial energy use. The accounting of trends in structural shift, material intensity, and technical energy efficiency

and their interactions can be extremely difficult. To understand trends in energy intensity it is important to analyze the structure of the industrial sector. Industrial energy use can be broken down into that of the energy-intensive industries (e.g., primary metals, pulp and paper, primary chemicals, oil refining, building materials) and the nonenergy intensive-industries (e.g., electronics and food). Reduction of energy intensity is closely linked to the definition of structure, structural change, and efficiency improvement. Decomposition analysis is used to distinguish the effects of structural change and efficiency improvement. Structural change can be broken down into intra-sectoral (e.g., a shift toward more recycled steel) and intersectoral (e.g., a shift from steel to aluminum within the basic metals industry). A wide body of literature describes decomposition analyses and explains the trends in energy intensities and efficiency improvement. Decomposition analyses of the aggregate manufacturing sector exist mainly for industrialized Western countries, but also for China; Taiwan; and selected other countries, including those in Eastern Europe. The results show that different patterns exist for various countries which may be due to specific conditions as well as differences in driving forces such as energy prices and other policies in these countries. More detailed analyses on the subsector level are needed to understand these trends better. Changes in energy intensities also can be disaggregated into structural changes and efficiency

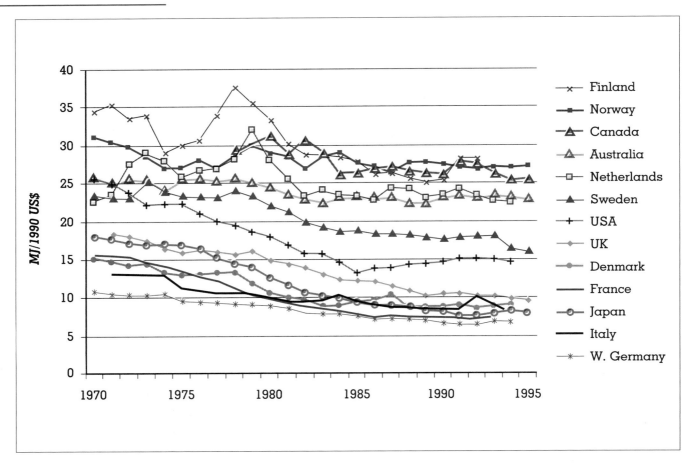

Figure 1.

Industrial sector economic energy intensity trends in selected industrialized countries, 1970–1995.

SOURCE: Lawrence Berkeley National Laboratory, Berkeley, California.

improvements at the subsector level. In the iron and steel industry, energy intensity is influenced by the raw materials used (i.e., iron ore, scrap) and the products produced (e.g., slabs, or thin rolled sheets). A recent study on the iron and steel industry used physical indicators for production to study trends in seven countries which together produced almost half of the world's steel. Figure 2 shows the trends in physical energy intensity in these countries, expressed as primary energy used per metric ton of crude steel. The large differences in intensity among the countries are shown, as well as the trends toward reduced intensity in most countries. Actual rates of energy efficiency improvement varied between 0.0 percent and 1.8 percent per year, while in the case of the restructuring economy of Poland, the energy intensity increased.

ENERGY SERVICES AND ENERGY EFFICIENCY

Energy is used to provide a service (e.g., a ton of steel) or to light a specified area. These services are called energy services. Energy efficiency improvement entails the provision of these services using less energy. About half of industrial energy use is for specific processes in the energy-intensive industries. On the other hand, various general energy conversion technologies and end uses can also be distinguished, such as steam production, motive power, and lighting. Hence, energy use in manufacturing can be broken down to various uses to provide a variety of services. A common breakdown distinguishes energy use for buildings, processes and utilities and boilers. The boilers provide steam and hot water to the processes, and the buildings. Due to the wide variety of indus-

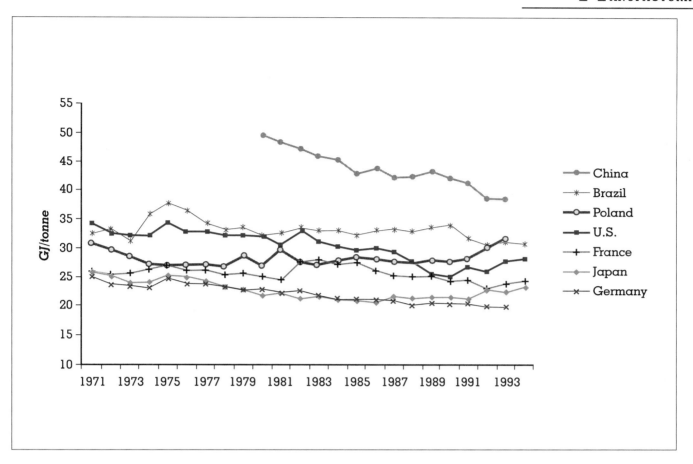

Figure 2.
Trends in physical energy intensity in seven countries between 1971 and 1994.
SOURCE: Lawrence Berkeley National Laboratory, Berkeley, California.

trial processes, we will limit our discussion to two energy intensive sectors—iron and steel, and the pulp and paper industries, as well as boilers, to illustrate important cross-cutting energy-consuming processes in industry.

Iron and Steel Industry

The first record of the use of iron goes back to 2500-2000 B.C.E., and the first deliberate production of iron began in about 1300 B.C.E. Small furnaces using charcoal were used. Evidence of such furnaces has been found in Africa, Asia, and Central Europe. The relatively low temperatures in the furnace lead to low-quality iron, and the slag had to be removed by hammering the iron. High-temperature processes started to be introduced in Germany in about 1300 C.E. The design of these furnaces is essentially the same of that of modern blast furnaces. The furnaces

still used charcoal, and in 1718 the first use of coke is reported in the United Kingdom. The higher strength of coke allowed larger furnaces to be built, increasing energy efficiency. By 1790 coke iron making contributed to 90 percent of British iron production. Between 1760 and 1800, energy use declined by about 2 percent per year, mainly through the use of steam engines permitting higher blast pressures. During the nineteenth century, coke demand was further reduced by 1 percent per year. The development of the modern blast furnace after World War II resulted in an annual reduction of energy intensity of 3 to 4 percent a year, due to the use of improved raw materials, ore agglomeration, larger blast furnaces, and higher air temperatures. Today the blast furnace is the main process to make iron, and provides the largest raw material stream in steelmaking.

Steel is made by reducing the carbon content in the iron to levels below 2 percent. This reduces the brittleness of the material and makes it easier to shape. The first steelmaking process was invented in 1855 by Bessemer, and was in commercial operation by 1860. In the Bessemer converter air was blown through the hot iron, which oxidizes the carbon. This principle is still followed in modern steelmaking processes. In the United States the last Bessemer converter was retired in the 1960s. In the late nineteenth century the open-hearth furnace (OHF) or Siemens-Martin furnace was invented, which uses preheated air and fuels to oxidize the carbon and melt the steel. This process is currently found only in developing countries and in Eastern Europe. The United States was one of the industrialized countries that phased out OHF at a very late stage. In the 1980s the dominant process became the basic oxygen furnace (BOF), using pure oxygen instead of air. The BOF process was developed in Austria in the 1950s. The productivity of this process is much higher than that of OHF, as is the energy efficiency. An alternative process is the electric arc furnace (EAF). The EAF process is mainly used to melt scrap. Performance of EAFs has improved tremendously; fuel and oxygen are starting to be used besides electricity. In the future it is expected that the BOF and EAF processes will follow similar developmental paths.

Liquid steel is cast into ingots or slabs and shaped in rolling mills to the final product. Although most energy use is concentrated in ironmaking and steelmaking, reduced material losses and productivity gains in casting and shaping (e.g., continuous casting and, currently thin slab casting) have contributed to dramatic increases in the energy efficiency of steelmaking.

Today, the U.S. iron and steel industry is made up of integrated steel mills which produce pig iron from raw materials (iron ore, coke) using a blast furnace and steel using a BOF, and secondary steel mills (minumills), which produce steel from scrap steel, pig iron, or direct reduced iron (DRI) using an EAF. The majority of steel produced in the United States is from integrated steel mills, although the share of minimills is increasing, growing from 15 percent of production in 1970 to 40 percent in 1995. There were 142 operating steel plants in the United States in 1997, of which 20 were integrated steelmills and 122 were minimills. The integrated mills are most often located near or with easy access to the primary resources; for example, in the United States these are concentrated in the Great Lakes region, near supplies of coal and iron ore and near key customers such as automobile manufacturers.

The worldwide average energy intensity of steelmaking is estimated at gigajoules (GJ) per metric tone, although large variations occur among countries and plants (see Figure 2). Today the most energy-efficient process would use 19 GJ/per metric ton for integrated steelmaking, and 7 GJ/per metric ton for making steel out of scrap. Analyses have shown that many technologies exist that could improve energy efficiency further. For example, in the United States the potential for energy efficiency improvement is estimated at 18 percent, using proven and cost-effective practices and technologies. Under development are new technolgies that could considerably lower the energy intensity of steelmaking. Smelt reduction in ironmaking would integrate the production of coke and agglomerated ore with that of ironmaking, leading to reductions in production costs and energy use. The development of direct casting techniques that abandon rolling would increase productivity while reducing energy use further. Combined, these technologies could reduce the energy intensity of primary steelmaking to 12.5 GJ/per metric ton of steel, and for secondary steelmaking to 3.5 GJ/ per metric ton reductions of 34 percent and 50 percent, respectively. In the highly competitive and globalizing steel industry, manufacturers must continuously look for ways to lower their energy intensity and costs.

Pulp and Paper Industry

Paper consists of aligned cellulosic fibers. The fibers may be from wood or other crops, or from recycling waste paper. Starting with wood fibers, the fibers need to be separated from the wood, which is done by pulping the wood. The separation can be done by chemicals, by heat, or by mechanical means. In the chemical pulping process chemicals and hot water are used to separate the cellulosis from the ligno-cellulosis. The amount of pulp produced is about half the amount of wood used. Chemical pulping results in high-quality paper. In the mechanical process the wood is ground under pressure, separating the fibers from each other. In mechanical pulping the ligno-cellulosis is generally not removed, resulting in a lower quality of paper (e.g., paper used for newsprint) but a higher recovery (about 90%) of the used wood. In chemical pulping a lot of steam is used to heat the water and concentrate the chemical by-products. However, recovery of the by-products to be recycled in the process can actually produce sufficient steam for the whole paper mill. The mechanical process uses large quantities of electricity, while some processes can recover steam from the grinding process. Chemical pulp can be bleached to produce white paper. Waste paper is being pulped by mixing with water, after which ink is removed and the pulp refined. Paper recycling reduces the energy needs of the pulping process. Waste paper use in the production of paper varies widely, due to different structures of the industry in different countries.

While energy efficiency improvement options do exist in the pulping step, greater opportunities exist in the chemical recovery step. The most common pulping process in the United States is the Kraft pulping process. Black liquor is produced as a by-product. The chemicals are currently recovered in a recovery boiler, combusting the ligno-cellulosis. Because of the high water content, the efficiency of the recovery boiler is not very high, and the steam is used to generate electricity in a steam turbine and steam for the processes. Gasification of black liquor would allow use of the generated gas at a higher efficiency. This would make a Kraft pulp mill an electricity exporter (Nilsson et al., 1995).

In papermaking the pulp is diluted with water at a ratio of about 1:100. This pulp is screened and refined. The solution with the refined fibers (or stock) is fed into the paper machine, where the water is removed. In the paper machine, the paper is formed into a sheet, and water is removed by dispersing over a wire screen. At the end of the forming section, 80 percent of the water has been removed. The rest of the water is removed in the pressing and drying sections. While only a small amount of water is removed in the drying section, most energy is used there. Hence, those using such energy efficiency opportunities try to reduce the water content by increasing the water removal by pressing. In a long nip press, the pressing area is enlarged. The larger pressing area results in extra water removal. New technologies are under development aiming to increase the drying efficiency considerably. One technology—impulse drying—uses a heated roll, pressing out most of the water in the sheet; this may reduce the steam consumption of the paper machine by 60 to 80 percent.

Energy Efficiency Measure	Typical Energy Savings (%)	Payback Estimate (years)
Distribution System		
Reducing Steam Leaks	3-5%	
Insulating Steam Pipes	5-10%	0.3-1.7 year
Condensate Return	10-20%	<1 year
Process Integration & Heat Recovery	5-40%	2-
System Operation and Maintenance		
Water Treatment	10-12%	
Load Control	3-5%	
Decentralization Steam Supply	<40%	<4 years
Hot Standby		
Boilers		
Boiler Tune-Up	1-2%	
Combustion Air Preheating	<12%	<5 years
Boiler Feed Preheating	2-10%	<4 years
New Low-NOx Boiler Type	>5%	n.a.
Monitoring and Control	1-4%	<3 years

Table 2.
Energy Efficiency Measures and Estimated Improvement Potentials for Steam Boilers.
SOURCE: Prindle et al., 1995; Jones, 1997; CADDET, 1999.

The pulp and paper industry uses approximately 6 to 8 EJ globally. Because energy consumption and intensity depend on the amount of wood pulped, the type of pulp produced, and the paper grades produced, there is a great range in energy intensities among industrialized countries of the world. In Europe energy use for papermaking varied between 16 GJ per metric ton and to 30 GJ per metric ton of paper in 1989. The Netherlands used the least energy per metric ton of paper, largely because most of the pulp was imported. Countries such as Sweden and the United States have much higher energy intensities due to the larger amount of pulp produced. Sweden and other net exporters of pulp also tend to show a higher energy intensity. Energy intensity is also influenced by the efficiency of the processes used. Many studies have shown considerable potentials for energy efficiency improvement with current technologies (Worrell et al., 1994; Nilsson et al., 1995; Farla et al., 1997), such as heat recovery and improved pressing technologies.

Cross-Cutting: Steam Production and Use

Besides the energy-intensive industries, many smaller and less energy-intensive, or light, industries exist. Light industries can include food processing, metal engineering, or electronics industries. In light industries energy is generally a small portion of the total production costs. There is a wide variety of processes used within these industries. Generally a large fraction of energy is used in space heating and cooling, in motors (e.g., fans and compressed air), and in boilers. Industrial boilers are used to produce steam or to heat water for space and process heating and for the generation of mechanical power and electricity. In some cases these boilers will have a dual function, such as the cogeneration of steam and electricity. The largest uses of industrial boilers by capacity are in paper, chemical, food production, and petroleum industry processes. Steam generated in the boiler may be used throughout a plant or a site. Total installed boiler capacity (not for cogeneration) in the United States is estimated at nearly 880 million megawatts. Total energy consumption for boilers in the United States is estimated at 9.9 EJ.

A systems approach may substantially reduce the steam needs, reduce emissions of air pollutants and greenhouse gases, and reduce operating costs of the facility. A systems approach assessing options throughout the steam system that incorporates a variety of measures and technologies is needed (Zeitz, 1997), and can help to find low-cost options. Improved efficiency of steam use reduces steam needs and may reduce the capital layout for expansion, reducing emissions and permitting procedures at the same time. Table 2 summarizes various options to reduce losses in the steam distribution and to improve system operation and the boiler itself. In specific cases, the steam boiler can be replaced almost totally by a heat pump (or mechanical vapor recompression) to generate low-pressure steam. This replaces the fuel use for steam generation by electricity. Emission reductions will depend on the type and efficiency of power generation.

Another option to reduce energy use for the steam system is cogeneration of heat and power (CHP) based on gas-turbine technology as a way to substantially reduce the primary energy needs for steam making. Low- and medium-pressure steam can be generated in a waste heat boiler using the flue gases of a gas turbine. Classic cogeneration systems are based on the use of a steam boiler and a back-pressure turbine. These systems have relatively low efficiency compared to a gas turbine system. Steam-turbine systems generally have a power-to-heat ratio between 0.15 (40 kWh/GJ) and 0.23 (60 kWh/GJ) (Nilsson et al., 1995). The power-to-heat ratio depends on the specific energy balance of plant as well as energy costs. A cogeneration plant is most often optimized to the steam load of the plant, exporting excess electricity to the grid. The costs of installing a steam turbine system strongly depend on the capacity of the installation. Gas-turbine-based cogeneration plants are relatively cheap. In many places (e.g., The Netherlands and Scandinavia) gas turbine cogeneration systems are standard in paper mills. The power-to-heat ratio is generally higher than for steam turbine systems. Aero-derivative gas turbines may have a power-to-heat ratio of 70 to 80 kWh/GJ . Aeroderivative turbines are available at low capacities, but specific costs of gas turbines sharply decrease with larger capacities.

POTENTIAL FOR ENERGY EFFICIENCY IMPROVEMENT

Much of the potential for improvement in technical energy efficiencies in industrial processes depends on how closely such processes have approached their thermodynamic limit. There are two types of energy efficiency measures: (1) more efficient use in existing equipment through improved operation, maintenance or retrofit of equipment and (2) use of more efficient new equipment by introducing more efficient processes and systems at the point of capital turnover or expansion of production. More efficient practices and new technologies exist for all industrial sectors. Table 2 outlines some examples of energy efficiency improvement techniques and practices.

A large number of energy-efficient technologies are available (see Table 3) in the steel industry, including continuous casting, energy recovery, and increased recycling. Large technical potentials ranging from 25 to 50 percent exist in most countries. New technologies are under development (e.g., smelt reduction and near net shape casting) that will reduce energy consumption as well as environmental pollution and capital costs. A few bulk chemicals such as ammonia and ethylene represent the bulk of energy use in this subsector. Potentials for energy savings in ammoniamaking are estimated to be up to 35 percent in Europe and between 20 percent and 30 percent in Southeast Asia.

Iron and Steel
Heat recovery for steam generation, pre-heating combustion air, and high efficiency burners

Adjustable speed drives, heat recovery coke oven gases, and dry coke quenching

Efficient hot blast stove operation, waste heat recovery for hot blast stove, top gas power recovery turbines, direct coal injection

Recovery BOF-gas, heat recovery of sensible heat BOF-gas, closed BOF-gas-system, optimized oxygen production, increase scrap use, efficient tundish preheating

UHP-process, Oxy-fuel injection for EAF plants, and scrap preheating

Heat recovery (steam generation), recovery of inert gases, efficient ladle preheating

Use of continuous casting, 'Hot connection' or direct rolling, recuperative burners

Heat recovery, efficient burners annealing and pickling line, continuous annealing operation

Chemicals
Process management and thermal integration (e.g. optimization of steam networks, heat cascading, low and high temperature heat recovery, heat transformers), mechanical vapor recompression

New compressor types

New catalysts

Adjustable speed drives

Selective steam cracking, membranes

High temperature cogeneration and heat pumps

Autothermal reforming

Petroleum Refining
Reflux overhead vapor recompression, staged crude pre-heat, mechanical vacuum pumps

Fluid coking to gasification, turbine power recovery train at the FCC, hydraulic turbine power recovery, membrane hydrogen purification, unit to hydrocracker recycle loop

Improved catalysts (reforming), and hydraulic turbine power recovery

Process management and integration

Pulp and Paper
Continuous digester, displacement heating/batch digesters, chemi-mechanical pulping

Black liquor gasification/gasturbine cogeneration

Oxygen predelignification, oxygen bleaching, displacement bleaching

Tampella recovery system, falling film black liquid evaporation, lime kiln modifications

Long nip press, impulse drying, and other advanced paper machines

Improved boiler design/operation (cogeneration), and distributed control systems

Cement
Improved grinding media and linings, roller mills, high-efficiency classifiers, wet process slurry

Dewatering with filter presses

Multi-stage preheating, pre-calciners, kiln combustion system improvements, enhancement of internal heat transfer in kiln, kiln shell loss reduction, optimize heat transfer in clinker cooler, use of waste fuels

Blended cements, cogeneration

Modified ball mill configuration, particle size distribution control, improved grinding media and linings, high-pressure roller press for clinker pre-grinding, high-efficiency classifiers, roller mills

Table 3.
Efficiency Improvement Measures in Energy Intensive Industry
SOURCE: Worrell, Levine, et al., 1997.

Energy savings in petroleum refining are possible through improved process integration, cogeneration, energy recovery, and improved catalysts. Compared to state-of-the-art technology, the savings in industrialized countries are estimated to be 15 to 20 percnet, and higher for developing countries. Large potentials for energy savings exist in nearly all process stages of pulp and paper production (e.g., improved dewatering technologies, energy and waste heat recovery, and new pulping technologies). Technical potentials are estimated at up to 40 percent, with higher long-term potentials (see above). Energy savings in cement production are possible through increased use of additives (replacing the energy-intensive clinker), use of dry process, and use of a large number of energy efficiency measures (such as reducing heat losses and use of waste as fuel). Energy savings potentials of up to 50 percent do exist in the cement industry in many countries through efficiency improvement and the use of wastes such as blast furnace slags and fly ash in cementmaking.

In the United States various studies have assessed the potential for energy efficiency improvement in industry. One study has assessed the technologies for various sectors and found potential economic energy savings of 7 to 13 percent over the business-as-usual trends (Brown et al., 1998) between 1990 and 2010. Technologies like the ones described above (see Table 2) are important in achieving these potentials.

However, barriers may partially block the uptake of those technologies. Barriers to efficiency improvement can include unwillingness to invest, lack of available and accessible information, economic disincentives, and organizational barriers. The degree to which a barrier limits efficiency improvement is strongly dependent on the situation of the actor (e.g., small companies, large industries). A range of policy instruments is available, and innovative approaches or combinations have been tried in some countries. Successful policy can contain regulations (e.g., product standards) and guidelines, economic instruments and incentives, voluntary agreements and actions, information, education and training, and research, development and demonstration policies. Successful polices with proven track records in several sectors include technology development, and utility/government programs and partnerships. Improved international cooperation to develop policy instruments and technologies to meet developing country needs will be

necessary, especially in light of the large anticipated growth of the manufacturing industry in this region.

SUMMARY

Manufacturing industry is a large energy user in almost all countries. About half of industrial energy use is in specific processes in the energy-intensive industries. On the other hand, various general energy conversion technologies and end uses can also be distinguished, such as steam production, motive power, and lighting. Opportunities and potentials exist for energy savings through energy efficiency improvement in all sectors and countries. Technology development, and policies aimed at dissemination and implementation of these technologies, can help to realize the potential benefits. Technologies do not now, nor will they in the foreseeable future, provide a limitation on continuing energy efficiency improvements.

Ernest Warrell

See also: Economically Efficient Energy Choices; Industry and Business Energy as a Factor of Production in; Industry and Business, Productivity and Energy Efficiency in.

BIBLIOGRAPHY

Ang, B. W. (1995) "Decomposition Methodology in Industrial Energy Demand Analysis." *Energy* 20(11):1081-1096.

Ang, B. W., and Pandiyan, G. (1997) "Decomposition of Energy-Induced CO_2 Emissions in Manufacturing." *Energy Economics* 19:363-374.

Brown, M. A.; Levine, M. D.; Romm, J. P.; Rosenfeld, A. H.; and Koomey, J. G. (1998). "Engineering-Economic Studies of Energy Technologies to Reduce Greenhouse Gas Emissions: Opportunities and Challenges." *Annual Review of Energy and the Environment* 23:287-385.

Center for the Analysis and Dissemination of Demonstrated Energy Technologies (CADDET). (1999). *CADDET Register on demonstration projects* Sittard, Neth.: Author/IEA.

De Beer, J.; Worrell, E.; and Blok, K. (1998). "Future Technologies for Energy Efficient Iron and Steel Making." *Annual Review of Energy and the Environment* 23:123-205.

De Beer, J.; Worrell, E.; and Blok, K. (1998). "Long-Term Energy-Efficiency Improvements in the Paper and Board Industries." *Energy* 23:21-42.

Farla, J.; Blok, J.; and Schipper, L. (1997). "Energy Efficiency Developments in the Pulp and Paper Industry." *Energy Policy* 25:745-758.

Howarth, R. B.; Schipper, L.; Duerr, P. A.; and Strom, S. (1991). "Manufacturing Energy Use in Eight OECD Countries, Decomposing the Impacts of Changes in Output, Industry Structure, and Energy Intensity." *Energy Economics* 13:135-142.

IEA. (1997). *Indicators of Energy Use and Efficiency: Understanding the Link Between Energy and Human Activity.* Paris: Author/OECD.

Jones, T. (1997). "Steam Partnership: Improving Steam System Efficiency Through Marketplace Partnerships." *Proceedings 1997 ACEEE Summer Study on Energy Efficiency in Industry*, Washington, DC: ACEE.

LBNL. (1998). *OECD Database.* Berkeley, CA, Lawrence Berkeley National Laboratory.

Li, J.-W.; Shrestha, R. M.; and Foel, W. K. (1990). "Structural Change and Energy Use: The Case of the Manufacturing Industry in Taiwan." *Energy Economics* 12:109-115.

Nilsson, L. J.; Larson, E. D.; Gilbreath, K. R.; and Gupta, A. (1995). "Energy Efficiency and the Pulp and Paper Industry." Washington DC: American Council for an Energy Efficient Economy.

Park, S.-H.; Dissmann, B.; and Nam, K.-Y. (1993). "A Cross-Country Decomposition Analysis of Manufacturing Energy Consumption." *Energy* 18:843-858.

Price, L.; Michaelis, L.; Worrell, E.; and Khrushch, M. (1998). "Sectoral Trends and Driving Forces of Global Energy Use and Greenhouse Gas Emissions." *Mitigation and Adaptation Strategies for Global Change* 3:263-319.

Prindle, W.; Farfomak, P.; and Jones, T. (1995). "Potential Energy Conservation from Insulation Improvements in U.S. Industrial Facilities." *Proceedings 1995 ACEEE Summer Study on Energy Efficiency in Industry.* Washington, DC: ACEEE.

Ross, M. H., and Steinmeyer, D. (1990). "Energy for Industry." *Scientific American* 263:89-98.

Schipper, L., and Meyers, S. (1992). *Energy Efficiency and Human Activity: Past Trends, Future Prospects.* New York: Cambridge University Press.

Sinton, J. E., and Levine, M. D. (1994). "Changing Energy Intensity in Chinese Industry." *Energy Policy* 21:239-255.

WEC. (1995). *Energy Efficiency Utilizing High Technology: An Assessment of Energy Use in Industry and Buildings*, prepared by M. D. Levine, E. Worrell, N. Martin, and L. Price. London: Author.

Worrell, E.; Cuelenaere, R. F. A.; Blok, K.; and Turkenburg, W. C. (1994). "Energy Consumption by Industrial Processes in the European Union." *Energy* 19:1113-1129.

Worrell, E.; Levine, M. D.; Price, L. K.; Martin, N. C.; van den Broek, R.; and Blok, K. (1997). *Potential and Policy Implications of Energy and Material Efficiency Improvement.* New York: UN Commission for Sustainable Development.

Worrell, E.; Price, L.; Martin, N.; Farla, J.; and Schaeffer, R. (1997). "Energy Intensity in the Iron and Steel Industry: A Comparison of Physical and Economic Indicators." *Energy Policy* 25(7-8):727-744.

Worrell, E.; Martin, N.; and Price, L. (1999). "Energy Efficiency and Carbon Emission Reduction Opportunities in the U.S. Iron and Steel Industry." Berkeley, CA: Lawrence Berkeley National Laboratory.

Zeitz, R. A., ed. (1997). *CIBO Energy Efficiency Handbook.* Burke, VA: Council of Industrial Boiler Owners.

MARKET IMPERFECTIONS

Modern economic theory provides a succinct description of the conditions under which the price system produces optimal outcomes in an idealized "laissez-faire" economy of perfectly competitive markets. This is the "First Welfare Theorem"; any competitive equilibrium is "Pareto optimal" (i.e., no agent in the economy can have his or her well-being increased except at the expense of another agent). Against this benchmark, the theory describes conditions under which policy interventions can, in principle, improve upon the performance of the unregulated market system. The possibility of such improvement arises from the existence of market "failures" or "imperfections," which are factors in or features of particular markets that cause private decision-making to produce less-than-optimal economic outcomes.

Historically, the key imperfection in energy markets was thought to be "economies of scale," or "declining average costs," in electric power generation. This means simply that this was the kind of industry where a single firm's costs of producing power would fall as its output was increased. Under this condition, unregulated market equilibrium with more than one competing firm would not result in economic efficiency, specifically the provision of power at minimum cost. Instead, power generation was a "natural monopoly"—a single firm could produce at lowest cost (in a given geographic area). This feature of power generation was the motivation for the U.S. system of privately owned, publicly regulated power companies, granted an exclusive license within a given service area with rates set by regulators. In the past several decades, however, technological change in electric power generation has resulted in a loss of economies of scale sufficient to motivate the deregulation or "restructuring" of this industry into a competitive form.

Currently, a different feature of energy markets having to do with the use of fossil fuels in energy production is recognized as entailing market failure. Among the assumptions required for the efficiency of competitive markets are that all commodities be both "rival" and "excludable." A rival commodity is one whose use by one precludes its use by another, while an excludable commodity is one whose production and consumption imposes no "side effects" on anyone not party to direct transactions involving the commodity. The market failures now most commonly ascribed to energy markets arise from the nonrivalry and nonexcludability of certain byproducts of energy consumption and production: so-called "environmental externalities," or side effects that have deleterious effects on health or welfare. Key examples are carbon monoxide emissions from vehicles, sulfur dioxide emissions from electric power generation, and emissions from power generation and from the use of fossil fuels. The creation of these by-products creates costs that are not reflected in market prices for energy products, so that in the absence of regulation or some other policy intervention there are no incentives to control them.

Sulfur dioxide emissions resulting from fossil fuel can have negative effects on urban air quality and create acid rain that harms aquatic life. These emissions are nonexcludable in that there is no private action that a particular individual can take to avoid this impact, and they are nonrival in that their effect on any one individual does not preclude or offset their effect on any other.

Another, more controversial, idea is that there may be market imperfections underlying the "energy-efficiency gap," the long-recognized apparent underinvestment by consumers and firms in energy-efficient technology. Beginning in the 1970s, energy analysts used the term "market barrier" to refer to any of the various possible reasons for such underinvestment. Since then, there has been some effort to distinguish those "barriers" that constitute market imperfections. Although no consensus has emerged, there is general agreement that the energy-efficiency gap may arise in part from informational problems involved in private investment decisions. It is now widely recognized that the nonrivalry of information can result in market imperfections. To the extent that

informational problems impeding optimal energy-efficiency investments are pervasive, the efficiency gap may be seen as another important example of an energy-related market imperfections.

POLICY RESPONSES

U.S. environmental regulations have for the most part tried to mitigate health and welfare effects from pollution through technological control strategies such as requiring installation of pollution reduction equipment. There has been a gradual increase of interest, however, in analyzing and attempting to correct environmental externalities within the paradigm of market failures or imperfections. In this paradigm, in the presence of market failures the government can under idealized assumptions reallocate resources to make some consumers better off while making none worse off—a "Pareto improvement." However, the recognition that in practice there will always be both winners and losers from a given policy led to the development of a less ambitious notion of the goals of policy. Thus, the focus of cost-benefit analysis is to determine how market failures justify policies in which some are better off, and these winners could in principle compensate the losers and still come out ahead. This is the "compensation principle."

We often think that solving the problem of environmental externalities means eliminating them altogether, but the cost-benefit approach applies a different criterion: emissions should be held to an optimal level, which is less than the unregulated level but in most cases not zero. The threshold is that the marginal damage from emissions should be equated to the marginal cost of abatement. This, in turn, naturally suggests the economic means of controlling emissions: the government should impose taxes or charges on them so that this marginal condition holds. This policy mechanism has been studied extensively as a way of reducing carbon emissions in an effort to mitigate global climate change. An alternative approach is for the government to issue permits to emitters that restrict the overall quantity of emissions; these permits would then be traded among emitters so that, overall, the cost of abatement would be minimized.

Alan H. Sanstad

See also: Government and the Energy Marketplace; Market Transformation.

MARKET TRANSFORMATION

The term "market transformation" first appeared in the literature in the early 1990s. The term emerged more as an abstraction than a concrete program strategy or model. Market transformation provides a vision of the ultimate objective of strategic interventions—markets that yield energy-efficient outcomes automatically as the result of normal market forces. Market transformation can be viewed as a catalyst for change—a means of intervening in imperfect markets to effect long-term changes to improve market performance with respect to energy efficiency.

While there is not a single, precise definition of market transformation in professional practice, the following definition captures the essential elements:

Market transformation is a strategic intervention to achieve a lasting, significant share of energy-efficient products and services in targeted markets. Market transformation is essentially synonymous with marketing strategy as used in the private business world. A key distinction is that market transformation is motivated by the social objectives of improving the performance of markets to yield greater energy efficiency. And like marketing strategies in the private sector, it often requires ongoing measures to achieve and sustain desired market outcomes. For example, market transformation will occur in the U.S. clothes washer market when energy-efficient washers (often horizontal axis machines that use about one-third the energy and water of old vertical-axis, agitator machines) become the norm as they already are in Europe. After introduction of new machines in the market, ongoing marketing and related support may be required to sustain a significant market share.

The overall goal of market transformation is to increase the share of energy-efficient products and services through fundamental, enduring changes in targeted markets. This goal also serves to improve the economy and reduce negative social and environmental effects that result from energy use. Intervention is needed because of market imperfections that do not allow the market on its own to provide an optimum level of energy-efficiency goods and services. In most cases intervention requires an

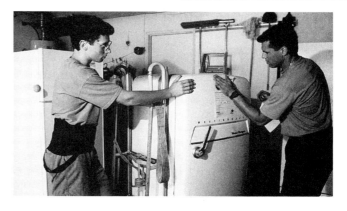

Jay Joseph and Don Gardner collect an old refrigerator after delivering a new one. In 1993, the Sacramento Municipal Utility District paid customers $100 for their old refrigerator plus an additional $100 rebate to buy an energy efficient model. (Corbis-Bettmann)

evolving mix of strategies and implementation measures over an extended period.

Market transformation as a strategy to improve market performance has become more important as most energy utility markets deregulate and restructure. Restructuring is resulting in more competition and reliance on market mechanisms in energy markets that have been highly regulated. Utility energy efficiency programs of the 1980s and 1990s, called demand-side management (DSM) programs, are being abandoned and market transformation is being introduced to fill the void. However, the roots of market transformation lie within the regulated energy industries.

The goal of most DSM programs has been relatively narrow: to reduce energy and power demand to avoid investments in new power plants or transmission and distribution systems. DSM was used within the context of integrated resource planning to yield the lowest cost of energy services by avoiding more costly construction and operation of supply-side power plants. DSM was considered a resource comparable and substitutable for supply-side resources (hence the name—integrated resource planning). Individual utilities have typically implemented DSM for their own customers, as ordered by public utility commissions or other regulatory bodies. All utility customers generally have shared the costs for DSM programs because regulators mandated such programs and consequently provided cost-recovery mechanisms for utilities.

Integrated resource planning and demand-side management have declined in the wake of the movement to restructure and deregulate energy markets. DSM has evolved to be more market-based, as program designers sought lasting change, and utilities reduced program costs and shifted some of the remaining costs to the direct beneficiaries of DSM programs.

As DSM evolved, program managers realized that their efforts could have much greater impact if they went beyond the service territories of single utilities to encompass regional and national markets. While not termed "market transformation," there were several early initiatives that took this approach, including the Manufactured Housing Acquisition Program in the Pacific Northwest and the Power Smart Program, which originated in British Columbia and was later adopted elsewhere. These and other state, regional, national and even international collaborations, with multiple parties contributing funds and expertise, have tried to change building practices, introduce new products and change market shares. Collaborations, such as the Consortium for Energy Efficiency (CEE) in the United States, have been making larger changes in the markets for target technologies. Examples include promoting super-efficient refrigerators, clothes washers, and motors. More broadly, the Energy Star Program of the United States Environmental Protection Agency (EPA) and Department of Energy (DOE) is an example of a market transformation effort that spans numerous household and commercial appliances and applications—from home computers and air-conditioners to energy ratings of commercial buildings. Energy Star is a labeling program that identifies the most energy-efficient technologies within a given appliance or application category.

Market transformation initiatives typically require collaboration among a diverse set of stakeholders, including manufacturers, retailers, utilities, research and development (R&D) organizations, government, and public-interest efficiency advocates. In recent years the need for coordination and collaboration among such a diverse set of actors to achieve consensus has led to the development of specific U.S. regional market transformation organizations, including the Northwest Energy Efficiency Alliance (NEEA), the Northeast Energy Efficiency Partnerships, Inc. (NEEP), and the Midwest Energy Efficiency Alliance (MEEA).

These organizations vary significantly in their structure, funding and operation. However, their overall approach to market transformation is similar. Market transformation typically includes the following steps (not necessarily in sequence):

- Identify needs through market analysis and research—the markets where opportunities exist to increase market share of products and services that respond to customer needs and deliver superior energy-efficient performance.
- Identify market participants (manufacturers, retailers, consumers) and stakeholders (such as consumer advocacy groups, trade organizations, and government).
- Form collaboratives and define roles among key market participants and stakeholders to lead and manage the market transformation initiative.
- Establish funding to cover costs of the initiative (program costs).
- Define program goals for target products or services within the chosen market.
- Establish market baselines against which intervention(s) will be evaluated.
- Design strategies and measures for the initiative, including a transition strategy that may be an exit or continued intervention such as advertising and education.
- Implement measures.
- Track market performance and evaluate results of the initiative.
- Continue, modify or end initiative as indicated by monitoring and evaluation results.

Implementation of market transformation programs requires adoption of coordinated measures targeted to various market participants over a fairly long period. The duration of market transformation programs depends on numerous factors, including the complexity of the market; customer response; support of manufacturers, distributors and retailers; and time required for manufacturers to change manufacturing operations. Experience with past DSM programs and early market transformation programs suggests that periods of five to ten years or more are needed to transform energy efficiency markets.

Typical measures used with market transformation programs may include the following:

- marketing, such as media advertisements, point-of-purchase displays, utility bill stuffers and other promotions

- labeling (a key example is the Energy Star program established and operated by the U.S. EPA and DOE)
- consumer education
- professional training (e.g., sales associates, skilled tradespeople, contractors, manufacturers)
- research and development in support of program needs
- codes and standards (to codify energy-efficient technologies by establishing minimum performance standards)
- consumer rebates or other incentives to increase consumer acceptance
- manufacturer and retailer incentives
- technology procurement (specifying required performance of technologies and aggregating customers to create sufficient demand for suppliers to respond to performance requirements)
- other types of bulk purchasing or buyer aggregation to create market pull
- design competitions based on desired performance.

Market transformation collaborations involve multiple parties—each with different motivations and objectives. The target markets are typically broad in geographic scope (regional and national markets). These two factors alone pose major challenges for market transformation programs. The complexity and dynamic nature of markets pose a different set of challenges. While the challenges of market transformation are many and complex, the advantages of market transformation versus traditional DSM intervention are substantial. Market transformation focuses on systemic, complementary measures for market improvement, whereas DSM programs typically were isolated efforts that addressed much narrower symptoms of market imperfections. For this reason, market transformation is growing rapidly as a dominant model for publicly and privately supported energy-efficiency programs.

Dan W. York
Mark E. Hanson

See also: Demand-Side Management; Efficiency of Energy Use, Labeling of.

BIBLIOGRAPHY

American Council for an Energy-Efficient Economy. (1998). *Proceedings from the 1998 American Council for an Energy-Efficient Economy Summer Study on Energy Efficiency in Buildings, Volume 7, Market Transformation.* Washington, DC: Author.

Eto, J.; Prahl, R.; and Schlegel, J. (1996). "A Scoping Study on Energy-Efficiency Market Transformation by California Utility DSM Programs," LBNL-39058. Berkeley, CA: Lawrence Berkeley National Laboratory.

Geller, H., and Nadel, S. (1994). "Market Transformation Strategies to Promote End-Use Efficiency." *Annual Review of Energy and the Environment* 19:301–346.

Nadel, S., and Latham, L. (1998). *The Role of Market Transformation Strategies in Achieving a More Sustainable Energy Future.* Washington, DC: American Council for an Energy-Efficient Economy.

Synergic Resources Corporation. (1996). *Market Trans-formation in a Changing Utility Environment.* Washington, DC: National Association of Regulatory Utility Com-missioners.

York, D., ed. (1999). "A Discussion and Critique of Market Transformation: Challenges and Perspectives," Report 186. Madison, WI: Energy Center of Wisconsin.

MASS TRANSIT

Mass transit is a transportation service available to the public for trips generally within metropolitan areas. As with all transportation services, an energy source is a critical input in the production of mass transit trips. Almost all of this energy consumption is derived from burning fossil fuels, a process that emits pollutants affecting human health, visibility, vegetation, and climate change. Transit service is increasingly being scrutinized for its pollutant emissions. Mass transit facilitates travel within densely developed, large urban areas. Because of the existence of mass transit, more intense development within an urban area can occur. Higher-density land use may enable reduced energy consumption when considering the settlement area as a whole.

FORMS OF MASS TRANSIT

The term "mass transit" generally refers to passenger vehicles that are common carriers in urban areas, as distinct from intercity travel. The terms "public transit" or simply "transit" also are frequently used. The major types of public transit are bus (rubber-tired vehicles), rail (running on tracks), and ferryboat. Within each type there are several subcategories.

Motor Bus

A rubber-tired, self-propelled transit vehicle using an internal combustion engine for power. Most use direct-ignition (diesel) engines, but gaso-

A seabus crosses a harbor in Vancouver, British Columbia. (Corbis-Bettmann)

Mode	Vehicle Distance Operated	Passenger Boardings
Bus	56%	58%
Rail rapid transit	19%	30%
Suburban rail	8%	5%
Streetcar/LRT	1%	3%
Demand response	12%	1%
Other*	3%	3%
Total	100%	100%

Table 1.
Service Supplied and Consumed, by Mode, in the United States, 1997.
NOTE: "Other" includes ferryboat, inclined plane, automated guideway, monorail, aerial tramway, and cable car.
SOURCE: U.S. Department of Transportation, 1999.

line, propane, natural gas, and other fuels are used as well. Buses are available in varying sizes and capacities, varying from 10 passengers to 150 or more passengers, including articulated and biarticulated vehicles. They can be designed for city service (fewer seats, more doors and standing room) or long-distance express service (often with no standees permitted). Buses can operate on most streets open to traffic, including expressways, and also can be operated on exclusive rights-of-way such as high-occupancy-vehicle (HOV) expressway lanes, bus lanes on city streets, or busways with on-line stations.

Electric Bus

When powered by an electric motor, generally taking power from overhead wires ("catenaries") such a vehicle is called an electric trolleybus or a trackless trolley. An electric bus taking power from current in the ground has been developed by Ansaldo-Breda. Battery-powered buses also have been developed, although all of these have been smaller than stan-dard-size buses. In the late 1990s, "hybrid" electric-internal-combustion buses were placed in service. These vehicles have a smaller than normal internal-combustion engine, which continually recharges a battery pack. The batteries power an electric motor, which provides supplementary power when needed. Fuel cells, which can be thought of as chemical batteries, also are used on an experimental basis to power transit buses.

Streetcar or Tram

The streetcar or tram, powered by electricity from overhead catenaries and running on rails on city streets, was the backbone of mass transit service from 1890 until the widespread deployment of the diesel bus in the early post-World War II period. In general, the only streetcar lines in North America that survived the transition to the bus were those in high-demand corridors, or those operating in their own right-of-way, such as a roadway median, tunnel, or subway. Since the 1970s, there has been increasing deployment of new streetcar service, now referred to as light rail transit (LRT). These services tend to use larger, articulated vehicles and frequently operate partly on exclusive rights-of-way such as disused rail corridors or expressway medians.

Rail Rapid Transit

Also known as metro, subway, elevated, underground, or heavy rail, this higher-capacity rail service is distinguished by its use of trained vehicles (several

	Public Transit Bus			Passenger Car		
	BTU per vehicle mile	BTU per passenger mile	passenger mile per vehicle mile	BTU per vehicle mile	BTU per passenger mile	passenger mile per vehicle mile
1970	31,796	2,472	12.9	9,301	4,896	1.9
1980	36,553	2,813	13.0	7,915	4,166	1.9
1990	36,647	3,735	9.8	6,183	3,864	1.6
1997	38,101	4,318	8.8	5,822	3,639	1.6
% change 1970-97	20%	75%	-31%	-37%	-26%	-16%

Table 2.
Average Energy Intensity and Occupancy Rates for U.S. Public Transit Buses and Passenger Automobiles, 1970–1997.
SOURCE: Davis, 1999.

cars attached together) and third-rail electric power (a live connection at grade). Because of its unprotected power source, rail rapid transit cannot be operated on city streets, and runs in elevated or underground rights-of-way. Unlike suburban rail, rail rapid transit service generally operates within cities and on a frequent schedule. Some of the new rail rapid transit systems, such as BART in San Francisco, have a metropolitan scope and distant stop spacing, and thus are similar to suburban rail.

Suburban Rail

Also known as commuter rail, this type of service generally provides long-distance—50 km (31 mi.) or more—routes to the far reaches of a metropolitan area using railroad rights-of-way. These operations use either diesel or electric power. Service is primarily oriented toward commuters from suburbs to central city locations and is concentrated in the peak commuting hours. Express bus operations also provide similar services, without the need for a separate right-of-way.

Ferryboat

In metropolitan areas adjacent to bodies of water, ferries can be used as mass transit—that is, providing frequent service useful for local travel. Geography limits ferryboat use to a select group of cities. Twenty-one cities in the United States and its territories have transit ferryboat service. Another example is Vancouver's SeaBus, which provides very frequent shuttle service between the center city and North Vancouver.

Table 1 shows the frequency distribution of U.S. mass transit service supplied (vehicle distance operated) and service consumed (passenger boardings) by transit mode in 1997. The official U.S. transit statistics also include the "demand response" mode. This type of transit consists of minibuses or vans operating by request, rather than on fixed routes, and typically available only to a portion of the public, such as elderly or disabled people. As shown in Table 1, demand response accounts for 12 percent of service operated but only 1 percent of boardings. Buses accounts for the majority of both service supplied and service consumed. Removing the few largest rail systems from the totals would reveal even greater significance of the bus mode.

IMPACT OF MASS TRANSIT ON ENERGY CONSUMPTION

The potential of mass transit to provide transportation services with low energy consumption relies on the high capacity of transit vehicles, since these vehicles have higher energy consumption per vehicle distance traveled compared to private motorized vehicles or nonmotorized modes. Therefore the occupancy rate of transit service is a key factor in determining its energy efficiency. This rate can be measured by the ratio of person distance traveled to vehicle distance traveled.

The energy efficiency of transit per passenger distance traveled depends both on its usage rate (person travel per vehicle travel) and its fuel efficiency (fuel consumption per vehicle travel). The transit vehicles with the greatest potential for energy efficiency gains

Year	Transit Bus	Transit Rail	Passenger Car	Intercity Rail	Intercity Bus
1970	2,472	2,453	4,896	3,677	1,051
1980	2,813	3,008	4,166	3,176	1,069
1990	3,735	3,453	3,864	2,609	944
1997	4,318	3,253	3,639	2,458	872
1970-97	75%	33%	-26%	-33%	-17%

Table 3.
Average Energy Intensity per Passenger Mile for Various U.S. Passenger Modes, 1970–1997.
SOURCE: Davis, 1999.

may not be those that can carry the most people per vehicle. Rather, the greatest potential gains can be had by matching vehicle capacity to travel demand. In fact, a high-capacity mode that is little used can increase energy consumption relative to the passenger automobile. On the other hand, a minibus with ten passengers may represent a significant reduction in energy use compared to the case of those ten passengers traveling by car.

The data in Table 2 dramatically illustrate the difference between vehicle fuel intensity and passenger fuel intensity from 1970 to 1977. (Fuel intensity, fuel use per distance traveled, is the inverse of fuel efficiency.) In 1970, the transit bus mode in the United States used 50 percent less energy per passenger mile than the automobile mode (passenger cars, not including light trucks). By 1997, transit buses used 19 percent *more* energy on average per passenger mile. How did this happen? The fuel efficiency of transit buses declined by 20 percent, while fuel efficiency improved by 37 percent for passenger cars. Occupancy declined in both cases, but twice as much for transit buses. The net result was a significant reduction in fuel use per passenger mile for cars despite declining occupancy, and an increase in fuel intensity for transit buses.

The evolution of transit bus compared to transit rail energy intensity in the United States from 1970 to 1997 is shown in Table 3. Transit bus and rail had similar energy intensities in 1970, but by 1997, energy use per passenger mile had increased by 33 percent for transit rail and by 75 percent for transit buses. This increase is yet more significant given that energy intensity for passenger cars declined 26 percent over the same period and given that intercity rail

and intercity bus operations also showed energy intensity reductions of 33 percent and 17 percent, respectively.

In 1997, the intercity bus was the least energy-intense of the modes shown in Table 3. Its energy intensity per passenger mile was one-fifth of transit bus energy intensity. The vehicles used for the two operations are similar, although operating conditions are different. Intercity buses operate mostly on expressways, and transit buses mostly on local streets. Still, the fact that transit bus energy intensity increased 75 percent while intercity bus energy intensity decreased 17 percent suggests that changes in operating patterns, rather than changes in technology, accounted for much of the difference in energy use.

In the United States in 1997, transit accounted for a small fraction of transportation energy consumption, largely because it played such a small role in the total travel market, representing only 1.8 percent of trips. Transit accounted for only 0.7 percent of transportation energy consumption (bus 0.4%, rail 0.2%, and suburban rail 0.1%).

Transit represents an increase in energy use compared to walking or bicycling. For countries with a significant amount of nonmotorized transport, increasing transit use may mean increasing energy use. However, the increased energy consumption may be associated with significant improvements in urban passenger transport, which can produce large economic benefits. It is generally easier for transit modes to compete successfully with nonmotorized modes than with the private automobile for new passengers. But as demand for higher-speed travel increases, those customers may switch to private motorized vehicles.

COMPARISON OF MASS TRANSIT MODES

In the largest metropolitan regions, such as in Tokyo, New York, London, and Moscow, the huge demand for transit makes rail transit both indispensable and energy-efficient. But because transit energy efficiency is so dependent on matching supply with demand, the greater flexibility of the bus mode may have an advantage in this aspect. Bus service characteristics can be easily altered as to time of service, location of service, and size of vehicle used. For rail transit, not much change is possible in the short run.

The rail modes require a significant investment of both fiscal resources and energy in infrastructure,

| Year | Diesel | Gasoline | Alternative Fuels | | | | | | Total |
			Compressed Natural Gas	Liquefied Natural Gas	Methanol	Propane (Liquid Petroleum Gas)	Other	Total	
1992	684,944	32,906	1,009	191	1,583	2,487	12	5,282	**728,414**
1993	678,511	37,928	1,579	474	4,975	2,098	197	9,323	**735,085**
1994	678,226	43,921	4,835	1,450	12,269	1,871	492	20,917	**763,981**
1995	678,286	42,769	10,740	2,236	11,174	3,686	865	28,701	**778,457**
1996	692,714	41,495	15,092	2,862	7,268	5,235	4,353	34,810	**803,829**
1997	716,952	41,547	23,906	4,030	965	5,150	7,771	41,822	**842,143**
1998	700,081	38,399	30,915	3,246	783	5,112	2,865	42,921	**824,322**

Table 4.
Mass Transit Fossil Fuel Consumption by Fuel (thousands of gallons).
SOURCE: American Public Transit Association, 2000.

including track, power supply, right-of-way, stations, and structures. This investment can produce a higher quality of service, but it commits the urban area to use of the rail mode and its relatively large vehicles.

The bus mode can use a broader range of vehicle sizes and does not require any separate infrastructure. Where right-of-way is shared with other vehicles, much of the energy used in producing the facilities may have been expended without consideration of use by mass transit. Where there is a need for exclusive bus infrastructure to bypass congested areas, busways can be constructed to improve service.

It is common outside of Europe and North America for bus service to be provided by private operators. The higher service efficiencies often achieved by private transit providers can translate into greater energy efficiency. While contracting out of transit service is possible for rail, it has been more common for bus operations.

Why has the load factor for transit decreased? Transit demand has declined due to higher incomes, higher automobile ownership, and a decrease in the share of jobs and population living in areas that are strong transit markets. At the same time, transit supply has increased, spurred by the growth of government subsidies for transit operation and capital investment. The long-standing problem of intense peaking of transit demand means that large vehicles are needed for only a few hours during the day, only to run nearly empty during off-peak hours. Transit agencies in the United States have been generally reluctant to use smaller vehicles. Privatized transit operations outside of London, England, were quick to adopt minibuses. Transit operation in the United States has largely become regionalized, with providers' operating areas spreading out over vast territories. Because all taxpayers in the region are typically required to contribute to transit via designated sales, property, or motor fuel taxes, transit agencies feel a political obligation to provide at least a minimum amount of service in all parts of the region, no matter how weak the transit demand.

Most urban rail service is electric-powered and most urban bus service is diesel-powered, although diesel rail and electric bus operations do exist, as noted above. The efficiency and environmental impacts of electricity depend greatly on the source of electric power. Although electric vehicles produce no tailpipe emissions, generation of electricity can produce significant emissions that can travel long distances. For example, coal-powered electricity plants produce particulate emissions that travel halfway across North America. Urban buses also can be powered by a variety of alternative fuels.

NEW TECHNOLOGIES

Applications of engineering and computer technology have the potential for reducing fuel use, greenhouse gas emissions, and toxic emissions from mass transit. They also may help in increasing transit load factors. Technology that enables greater operating efficiency of transit also has the potential to reduce fuel consumption per passenger distance traveled. The introduction of electronic fare cards in New York City prompted the development of new fare policies, including free transfer from bus to rail. The result was a 36 percent increase in bus use between 1997 and 1999. With only a 9 percent increase in bus service during the same period, the fuel efficiency of transit increased dramatically.

Traffic signal preemption and other technologies that seek to speed transit can have a similar effect. Faster bus travel means less fuel consumption per vehicle distance traveled. However, faster transit service increases demand for transit, potentially increasing the transit load factor.

The "conventional" fuels used for transit applications include gasoline, diesel fuel, and electricity. Alternatives to these fuels have been sought to reduce energy consumption, pollutant emissions, greenhouse gas emissions, and use of imported fuels. The conventional fuels for internal-combustion engines are the most energy-dense fuels: petroleum and diesel fuel.

Alternative fuels often are handicapped by the need to develop an alternative fueling infrastructure. Most transit fleets have their own fueling stations, so they are prime candidates for early introduction of new fuels. And transit agencies have experimented with many different fuels. Alcohol-fueled engines proved unreliable for heavy-vehicle applications. Los Angeles found that methanol and ethanol corroded bus engines, scrapped some of the vehicles, and converted the remainder to diesel in 1998-1999. The most popular alternative fuel has been compressed natural gas (CNG). This fuel offers lower pollutant emissions than diesel, but at a higher total cost than diesel. In addition to requiring new vehicles with natural-gas-powered engines and storage areas for fuel cylinders, using CNG requires an investment in new fueling and compression facilities and modifications to depots and maintenance facilities. Considering fueling, inspection, and maintenance, operating costs typically are higher for CNG than for comparable diesel vehicles.

Between 1992 and 1998, alternative fuels increased from less than 1 percent to more than 5 percent of total mass transit fossil fuel consumption in the United States (see Table 4). The share of alternative fuel consumption that was CNG increased from 19 percent to 72 percent over the same period.

One promising new technology is the hybrid electric vehicle, which combines an internal-combustion engine and an electric motor powered by batteries. The engine recharges the batteries, and the electric motor provides additional power during acceleration. These vehicles do not have the severe range and capacity limitations of battery-powered electric vehicles and do not have long refueling (recharging) times. Hybrids equipped with regenerative braking are well suited for city bus routes requiring extensive stop-and-go driving.

Ambient particulate matter (soot) has been linked to increased mortality. Diesel engines are high emitters of particulates. Particulate is also formed in the atmosphere from other pollutants, including those emitted by gasoline engines. As the public has become more aware of this problem, diesel engines have become much cleaner. By adding emissions control devices and electronic controls and by modifying engine designs, a large reduction in the emissions of particulate matter and nitrous oxides (precursors to both soot and smog) has already been achieved. In 2000, the U.S. Environmental Protection Agency proposed new regulations requiring very-low-sulfur diesel fuel and further large reductions in particulate and nitrous oxide emissions. Low-sulfur fuel reduces sulfate particulate emissions and also enables the introduction of emissions control devices that would be damaged by high-sulfur fuel. These regulations require diesel vehicles that have an emissions profile similar to today's CNG buses.

In the longer term, more exotic technologies, such as fuel cells powered by hydrogen, may be feasible. These technologies are far from being economically feasible, but rapid progress is being made. However, as conventional vehicles become cleaner, the relative emissions-reduction benefits from alternative fuels declines.

LAND USE AND SOCIAL AND POLITICAL FACTORS

Matching the supply of transit service to its demand is the primary determinant of the energy efficiency of transit. The demand for transit depends on income levels and land use. As average income rises, so does the value of travel time, and therefore the cost of time spent traveling. Higher incomes also make automobile use more obtainable.

Land use is the other key determinant of transit use. Public transit requires a concentration of trips in the same time and place. A concentration of residential and commercial land use, such as that typically

A conductor on the number five train at Grand Central Terminal in New York on the eve of a possible strike by the subway and bus workers, December 1999. (Corbis Corporation)

found in cities, is necessary to generate significant transit demand. Concentrated land use has another, perhaps more important effect. To function efficiently, the private automobile must be provided with a significant amount of space for both storage and movement. In many cities there is insufficient space for the automobile, making automobile travel expensive. The largest costs of urban automobile travel are storage (parking) costs and travel time costs, due to traffic congestion. Under these circumstances, mass transit modes can compete effectively. They provide travelers with similar or lower total travel time (including time spent walking to and waiting for transit vehicles), and the avoided cost of parking may be several times greater than the cost of the transit fare.

However, most North American urban development since 1945 has been designed around the space needs of the automobile. A generous quantity of off-street parking is required for all new residential and commercial development. Streets are designed to be wide enough to accommodate free-flowing traffic, even in peak periods. Expressways have been constructed to connect all parts of the metropolitan region. The result has been the development of urban areas that are convenient for motoring. The same set of changes also has made them inhospitable to any mode of daily transport other than driving. The low population density of residential development makes postwar neighborhoods difficult to serve by public transit. More important, the prevalence in North America of widely scattered office parks with ample free parking has ensured that transit cannot compete for trips to those destinations, since driving is inexpensive and transit is either not available or takes up to twice as long.

Cities that were large (1 million or more) before 1920 have cores with a large concentration of jobs, expensive parking, and a large transit share of trips. The New York metropolitan area is the premier example, accounting for nearly 40 percent of all U.S. transit trips. Although newer areas have been able to increase transit ridership, none has transit shares approaching what is still common in the older large cities.

Per capita transportation energy consumption in a city such as New York is much lower than the U.S. average as a result of much lower car use. Although high-quality public transit service is one explanation for this result, parking costs, bridge and tunnel tolls, and the convenience of walking are equally important. Public transit is vital for the transportation needs of New York City residents. But many of the trips that residents of the typical U.S. metropolitan area take by auto are taken by New Yorkers on foot, or not at all. It is the lower number of auto trips, rather than a one-for-one substitution of transit trips for auto trips, that has a large impact on reducing energy use. Transit service in New York City is well used, making it energy-efficient despite the slow speed of surface transit travel.

In the United States, outside of the core areas of older major cities, transit has become the transport of last resort. There is a substantial social stigma attached to using transit, due to the low income levels of transit patrons in most U.S. cities. Transit customers also sometimes fear their vulnerability to crime, especially while waiting at bus stops.

INTERMODAL TRAVEL

Many transit trips involve transfers between transit vehicles. All else being equal, passengers would prefer not to transfer. However, the mode of access to transit is of interest. The vast majority of transit customers walk to transit and then walk again to their final destination. It is difficult to develop high transit demand in an urban environment that is not conducive to walking.

Park and Ride

A recent trend is the development of park and ride areas for transit passengers to drive to transit. These facilities help make transit accessible to a wide geographic area. Because emissions control devices are ineffective when cold, the auto trip to the transit station may produce nearly as many pollutants as a direct trip to the final destination. Further, park and ride lots occupy a lot of space, making transit stations

unfriendly to pedestrians and reducing the opportunities for station-area real estate development. Finally, those driving to transit must still pay the full costs of automobile ownership. In fact, the only reason they are likely to take transit is if the cost of station-area parking and transit fare is less than the parking cost at the final destination. This explains why park and ride areas are most successful on transit lines serving urban cores with high parking costs. Most workers in North America, however, pay nothing out of pocket for parking at work.

Bicycling and Transit

Bicycling to transit is another solution for providing access. With no fossil fuel consumption and no pollution, bicycling to transit is attractive from an environmental point of view. Personal fitness is a major motivation for many bicyclists. Public health authorities concerned about physical inactivity have started to promote bicycling. Bicycle access to train stations is very popular in Japan and the Netherlands and some places in Germany. Since bicyclists can travel faster than local transit on some routes, the bike and ride option is most attractive for accessing express services such as suburban rail or express bus. Bike stations that provide bike storage, rental, and repair services and changing facilities adjacent to transit stations have been developed in three California cities.

Permitting bicyclists to take bicycles aboard transit vehicles allows them to use the bicycle for the trip to their final destination as well. Although many transit agencies are reluctant to make room for bicycles during crowded peak hours, railcars designed with bicycle storage have been deployed in Europe and North America. In the 1990s, hundreds of buses in North America were equipped with bicycle-carrying racks.

Despite the energy, emissions, and public health benefits of combining bicycle use with transit, few people in North America take advantage of this combination. Many urban roads are designed for high-speed operation and have narrow lanes. These designs often scare bicyclists from using the roads, a fear sometimes reinforced by motorists and police who do not believe that bicyclists have a right to use roads. Secure, sheltered bicycle storage at transit stations is rare, and therefore the threat of bicycle vandalism and theft deters others from using bicycling as an access mode to mass transit.

Paul M. Schimek

See also: Batteries; Behavior; Bicycling; Diesel Cycle Engines; Diesel Fuel; Electric Vehicles; Emission Control, Vehicle; Engines; Fuel Cells; Fuel Cell Vehicles; Gasoline and Additives; Gasoline Engines; Hybrid Vehicles; Locomotive Technology; Petroleum Consumption; Railway Passenger Service; Traffic Flow Management; Transportation, Evolution of Energy Use and.

BIBLIOGRAPHY

American Public Transportation Association. (2000). *Public Transportation Fact Book 2000.* Washington, DC: Author. <http://www.apta.com/stats/>.

Davis, S. C. (1999). *Transportation Energy Data Book: Edition 19.* ORNL-6958. Oak Ridge, TN: Oak Ridge National Laboratory. <http://www-cta.ornl.gov/Publications/Tedb.html>.

Kenworthy, J. R.; Laube, F. B.; Barter, P.; Newman, P.; Raad, T.; Poboon, C.; and Guia, B. (2000). *An International Sourcebook of Automobile Dependence in Cities, 1960-1990.* Boulder, CO: University Press of Colorado.

Parcells, H., and Replongle, M. (1992). *Linking Bicycle/Pedestrian Facilities With Transit (National Bicycling and Walking Study. Case Study No. 9).* Washington, DC: U.S. Department of Transportation, Federal Highway Adminstration.

Pickrell, D. (1999). "Transportation and Land Use." In *Essays in Transportation Economics and Policy: A Handbook in Honor of John R. Meyer,* ed. J. A. Gomez-Ibanez, W. B. Tye, and C. Winston. Washington, DC: Brookings Institution Press.

Shoup, D. C. (1995). "An Opportunity to Reduce Minimum Parking Requirements." *Journal of the American Planning Association* 61:14-28.

U.S. Department of Transportation, Federal Transit Administration (1999). *National Transit Database 1997.* Washington, DC: Author. <http://www.fta.dot.gov/ntl/database.html>.

MATERIALS

Energy and materials were developed together in early civilizations, beyond the use of fire for cooling and heating. In the courtyard of the stepped pyramid

---

MATERIALS IN AUTOMOBILES AND TRUCKS

Advances in materials science have been a leading factor contributing to improvements in vehicle fuel economy and performance since the 1973 oil embargo. Better tire materials have reduced rolling resistance, better materials for moving parts have reduced friction, and most importantly lighter body parts have reduced the overall weight of vehicles.

The average 1993 model American car weighed about 3,200 pounds (1,450 kg) and delivered 28 miles per gallon of fuel (171 km/liter), representing a 19 percent reduction in weight and a 44 percent reduction in fuel consumption as compared to the average 1975 model. This reduction was achieved by substituting polymers for interior metallic trim, aluminum engines for cast iron, and careful design attention to weight factors. Weight reduction and improved engine design contributed greatly to better fuel economy. There is further potential for weight reduction through increased substitution of polymer matrix composites, aluminum, and ceramics for metals in vehicles.

---

at Sakahra, built near Cairo, Egypt, in about 2600 B.C.E., there are a series of carved relief panels in stone, showing that the ancient Egyptians had mastered the smelting and working of metals with heat energy (fire), as well as many other technical skills. Glass melting was discovered at least 10,000 years ago, and fired ceramics (pottery) even much earlier.

Solid materials are essential to the production and transmission of energy; in the next section gives examples of materials used in these processes. The article then focuses on the large energy requirements needed to produce the metals, ceramics, glasses, and electronic materials (silicon and germanium) that our technological civilization demands, and finishes with an overview of the environmental problems encountered in trying to satisfy this vast demand for materials.

Metal	Form or Process	Energy 10^{12} J/kg
Iron	Steel	28
Bismuth	Bulk	30
Lead	Ingot	31
Thallium	Sponge	36
Zinc	Electrolytic	70
Sodium	Bulk	107
Copper	Refined Bar	128
Cobalt	Electrolytic	144
Nickel	Electrolytic	167
Cadmium	Electrolytic	178
Tin	Ingot	221
Aluminum	Electrolytic, ingot	284
Tantalum	Powder	289
Tungsten	Powder	402
Magnesium	Electrolytic	416
Mercury	Liquid	459
Titanium	Sponge	474
Indium	Bulk	590
Cesium	Bulk	612
Hafnium	Sponge	768
Zirconium	Sponge	1390
Silver	Bars	1710
Rhenium	Powder	3600
Beryllium	Cast	6100
Gallium	Electrolytic	13,500
Gold	Bars	68,400

Table 1.
Energy Required to Produce Purified Metals

MATERIALS IN ENERGY PRODUCTION, TRANSMISSION, AND STORAGE

Metals are of overwhelming importance in these applications. Steel, which is iron containing carbon and many different metallic additions, is still the most used metal. Aluminum is increasingly used in many applications because of its light weight and resistance to chemical attack. Copper is required for transmission lines, wiring, and generators because of its high electrical conductivity. There are many specialized uses of other metals: manganese, vanadium, and molybdenum as alloying elements to improve strength and chemical durability of steel; uranium and plutonium in nuclear reactors; silver in electrical contacts; and tungsten as filaments in lamps.

Ceramics, including concrete, are useful especially in structures, reactors, as refractories in combustion of fuels, and as nuclear fuel. Porcelain insulators on transmission lines are an example of a specialized application of ceramics.

Electronic materials are needed for computers and control devices; purified silicon is the basic material for these applications. In addition silica glass (SiO_2) is an insulator, aluminum an electrical conductor, and polymers are reactive materials for patterning in these devices. Control of every step of energy production and transmission is now completely dependent on electronics.

ENERGY USE: REFERENCES AND METHODS

The energy use data in the tables come from three Battelle-Columbus reports and an article by H. H. Kellogg on "Energy Considerations in Metals Production in the Encyclopedia of Materials Science and Engineering." The Battelle-Columbus reports have detailed descriptions of the processing of all the materials listed in the tables, with methods from mining through separation to purification, and cost estimates at each processing step. There is a remarkable amount of valuable information on each material in these reports. The section on metals also relied heavily on the article by Kellogg.

The estimates for energy use can be separated into the following components:

$$energy\ use = F + E + S - B$$

where F is the heating values of fuels used, E is the fuel equivalent of electrical energy (from the U.S. average fuel equivalent of $11.1(10)^6$ joules per kilowatt hour), S is the fuel equivalent of supplies and chemical reagents consumed in the processing, and B is the fuel equivalent of by products and surplus steam. The units of energy use are given in joules per kilogram of final material produced in the tables; the conversion to English units is: divide J/kg by 1.16 to get Btu/ton (British thermal units per 2,000 pounds).

The energy equivalent of one barrel (159 liters) of crude petroleum is about $6.6(10)^9$ joules. Table 1 shows that the production of a metric ton (1,000 kg or 2,200 pounds) of steel requires about four barrels of oil; a ton of aluminum requires about forty barrels

Metal	Energy 10^{12} J/kg
Manganese	58
Chromium	71
Silicon	89
Niobium	220
Vanadium	570

Table 2.
Energy to Produce Metal for Alloying with Iron (Ferrometals)

Metal	Consumption in millions of kilograms	Fraction as scrap
Iron and steel	97	0.35
Aluminum	4.0	0.34
Copper	2.9	0.45
Lead	1.1	0.55
Zinc	0.95	0.29
Nickel	0.13	0.27
Platinum group	0.10	0.12
Gold	0.087	0.57
Magnesium	0.08	0.09
Titanium	0.036	0.34
Silver	0.033	0.72

Table 3.
Total Consumption of Certain Metals in the United States in 1981

of oil, and a ton of gold more than ten thousand barrels of oil. Table 2 shows the energy needed to produce metals for alloying with iron to produce steel.

ENERGY FOR MINING

The amount of energy required for mining the ores and minerals needed to make materials depends on their depth in the ground, and processing and separation methods. Ore and minerals lying near the surface need only be excavated by shovel or dredge, and thus require low energy expenditure (10^{11} J/kg). Fine grinding (0.1 mm) requires up to $3(10)^{11}$ J/kg. Loading, elevation out of a mine, and transporting the ore can require up to $5(10)^{11}$ J/kg.

If the ore consists of separate grains containing the desired material, it can be separated from undesired minerals by physical methods such as flotation, sedimentation, or magnetic separation. For metals this step can lead to 80 to 95 percent concentration of the value of the ore. Ceramic raw materials such as sand and clay can often be found pure enough in nature so that no concentration is needed.

If the desired material is not in separate grains, chemical treatment of the ore is required for metals, and for purification of ceramics.

ENERGY USE FOR PROCESSING OF METALS

Table 1 shows estimated values of energy to produce a kilogram of reasonably pure metals, or for metal useful for practical applications. The range of a factor of more than 2,000 between steel and gold depends on the concentration in ore, the chemical processing needed, and amount of technology development for the particular metal. Steel production has been developed during about three millennia, and iron ore is highly concentrated and can be reacted directly to steel

alloys. Gold is widely dispersed in low concentrations, and requires intensive chemical treatment of ores.

Table 3 shows the total consumption of some metals in the United States.

If the metallic compound in the ore can be selectively leached by acid or base without dissolving much of the remaining ore, then the energy requirement is only about 10^{12} J/kg. Examples are leaching of oxide ores of copper, zinc, or uranium with sulfuric acid.

Much more energy is needed if the entire rock matrix must be chemically dissolved to free the metallic compounds. Examples are the separation of aluminum oxide (Al_2O_3) from bauxite ore by dissolution by strong base at elevated temperature and pressure (Bayer process), with an energy requirement of about $8(10)^{12}$ J/kg, and smelting of nickel ores by heating to produce a molten nickel-iron alloy and oxide slag, with energy up to $(10)^{13}$ J/kg needed. These high energy requirements result from the fuels needed to heat furnaces or reactors to the high temperatures of these chemical reactions, and for the energy equivalent to make the chemical reagents employed, such as acids, bases and iron alloys.

The grade of metallic ore is the percentage of metal (native and chemically combined) in the ore. For example, high grade iron ore can contain up to 65 percent iron, whereas usual gold ores contain less than 0.001 percent gold. The weight of ore that must be processed is inversely proportional to the grade of the ore. For example, about $7(10)^{13}$ J/kg are required for mining and concentration of hard rock ore containing 0.6 percent to copper, and $4(10)^{16}$ J/kg are needed to mine and concentrate a gold ore containing 0.001 percent gold. Thus, the grade of ore is a major factor in energy consumption.

High grade ores are used first, and are already substantially depleted for most metals. With time one might expect the energy use in Table 1 to increase. Improved technology can offset this increase somewhat; belt conveyors in mines and computer control of grinding are examples. Ordinary rocks contain small quantities (parts per million by weight) of many different metals, and have been suggested as a source for rare metals and those with depleted ores. For example, many rocks contain about 0.01 percent copper; it would require about one thousand times the energy to recover copper from these ores as from presently-available ores (Table 1). The large energy demand precludes the use of these rocks for producing copper, because of the low price of copper. Gold, however, is more valuable, and would justify a larger energy input.

In the concentrated ores most metals are in chemical compounds, as oxides or sulfides. Reducing these compounds to the metallic state in the final stage in producing metal can be accomplished by chemical processes or electrolysis. Two examples of chemical reduction are

Copper sulfide

$$CuS + O_2 = Cu + SO_2 \qquad (1)$$

Steel blast furnace

$$2Fe_2O_3 + 4C + O_2 = 4Fe + 4CO_2 + heat \qquad (2)$$

where $2Fe_2O_3$ is hematite coke ore. Because the oxides and sulfides of many metals are stable, their chemical reduction is difficult, and they are reduced to metal by electrolysis (electrowinning); examples are zinc, aluminum, and magnesium.

The electrolytic processing of concentrated ore to form the metal depends on the specific chemical properties of the metallic compound. To produce aluminum about 2 to 6 percent of purified aluminum oxide is dissolved in cryolite (sodium alumino-fluoride, Na_3AlF_6) at about 960°C. The reduction of the alumina occurs at a carbon (graphite) anode:

$$2Al_2O_3 + 3C = 4Al + 3 CO_2 \qquad (3)$$

Magnesium is reduced from a mixture of magnesium, calcium, and sodium chlorides. Electrolysis from aqueous solution is also possible: zinc, copper, and manganese dissolved as sulfates in water can be reduced electrolytically from aqueous solution.

Process energies are found by subtracting energies for mining and concentrating from the values in Table 3. The free energies of formation of the metal oxides are a measure of the total (theoretical) energy required to reduce the metal from the oxide. The ratio of the actual process energy to the free energy of formation is a rough measure of the efficiency of the reduction process. The free energies of formation are a measure of the chemical stabilities of the oxides; stable oxides such as aluminum, magnesium, and titanium intrinsically require more energy for reduction than from less stable oxides (or sulfides) of copper, lead, and nickel.

The most efficient processes in Table 1 are for steel and aluminum, mainly because these metals are produced in large amounts, and much technological development has been lavished on them. Magnesium and titanium require chloride intermediates, decreasing their efficiencies of production; lead, copper, and nickel require extra processing to remove unwanted impurities. Sulfide ores produce sulfur dioxide (SO_2), a pollutant, which must be removed from smokestack gases. For example, in copper production the removal of SO_2 and its conversion to sulfuric acid adds up to $8(10)^{12}$ J/kg of additional process energy consumption. In aluminum production disposal of waste cryolite must be controlled because of possible fluoride contamination.

As global warming develops, the formation of large quantities of carbon dioxide (see Equations 1 and 3) may become a problem in metals production. There is no simple or inexpensive way to reduce these emissions. Switching to electrolysis processing of steel, or a different electrode reaction for aluminum, would involve unacceptably large energy use and cost. Chemical absorption of carbon dioxide may be necessary in a wide variety of chemical and energy-producing processes, at enormous cost.

The scale of production also influences efficiency. Small-scale batch processing for metals such as titanium, tungsten, and zirconium leads to higher energy use and costs.

RECYCLING OF METALS

Reuse of waste metals generated from metal fabrication and from discarded products (scrap) can save large amounts of energy, particularly for metals that have high energy use in production, such as aluminum. The low fractions of energy used to produce metals from scrap for aluminum, certain sources of copper, and nickel show the value of recycling these metals.

The purity of the scrap mainly determines the fraction of energy needed to produce metal from it, and the value of recycling. Clean copper scrap need only be remelted and cast to form recycled copper; if the copper is contaminated with organic materials and other metals, more complex separation processes are needed that are similar to production from ores. It is easier to remelt the steel of a car driven in Arizona compared to one rusted by the road salt in snowy areas. Scrap that is produced as a by-product of metal processing can be easily recycled, and it can be collected from relatively few locations. There has been a strong effort to educate both householders and industrial users to separate scrap and return it to waste collectors, leading to a supply of reasonably separated scrap.

Despite the efforts of many communities to encourage recycling, there is still a large amount of metal that is not recycled. Only an estimated 30 percent of aluminum is recycled, as compared with up to 50 percent for precious metals. Landfills contain large amounts of metals, especially large use metals (Table 3) such as iron, aluminum, and copper, and more metals continue to accumulate in landfills. As the cost of disposal increases and ores of metals such as copper become of lower grade, it may be economically feasible to "mine" landfills for metals. Development of new technologies for treatment and separation of waste materials is needed to make this mining economical.

ENERGY USE IN PRODUCING CERAMICS

Traditionally ceramic raw materials have been dug out of the ground and used with little or no treatment or purification. Sand, fireclay, talc, and gypsum are examples. The energy expenditure for producing these materials is therefore small. Some of these materials can be found naturally in high purity. Silica sands (SiO_2) with less than 100 ppm (parts per million by weight) of impurities are known, and some clay deposits are nearly pure kaolin. Minerals such as feldspar, kyanite, and kaolinite (clay) can be purified by washing or solution treatments at near ambient temperatures, with low energy expenditure.

Many ceramic products require firing at high temperatures, and the fuels required to reach and sustain these temperatures are major factors in the energy consumed to make these products. Portland cement is made by firing a mixture of compounds, mainly carbonates, sulfates, and silicates to form the desired calcium silicate products. The firing is done in a rotary furnace or kiln, so that a fraction of the raw materials become liquid. As the resultant calcium silicates cool, they go through a large volume change that causes the cement particles to break into smaller sizes.

Concrete is made from a mixture of about equal parts of sand, gravel, and cement, plus some added water to give a mixture that flows. The low energy expenditures to make these raw materials mean that concrete is a material that requires very low energy; the only additional energy is a small amount for transport and for mixing the constituents. Concrete requires about one-third the energy expenditure for steel, and one thirtieth that for aluminum (Table 1). In Western Europe concrete has replaced metals and wood in many applications, because forests are depleted and energy costs are higher than in the United States. Examples are in building; American homes still use a wood frame, but in Western Europe almost all homes are made from concrete or stone. Electrical transmission poles in the United States are made of wood or aluminum, and in most of the rest of the world these poles are made of concrete because of their lower energy requirements. Considerable energy savings could occur by substituting concrete for metals in a variety of applications. Concrete has excellent compressive strength but is weak in tension or bending. By reinforcing concrete with steel bars the concrete building or structure has good strength in bending and tension as well as compression.

Other applications of ceramics require clay, either raw or purified, sand, and feldspar. Brick, porcelain, and white wares are made from these raw materials; the main expenditure in making these products is in firing the mixtures of powders to a dense solid. Ordinary brick made from fire-clay requires a small amount of energy; even refractory brick for high temperatures and chemical durability, made partly from purified oxides such as alumina or chrome ore, requires only about the same energy to make as an equivalent weight of steel.

Glass for containers is made continuously in a large tank or furnace. Raw materials (sand, soda-ash, limestone) are fed in at one end of a gas-fired furnace. The molten glass slowly passes through the furnace at high temperature (1,200°C-1,300°C) to homogenize it and remove bubbles. At the exit end the molten glass is fed into special molds in two stages to make containers of desired sizes and shapes. Flat glass (windows) is also made in a continuous furnace; a glass layer leaving the furnace is spread onto a bath of molten tin (float glass) to provide smooth surfaces. Lamp bulbs are made from a continuous furnace; a ribbon of glass is fed to blowers that blow the bulbs into a mold (ribbon machine). All these processes are continuous with large furnaces for melting, and so are energy efficient, using about $20(10)^{12}$ J/kg, a factor more the ten lower than the energy required to make an equivalent weight of aluminum. Nevertheless, aluminum containers have replaced glass for many purposes, because aluminum is easier to handle, and harder to break. Polymer (plastic) containers are also popular because of low cost, chemical durability, and ease of handling. As energy costs increase, aluminum containers will become less attractive than glass; the raw material (petroleum) for polymers may also become more expensive, leading to a return of glass as the primary container material.

For many specialized uses glass is made in small batches, so the energy costs are much higher than for the continuous furnaces. Special processes, such as for drawing fibers, casting optical components, and making laser glass, require highly purified or controlled raw materials, leading to much higher energy requirements than for continuously made glass.

Many ceramic applications are high value and small volume, so energy expenditure is high. Ferroelectric magnets, electronic substrates, electro-optics, abrasives such as silicon carbide and diamond, are examples. Diamond is found naturally, and made synthetically by the General Electric Company at high pressure and temperature. Synthetic diamonds for abrasives require less energy to make than the value in Table 4; nevertheless, the market is carefully divided between natural and synthetic diamonds.

Large quantities of uranium oxide are required for nuclear reactor fuel. The uranium ore must be carefully purified and processed to desired shapes, causing high energy expenditure.

Single crystals of synthetic quartz are made by crystallization from aqueous solution at temperatures and pressures well above ambient. The crystallization is slow and carefully controlled, so energy costs are high.

The energies for producing some gases are listed in Table 5 for comparison with those for other materials.

RECYCLING OF CERAMICS

Bulk ceramics such as building materials, porcelain, and concrete are not recycled, because of the low energy required to make them and the difficulty of collecting, transporting, and reforming them into useful shapes. Some glass is recycled in the form of "cullet," which is waste glass. The amount of cullet in a glass furnace is rigidly controlled, because the final product of the furnace must have just the right viscosity for the automatic machinery (container mold, tin bath, or ribbon machine) that forms the glass. Glass manufacturers are unwilling to build tanks to accept waste glass, because its variable composition leads to uncontrollable variations in the viscosity of the glass. Viscosities of silicate glasses are highly sensitive to impurities, especially water and alkali (sodium and potassium) compounds.

ELECTRONIC MATERIALS

Silicon wafers are the basis for electronic circuits. The silicon must be highly purified, then grown as a single crystal containing a small amount (a few parts per million) of additions to give either negative carriers (electrons from phosphorous or arsenic) or positive carriers (holes, from boron or aluminum). These processes require temperatures above the melting point of silicon (1,414°C) and careful control of several processing steps. The energy expenditure for making silicon for wafers (chips) is about the same as that for germanium, given in Table 4. The energy of about $2,500(10)^{12}$ J/kg is greater than that required to make such a valuable metal as silver, or zirconium, which is strongly bonded in compounds, because of the highly complex processing and high purity required for the semiconductors. Subsequent processing of silicon wafers to form devices on the wafers for practical use is highly specialized, carefully controlled, and expensive in cost and energy.

Material	Form or process	Energy 10^{12} J/kg
Sulfur	Frasch	2.2
Antimony	Bulk	7
Graphite (carbon)	Refined, bulk	40
Tellurium	Bulk	96
Phosphorous	Bulk elemental	200
Selenium	Bulk solid	340
Iodine	Solid	620
Germanium	Semiconductor grade	2500
Diamond	Natural	830,000

Table 4.
Energy to Produce Semi-Metals and Semiconductors

Gas	Energy 10^{12} J/kg
Oxygen	4.2
Argon (liquid)	4.9
Bromine	17
Chlorine	21
Ammonia	45

Table 5.
Energy to Produce Commercial Gases

ENVIRONMENTAL CONCERNS

There is great interest in the more energy efficient production of materials to reduce costs and environmental damage. The products of materials production receiving the greatest attention are sulfur dioxide, fluorides, and carbon dioxide. The production of energy from fossil fuels, especially coal and oil, leads to production of sulfur dioxide, which causes much damage locally and at long distances. It leads to respiratory problems and damage to plants, especially trees, and can acidify soils and lakes, damaging them for growing plants and animals. Sulfur dioxide can be scrubbed from flue glass at considerable expense, but much of it is still discharged into the atmosphere. Sulfur dioxide is a by-product in much of the materials production discussed here; sulfide ores (copper) when oxidized produce sulfur dioxide (Equation 2), and some raw materials for cement contain sulfates.

Fluorides are used in many materials processes, and can poison the environment when they are discarded. Examples are cryolite (sodium aluminofluoride, Na_3AlF_6) used to dissolve aluminum oxides for electrolysis, and hydrofluoric acid (HF) used in etching lamp bulbs and semi-conducting circuits. Today lamp bulbs are etched much less than they used to be to reduce fluoride disposal; not much has been done to reduce the amount of cryolite for aluminum production.

Some heavy metals and semi-metals are quite toxic (chromium, lead, and antimony) and expensive care is needed to prevent them from being dispersed in the environment. Lead in gasoline and paint has been almost completely eliminated; its use in storage batteries has resisted efforts to find a suitable substitute.

The discharge of carbon dioxide from combustion of fuels from vehicles, and from processes such as steelmaking, cement production, and much other materials production has increased the concentration of carbon dioxide in the atmosphere. Some computer models demonstrate that this increase is responsible for an increase in the mean temperature of the surface of the Earth, and there are numerous predictions of further temperature increases as more carbon dioxide is discharged into the atmosphere. There are other claims that this result is not proven. Reduction of carbon dioxide emissions is highly difficult and expensive. If the connection between carbon dioxide emissions and global warming is proven more conclusively and a carbon dioxide reduction plan is instituted, materials industries will feel a great impact because they consume about 20 percent of all industrial energy.

Reduction of overall energy use is one solution to the above problems. It requires money, technical advances, political power, and courage; some reduction has been achieved, but much more is needed to reduce emissions of gases. One solution being advanced is use of processes to produce energy that do not emit gases. Hydropower has been exploited about as fully as possible, and supplies only a small fraction of total energy needs. Other sources such as wind and solar power are still much too expensive.

One energy source that first appeared to be highly attractive was nuclear power. The problem with nuclear power is that some costs were hidden in its initial development. Especially pernicious is the disposal of uranium oxide fuel after it has become depleted. It can be reprocessed, but at considerable expense, and the product plutonium can be used for weapons. In the United States the plan is to bury depleted uranium from reactors, but many persons are not convinced that burial is safe. Much work has been done on encapsulation of radioactive waste in glass; the problem of reactor waste remains.

SUMMARY

The energy required to produce materials varies widely, gold requires more than two thousand times the energy to produce the same weight of steel, and diamonds two hundred thousand times the energy required to make ordinary brick. Factors in energy use are the quality (concentration) of ore, the complexity of processing, and the technological development of processing.

Some recycling of metals occurs; much more is possible, and substitution of materials requiring less energy for those requiring more has much potential.

Reduction of environmental pollution requires lower energy use and new technology to decrease emission of gases such as sulfur dioxide and carbon dioxide, and to prevent toxic fluoride, heavy metal, and radioactive wastes from discharging into the environment.

Robert H. Doremus

See also: Building Design; Climatic Effects; Drilling for Gas and Oil.

BIBLIOGRAPHY

Battelle-Columbus Laboratories, Energy Use Patterns in Metallurgical and Nonmetallic Mineral Processing. (1975). *High Priority Commodities*, PB 245 759; (1975). *Intermediate Priority Commodities* PB 246 357; (1976). *Low-Priority Commodities*, PB 261 150.

Doremus, R. H. (1994). *Glass Science*. New York: Wiley.

Fine, H. A., and Geiger, G. H. (1993). *Handbook on Material and Energy Balance Calculations in Metallurgical Processes*. Warrendale, PA: TMS.

Kaplan, R. S., and Ness, H. (1986). "Recycling of Metals: Technology." In *Encyclopedia of Materials Science and Engineering*, ed. M. B. Bever. Cambridge, MA: MIT Press.

Kellogg, H. H. (1986). "Energy Considerations in Metals Production." In *Encyclopedia of Materials Science and Engineering*, ed. M. B. Bever. Cambridge, MA: MIT Press.

Kingery, W. D.; Bowen, H. K.; and Uhlmann, D. R. (1976). *Introduction to Ceramics*. New York: Wiley.

Mayer, J. W., and Lau, S. S. (1990). *Electronic Materials Science*. New York: Macmillan.

Shackelford, J. F. (1997). *Introduction to Materials Science for Engineers*. New York: Macmillan.

MATTER AND ENERGY

The entire observable universe, of which the Earth is a very tiny part, contains matter in the form of stars, planets, and other objects scattered in space, such as particles of dust, molecules, protons, and electrons. In addition to containing matter, space also is filled with energy, part of it in the form of microwave radiation.

INERTIA, MASS, AND ACCELERATION

Matter itself has energy, called "rest energy." What distinguishes matter-energy from other forms of energy is that all matter has inertia and is subject to the force of gravity when at rest as well as when in motion. Inertia measures the resistance of an object to being accelerated by a force, and the inertia of an object at rest is proportional to its mass.

According to a law of physics first formulated by Isaac Newton and later modified by Einstein in his general theory or relativity, any object with mass can be accelerated by applying a force to it. Physicists use the term "acceleration" not only to describe the speeding up or slowing down of an object but also for changing its direction. A car going around a curve at constant speed is accelerating because its direction changes.

If you flick a small plastic ball on a table top with your finger, thereby exerting a small force on the ball, you will see it move rapidly from its resting position. But if you do the same with a steel ball of the same size, the same flick of the finger (the same force) will produce noticeably less motion. The steel ball has greater mass and therefore greater resistance to being accelerated. The ratio of the accelerations of two objects experiencing the same force is equal to the ratio of their masses.

The basic unit of mass is the kilogram, which is the mass of a standard platinum cylinder located in the city of Paris. A kilogram has a weight of 2.2 pounds. The basic unit of energy is the joule, which is equal to the kinetic energy that a one-kilogram object has when it is moving at a speed of 1.41 meters per second or the amount of potential energy the object has when lifted to a height of 0.102 meters.

PROPERTIES OF MATTER

Matter on earth commonly takes the form of solids, liquids, or gases, but may also be in the form of plasmas, which are "ionized" gases, that is, gases in which some of the atoms of the gas have lost one or more of their electrons. These electrons move within the plasma.

Gases, liquids, and solids have different physical properties. A gas fills its container, so that if a certain amount of gas is transferred from a small container into a large one, the gas will expand to fill the new container. If there is a hole in the top of a container filled with gas, the gas will escape. A liquid keeps the same volume when transferred from one container to another, but takes the shape of the new container. On Earth, a liquid has a flat, horizontal surface, If there is a hole in its container below that surface, the liquid will spill out. A solid keeps both its shape and its volume when transferred from one container to another.

Solids, liquids, and gases all change their volumes when the temperature is changed. All gases and nearly all solids and liquids tend to expand when their temperature is raised. When heat is applied to a solid, its temperature normally goes up, but at a certain temperature it can change its state (phase) to a liquid while the temperature remains constant. Similarly, as heat is applied to a liquid, its temperature normally rises until a certain temperature, when it changes its state to a gas. On still further heating, the gas will expand if the container has a movable piston to let it expand; otherwise, the pressure of the gas will increase. Eventually, if enough heat is applied, the gas can become partially ionized, that is, it can turn into a plasma.

The sun is a partially ionized gas or plasma. It has no container to hold it together. Instead, the enormous gravitational forces in the sun do the job. Even on Earth, the atmosphere (a gas) does not escape to outer space because of gravity. On the surface of the moon, gravity is only one-sixth as strong as on the surface of the earth. This is not strong enough to hold gases on the moon, so the moon has no atmosphere.

According to Albert Einstein's theory of relativity, no object with mass can travel as fast as the speed of light in empty space (in vacuum). So another definition of matter is anything that is subject to gravity and is either at rest or traveling slower than the speed of light in vacuum. The speed of light in vacuum, denoted by the special symbol c, is a constant speed (186,000 miles per second, or 300,000 kilometers per second) and is the speed in vacuum of any quantum (packet) of energy that has no mass. (At this speed,

Is the universe matter dominated or energy dominated? The very early universe was "energy dominated" in the sense that only a small fraction of the energy of the universe at that time was in the form of the rest energy of matter and antimatter. Much of the energy of the universe was in the form of electromagnetic radiation. It is not known whether the universe originally had an excess of matter over antimatter or whether the excess of matter developed as time went on. Most cosmologists favor the idea that reactions in the early universe led to the excess of matter, but the mechanism is not known. As the universe expanded it cooled, until, about 300,000 years after the Big Bang, the quanta of radiation (photons) no longer had enough energy to create electrons and positrons. As the universe continued to expand, the radiation cooled still further, until today all that is left of it is the microwave background radiation at a temperature of 2.7 K. The amount of energy in the background radiation plus all the light energy from stars is only a tiny fraction of the rest energy of matter. Therefore, today, the visible universe appears to be matter dominated. However, it is an open question whether other, as yet unseen, forms of energy exist that would make the universe energy dominated.

light can travel seven times around the world in a little under a second.) Light, and anything else that travels at the speed of light in a vacuum, is not considered to be matter. However, all things that travel at the speed of light, including light itself, possess kinetic energy.

The theory of quantum mechanics says light has some properties of a wave (for example, a wavelength), but its energy is concentrated in little packets called photons. These quanta, or particles of light, do not have mass and always travel at the speed of light in vacuum. Light travels slower in a material medium (such as glass) than in a vacuum because the photons get absorbed and reemitted by the atoms of the medium, thereby slowing down the progress of the light wave.

Forces, such as the gravitational force, are transmitted by means of quantities that scientists call "fields." All matter is influenced by forces carried by gravitational fields. Electrically charged particles are influenced by forces carried by electromagnetic fields. The gravitational and electromagnetic fields are not matter, but they have energy.

As far as is known, ordinary matter is made of tiny building blocks called elementary particles. For example, an atom is made up of a nucleus surrounded by one or more electrons. As far as scientists have been able to determine, the electrons are elementary particles, not made of anything simpler. However, an atomic nucleus is not elementary, but is a composite particle made up of simpler particles called protons and neutrons. (The lightest nucleus is the nucleus of ordinary hydrogen, which consists of only a single proton.) Today, physicists believe that even protons and neutrons are not elementary but are composite particles made up of still simpler building blocks called quarks.

At the present time, quarks are believed to be elementary particles. All the particles in an atom, whether elementary or not, are particles of matter and possess mass. Electrons, protons, and neutrons can also exist outside of atoms.

In addition to ordinary matter, scientists have evidence for the existence in the universe of "dark matter." Some of the dark matter is ordinary matter, such as dust in outer space and planets going around other stars. Astronomers cannot see ordinary dark matter because any light coming from such matter is too faint to be observed in telescopes. However, most of the dark matter in the universe is believed not to be ordinary matter. At the present time it is not known what this mysterious dark matter is, or what it is made of. Scientists know that this dark matter exists because it exerts a gravitational force on stars (which are made of ordinary matter), causing the stars to move faster than they otherwise would. According to present estimates, there is perhaps five times as much dark matter in the universe as ordinary matter.

GRAVITY

Gravity is a force that acts not only on matter but on anything, such as light, that possesses energy. A gravitational force cannot speed up light or slow it down, but it can accelerate light by changing its direction.

Light from stars directly behind the sun can be seen on earth during an eclipse of the sun because the sun's gravity bends some starlight around it. An observation of this effect was first made in 1919 during a solar eclipse, and the amount of bending observed was in agreement with the predictions of general relativity.

Gravity can also give energy to light or take energy away from light. If light from a laser is directed down from the top of a building to the ground, the light will gain a small amount of energy by "falling" in the gravitational field. Scientists can measure this slight energy gain as an increase in the frequency (decrease in the wavelength) of the light. Conversely, if the laser light shines upward, the light loses energy, and its frequency slightly decreases.

According to the theory of relativity, all matter has a kind of energy, called rest energy, denoted by the symbol E. If an object at rest has a mass denoted by m, its rest energy E is given by Einstein's famous formula $E = mc^2$. Because the speed of light c is such a large number, a small amount of matter contains a large amount of rest energy.

It has been noted that a one-kilogram object moving at 1.41 meters per second has a kinetic energy of 1 joule. The rest energy of the object is 90 million billion times as great. In fact, the rest energy of ordinary objects is so large that some people dream of unlocking that energy and converting it into more useful forms of energy, such as kinetic energy and heat. The laws of physics do allow matter to be converted into energy and energy into matter. However, at present, no way is known to convert the rest energy of matter entirely into energy except by "annihilation" in a collision with a form of matter known as antimatter.

PROPERTIES OF ANTIMATTER

Earth and the sun, and, as far as is known, the stars and planets in the rest of the visible universe, are made of ordinary matter. However, according to a theory first proposed by Paul Dirac in 1928, for every kind of particle of ordinary matter that exists in nature, there can exist an antiparticle made of antimatter. Some antiparticles have been discovered: for example, the antiparticle of the electron, called the positron, was discovered in 1932 in cosmic rays falling on earth and have also been created in experiments performed in the laboratory. Antimatter is very simi-

Lasers are focused on a small pellet of fuel. This is an attempt to create a nuclear fusion reaction for the purpose of producing energy. (Corbis-Bettmann)

lar in some of its properties to ordinary matter, while other properties are quite different. For example, an electron is a particle of ordinary matter with negative electric charge. However, although the positron has a mass equal to that of an electron, it has positive electric charge. The mysterious dark matter of the universe is not the antimatter of ordinary matter.

According to general relativity, both matter and antimatter are attracted by gravitational forces. However, as yet no experiment has succeeded in showing that antimatter falls under Earth's gravity. The reason is that only small particles of antimatter, such as antiprotons, antineutrons, and positrons, have been created in the laboratory. The electric and magnetic forces acting on these particles are much stronger than the gravitational forces and mask the effects of gravity.

The most spectacular difference between a particle and an antiparticle is that, as the result of a collision, the particle and antiparticle can both annihilate into pure energy. For example, if an electron and a positron collide, they may destroy each other (annihilate) into radiation. This is an example in which the rest energy of matter and antimatter is converted entirely into another form of energy. Conversely, under some conditions the kinetic energy of rapidly moving particles can be converted into new particles of matter, usually together with particles of antimatter. Because antimatter is rare in the universe, nobody has to worry about our earth colliding with enough antimatter to annihilate the earth, although some particles on earth are annihilated by antiparticles from outer space.

It is because of the rarity of antimatter that we cannot use annihilation of matter as a source of kinetic energy, heat, light, and other forms of energy. Of course, scientists can create antimatter, but they have to supply the energy to create it. When the created antimatter annihilates, the scientists get back only the energy that they put in. It is actually much worse than that, because creation of antimatter is a very inefficient process, and most of the input energy is wasted. Furthermore, it is very difficult to store antimatter. It cannot be stored in any container made of matter, as it will annihilate with the walls of the container. Antimatter has to be contained by electromagnetic forces in a vacuum.

CONVERTING MATTER TO ENERGY

Although it is impractical to convert the rest energy of matter entirely into other forms of energy, nevertheless, a small fraction of rest energy is converted in chemical and nuclear reactions. For example, if hydrogen is burned in oxygen (a chemical reaction), the product is water plus heat and light. A scientist can describe this process by saying that burning converts chemical energy into heat and light. However, the process can be looked at in another way. If careful measurements are made, it is found that the mass of the water is slightly less than the sum of the masses of the original hydrogen and oxygen. So it can also be said that "the burning process" converts a small amount of rest energy of the hydrogen and oxygen into heat and light.

Normally, electrons in an atom are "bound" in the atom by the attractive electrical forces between the electrons and the atomic nucleus. A certain amount of energy must be applied to an atom to release an electron from the atom, thereby ionizing it. This amount of energy is called the "binding energy" of the electron in the atom. The binding energy of the electrons in hydrogen and oxygen is slightly different from the binding energy of the same electrons when the oxygen and hydrogen are combined in water. A change in binding energy causes a change in rest energy of the same amount. This difference in binding energy (or rest energy) is the source of the heat and light when hydrogen is burned.

Just as electrons are bound in atoms, so are protons and neutrons bound in atomic nuclei, but the binding energy of the protons and neutrons is far greater. Consequently, changes in the binding energy in nuclei are a much greater source of heat and light than changes in the binding energy of electrons. More than a million times as much matter can be converted into energy in a nuclear reaction as in a chemical reaction, and even in such a nuclear reaction only about 0.1 percent of the matter is converted into energy.

Most of the energy of the sun comes from changes in binding energy when hydrogen is converted into helium in nuclear reactions. When very light nuclei, such as hydrogen nuclei, are combined to produce nuclei having less total mass than the very light nuclei, energy is released. The process is called "nuclear fusion." The energy released in the sun in nuclear fusion is what causes the sun to shine.

When very heavy nuclei, such as those of uranium and plutonium, are split into lighter nuclei having less total mass than the very heavy nuclei, energy is released. The process is called "nuclear fission." In either nuclear fission or nuclear fusion, much of the converted rest energy emerges as kinetic energy, heat, and light.

The explosion of an atom bomb (a uranium or a plutonium bomb) and the operation of a nuclear reactor are cases of energy released in nuclear fission, the first in a very fast process and the second in a slower, controlled way. In a nuclear reactor, a small fraction of the rest energy of the uranium or plutonium is converted into heat. The heat is then used to turn water into steam, which drives a turbine attached to an electric generator in order to generate electricity (electrical energy). In a hydrogen bomb, most of the energy released comes from nuclear fusion. Scientists have tried for almost fifty years to build a fusion reactor. Although scientists have been able to generate a small amount of heat by controlled fusion, they have not succeeded in generating large amounts of heat from controlled fusion in a profitable way.

Don Lichtenberg

See also: Einstein, Albert; Hydrogen; Molecular Energy; Nuclear Energy; Nuclear Fission; Nuclear Fusion; Thermodynamics; Units of Energy.

BIBLIOGRAPHY

Clark, J. O. W. (1994). *Matter and Energy: Physics in Action.* New York: Oxford University Press.

Einstein, A. (1961). *Relativity,* tr. R. Lawson. New York: Three Rivers Press.

Schrödinger, E. (1953). "What Is Matter?" *Scientific American* 189:52-57.

MAXWELL, JAMES CLERK (1831-1879)

James Clerk Maxwell is the one theoretical physicist between Isaac Newton and Albert Einstein of a stature comparable to theirs. Maxwell's contributions to science ranged over many areas, of which the two greatest were his creation of the electromagnetic theory of light, and his work on molecular physics, gas theory, and statistical mechanics. He entered the scientific scene in the early 1850s, immediately after the principle of conservation of energy had been established. Its impact is seen everywhere in his work.

A descendant of a distinguished Scottish family, the Clerks of Penicuik, and, by an illegitimate line, of the ninth Lord Maxwell, he was born in Edinburgh but lived much of his life at his estate in Galloway in southwest Scotland, where he inherited 2,000 acres of rich farmland. From the ages of ten to nineteen he was educated in Edinburgh, entering the University of Edinburgh in 1847. At nineteen he went on to Cambridge University to take the rigorously severe mathematical *tripos*, from which he graduated in 1854 second in order of merit. He became Fellow of Trinity College in the following year and then, in 1856, at twenty-five, was appointed professor of natural philosophy at Marischal College, Aberdeen. In 1858 he married Katherine Mary Dewar, daughter of the principal. Though lacking any prior scientific training, she became an enthusiast in experimental research and worked closely with Maxwell on several experiments, first in color vision and then in physics. They had no children.

In 1860 Maxwell became a professor at King's College, London, where he served for five years. He retired from professorial life in 1866, at the age of thirty-five, to spend six years writing his famous *Treatise on Electricity and Magnetism* (1873). During the same time he also produced his small but important *Theory of Heat* (1871). In 1871 he was appointed Cavendish Professor of Experimental Physics at Cambridge and was responsible for designing and setting up the Cavendish Laboratory. Maxwell died of abdominal cancer in 1879 at age forty-eight.

James Clerk Maxwell. (Library of Congress)

ELECTROMAGNETIC THEORY

When Maxwell began studying electricity and magnetism in 1854, the field was in a state of confusion. The laws of electric and magnetic force had been established by Charles Augustin de Coulomb in the 1780s, and impressive mathematical structures had been built on them. However, the triumph was unsettled by Hans Christian Oersted's discovery in 1820 of electromagnetism—a peculiar twisting action exerted by an electric current on a magnet. This departure from Newtonian attractions and repulsions met two contrasting reactions. André Marie Ampère sought to reinterpret Oersted's force as a disguised form of attraction. Michael Faraday treated it as primary and related it geometrically to properties of lines of magnetic and electric force.

It is wrong to see Maxwell's achievement as one of merely translating Faraday's ideas into precise mathematical language. Though he once described Faraday as "the nucleus of everything electric since 1830," two other men, William Thomson (Lord Kelvin) and Wilhelm Weber, were equally influential.

From Faraday Maxwell gained a way of thinking; from Thomson, the first mathematizations of Faraday's ideas and several groundbreaking connections to the concept of energy; from Weber, the remarkable insight that the ratio of the two kinds of force, electrostatic and electromagnetic, somehow involves a velocity.

Between 1855 and 1868 Maxwell devoted great effort (five substantial papers) to clearing up the confusions in electromagnetism. The outcome was the dramatic discovery that light is an electromagnetic phenomenon, and the prediction—twenty-seven years before they were detected by Heinrich Hertz—of radio waves. Crucial was Maxwell's devising in 1861 of a speculative "ether" transmitting Faraday's lines of magnetic force. To his astonishment he found that this ether would transmit waves. Using some measurements by Weber and Friedrich Kohlrausch, Maxwell then calculated their velocity and found, to his even greater astonishment, that it was just equal to the velocity of light. Thus the great discovery was made and thus began the great intellectual metamorphosis, shaped by Maxwell and Einstein, in which the velocity of light was transformed from an isolated quantity into a universal fundamental constant influencing every part of physics.

The essence of Maxwell's later development of his theory was in the electromagnetic equations and the idea that electric and magnetic energies, instead of being located on charged bodies, are disseminated through space. That he could so quickly discard his ether model was closely related to the new doctrines of energy. Rather than attempt to explain light or electromagnetism in terms of a mechanism, Maxwell demonstrated that one set of unexplained equations describes both. Philosophically, the theory became a theory of relations. In this line of thought, Maxwell was strongly influenced by his mentor at Edinburgh, Sir William Hamilton, who held that all human knowledge is of relations rather than absolutes.

Maxwell's *Treatise on Electricity and Magnetism* (1873) covered every branch of the science and was a source of ideas and discoveries for fifty years to come.

GASES, MOLECULES, AND STATISTICS

In 1859 Maxwell, who had just completed a famous essay on the structure of the rings of Saturn, chanced to read a paper by Rudolph Clausius on gas theory. Maxwell had proved that the rings had to be composed of large numbers of independent bodies constantly colliding with each other. Clausius, expanding on earlier work by James Prescott Joule and August Karl Krönig, proposed that in a gas the rapidly moving molecules are constantly colliding. His interest at once aroused, Maxwell in a few months had written the first of several papers that created the modern kinetic theory of gases.

Maxwell's and Clausius's innovations were of two kinds, mathematical and physical. Mathematically, the key to dealing with large numbers of molecules was statistics, used not as a means of processing scientific data but as a fundamental explanatory idea. Clausius recognized that molecules must travel a certain average distance between collisions—the mean free path—but restrictively assumed that they all have the same speed. Maxwell transformed the discussion by introducing his velocity distribution function, giving the proportion of molecules traveling with a particular speed. Armed with this mathematical weapon, he could attack many previously intractable physical phenomena. He obtained theoretical formulas for viscosity, diffusion, and heat conduction in gases that then could be compared with experimental data. One startling consequence was that the viscosity of a gas should be independent of its pressure. When this was confirmed in independent experiments by Oskar Emil Meyer and by Maxwell and his wife, it added tremendous credibility to the theory.

The work on gas theory had many extensions. In 1865 Johann Josef Loschmidt used estimates of the mean free path to make the first generally accepted estimate of atomic diameters. In later papers Maxwell, Ludwig Boltzmann, and Josiah Willard Gibbs extended the mathematics beyond gas theory to a new generalized science of statistical mechanics. When joined to quantum mechanics, this became the foundation of much of modern theoretical condensed matter physics.

Through his famous "demon" Maxwell addressed one mystery of energy physics: the relation between the first law of thermodynamics, which states that energy as a whole is conserved, and the second law, which states that mechanical energy will be gradually dissipated. Maxwell was the first person to realize and forcefully argue that the second law is a statistical rather than a dynamical truth. Following this clue, Boltzmann in 1872 found the exact formal expression relating entropy to probability. Their work, together with earlier reflections by Kelvin, framed a discussion of irreversibility in physics, embracing even the nature of time, that has continued to this day.

C. W. F. Everitt

See also: Ampère, André-Marie; Clausius, Rudolf Julius Emmanuel; Electricity; Electricity, History of; Faraday, Michael; Gibbs, Josiah Willard; Magnetism and Magnets; Molecular Energy; Oersted, Hans Christian; Thomson, William.

BIBLIOGRAPHY

Brush, S. G. (1976). *The Kind of Motion We Call Heat*, Vols. 1-2. Amsterdam: North-Holland.

Buchwald, J. T. (1985). *From Maxwell to Microphysics*. Chicago: University of Chicago Press.

Campbell, L., and Garnett, W. (1882). *The Life of James Clerk Maxwell*. London: Macmillan.

Everitt, C. W. F. (1975). *James Clerk Maxwell, Physicist and Natural Philosopher*. New York: Scribner.

Harman, P. (1998). *The Natural Philosophy of James Clerk Maxwell*. Cambridge, Eng.: Cambridge University Press.

Siegel, D. M. (1991). *Innovation in Maxwell's Electromagnetic Theory*. Cambridge, Eng.: Cambridge University Press.

Whittaker, E. T. (1954). *History of the Theories of Aether and Electricity*, Vols. 1-2. New York: Philosophical Library.

Julius Robert Mayer. (Library of Congress)

MAYER, JULIUS ROBERT (1814–1878)

In the early nineteenth century many scientists had glimmerings of the conservation-of-energy principle. The three most important among these were the Frenchman Marc Séguin, the American-born, well-travelled soldier of fortune Benjamin Thompson, and the chief engineer of the city of Copenhagen, Ludwig Colding.

The three men whose work later in the nineteenth century was crucial in bringing clarity to this principle were two Germans, the physician Julius Robert Mayer and the great polymath Hermann von Helmholtz, and the British amateur scientist James Joule. In a lecture delivered by Helmholtz on February 7, 1854, in Königsberg on "The Interaction of Natural Forces,"

he referred to Mayer as "the founder" in 1842 of the principle of conservation of energy and acknowledged Mayer's priority in this discovery over Colding (1843), Joule (1843), and Helmholtz himself (1847). Rudolf Clausius agreed with Helmholtz and put Mayer in touch with the British physicist John Tyndall, who quickly became Mayer's English champion in his long-drawn-out priority dispute with Joule and his British supporters, William Thomson (Lord Kelvin) and George G. Stokes.

MAYER'S LIFE AND CONTRIBUTIONS TO SCIENCE

Julius Robert Von Mayer was born in 1814 in Heilbronn, a small town on the Neckar river, halfway between Heidelberg and Stuttgart. Interested in science as a youth, he decided on a career in medicine, and in 1832 began his medical studies at the University of Tübingen. After completing his studies there in 1838, he received his M.D. degree.

In February 1840, Mayer embarked for a year as physician on a freighter carrying cargo to Jakarta in the East Indies, just south of the equator. In Jakarta he noticed that sailors' venous blood was a much brighter red than he had observed in patients back in Germany. He surmised that this change was due to the hot climate and the reduced oxidation needed to preserve normal body temperature. This stimulated him to thinking more generally about how heat affects human metabolism. This interest in heat, work, and what is now called energy became the passion of Mayer's life.

Mayer returned to Heilbronn in 1841, began his medical practice, and eventually became chief surgeon of the town. In his free time he did some experiments and struggled with difficult, abstract concepts in an attempt to understand the nature of energy. He knew so little physics from his one semester of the subject at Tübingen that many of the papers he submitted for publication were rejected as incompetent by the important scientific journals of the day. He was forced to publish most of his writings at his own expense, and so their circulation was confined primarily to Heilbronn residents.

The first published results of Mayer's work appeared in the March 1842 issue of *Liebig's Annalen der Chemie und Pharmacie* in an article entitled "Remarks on the Forces of Inanimate Nature." It contained a great deal of philosophy and some inaccurate science, and yet buried among these distracting elements was the essence of the conservation-of-energy principle. Perhaps more important to scientists, the article contained the first quantitative attempt to determine the relationship of the "calorie" (a unit of heat) to what is now called the "joule" (a unit of energy).

Mayer's calculations were based on his conviction that there is a definite quantitative relationship between the height from which a mass falls to the ground and the heat generated when it strikes the ground. Here he was stating his conviction that energy is conserved in this process if heat is considered a form of energy. He calculated the ratio of the joule to the calorie, and found that it was equal to 3.59 J/cal, which differs considerably from today's accepted value of 4.18 J/cal. His whole approach to the problem was correct, however, and his numerical result was certainly of the right order-of-magnitude, but he had used inaccurate values for some constants needed in his calculation.

Few details of his research were given in Mayer's 1842 paper, but in 1845 he published (at his own expense) his most original and comprehensive paper, "Organic Motion and Its Relation to Metabolism," in which he gave full details of his earlier work. In 1848 he also published, again privately, "Contributions to Celestial Dynamics," in which he made the interesting conjecture that the source of the energy radiated by the sun was its constant bombardment by high-energy meteors.

Mayer had married in 1842, and for the first few years his marriage was a very happy one, but then his life began to fall apart. Between May 1845 and August 1848 three of his children died. A nasty priority controversy with Joule became public when their claims were read and discussed by the Academy of Science in Paris. Mayer was upset that his writings on energy conservation were not more widely read, and that they often were not appreciated by the few scientists who did read them. And, as the final blow, he was accused by local scientists of being more a mad philosopher than a competent scientist.

Finally on May 28, 1850, in a fit of despair, Mayer threw himself out of his bedroom window to the street thirty feet below, but escaped without serious injury. He spent three years in mental hospitals and did little scientific work of value after his release in 1853, although he was able to return to limited service as a physician in Heilbronn.

Physicists around the world gradually came to appreciate Mayer's scientific work, but by this time they were unsure whether he was still alive and, if so, what his mental condition was. In his later years he finally reaped some fruit from his scientific labors. In 1859 he received an honorary doctorate from the University of Tübingen. This was followed in 1871 by his reception of the Copley Medal from the Royal Society of London, and then the Prix Poncelet from the Paris Academy of Sciences. It is unknown how appreciative Mayer was of this belated notoriety when he died of tuberculosis in 1878.

Joseph F. Mulligan

BIBLIOGRAPHY

Caneva, K. L. (1993). *Robert Mayer and the Conservation of Energy*. Princeton, NJ: Princeton University Press.

Friedländer, S. (1905). *Julius Robert Mayer*. Leipzig: T. Thomas.

Kuhn, T. S. (1977). "Energy Conservation as an Example of Simultaneous Discovery." In *T. S. Kuhn: The Essential Tension*. Chicago: University of Chicago Press.

Lindsay, R. B. (1973). *Robert Mayer, Prophet of Energy.* Elmsford, NY: Pergamon Press.

Schmolz, H., and Weckbach, H. (1964). *Robert Mayer, Sein Leben und Werk in Documenten.* Heilbronn: Anton H. Konrad Verlag.

Turner, R. S. (1974). "Mayer, Julius Robert." In *Dictionary of Scientific Biography, Vol. 9*, ed. C. C. Gillispie. New York: Scribner.

MECHANICAL TRANSMISSION OF ENERGY

Mechanical devices are used to magnify the applied force (mechanical advantage), to magnify the distance moved, or to change the direction of the applied force. They of course cannot decrease the amount of work (force × distance) necessary to do a job; they only make it more convenient to do it. In many cases, without a machine, the job would be impossible.

There is generally considered to be five distinct simple machines: lever, wedge, wheel and axle, pulley, and screw. The transmission of energy by these simple machines is so basic that people use them with little understanding of the physical principles involved. Most learn their use intuitively, through experience, and consider their application just plain common sense.

THE BASICS

The history of the origin of simple machines is largely conjectural, but there also exists documentation of the ancient Egyptians using simple machines to build pyramids nearly 5,000 years ago. An inscription in a 4,000-year-old tomb tells of 2,000 men pulling a statue estimated at 132 tons into place. The mass of the 2,000 men would be about the same as the mass of the statue, and it would probably take that many because they moved it on sledges without wheels.

The use of simple machines has sometimes been taken as a definition of what separates humans from animals; however, some primates have been observed fashioning probes out of sticks to pry out or to reach food. One of the most powerful images depicting the use of tools as defining humanity is the opening scene in the movie *2001*. An ape has discovered the club and is bashing some bones. One of the bones flies upward and in slow motion morphs into a spaceship. The club or the hammer is such a basic tool that it does not even make it into the classical listing of the five simple machines. However, it is also a mechanical device that multiplies force and transmits energy.

In the transmission of energy by these simple machines, the conservation law always applies: The work input equals the work output. When work is done *by* a system, energy is transferred out of it; and when work is done *on* a system, energy is transferred into it. When two objects interact by way of a machine (e.g. a lever), the work out of one object equals the work into the other. The work done by a person forcing one end of a lever downward equals the work done lifting a load at the other end as the lever moves upward. In any practical situation, the frictional forces resisting motion will always increase the amount of force (and work) required to do a job.

The amount of work done on an object is determined by the force exerted on it multiplied by the distance it moves in the direction of the force. Therefore the key to figuring out how much the force is magnified by a simple machine is to compare distances moved. For example, if the end of a lever under a stone weighing 2,000 newtons moves upward 1 meter, the amount of work done lifting the stone is 1 meter × 2,000 newtons = 2,000 joules. An equal amount of work must be done by a person on the other end of the lever. If that end of the lever is pushed downward 2 meters, then the person needs to apply a force of only 1,000 newtons to do the same amount of work (2,000 joules) and lift the 2,000 newton stone.

In brief, the mechanical advantage of a lever or any machine equals the ratio of the distance the applied force moves to the distance the load moves.

THE LEVER

The lever is such a part of everyday activity that its application usually requires no conscious thought: the pop top on a soda can, a doorknob, a wrench, pliers, a fishing pole, a faucet with a handle that lifts, a wheelbarrow, fingernail clippers, and so on. The crank and winch used to pull a heavy boat out of the water up onto a trailer can be thought of as a lever in circular form.

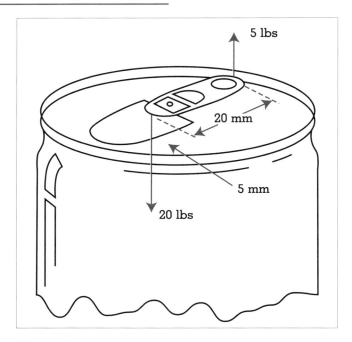

Figure 1.
Standard design of a can with a pop top.

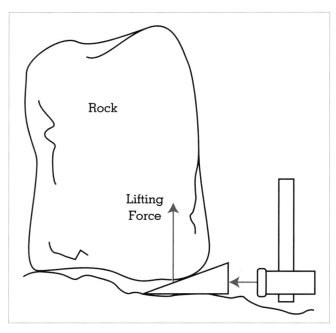

Figure 2.
Forces involved in the use of a wedge and hammer.

Basically, a lever is a solid object with an axis about which it rotates (fulcrum). As the lever rotates about its fulcrum, a point on the lever farther from the fulcrum moves a greater distance. The conservation of energy applied to the lever results in the fact that the output force times its distance from the fulcrum equals the input force times its distance from the fulcrum. A little experience lifting heavy loads with a lever soon teaches one that to maximize the output force, the load should be placed as close to the fulcrum as possible and the input force as far from the fulcrum as possible. To dramatize the nearly infinite possibility of the lever to magnify force, Archimedes said that if he had a lever long enough and somewhere to stand, he could move Earth.

As an illustration of the lever, consider the current design of the pop top on a popular brand of cola (Figure 1). The built-in opener is a piece of aluminum 25 mm long, made rigid by crimping its edges. The fulcrum is 5 mm from the end which presses into the top of the can to "pop" it open. This leaves 20 mm from the fulcrum to the end, which the user lifts up on with a force of about 5 pounds to open the can. (Since ratios of quantities are involved, there is no problem with mixing English and metric units.) The ratio of distances is 20 mm/5 mm = 4,

which means that the opener presses into the top of the can with a force of 4 × 5 pounds = 20 pounds. If the built-in opener is missing, it is necessary to open the can by pressing a small object such as a key into the top with a force of 20 pounds. It is not impossible but it is awkward, and a slip could be messy.

THE WEDGE, THE INCLINED PLANE, AND THE SCREW

These three simple machines change the direction of the applied force as well as magnify it. Each one's operation can be understood by nearly the same physical principles.

A wedge is fairly easy to understand. One side of a heavy rock can be lifted a small amount by pounding a wedge under it, as illustrated in Figure 2. If friction is neglected, the force pushing the wedge under the rock is magnified by the ratio of the distance the wedge moves to the amount the rock is lifted. This follows from requiring the work done by the driving force to equal the work done in lifting the rock. This magnification is the ratio of the length of the wedge to its width, and is obviously greater the smaller the angle of the wedge. The friction force of the wedge against the rock decreases the available lifting force, but it is to

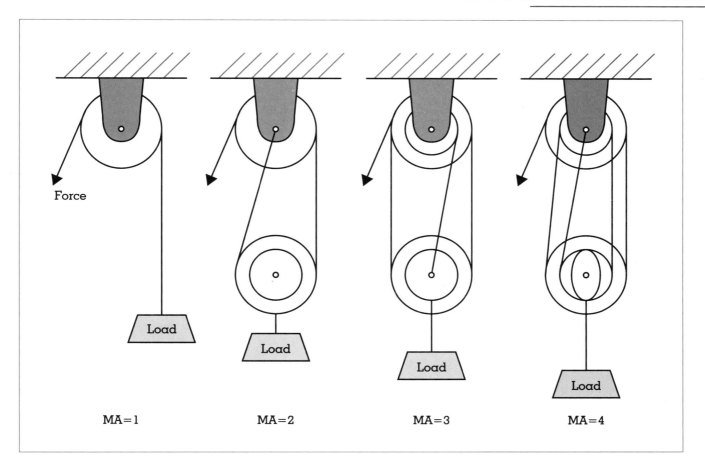

Figure 3.
Pulley systems with varying mechanical advantages.

some extent an advantage. It holds the wedge in place even when no driving force is applied. A wedge is usually used in combination with a hammer.

The hammer itself is an important simple machine used to magnify force. The ape in *2001* was shown inventing the hammer or the club. The user applies a force on the hammer through a relatively large distance, giving it kinetic energy. The hammer comes into contact with an object, compressing it or moving it some short distance. The average delivered force depends on how much compression or motion of the object there is and on whether the hammer rebounds off the object. The mechanical advantage is approximately the ratio of the distance the hammer moves to the distance the wedge, for instance, moves. A person might apply a relatively small force to a hammer for a 1-meter swing. If it drives the wedge 1 millimeter, then the

mechanical advantage is 1,000. No wonder that striking a wedge with a hammer can lift an object weighing several tons.

The inclined plane is stationary, and the load to be lifted is moved; whereas it is the other way around for the wedge. The force is magnified by the ratio of the distance the load is pushed to the height through which it is lifted. Again friction will increase the amount of force that has to be applied. If the coefficient of friction is large enough (at least as great as the tangent of the angle of incline), then the load will not slip back down the incline when the pushing force is released. For most materials (coefficient of friction greater than, 0.6 say), an angle less than 30 degrees will keep the load from slipping back on its own. This allows the pushers to take a break to regain their strength before finishing the job. Or they can push the load up the incline a little at a time by lunging at it. In this way they play the role of the hammer relative to the wedge as they collide with the heavy load.

The screw is an inclined plane that is conveniently wrapped around a circular cylinder. The incline of the screw is in the form of a helix similar to a spiral staircase. The mechanical advantage is the ratio of the distance the driving force moves in a circle as it rotates the screw to the distance the load is lifted. A screwdriver with a large handle will provide a larger mechanical advantage, as will a screw with threads closer together. By using a screw jack with a long enough handle, a person can easily lift an automobile or a house. Again, friction conveniently keeps the screw from backing up on its own so that the user can just leave the jack supporting the load, assured that the jack will stay in place.

THE PULLEY

A fixed pulley is a device for changing the direction of an applied force. A common form is a mounted wheel with a rim around which a rope passes. In a very primitive form it could be a vine looped over a tree branch. A pull downward on the rope (vine) results in lifting a load on the other end. Neglecting friction, the mechanical advantage of the single fixed pulley is 1; the load moves the same distance as the applied force.

A combination of fixed and movable pulleys can achieve larger mechanical advantages. The diagrams in Figure 3 show systems of pulleys with mechanical advantages of up to 4. The mechanical advantage is always the ratio of the distance the applied force moves to the distance the load moves. Another, and perhaps easier, way to determine the mechanical advantage of a system of pulleys is to count the number of rope segments that support the load. That number is the mechanical advantage.

Friction is a very important factor in the actual mechanical advantage of a system of pulleys because the frictional losses are compounded each time the rope passes over a pulley. For a typical coefficient of friction of 0.03 (greased shaft with no ball bearings), the system with a theoretical mechanical advantage of 4 would be decreased by friction to an actual mechanical advantage of about 3.5. The largest mechanical advantage that could be obtained with pulleys having a coefficient of friction of 0.03 is about 16 no matter how many pulleys are used. A coefficient of friction of more than

0.333 (such as a vine over a tree branch) results in an actual mechanical advantage of less than 1 for any system of pulleys.

The friction coefficient of 0.03 used above is so low due to the advantage of a wheel on an axle. The friction force at the axle of the pulley wheel is overcome by a smaller force applied at the rim in accord with the principle of the lever.

The friction forces in a pulley system will never hold the load by themselves, but the effort required can be quite small. In the case of a block and tackle used to lift an engine out of an automobile, the weight of the chain hanging from the last pulley may be enough to keep the engine in place. By wrapping a rope once around a post, a cowboy can hold a raging bull in check.

GEARS

Gears are used almost entirely in rotary motion applications, and as such it is easier to discuss the mechanical advantage as a multiplication of torque rather than as a multiplication of force. The work involved in rotary motion is torque times angle; whereas for the linear motion discussed above, it is force times distance.

Torque arises when a force is applied so that it tends to rotate an object about an axis. The force must have a component at right angles to the axis and at some distance from the axis. The torque produced is the product of the force component and its perpendicular distance from the axis. The units of torque turn out to be the same as the units of energy: force × distance. However, torque is not energy. An angle is the ratio of two lengths (arc length/radius for the angle in radians) and has no units; thus the work in rotary motion (the product of torque and angle) has the appropriate units for work.

Upon comparing the work input with the work output for a gear system similar to the one shown in Figure 4, the mechanical advantage is found from the ratio of the angles turned by the respective shafts as the gears engage. This ratio in turn is equal to the ratio of the numbers of teeth on each gear. For example, if a gear (pinion) with 10 teeth drives a gear with 40 teeth, the mechanical advantage is 4—that is, the torque imparted to the large gear is 4 times the torque input by the small gear. There is a commensurate speed reduction; the large gear will rotate once for each four revolutions of the small gear.

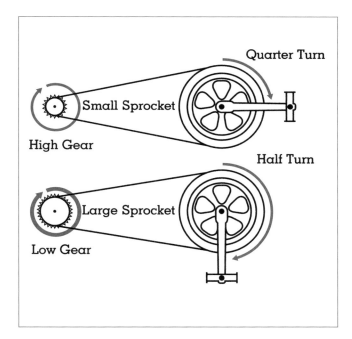

Figure 4.
Gear set with a 12/24 ratio.

Gears may be used to increase the available torque, as in most applications of electric or internal- combustion engines, or to increase the amount of motion, as in a bicycle.

In an automobile, except for the highest gear, the transmission reduces the rotation rate of the drive shaft relative to the engine speed. The differential gears further reduce the rate of rotation. For a typical automobile cruising at 60 mph, the engine runs at about 2,000 rpm while the wheels turn at about 800 rpm giving a mechanical advantage of 2,000/800 = 2.5.

For a typical 21-speed bicycle, on the other hand, the lowest gear ratio is 1.0 and the highest is nearly 3.5. In the highest gear the wheels rotate 3.5 times for each rotation of the pedals. The torque with which the rear wheel propels the bicycle is less than the torque exerted on the pedals by a factor of 3.5. The propulsion force is further reduced by the fact that the radius of the pedals is less than the radius of the rear wheel. The advantage of a bicycle is basically that the wheels move faster than the pedals, coupled with the fact that it takes very little force to overcome the rolling friction of the wheels on a hard, smooth surface. This allows the rider to move faster than a pedestrian using the same energy. However, there is no advantage in force even in the lowest gear. To climb a really steep incline with a bicycle, it is better to get off and walk. From a standing start the greatest acceleration is achieved by pushing the bicycle.

BELTS AND CHAINS

Another function of the transmission of energy by mechanical means is to transfer the motion of a rotating shaft to another shaft at a distant location. A historical application was to get the rotation of a water-wheel coupled to a mill located safely away from the stream in order to grind grain. Another was to drive several rotating machines in a factory from a single large steam engine. Before the invention of the electric generator and the electric motor this type of problem was solved by the use of belts. Today it would be ridiculous to use belts in this way. Think how complicated it would be to connect the power company's spinning steam turbine to a refrigerator in a home by means of belts. Electricity makes it easy. The power is distributed electrically through relatively small wires to drive individual small electric motors.

Today belts are used in automobiles to drive auxiliary devices such as air conditioning, power brakes, power steering, the alternator, and the cooling fluid pump. Belts also can be found in household appliances such as vacuum sweepers, on lathes in machine shops, or inside copying machines.

A belt drive is an inexpensive solution for transmitting mechanical energy over short distances. Compared with machining precision gears, fashioning two pulleys to be linked by a belt is not technologically demanding. The mechanical advantage is the ratio of the diameter of the pulley on the load to the pulley on the driver. The length of the belt makes no difference. By adjusting the sizes of the two pulleys, the mechanical advantage can be conveniently changed. Also, one twist in the belt will reverse the direction of rotation. One disadvantage of a belt is that the amount of torque it can deliver is limited to the product of three factors: the coefficient of friction between the belt and the pulley, the force producing tension in the belt, and the radius of the load pulley.

A V-belt greatly increases the deliverable torque, since the wedging of the belt in the sheave groove increases the force of contact between the surfaces (N) far above the tension force (P). The driving action takes place through the sides of the belt rather than the bottom, which normally is not in contact with the sheave at all. This is yet another example of the application of the wedge as a force multiplier.

For a situation where large torques are involved, such as a bicycle drive, a chain linkage is superior to a belt. A person putting all his or her weight on a pedal probably would make most belt systems slip. Another advantage of a chain over a belt is that a chain is more efficient, mainly because it does not require any ambient tension. The return side of a chain drive has only enough tension to support itself. Furthermore, the chain links are equipped with rollers, which can rotate as they contact the teeth, reducing the frictional forces and wear.

The mechanical advantage of a chain linkage can be calculated by counting the teeth on the load sprocket and the drive sprocket. The output torque is found by multiplying the input torque by the ratio of the number of load teeth to the number of drive teeth. A chain drive is also compact compared to a belt. Imagine trying to arrange 21 speeds on a bicycle derailleur using belts.

A serious disavantage of belts and chains for transmitting energy is that they can be quite dangerous. Whereas the low torque of a bicycle chain rarely results in bad injuries when trouser cuffs or shoestrings get caught in the chain, high-torque industrial and agricultural machinery (such as mechannical reapers with numerous belts and chain drives) is another matter. They have caused many grave inuries and loss of limbs because of the tremendous torque that engines and motors transmit to belts and chains. Therefore, as a safety measure, almost all new equipment comes equipped with guards to prevent accidental injuries either from broken belts and chains whipping out with tremendous force, or the careless actions of workers.

Don C. Hopkins

See also: Bicycling; Drivetrains; Electric Power, Generation of; Electric Power Transmission and Distributive Systems; Engines; Flywheels; Kinetic Energy; Propellers; Steam Engines.

BIBLIOGRAPHY

Adkins, J. (1980). *Moving Heavy Things.* Boston: Houghton Mifflin.

Barnes, M.; Brightwell, R.; Von Hagen, A. L.; and Page, C. (1996). *Secrets of Lost Empires.* New York: Sterling.

National Geographic Society. (1986). *Builders of the Ancient World.* Washington, DC: Author.

Macaulay, D. (1988). *The Way Things Work.* Boston: Houghton Mifflin.

"Machines and Machine Components." (1973). *Encyclopaedia Britannica, Macropeadia,* Vol. 11, pp. 230-259. Chicago: University of Chicago.

MEITNER, LISE (1878–1968)

In a 1959 lecture at Bryn Mawr College in Pennsylvania, Lise Meitner reflected that "Life need not be easy, provided that it is not empty." Life was not easy for any Jewish woman scientist in Germany in the first half of the twentieth century, and Meitner certainly had her own experience in mind when she made this statement.

Lise Meitner grew up in the Vienna of Emperor Franz-Josef and horsedrawn trolley cars. She was born there in 1878 into a well-to-do Jewish family and decided at an early age that she wanted to be a scientist like Madame Curie. (Later Albert Einstein would call her "the German Madame Curie.") In 1901, she entered the University of Vienna. There, where serious women students were considered odd, she was treated rudely by many of her fellow students. In 1905 she was only the second woman in the university's history to receive a Ph.D. in science.

In 1907, she went to Berlin to study under Max Planck, promising her devoted parents that she would return to Vienna in six months at the most. She stayed in Berlin for thirty-one years. In Berlin, Meitner met Otto Hahn, a professor of chemistry, and took an unpaid position assisting Hahn with his research on the chemistry of radioactive substances. At that time women were not allowed to work in the Chemical Institute, and she had to set up her laboratory in a carpenter's workshop outside the Institute.

While continuing work with Hahn at the new Kaiser Wilhelm Institute for Chemistry in Berlin-Dahlem, beginning in 1912 Meitner served as assistant to Max Planck at the Institute for Theoretical Physics at the University of Berlin, and in 1918 was appointed head of the physics department at the Kaiser-Wilhelm Institute.

Lise Meitner. (Library of Congress)

MEITNER'S CONTRIBUTIONS TO NUCLEAR ENERGY

In their years together Hahn and Meitner did significant research on beta- and gamma-ray spectra. They discovered the new element protoactinium-91 and, at Meitner's suggestion, took up, and made great progress with, work on neutron bombardment of nuclei that Enrico Fermi had commenced in Rome. In 1938, this research was suspended when Adolph Hitler annexed Austria and Meitner had to flee Germany.

Based on the strong recommendations of her German physics colleagues, Meitner received a research position in the Stockholm laboratory of Manne Siegbahn, the Swedish physicist who had received the 1924 Nobel Prize in Physics for his precision measurements on X-ray spectra. Siegbahn provided laboratory space for Meitner, but no suitable equipment for her to continue the research she had started in Berlin, and little encouragement for her work.

Meitner was left very much to herself in the Stockholm Physics Institute, which had little

research in progress when she arrived in 1938. Meitner's stipend from the Nobel Foundation was a pittance, and she was forced to do without many of the little comforts she had grown used to in Berlin. It was a difficult time for her, relieved only by regular letters from Hahn containing news of what had once been their joint research effort.

At the end of 1938, Hahn sent her a description of his experiments on the interaction of neutrons with uranium. He and a young chemist, Fritz Strassman, had determined that one of the reaction products was clearly barium. Meitner was so excited about this that she showed Hahn's letter to her nephew, physicist Otto Frisch. Their discussions on the topic gave birth to the idea of nuclear fission.

Frisch then demonstrated in his laboratory the tremendous release of energy accompanying fission, and a short paper by Meitner and Frisch in the British journal *Nature* in 1939 revealed the momentous concept of nuclear fission to the scientific world. It provided a new source of energy for the Earth, while at the same time introducing the possibility of a new weapon capable of unbelievable destructive power.

The step from nuclear fission to a nuclear chain reaction and the atomic bomb was, in principle, quite straightforward. In practice, however, it consumed more time and money than was ever foreseen. Although it was her basic insight that eventually led to the fission bomb dropped on Hiroshima, Meitner refused to work on the bomb and, for humanitarian reasons, hoped that it would not work.

When the question of the award of a Nobel Prize in Physics for the discovery of nuclear fission arose at the end of World War II, it was complicated by the fact that both Hahn and Strassmann were chemists. Another complication was that the Nobel Prize Committee had always considered radioactivity and radioactive atoms the responsibility of their chemistry committee—despite the fact that the discovery of fission had been interdisciplinary from beginning to end. The Swedish Academy of Science was divided on whether the Chemistry Prize should be given jointly to Hahn and Meitner, or to Hahn alone. Finally they decided by a close vote to give the 1945 chemistry prize solely to Otto Hahn.

The physics prize was still in question, and many nominators were strong in their support of Meitner as the recipient. She had continued to correspond with Hahn and advise him from afar on experiments to be performed in Berlin, but when Hahn and Strassmann published their paper on barium as a reaction product of neutron bombardment of uranium, Meitner's name was not included as a coauthor. The Nobel Committee finally decided to award no physics prize for the discovery of nuclear fission in 1945, and gave the prize for that year to Wolfgang Pauli for his theoretical discovery of the Exclusion Principle.

After the war, although now famous, Meitner continued her research in Stockholm, interrupted only by trips to receive honorary degrees and other scientific accolades. She shared in the prestigious Enrico Fermi Prize awarded by the U.S. Atomic Energy Committee in 1966. She retired to Cambridge, England, in 1960, to be near her nephew, Otto Frisch, and died there in 1968 at the age of ninety. Like so many people all over the world during the Hitler period, Meitner's life had been far from easy, but no reasonable person would ever be tempted to call her life empty.

Joseph F. Mulligan

BIBLIOGRAPHY

Crawford, D. (1969). *Lise Meitner, Atomic Pioneer*. New York: Crown Publishers.

Crawford, E.; Sime, R. L.; and Walker, M. (1996). "A Nobel Tale of Wartime Injustice." *Nature* 382:393-95.

Frisch, O. R. (1970). "Lise Meitner." *Biographical Memoirs of Fellows of the Royal Society* 16:405-420.

Frisch, O. R. (1974). "Lise Meitner." In *Dictionary of Scientific Biography*, Vol. 9, ed. C. C. Gillispie. New York: Scribner.

Meitner, L. (1964). "Looking Back." *Bulletin of the Atomic Scientists* 20:2-7.

Meitner, L., and Frisch, O. R. (1939). "Disintegration of Uranium by Neutrons: A New Type of Nuclear Reaction." *Nature* 143:239.

Sime, R. L. (1996). *Lise Meitner: A Life in Physics*. Los Angeles: University of California Press.

METHANE

The hydrocarbon methane (CH_4) is the major component of natural gas (around 90 percent) that is found in oil and gas wells throughout the world. Since the beginning of time, methane has also been produced by a number of biological sources—both natural and human—by the decomposition of organ-

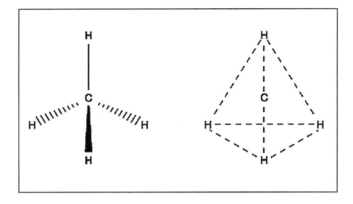

Molecular structure of methane. (Gale Group)

ic material. From 1800 to 2000, atmospheric concentrations of methane, which are approximately 0.00017 percent, have grown around 150 percent. However, the patterns of methane emission is highly irregular and, for reasons yet unclear, the rate of increase slowed considerably from 1980 to 2000. The major natural releases of methane are from wetlands (marsh gas) and termites; the major human releases are from energy use, rice paddies, gaseous emissions from animals, human/animal wastes, landfills and biomass burning. Methane research is proceeding in two major directions: the energy course, looking for ways to make bioconversion of wastes to methane more economically attractive as an alternative fuel; and the environmental course, looking for ways to limit its release into the atmosphere since it is a much more potent greenhouse gas than carbon dioxide—a thermogenic effect four to six times that of carbon dioxide. The shared goal is finding ways to "harvest" for energy production much of the methane now being released into the atmosphere.

In the energy sector, many coal mines are looking at ways to put the methane produced as a result of the mining process to work instead of venting it into the atmosphere. The U.S. Environmental Protection Agency estimates that up to 40 percent of the methane that migrates to the atmosphere can be used for power generation (electricity and heat), injection into pipeline systems, methanol production, or onsite applications like coal drying. Besides methane sales revenue and greenhouse gas reductions, the removal of methane from coal seams could serve the vital function of decreasing the risk to workers of firedamp—methane-air mixtures igniting inadvertently.

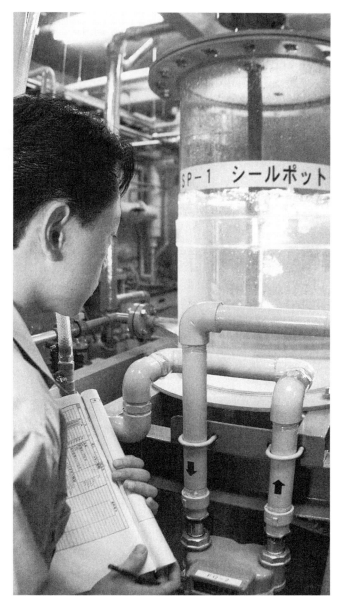

A researcher checks a sealed poy of methane in Kajima Corp.'s newly developed system designed to break down organic waste. (Corbis-Bettmann)

Other energy sector concerns are methane emissions from unburned fuel, and from natural gas leaks at various stages of natural gas production, transmission and distribution. The curtailment of venting and flaring stranded gas (remotely located natural gas sources that are not economical to produce liquefied natural gas or methanol), and more efficient use of natural gas have significantly reduced atmospheric release. But growth in natural gas production and consumption may reverse this trend. Methane has the highest ratio of hydrogen to carbon of any fossil fuel (4:1), so switching to natural gas is increasingly seen as an attractive option for cleaner air and carbon dioxide reduction.

Unlike natural gas production, the biological production of methane is attractive from a sustainability perspective. Whereas the methane that sits in underground natural gas reservoirs is finite, the methane production from biological sources is potentially very large as well as renewable. Anaerobic bacteria digestion of organic materials, in the absence of air, can produce a biogas that is 60 to 70 percent methane (state-of-the-art systems have reported producing 95 percent pure methane), the rest being carbon dioxide and other trace gases. Burning this gas can provide energy for cooking and space heating, or electricity generation.

Bioconversion of manure-to-methane is accomplished with biogas devices called digesters: organic material fed into the digester tank is heated to increase the natural decomposition rate by microorganisms, and then a pipe carries the biogas to where it will be used. There is an outlet for digested residues that usually are used as fertilizer. Several demonstration projects are taking place that are showcasing the technology. The Mason Dixon Dairy located in Gettysburg, Pennsylvania, produces enough methane from cow manure to meet its power needs, with excess power sold to the local utility. However, wide-scale development is unlikely because the low margins characteristic of farming do not support the additional capital investment to build such an operation.

Landfill gas-to-energy is another promising way to reduce atmospheric release and provide inexpensive energy to local industry and communities. Boilers for steam heat, hot water and the generation of electricity can be designed to burn a blended fuel or fueled exclusively on landfill gas. Because large-scale landfill methane recovery projects are most economical when an industrial facility is located nearby, landfill gas-to-energy projects are few and can only provide a limited amount of energy.

Controlling methane release from wetland, rice paddies and gaseous emissions from animals is more problematic. The release from rice paddies and wet lands is slow, intermittent and takes place over a wide geographic area, and thus very difficult to control. Gaseous emissions from agricultural animals contribute to atmospheric accumulation of methane due to fermentative digestion that produces methane in the rumen (stomach). Although the rate of release is highly variable, affected by factors such as quantity and quality of feed, body weight, age and exercise, beef cattle and draught animals are suspected of contributing around 50 percent, dairy cows around 20 percent, and sheep around 10 percent. Higher quality feed standards, which increase the efficiency of nutrient use, are being recommended as a way to curtail these emissions.

Methane from renewable biological sources will never be a major energy resource, yet it can be a valuable addition to the energy supply mix. Nevertheless, whether methane comes from fossil fuel reservoirs or from bioconversions, it is certain to provide useful energy for many years to come.

John Zumerchik

See also: Biofuels; Capital Investment Decisions; Climatic Effects; Natural Gas, Consumption of; Natural Gas, Processing and Conversion of; Natural Gas, Transportation, Distribution, and Storage of.

BIBLIOGRAPHY

Buell, P., and Girard, J. (1994). *Chemistry: An Environmental Perspective.* Englewood Cliffs, NJ: Prentice-Hall.

Leng, R. A. (1993). "Quantitative Ruminant Nutrition—A Green Science." *Australian Journal of Agricultural Research* 44:363-80.

Olah, G. A., and Molnar, A. (1995) *Hydrocarbon Chemistry.* New York: John Wiley.

METHANOL

Robert Boyle, an Irish chemist noted for his pioneering experiments on the properties of gases, discovered methanol (CH_3OH) in 1661. For many years methanol, known as wood alcohol, was produced by heating hardwoods such as maple, birch, and hickory to high temperatures in the absence of air. The most popular modern method of producing methanol, which is also the least costly, is from natural gas (methane) by the direct combination of carbon monoxide gas and hydrogen in the presence of a catalyst. Methanol also can be produced more expensively from oil, coal, and biomass.

Methanol accounted for less than one hundredth of one percent of total transportation fuel consump-

tion in 1999. However, it is a high-octane, clean-burning liquid fuel that has been an important component of rocket fuel, and has been the fuel of choice for Indianapolis-500 race cars since 1965. Internal combustion engine vehicles can operate on methanol alone, but M85 (85% methanol and 15% unleaded gasoline), with an octane rating of 102, is a more common automotive fuel since it requires less dramatic vehicle modifications.

Automobiles that are designed to run on methanol need a few modifications to become flexible fuel vehicles (vehicles that run on either gasoline or methanol). First, for the fuel tank, fuel lines and fuel-injection equipment, the vehicle needs noncorrosive materials such as stainless steel and high-fluorine elastomers. Second, since methanol is a lower energy density fuel, fuel injectors must be larger to provide greater volumes of fuel, and vehicles must be equipped with larger fuel tanks to achieve a range comparable to a gasoline vehicle. Third, a fuel sensor that detects fuel composition is needed to relay information to the on-board computer. And finally, the lower volatility and higher heat vaporization of methanol requires a special starting system for convenient cold weather start-ups.

By 1996, California had about 13,000 cars and 500 buses and trucks running on methanol, which was more methanol vehicles than the rest of the United States combined. The driving force for methanol vehicle growth was stringent California emission regulations designed to improve air quality. Since methanol vehicles cost only marginally more than gasoline-only vehicles, and the combustion of M85 produces less smog-forming and toxic air pollutants than comparable gasoline-powered vehicle—around 30 percent of the non-methane hydrocarbons (HC), 20 percent of carbon monoxide (CO), and 10 percent of the nitrous oxide (NO_x) of conventional gasoline engines—fuel-flexible methanol vehicles were seen as a very cost-effective means of improving air quality. However, the near-term future for methanol vehicles looks bleak. Los Angeles had to scrap its methanol transit fleet in the late 1990s because of the corrosive effect of the fuel on the heavy-duty diesel engines. The oil companies also responded to the methanol threat by developing much cleaner-burning reformulated gasoline. Combined with the tremendous advances in vehicle technology, such as improved fuel injection, advanced computer controls, variable valve timing, and electrically heated catalysts, some standard model year 2000

THE UNTAPPED METHANE RESERVES: HYDRATES

Since methane is almost always a byproduct of organic decay, it is not surprising that vast potential reserves of methane have been found trapped in ocean floor sediments. Methane forms continually by tiny bacteria breaking down the remains of sea life. In the early 1970s it was discovered that this methane can dissolve under the enormous pressure and cold temperatures found at the ocean bottom. It becomes locked in a cage of water molecules to form a methane hydrate (methane weakly combined chemically with water). This "stored" methane is a resource often extending hundreds of meters down from the sea floor.

Some petroleum geologists believe that there may be more methane trapped in hydrates than what is associated with natural gas reserves. However, as an energy source, there is considerable uncertainty whether this methane can ever be recovered safely, economically, and with minimal environmental impact. The Russians have experimented with the use of antifreeze to break down hydrates at some onshore locations in Siberia. But perhaps a more promising approach would be to pipe warm surface water to the bottom to melt the hydrates, with a collector positioned to convey the gas to the surface. Another approach might be to free methane by somehow reducing the pressure on the methane hydrates.

Regardless of the "harvesting" method, before these vast methane hydrate reserves can become a viable energy source, ways must be found to minimize the impact to the ocean floor and ocean-bottom ecosystems, and to limit the amount of methane escaping into the atmosphere.

gasoline-powered automobiles not only can meet the California Air Resource Board's Ultra Low Emission Vehicle standards (HC: 0.040, CO: 1.7, NO$_x$: 0.040

grams/mile), but also the proposed Super Ultra Low Emission Vehicle standards (HC: 0.02, CO: 1.0, NO$_x$: 0.010 grams/mile).

In the rest of the United States, the primary use of methanol is as a chemical feedstock and in the synthesis of methyl-t-butyl ether (MTBE), the most widely used gasoline additive that boosts octane and reduces the level of emissions. To reduce carbon monoxide emissions, the Clean Air Act of 1992 designated thirty-nine areas in the United States where gasoline sold from November through February must contain 2.7 percent oxygen. MTBE, with an octane rating of 116, is the primary oxygenate additive. Oil companies favored MTBE over ethanol, which has an octane rating of 108, since gasoline formulated with MTBE runs more smoothly with less knock because of its higher boiling point, and MTBE does not vaporize out of the gas tank as quickly as ethanol. The downside for MTBE, however, is that it adds five to ten cents a gallon to the cost of gasoline and reduces gas mileage by around 10 percent. There have also been concerns about the toxicity of any spilled MTBE that may get into the water supply.

As a fuel, none of the four ways of producing methanol can compete with petroleum at its 1999–2000 price of $15 to $30 a barrel. The per-barrel price of oil would have to surpass $40 for methanol made from natural gas to become a competitive alternative, over $50 if the methanol was produced from coal, and close to $70 if it was made from wood. Although methanol is a cleaner burning fuel than gasoline and diesel fuel, when produced from natural gas, it generates more carbon dioxide emissions (which is suspected of contributing to global warming) than diesel fuel.

Although methanol's future is bleak as a fuel for internal combustion engines, its future is much brighter for fuel cell vehicles. Fuel cell vehicles run on hydrogen, yet due to hydrogen's low energy density, it is expensive to transport and store. Thus, auto makers are looking at ways to extract hydrogen from methanol through a device called a steam reformer. Steam reformers combine methanol with steam and heat to produce hydrogen, carbon dioxide and trace amounts of carbon monoxide (that must be removed by an oxidation reactor downstream of the reformer):

$$CH_3OH + H_2O \rightarrow XO_2 + 3H_2$$

By 1999, General Motors, Daimler-Chrysler, Toyota, and Nissan all had demonstration fuel cell vehicles operating on methanol, with plans to start introducing vehicles into the market by 2005. Auto makers have shown a preference for methanol over gasoline primarily because of the likelihood of the sulfur content in gasoline poisoning some of the catalysts used in the fuel cell.

Nevertheless, methanol still faces a major technological hurdle: the endothermic (requires heat) nature of the methanol steam reformer. Heat must come from an additional reactor to burn some of the fuel or exhaust gases from the fuel cell stack. It also takes time to reach operating temperature, which may be an unacceptable compromise for potential auto buyers accustomed to start-and-go vehicles. Additionally, if methanol for fuel cell vehicles began to capture a significant share of the transportation fuel market, it would require major investments in infrastructure. Gasoline and methanol cannot share the same distribution system since methanol can be highly corrosive and presents a greater fire hazard. And since methanol is much more toxic than gasoline, colorless and nearly odorless, greater precautions are needed to lower the risk of groundwater contamination from leaking storage tanks.

John Zumerchik

See also: Alternative Fuels and Vehicles; Biofuels; Climatic Effects; Emission Control, Vehicle; Methane; Synthetic Fuel.

BIBLIOGRAPHY

Greene, D. L. (1996). *Transportation and Energy*. Landsdowne, VA: Eno Transportation Foundation, Inc.

Olah, G. A., and Molnar, A. (1995). *Hydrocarbon Chemistry*. New York: John Wiley.

MILITARY ENERGY USE, HISTORICAL ASPECTS

According to historian Vaclav Smil, the destructive energy of military weapons has increased by sixteen orders of magnitude over the past five thousand years. The exploitation of inanimate energy sources has also resulted in the increased technical specialization of men-at-arms, with corresponding changes in the relationships between industry, the state, and military forces. Conquests were once limited by the availability of food for the pillaging troops; naval exploits were at the mercy of the winds, tides, and aggregate human energy of rowing crews. The discovery and subsequent harnessing of ever more efficient sources of energy unshackled the militaries of the Western world, but forced soldiers, sailors, and civilians to reexamine and redefine the place of the military in larger society. By the late nineteenth century, Western militaries were highly industrialized and bureaucratized institutions with intricate divisions of labor. Scientific enterprises, the purview of lone investigators and small collectives of enthusiasts before the nineteenth century, increasingly came under the direct control of state authorities.

ANCIENT HISTORY

Organized military aggression was limited to the use of human muscle power for thousands of years. Warfare was a matter of close-quarters combat, much of it hand-to-hand, and the energy expended in this activity was limited to the physical endurance of the participants. Personal kinetic energy weapons—slings, bows, and crossbows—increased slightly the fighting range of the combatants, but again the energy expended to kill others was limited to the strength of the warrior. The use of the horse as a combat mount increased the destructive energy available to the military, as cavalry were used to bear down on unprotected footsoldiers, although special weapons such as the pike, or architectural protections such as obstructions or fort walls, could limit the impact of the mass of the charging horse and rider.

Siege ballistic weapons, the largest of the early kinetic energy weapons, were developed and deployed to counter the protective walls and other stubborn defenses of cities and fortifications. The most dramatic early example of this kind of device was the catapult, invented in 399 B.C.E. in the siege workshops organized by Dionysus I of Syracuse. The early catapult used mechanical elastic energy to hurl projectiles and smash high fortification walls or disrupt formations of men. Although the catapult did not significantly extend the destructive range of the attacker, this device did allow for the use of projec-

Early example of a catapult. A simple lever is turned into a potent long-distance weapon. (Corbis Corporation)

tiles of ever greater mass, with a corresponding increase in the destructive kinetic energy applied to an opponent. Other offensive devices employed by Dionysus included the battering ram and siege tower, elements of the so-called "offensive-defensive inventive cycle," a kind of arms race that pitted fortress and city defenders throughout Europe and East Asia against marauding attackers. Although the application of kinetic energy to fortress walls could influence the course of a siege, the availability of biomass energy (food) for the consumption of attackers or defenders often determined the ultimate outcome of these stalemates that were so common in European and Near Eastern warfare before the fifteenth century.

The immediate availability of biomass energy, in the form of forage and food, placed serious constraints on military operations before 1850. Feudal political arrangements in Europe and Asia facilitated the growth of large armies capable of protracted war-

fare, but only at great human and monetary expense. The long-standing constraints on marching range and military strength were gradually addressed as emerging national governments created more effective systems of taxation and resource allocation to support military forces. Logistics, a system of energy distribution for the benefit of military forces, developed into a kind of science as emerging states recruited large standing armies and engaged in open warfare against other organized political groups. Although troops relied heavily on local food supplies (often with disastrous consequences for the local civilian population, as was the case during the Thirty Years' War in Central Europe from 1618 to 1648), the growth of logistics support institutions in the centuries that followed relieved some of the energy-demand burden from non combatants. The presence of logisticians, either civilian contractors or soldiers consigned to a support role, precipitated wider military reforms and the changing role of the army and navy in society at large.

Harnessing the energies of explosive materials, first in the form of gunpowder, produced a revolution in warfare, albeit a very slow one. Explosive powders were commonplace in Asia before the fifteenth century, although dynastic China did not use this material in an effective military capacity. Early hand-cannons, sometimes made of leather-wrapped iron cylinders, were little more than launch tubes for unrefined (and inaccurate) projectiles. The military firearm evolved slowly, beginning with the invention in the fifteenth century of the matchlock (or arquebus), followed by the invention of the German wheel-lock in 1515. The flintlock, subject to various mechanical improvements until the weapon was eclipsed in the nineteenth century by the breech-loading rifle, was a standard infantry arm by the close of the eighteenth century.

With subsequent improvements of these weapons, the explosive energy released by the combustion of gunpowder was harnessed to yield longer projectile ranges with more accuracy. State interest in firearms technology led to government sponsorship of scientific research into aspects of chemistry and ballistics in an effort to better understand the mechanics of explosive energy. The alignment of guilds and other productive tradesmen with military institutions in the sixteenth century foreshadowed a restructuring of the relationship between armies and industry in the years to come.

Gunpowder technologies heightened the competition between defense and offense, and raised the stakes for sovereign polities. Vauban, a French military engineer in the employ of Louis XIV, devised an innovative fortification system to respond to the threat of new siege techniques developed since 1400. Among these new threats was the increasing power of portable artillery and, more ominously, an improved form of "sapping," a process of undermining the walls of a fortification with explosives deposited by tunneling enemies. In the 1490s, the first portable siege guns were deployed as part of a larger military expedition organized by Charles VIII; within a century, small cannons could be found in the arsenals of every European power, as well as various kingdoms in India and Asia. Vauban's geometric fortifications were designed to maximize the firepower of the defenders with ramparts that would effect devastating cross-fires on any

enemy bold enough to approach the high walls. The importance of specialists, such as miners, to the military endeavor stratified armies and increased the costs of warfare, but the effectiveness of new destructive energies made this kind of investment worthwhile.

THE AGE OF STEAM

One of the hallmarks of the industrial revolution, the steam engine, had important military uses. After nearly two centuries of labor-augmenting industrial use, steam power matured into a viable energy source for military applications in the nineteenth century. The same rails that were used to transport goods between the rural border regions and the urban centers of the Western world were also used for the rapid transportation of troops. In India, the "famine and security" lines built with the assistance of the British government in the late nineteenth century linked provincial territories with the administrative centers where military forces were housed. These rail lines were used to move food and other vital supplies to impoverished areas for relief of the population, or to move troops to quell civil unrest. Railroads proved to be important military assets during the American Civil War (1861–1865) and the Wars of German Unification (1864–1871); the rails served as arteries of support for the combatants. The rapid mobilization of troops in 1914, which eventually led to the horrors of stalemate on the Western Front, was facilitated by the efficiency of Western European railroads.

Guncotton (nitrocellulose) and nitroglycerine, substances that exponentially increased the explosive energy available to mankind, were developed in 1840. In 1867 Alfred Nobel found that nitroglycerine liquid could safely be absorbed in a clay-like substance called kieselguhr. This solid and relatively safe form of explosive became known as dynamite. In 1875, Nobel mixed nitroglycerine liquid with collodion cotton to make an explosive gelatin with both mining and military applications. These early forms of high explosive were refined into more powerful and destructive substances, including French Poudre B (1884) and ballistite (1888). By the 1880s, nitroglycerine (in various forms) and other nitrated organic compounds were important components of munitions manufacturing in Europe and the United States.

The development of high-energy explosives in the nineteenth century corresponded to a period of great

The war of 1854–1856 was fought mainly in the Crimean peninsula between Russia on the one hand and Great Britain, France and Turkey. Shown are cannonballs used during the war. (Corbis-Bettmann)

innovation in artillery design and manufacturing, as more powerful and longer-range guns were introduced into the arsenals of the West. The ranges of these weapons increased from about two kilometers in the 1850s to more than thirty kilometers in the 1890s. The production of the steel and powder nec-essary for the deployment of these weapons required a tremendous amount of energy, most of it derived from fossil fuels.

The expansion of the colonial empires between 1870–1900 involved the deployment of troops and military hardware to the far reaches of the world, and placed enormous energy demands on the economies of the imperial powers. The so-called tools of empire—steamships, the railroads, and canals—were energy-intensive projects with important military uses. The first oceangoing steam-powered naval vessels were commissioned in the 1850s, and were quickly demonstrated as important military tools. During the Crimean War, the British government nationalized a fleet of private steamships to transport men and supplies to the Ottoman Empire, an arrangement made possible by a special "militarization" clause inserted in the charter agreements between the British government and commercial shipping companies. Private and military fleets grew substantially in the decades that followed; during the period from 1870 to 1910, steam vessels became increasingly more energy-efficient and powerful. Around 1880, the triple-expansion steam engine was invented, followed by the introduction of the quadruple-expansion steam engine in the 1890s. These innovations made warships more fuel efficient with longer cruising ranges. This also meant that remote colonial possessions became more strategically important as coal refueling stations for modern naval forces.

The harnessing of electrical energy had a profound effect on the conduct of military operations. Perhaps most importantly electricity revolutionized communications, with consequences for the conduct of military affairs. The British telegraph network of the late nineteenth century connected the vast reaches of the empire to London, and other imperial powers were often beholden to the British system for news and diplomatic correspondence. In 1901, the first trans-Atlantic "wireless" communications system was demonstrated by Guglielmo Marconi, who promptly approached representatives of the American, British, and Italian navies with the hope of selling his invention. Wire-based telegraph and telephone systems, as well as the "wireless" radio, were adopted by the world's armed forces between 1900 and 1914. The defeat of a Russian expeditionary force in Eastern Asia at the hands of the Japanese army (1904–1905) was widely attributed to the latter's decisive use of the telephone and telegraph to coordinate the movement of troops and the distribution of supplies. The availability and proper application of inanimate energy, it seemed, was vital for the correct distribution of biomass during protracted campaigns, a necessary condition for victory.

Despite the obvious implications that new and more efficient inanimate energy resources held for the world's armies and navies, it was some time before military planners began to consider the importance of energy resources and infrastructure. In 1830 revolutionaries in Paris attempted to paralyze the state government by plunging the city into darkness by attacking the gasworks, the principle distribution center for the gas supply that fed the city's street lamps. The growing popularity of central electric plants in the United States and Europe around 1900 raised questions about the vulnerability and strategic importance of these facilities. By 1914, electric power plants, which supplied crucial power to the war industries, were being protected from sabotage by militiamen and professional soldiers. When called upon, military forces were deployed to disrupt the activities of striking coal miners and others who threatened economic stability or national security. In other instances, military forces were sent abroad to protect energy resources from the predations of other states.

OIL AND THE INTERNAL COMBUSTION ENGINE

Large-scale crude oil exploitation began in the late nineteenth century. Internal combustion engines, which make use of the heat and kinetic energy of controlled explosions in a combustion chamber, were developed at approximately the same time. The pioneers in this field were Nikolaus Otto and Gottlieb Daimler. These devices were rapidly adapted to military purposes. Small internal-combustion motors were used to drive dynamos to provide electric power to fortifications in Europe and the United States before the outbreak of World War I. Several armies experimented with automobile transportation before 1914. The growing demand for fossil fuels in the early decades of the twentieth century was exacerbated by the modernizing armies that slowly introduced mechanization into their orders of battle. The traditional companions of the soldier, the horse and mule, were slowly replaced by the armored car and the truck in the early twentieth century.

Internal combustion and electricity wrought a number of changes in naval architecture. Electric power was introduced on warships in the 1880s, allowing for the construction of larger ships, with better light and ventilation, deeper decks, improved artillery fire control, and improved efficiency in han-

British pilots on the Western Front swoop down on a German formation during actual combat in World War I. (Corbis Corporation)

dling ammunition and steering the ever-larger gun turrets. Highly energy-efficient oil-burning turbine engines were introduced on naval vessels about 1900. In 1906, the British navy commissioned the Dreadnaught, widely hailed as a revolutionary warship design, powered by turbine engines, equipped with heavy steel plate armor, and fitted with powerful long-range cannons. Twentieth-century battleships dwarfed the steam-powered vessels of the previous century in terms of displacement and destructive power.

Various schemes of powered flight were finally realized at Kitty Hawk, North Carolina, in 1903 when the Wright brothers successfully demonstrated a self-propelling flying machine. The Wright brother's plane was a breakthrough because of its lightweight aluminum internal combustion engine. Within a few years, strategic planners were speculating about the military importance of the airplane, although there was some debate as to whether the machines would ever pose a threat to static fortifica-

tions or field armies. Aircraft were used by the belligerents during World War I, ordered to perform reconnaissance and harassing work. Improvements in engine design, metallurgical science, and aerodynamic theory resulted in net increases in the energy efficiency of aircraft. By the mid-1930s, both airships and airplanes could perform a range of military functions, serving as interceptors, bombers, transports, and reconnaissance craft.

MILITARY ENERGY USE IN THE TWENTIETH CENTURY

The great demand for fossil fuels, a consequence of the rapid industrialization of emerging world powers in the early twentieth century, created political frictions between states vying for energy supplies. The United States, Britain, France, and Russia benefited from direct control of oil- and coal-producing territories, while Germany, Japan, and other modernizing nations were forced to import great quantities of precious energy. Historians have suggested that Japanese expansionism in the 1930s was an expression of energy insecurity; regardless of the causes, Japanese aggression in the Pacific, culminating with the attack on Pearl Harbor, Hawaii, on December 7, 1941, resulted in a protracted war, during which an unprecedented amount of energy was consumed by the combatants. The importance of energy resources, especially fossil fuels, affected the military strategies of all involved. German campaigns in North Africa were planned with the intention of liberating the Suez Canal from British control and permitting the Nazis access to Middle Eastern oil supplies. Allied bombing raids on Ploetsi, Romania, struck the oilfields of that region, a move designed to deny the Germans and Italians access to that energy. The German army's drive to seize the Caspian oilfields in 1941, an operation that compromised the effectiveness of the siege of Stalingrad, was an attempt to secure energy resources for the war effort. The Japanese invasion of Dutch Indonesia was undertaken to obtain the precious oil reserves found among the islands of the archipelago.

World War II was ultimately a contest between economies, and victories were a direct result of effective resource mobilization. The atomic bombs dropped on Hiroshima and Nagasaki in August 1945 released a tremendous amount of energy in the form of heat and radiation; the development of that weapon required a substantial economic and energy investment. After 1945, states went about building up strategic reserves of important natural resources; renewed national environmental and conservation efforts began with concerns about security. In some instances, the interests of the state, the military, and industry aligned, resulting in the execution of energy policies designed to protect the stability and security of the state. In the 1960s, for example, France undertook an aggressive nuclear energy development program in response to the agitations of the domestic coal-mining unions; it was feared that the miners had developed close ties with the French Communist Party. Civil nuclear power was the crossover manifestation of a technology originally developed for military purposes, but adapted for civilian use. Fuel cells and other high-yield, portable power-generation devices, developed in the mid-twentieth century and designed for use in space or other hostile and isolated environments, have both civilian and military applications.

The availability of ever-more-efficient kinds of inanimate energy has made the coordination of men and machines easier. With more efficient and powerful forms of energy generation, the coordination of the movements of man and machines became more effective. Two hundred years ago, soldiers were dependent upon verbal and visual signals for direction; the effective use of electrical communications technologies has resulted in the increased scope and scale of military operations. Digital computer technology, developed in the latter half of the twentieth century, found a ready audience among military officers seeking to maximize their control over subordinates and improve the collection and distribution of intelligence information.

The improvement of human control over inanimate forms of energy, put to use to military ends, has improved the logistics and coordination aspects of armies and navies, and increased the overall destructive capacity of humanity. Energy-efficient propulsion systems have reduced the costs and increased the ranges of various forms of transportation, both military and civilian. For the military, energy is both a blessing and a vulnerability, requiring ever-more-specialized soldiers and more expensive equipment to remain effective in the face of competition from other modern military forces.

Shannon A. Brown

See also: Communications and Energy.

BIBLIOGRAPHY

Brodie, B., and Fawn, M. (1962). *From Crossbow to H-Bomb*. Bloomington, IN: Indiana University Press.
Cipolla, C. M. (1966). *Guns, Sails and Empires*. New York: Pantheon Books.
Ferrill, A. (1985). *The Origins of War: From the Stone Age to Alexander the Great*. London: Thames and Hudson.
Gibbs-Smith, C. H. (1970). *Aviation: An Historical Survey*. London: HMSO.
Hall, B. S. (1997). *Weapons and Warfare in Renaissance Europe*. Baltimore: Johns Hopkins University Press.
Headrick, D. R. (1981). *The Tools of Empire*. New York: Oxford University Press.
Headrick, D. R. (1988). *The Tentacles of Progress*. New York: Oxford University Press.
McNeill, W. H. (1982). *The Pursuit of Power*. Chicago: University of Chicago Press.
McNeill, W. H. (1989). *The Age of Gunpowder Empires, 1450–1800*. Washington, DC: American Historical Association.
Mumford, L. (1934). *Technics and Civilization*. San Diego: Harcourt, Brace and Company.
Nye, D. E. (1998). *Consuming Power: A Social History of American Energies*. Cambridge, MA: MIT Press.
Pacey, A. (1974). *The Maze of Ingenuity*. Cambridge, MA: MIT Press.
Pacey, A. (1990). *Technology and World Civilization*. Cambridge, MA: MIT Press.
Parker, G. (1988). *The Military Revolution*. Cambridge: Cambridge University Press.
Pick, D.. (1993). *War Machine: The Rationalisation of Slaughter in the Modern Age*. New Haven, CT: Yale University Press.
Smil, V. (1994). *Energy in World History*. Boulder, CO: Westview Press.
White, L. (1978). *Medieval Religion and Technology*. Berkeley: University of California Press.

MOLECULAR ENERGY

The fusion reactions that power the stars are nuclear reactions, in which one type of atom is converted into another. The extremely high heat of our Sun provides energy so intense that when atoms collide, they may collide with enough force to allow their nuclei to fuse together, a process that in itself releases tremendous energy. At moderate temperatures, as on our temperate planet Earth, when two elements bump into each other, they do not ordinarily contain enough energy to fuse. However, under the right circumstances, they may bind together to form a new compound. What

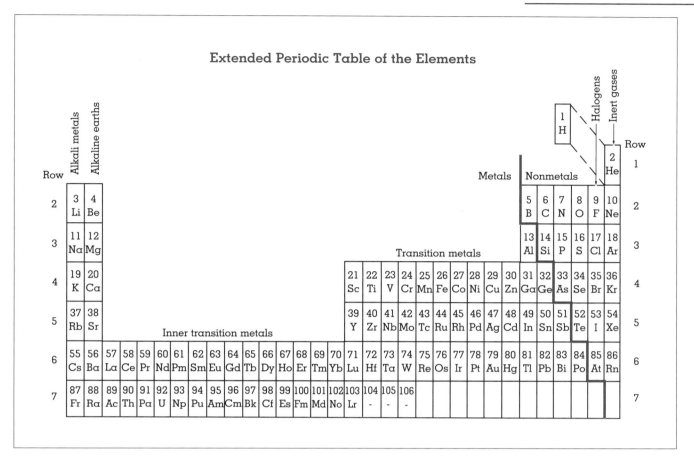

Figure 1.
When the elements are arranged in order of atomic number, a remarkable repetition of chemical properties occurs. The periodic table arranges the elements so that those in any one vertical column possess similar properties.

holds the two elements together is called a chemical bond. The energy involved in forming a typical chemical bond is ten million times smaller than that involved in nuclear reactions. Yet it is chemical bonding that allows for the existence of the highly organized and complex collection of chemical entities found here on Earth, even for life itself.

ELECTRONIC STRUCTURE OF THE ATOM

An atom is composed of a nucleus, which contains two types of relatively massive particles: the positively charged proton and the neutral neutron. The nucleus is surrounded by very light, negatively charged electrons equal in number to the number of protons, so that the overall charge on the atom is neutral. The number of protons in an atom, its atomic number,

defines it as a chemical element. There are 105 elements, each with a different number of protons in its nucleus, and hence a unique atomic number. For example, all atoms with six protons are by definition carbon; those with seventy-nine protons, gold. The electrons surround the nucleus in distinct spatial patterns and possess specific energies. These energies and spatial patterns define the element's chemical behavior. They are in large part responsible for determining which elements will react with each other to form molecules. When atoms come close enough for their outer electrons to interact, attractions among the atoms may occur that are strong enough to hold them together to form a chemical bond.

In the mid-seventeenth century, when accurate atomic weights for a large number of elements were

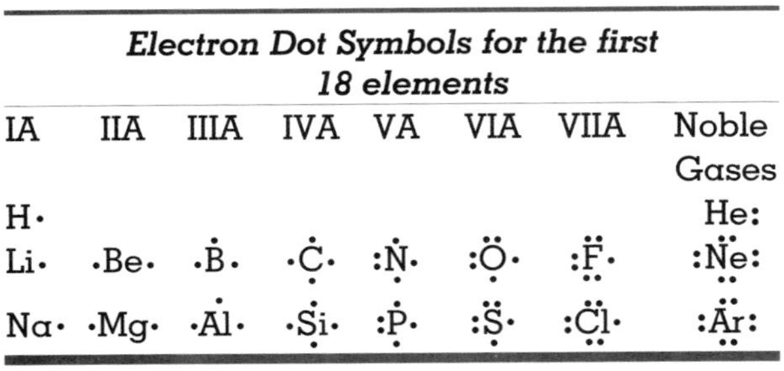

Electron Dot Symbols for the first 18 elements							
IA	IIA	IIIA	IVA	VA	VIA	VIIA	Noble Gases
H·							He:
Li·	·Be·	·Ḃ·	·Ċ·	:Ṅ·	:Ö·	:F̈·	:N̈e:
Na·	·Mg·	·Äl·	·Ṡi·	:Ṗ·	:S̈·	:C̈l·	:Är:

Figure 2.
Electron-dot symbols for the first eighteen elements. This scheme, invented in the early twentieth century by G. N. Lewis, provides a rough but useful tool for predicting the availability of an atom's valence electrons for chemical bonding.

becoming increasingly available, it was noted that there is a periodic recurrence of similar properties among the elements, both in terms of their characteristics as elements (are they, for example, metallic or nonmetallic?) and their behavior in chemical reactions (reactive or stable?). (See Figure 1.) Through the latter half of the nineteenth century natural philosophers amassed an ever-increasing collection of empirical rules, called "valence rules," to describe the formation of chemical bonds and to systematize the characterization, synthesis, and use of new compounds. The periodic table, developed in the mid-nineteenth century, provided a basis for predicting the nature of chemical bonds and the likely composition of a growing list of compounds. However, the approach was largely intuitive and descriptive. No underlying theoretical principle existed to explain chemical bonding.

In the early part of the twentieth century, G. N. Lewis observed that chemical bonding seemed to favor a state in which the atoms in stable compounds, by "sharing" electrons, achieved the stable electron distribution exhibited by the nonreactive noble gases, so-called because they are almost always found as pure elements in the gas state. He proposed that the electrons shared between two elements act as an electromagnetic "glue" to hold the two atoms together. The positive nuclei are attracted to the negative electrons; the electrons spend most of their

time between the two nuclei, where the negative charge on the electrons screens the repulsive forces between the two nuclei. This attraction between positive and negative charges and the shielding of the positive nuclear charges by the electrons create a lower-energy state for the two atoms. The energy required to pull these two atoms apart again is called the bond energy. To pull apart a mole of H_2 molecules into two moles of H atoms requires 432 kJ of energy. Lewis's schemes for predicting bond formation among elements, the octet rule and electron-dot diagram, although incomplete, provided a strong predictive tool for describing covalent bonding and are still useful. (See Figure 2.)

The advent of quantum mechanics—developed over the first half of the twentieth century—provided the theoretical framework for our modern understanding of the chemical bond. This body of work predicates that light, ordinarily described in terms of electromagnetic waves, also can be understood as a collection of quanta. And, likewise, moving particles can possess wave-like properties. (This interchangeability is known as "wave-particle duality.") Infrared light quanta are similar in scale (and energy) to molecular bonds; the wavelengths of electrons in atoms and molecules are similar in scale (and energy) to the wavelengths of ultraviolet and visible light.

At the turn of the twentieth century, the planetary model for the atom, with electrons orbiting the nucleus, held sway, despite its fatal flaws: The laws of classical physics stipulate that the orbiting, charged electron should radiate energy and gradually spiral and crash into the nucleus. In 1913 Niels Bohr proposed discarding this classical notion and limiting the electron orbits to those having angular momenta that are integer multiples of Planck's constant h divided by 2π. Although not completely correct, Bohr's theory of the hydrogen atom paved the way for quantum mechanics and won him a revered place in science history.

In 1926, Austrian physicist Erwin Schrödinger made the next leap. He expanded the suggestion by Louis de Broglie (1924) that likened the wave behavior of the hydrogen electron, with its fixed energy values, to a guitar string. The string vibrates at a distinct frequency and has harmonics at wavelengths related to string length by integer numbers, indicating that the guitar string's vibrations are "quantized": The string can vibrate only at those frequencies characterized by integer numbers. To set up an analogous mathematical description of systems such as the electron in the hydrogen atom, Schrödinger described a nonclassical wave function that contains information on the whereabouts of an electron in an atom (or molecule). The wave function (actually, its square) tells us the *probability* of finding the electron at a given point in space. The probability distribution for an electron in a given energy level is known as an orbital.

We now know that electrons in atoms can hold only particular energies and that their probable whereabouts are described by Schrödinger's wave function. The energies and probable locations depend on integer numbers, or "quantum numbers." Quantum numbers describe the energy and geometry of the possible electronic states of an atom. These states, in turn, determine the chemical behavior of the elements—that is, how chemical bonds can form.

The periodicity of the elements derives from the rules dictating the manner in which electrons in atoms fill the energy levels. These levels can be roughly collected as a series of "shells," which one might think of as floors in a multistory building. Each floor, or shell, lies at a discrete energy; higher floors lie at higher energies. Quantum numbers dictate the number of orbitals in each shell. Each orbital can hold no more than two electrons. Electrons begin occupying the lowest-energy shell first; once a shell is filled, the additional electrons must go into the next, higher-energy shell. The most stable (i.e., the lowest-energy) configuration for an atom is to have completely filled electron shells. An atom with one or two vacancies in its highest shell is stabilized by the addition of one or two electrons. An atom with only one electron in its highest shell is stabilized by the release of that lone electron. Atoms interact to form chemical bonds by capturing, losing, or sharing electrons. The shared electrons hold groups of atoms together to form molecules of distinct geometry and character.

The ionization energy, electron affinity, and orbital occupancy determine the chemical behavior, or reactivity, of the elements. The uppermost (highest-energy) occupied orbitals are called the valence orbitals; the electrons occupying them are the valence electrons. An element's ionization energy, the energy required to remove an electron from a neutral atom, is related to its reactivity: A low ionization energy means that the valence electron is readily removed, and the element is likely to become involved in

Electron Configurations for the First 20 Elements		
Name	Atomic Number	Electron Structure
Hydrogen	1	$1s^1$
Helium	2	$1s^2$
Lithium	3	$1s^2 2s^1$
Beryllium	4	$1s^2 2s^2$
Boron	5	$1s^2 2s^2 2p^1$
Carbon	6	$1s^2 2s^2 2p^2$
Nitrogen	7	$1s^2 2s^2 2p^3$
Oxygen	8	$1s^2 2s^2 2p^4$
Fluorine	9	$1s^2 2s^2 2p^5$
Neon	10	$1s^2 2s^2 2p^6$
Sodium	11	$1s^2 2s^2 2p^6 3s^1$
Magnesium	12	$1s^2 2s^2 2p^6 3s^2$
Aluminum	13	$1s^2 2s^2 2p^6 3s^2 3p^1$
Silicon	14	$1s^2 2s^2 2p^6 3s^2 3p^2$
Phosphorus	15	$1s^2 2s^2 2p^6 3s^2 3p^3$
Sulfur	16	$1s^2 2s^2 2p^6 3s^2 3p^4$
Chlorine	17	$1s^2 2s^2 2p^6 3s^2 3p^5$
Argon	18	$1s^2 2s^2 2p^6 3s^2 3p^6$
Potassium	19	$1s^2 2s^2 2p^6 3s^2 3p^6 4s^1$
Calcium	20	$1s^2 2s^2 2p^6 3s^2 3p^6 4s^2$

Table 1.

Bohr diagrams showing the electronic configurations for the first ten elements, along with representative bond formation.

bonding with another atom by losing an electron to it. Electron affinity, the energy released upon capture of an electron by a neutral atom, is an important characteristic with respect to ionic bonding: The energy of an atom with high electron affinity is lowered as another electron approaches. If that approaching electron is attached to another atom, the two atoms can form a bond.

Atoms are held together to form molecules either by the attraction between oppositely charged atoms (ionic bond) or by the sharing of electrons among positively charged atomic nuclei (covalent bond). In the language of quantum mechanics, a "bonding molecular orbital" is formed between the atoms in a molecule by the combination of atomic orbitals in a way that increases the likelihood of finding electrons between the nuclei. Both the attraction between opposite charges and the screening of repulsive forces between the positive nuclei by the negative electrons lower the overall energy of the system. The bonding

1. Electrons in an atom differ in energy depending on the orbital they occupy—that is, their average location in space relative to the nucleus.
2. Electrons closest to the nucleus have the lowest energy.
3. Electrons can obtain only certain discrete energy values in a stable atomic state.

molecular orbitals are lower in energy than the isolated atomic orbitals from which they arise.

BOND FORMATION BY THE FIRST TEN ATOMS

The "first floor" of the atom can hold only two electrons. When two hydrogen atoms, each with one electron, come together, each nucleus is in effect "sharing" two electrons, and as far as each nucleus is concerned, its shell is filled. The helium atom already contains two electrons; its first floor is filled. That is why helium is unreactive. When two helium atoms collide, rather than stick together, they simply bounce apart. (See Table 1.)

The second shell can hold up to eight electrons. Lithium (Li), the first element in the second row of the periodic table, has atomic number 3. With three electrons, lithium's first shell is filled and its second shell has only one electron. Lithium is highly reactive and readily gives up its lone second-shell electron. In fact, all elements in the same column as lithium, called the alkali metals, have a single valence electron and share these properties. Because an alkali metal gives up its valence electron readily, the bonds formed between alkali metals and other elements are ionic: The electrons are not shared equally. In sodium chloride, $NaCl$, sodium carries a positive charge and chlorine a negative charge.

Electrons are not only charged, they also have a characteristic physicists call "spin." Pairing two electrons by spin, which has two possible values, "up" or "down," confers additional stability. Beryllium (Be, atomic number 4) has two spin-paired electrons in its second shell that are easily given up in chemical reactions. Beryllium shares this characteristic with other elements in column two, the alkaline earth metals. These atoms also generally form ionic bonds. Boron

(B, atomic number 5) will readily give up three electrons.

Carbon (C, atomic number 6) is the most versatile of all elements and is found abundantly in compounds produced by living organisms. With four unpaired electrons, it can make as many as four covalent bonds to fill its outer shell to eight, producing extremely stable molecules. Carbon can form branched and cross-linked chains, and it can form especially stable double bonds and "aromatic" bonds (highly stable, delocalized bonds found in planar ring compounds, like benzene) that can absorb energy in the visible and ultraviolet range of the electromagnetic spectrum.

Nitrogen (N, atomic number 7) has five outer electrons, two of which are paired. It therefore has three electrons to share in chemical bonds. Oxygen (O, atomic number 8), with six outer electrons, has two paired and two single electrons. Oxygen combines readily with other elements to gain two electrons to fill its outer shell, in a process called oxidation. Reactions between oxygen and carbon are especially favorable. (Carbon-oxygen compounds, particularly carbon dioxide, CO_2, are stabilized to a very low energy.)

Fluorine (F, atomic number 9) has seven outer electrons, one unpaired. Because it needs to obtain only one electron to fill its outer shell and gain stability, it is highly reactive. Neon (Ne, atomic number 10), on the other hand, has a filled outer shell. Like helium and its other column mates, the noble gases, neon does not readily react with any element.

ENERGY AND CHEMICAL CHANGE

In chemical reactions, when the atomic configurations of molecules are changed, matter is neither created nor destroyed (Law of Conservation of Matter). The identity and number of atoms remain unchanged. When methane gas (CH_4) is burned, its atoms don't disappear; they combine with oxygen (O_2) in the air and are transformed into carbon dioxide (CO_2) and water vapor (H_2O):

$$CH_4 + 2\,O_2 \rightarrow CO_2 + 2\,H_2O + heat$$

Likewise, in chemical change, energy is neither created nor destroyed (Law of Conservation of Energy). When charcoal is burned the potential energy stored in the carbon-carbon bonds is released as heat. Although in this reaction the *forms* of matter and energy are changed, both are ultimately conserved; the number of atoms and the amount of energy in the universe remain unchanged.

Energy is the capacity to do work. Potential energy is the energy possessed by an object as a result of its position. Heat, another form of energy, ran be thought of in molecular terms as frictional losses of the uncoordinated motion of molecules.

In a heat-producing (exothermic) reaction the molecular energy of the products is lower than that of the reactants. In the combustion of methane, the energy stored in the bonds of the molecules CO_2 plus two molecules of H_2O is less than that in CH_4 plus two molecules of O_2. If the molecular energy of the products is greater than that of the products (an endothermic reaction), energy (often in the form of heat) must be added for the reaction to occur.

Oxidation reactions release energy because oxidized compounds have lower energy than the reduced compounds from which they were formed. This energy difference derives from the strong "greediness" of oxygen for electrons. Oxygen has a high electronegativity, the attraction of an atom for an electron in chemical bonding. The electronegativities of carbon and hydrogen are roughly the same. Similarly, in the oxygen molecule, the two oxygen atoms share their bonding electrons equally. In carbon dioxide, however, oxygen, which is more highly electronegative than carbon, holds its electrons much more tightly than carbon. The same is true for the other product molecule: water. It is a more stable (lower-energy) situation for the electrons to be held closer to the nucleus of the highly electronegative atom, oxygen, than it is for the electrons to be shared between atoms (carbon and hydrogen) that are less electronegative. Oxygen "wants" the electrons, while carbon and hydrogen are somewhat more "ambivalent" about holding on to them.

SPONTANEITY AND RANDOMNESS

Like a ball at the top of an incline, chemical reactions seem to have a tendency to naturally "roll downhill" from a state of higher to a state of lower potential energy, with the potential energy converted to kinetic energy along the way (motion for the ball, heat for the chemical reaction). However, while most spontaneous chemical reactions are indeed exothermic, many spontaneously occurring chemical changes actually absorb heat, notably ordinary table salt dissolving in water.

In determining if chemical change will occur spontaneously, one must consider not only whether heat is released (the reaction rolls downhill energetically), but also whether it creates order or disorder. Both the lowering of energy and an increase in disorder are changes that tend to occur spontaneously. The measure of the randomness of a system, its "entropy," increases, for example, when a solid or liquid is converted to a gas, when a solid or a liquid is dissolved in solution, with increasing weakness between bonds, or with an increasing number of particles. Sodium chloride (table salt) dissolves spontaneously in water. Energy must be added to break the salt's crystalline bonds, but the increase in entropy (disorder) produced as hydrated sodium and chloride ions are freed from the ordered NaCl crystal into solution drives the reaction forward. The importance of entropy in determining whether a chemical reaction will proceed spontaneously increases as the temperature is increased. For example, ice is the lowest-potential energy state for water. However, at temperatures above 0°C entropy becomes a driving force for melting, and water goes from a low-entropy crystalline phase to a more disordered liquid phase in which the water molecules can move more freely.

OXIDATION AND REDUCTION

Reactions that involve the loss and gain of electrons are termed reduction-oxidation, or Aredox," reactions. The two processes of reduction (loss of an electron) and oxidation (gain of an electron) are complementary and always occur together: One substance's loss is another's gain. Reduced forms of matter—often fuels such as sugar, fat, gasoline, and charcoal—are high in potential energy. Oxidized forms, such as carbon dioxide and water, are low in potential energy.

The term "oxidation" derives from its original use to describe reactions involving combination with oxygen. The broader definition includes reactions in which an element or compound gains oxygen (in the combustion of methane, the carbon gains oxygen atoms and is thus "oxidized"), loses hydrogen, or loses electrons. The working definition for reduction includes reactions in which an element or a compound loses oxygen atoms, gains hydrogen atoms, or gains electrons. The oxidation of fuels—by engine, stove, or living organism—releases energy, often in the form of heat, that can be used to perform work. Green plants convert the light energy of the sun, by reducing carbon dioxide during photosynthesis, into potential energy stored in hydrocarbon bonds:

$$6\ CO_2 + 6\ H_2O + \text{solar energy} \rightarrow$$
$$C_6H_{12}O_6 (\text{glucose}) + 6\ O_2$$

Redox reactions occur in the reduction of ores (metal oxides) into pure metals and the corrosion (oxidation) of pure metals in the presence of oxygen and water. Rusting iron, $4Fe + 3O_2 + 6H_2O \rightarrow 4Fe(OH)_3$, is a good example of metal oxidation. Strong oxidizing agents can be used as antiseptics (hydrogen peroxide, H_2O_2) or bleaches (sodium hypochlorite, $NaOCl$).

A voltaic cell produces electrical energy through spontaneous redox chemical reactions. When zinc metal is placed in a solution of copper sulfate, an electron transfer takes place between the zinc metal and copper ions. The driving force for the reaction is the greater attraction of the copper ions for electrons:

$$Zn + Cu^{2+} \rightarrow Zn^{2+} + Cu + \text{energy}$$

To exploit the energy produced in this reaction, the "half reactions" are separated. The oxidation reaction is carried out at a zinc electrode ($Zn \rightarrow Zn^{2+} + 2$ electrons) and the reduction reaction is carried out at a copper electrode ($Cu^{2+} + 2$ electrons $\rightarrow Cu$ metal). Electrons flow through a metal wire from the oxidizing electrode (anode) to the reducing electrode (cathode), creating electric current that can be harnessed, for example, to light a tungsten bulb.

BIOENERGETICS

Biological systems ultimately rely on energy from the Sun to build complex, high-potential energy molecules used to store fuel (glucose), provide structure (collagen), transmit signals (hormones, neurotransmitters), or carry information (DNA). Plants harvest the Sun's energy directly through photosynthesis to produce glucose from carbon dioxide and water. Whether by means of a vegetarian, carnivorous, or omnivorous diet, animals, too, benefit from the Sun's energy, albeit less directly. Living organisms couple energy-releasing reactions to energetically unfavorable reactions in synthesizing complex compounds that are high in potential energy. The primary energy-producing reaction in

living organisms (those that use oxygen) is the combustion of sugar:

$$C_6H_{12}O_6 (\text{glucose}) + 6\ O_2 (\text{oxygen gas}) \rightarrow$$
$$6\ CO_2 (\text{carbon dioxide}) + 6\ H_2O + \text{energy}$$

Biological glucose combustion is coupled through a complex set of reactions to the synthesis of the energy-storing molecule adenosine triphosphate, ATP. This elaborate metabolic pathway ensures that much of the energy released, 686 kilocalories per mole of glucose, is stored in small parcels of useful energy in the phosphate bonds of ATP, while allowing as little energy as possible to be wasted as heat. Glucose metabolism can be broken down into three parts: glycolysis, the citric acid cycle, and the respiratory chain. In glycolysis, also known as anaerobic fermentation, glucose is broken down into pyruvic acid in about ten steps, none of which requires oxygen. Over the course of these steps a significant portion of the energy released is taken up by the formation of ATP or by the reduction of special molecular intermediaries, nicotinamide adenine dinucleotide, NAD^+, and flavin adenine dinucleotide, FAD:

$$NAD^+ + H_2O + 52.7\ \text{kcal/mole} \rightarrow$$
$$NADH + H^+ + \tfrac{1}{2}O_2$$
$$FAD + H_2O + 36.2\ \text{kcal/mole} \rightarrow FADH_2 + \tfrac{1}{2}O_2$$

The citric acid cycle, a nine-step process, also diverts chemical energy to the production of ATP and the reduction of NAD^+ and FAD. In each step of the citric acid cycle (also known as the Krebs cycle) a glucose metabolite is oxidized while one of the carrier molecules, NAD^+ or FAD, is reduced. Enzymes, nature's chemical catalysts, do a remarkable job of coupling the oxidation and reduction reactions so that energy is transferred with great efficiency.

The reduced carrier molecules NADH and $FADH_2$ are recycled in the respiratory chain. The reoxidation of these intermediaries releases additional energy for use in the production of more ATP:

$$NADH + H^+ + \tfrac{1}{2}O_2 \rightarrow NAD^+ + H_2O + 52.7\ \text{kcal/mole}$$
$$FADH_2 + \tfrac{1}{2}O_2 \rightarrow FAD + H_2O + 36.2\ \text{kcal/mole}$$

Photosynthesis uses many of the same enzyme-driven steps found in glucose metabolism, only in reverse. The stages of photosynthesis termed the "dark reactions" are fueled not by the Sun but by ATP and reduced nicotinamide adenine dinucleotide phosphate, NADPH. This series of reactions, which takes place in the chloroplasts of the plant cell, synthesizes glucose from carbon dioxide. In the "light reactions," solar energy trapped by chlorophyll molecules is used to produce the continuous supply of ATP and NADPH (NADH in bacteria) necessary to power the dark reactions. The light-harvesting apparatus is composed of chlorophylls and beta-carotene, aromatic molecules with molecular orbitals that are highly Adelocalized," meaning that they are spread over a major portion of these large molecules. These delocalized molecular orbitals have energy spacings that are similar in energy to light in the visible range: Chlorophyll-*a* absorbs violet light; chlorophyll-*b* absorbs blue and, to a lesser extent, red; and the carotenes are long, antenna-like molecules that harvest light at wavelengths not picked up by the chlorophylls, in the orange and yellow portion of the visible spectrum. Green light is not absorbed, but is reflected, giving green plants their characteristic color.

Ellen J. Zeman

See also: Animal and Human Energy; Batteries; Conservation of Energy; Nuclear Energy; Solar Energy; Thermodynamics; Units of Energy.

BIBLIOGRAPHY

Asimov, I. (1965). *A Short History of Chemistry.* Garden City, NY: Doubleday.
Companion, A. (1979). *Chemical Bonding,* 2nd ed. New York: McGraw-Hill.
Dickerson, R., and Geis, I. (1976). *Chemistry, Matter, and the Universe.* Reading, MA: W. A. Benjamin.
Dickerson, R. E.; Gray, H. B.; and Haight, G. P., Jr. (1979). *Chemical Principles,* 3rd ed. Reading, MA: Benjamin/Cummings.
Hill, J. W. (1992). *Chemistry for Changing Times,* 6th ed. New York: Macmillan.
Hill, J. W., and Feigl, D. M. (1987). *The Chemistry of Life,* 3rd ed. New York: Macmillan.
Pauling, L., and Hayward, R. (1964). *The Architecture of Molecules.* San Francisco: W. H. Freeman.
Snyder, C. H. (1998). *The Extraordinary Chemistry of Ordinary Things,* 3rd ed. New York: John Wiley & Sons.

MONOPOLY, NATURAL

See: Regulation and Rates for Energy

MOTOR FREIGHT

See: Freight Movement

N

NANOTECHNOLOGIES

Nanotechnology is the technology of building things at a molecular scale, that is, where objects are measured in nanometers. Nanotechnology has the potential to improve both energy production technology, and the efficiency of end–use technology. Before this is possible, scientists must develop techniques to manipulate atoms and molecules well enough to build machines and structures. This means that the design of an object must specify each atom in the object, and all the chemical bonds between them. Such a design or object is called "atomically precise." In plants and animals, molecular biology has this capability, with certain substantial limitations. In the laboratory, atoms and molecules can be moved with the tip of a scanning probe, such as an STM (scanning tunneling microscope) or an AFM (atomic force microscope), but with other limitations. Other approaches include manipulation with electron beams, light waves ("optical tweezers"), and deposition in thin films (sputtering, molecular beam epitaxy).

HISTORY

In the 1950s, biologists (notably Francis Crick and James Watson) discovered the molecular basis for information coding in DNA and established that the workings of cells were molecular machines with understandable structure and function. Mathematician John von Neuman developed a mathematical theory of self-reproducing machines based on the biological theories.

In the 1960s, physicists such as Richard Feynman and Carver Mead showed that the speed and power efficiency of machines can improve with decreasing scale, particularly for solid-state electronics. A trend toward miniaturization in electronics was accelerated by the space program. Biologists began to decipher the genetic code.

In the 1970s, computers on a chip (microprocessors) appeared. Electron microscopes neared molecular resolution. By the end of the decade, transistors could be made the size of a human cell (10 microns).

In the 1980s, biologists began manipulating ("recombining") DNA–they could now read and write information at the molecular scale. Physicists (notably Gerd Binnig and Heinrich Rohrer) invented "scanning probe microscopes" able to image individual atoms. Technologists (notably William Trimmer) began to build "micromachines" using techniques from microelectronics. Futurists (notably Eric Drexler) began describing nanomachines and using the term "nanotechnology." At the end of the decade, transistors could be made the size of a bacterium (1 micron).

In the 1990s, nanotechnology emerged as a distinct field, with its own journals, conferences, and funding programs. MIT conferred the first Ph.D. in nanotechnology (to Drexler). Micromechanics became a burgeoning commercial field. Although most of the activity was still nanoscience as opposed to nanoengineering, the explosion of computational power made designing and simulating molecular machines feasible. Drugs and industrial catalysts were routinely designed and simulated in this fashion. Genetically modified organisms were commercial technology. Nanotechnologists (notably Nadrian Seeman) built controllable, albeit very rudimentary, molecular machines. By the end of the decade, commercial electronics had one-tenth micron (100 nanometer) transistors, but switching with single molecules had been done in laboratories.

ENERGY

As the field of nanotechnology matures, the ability to manipulate atoms and molecules will develop into the ability to build machines that do so. This will revolutionize the generation, storage, transmission, and use of energy. For example, a major source of energy is the oxidation of fossil fuels. If the fuel is burned, the energy is thermalized, and must be recovered by a heat engine. The laws of thermodynamics together with engineering constraints give this process a poor efficiency, typically 30 to 50 percent. Efficiency is even worse for small (e.g., hand-held) engines. Nanotechnology gives us the prospect of a different approach. Molecules of oxygen and fuel could be positioned rigidly and moved mechanically through the oxidation process, so that it becomes reversible (both practically and in the technical thermodynamic sense). This would raise the efficiency to near 100 percent, even in small engines.

Reversing the process yields an efficient method of energy storage. The process inputs combustion products such as carbon dioxide and water, and energy in the form of electricity or shaft power, and outputs oxygen and fuel (typically hydrogen or hydrocarbons).

Reversible fuel oxidation is as yet not well understood, and can be considered a major goal of nanotechnology in the energy area. Other forms of reversible conversion, such as shaft power to electricity and vice versa (generators and motors), are better understood and there are existing designs awaiting only construction capability. When of reversible oxidation becomes a reality, nanotechnology will enable the construction of heat engines that are small, powerful, clean, and as efficient as thermodynamics permits. One early application that is the subject of research is storage of hydrogen as a fuel.

The understanding of the electronic properties of nanostructures is one of the most rapidly advancing areas in science. This has two major implications: first, it will lead to the construction of nanocircuitry and nanocomputers that will use considerably less power than current computers while being faster and smaller; and second, it will lead to increasing efficiency and decreasing cost of photovoltaic power conversion ("solar energy").

ATOMIC IMAGING AND MANIPULATION

A scanning-probe microscope consists of a sharply pointed object, preferably so sharp that its tip is a single atom, mounted on a block of piezoelectric material such as a quartz crystal. A voltage is applied to the block, which warps in response. This warping can be controlled to atomic dimensions, allowing the tip to be steered across a molecular sample. The proximity of the tip to atoms of the sample can be sensed by various means, allowing a computer to build up a picture of the sample by scanning the tip across it. Kinds of scanning probe microscope include STM (scanning tunneling microscope) in which a current tunneling from the sample to the probe is measured, the AFM (atomic force microscope) in which the probe presses on the sample and the resulting force is measured, near-field optics in which the probe is an optical funnel focusing or detecting photons with much greater precision than their free-space behavior would allow, and many others.

Scanning probes can also be used to manipulate atoms and molecules individually, placing the tip in contact with the subject atom and pushing or pulling (atoms stick to the tip by virtue of the van der Waals force).

COMPUTATIONAL NANOTECHNOLOGY

Computer-aided design (CAD) and simulation of molecular structures is a rapidly advancing and widely applicable field. Molecules can be designed with molecular CAD programs; early programs allowed the user to specify the type and place of each atom or of substructures ("moieties") from a library. Research in molecular CAD is now focusing on the automation of parts of the process, following in the footsteps of a similar development in design software for digital electronics (e.g., for microprocessors).

Simulation of molecules can be done at the quantum mechanical level, as is necessary to determine the electronic properties of molecules, to analyze covalent bonds or simulate bond formation and breaking. However, quantum mechanical simulation is extremely computationally intensive and is too time-consuming for all but the smallest molecular systems.

A more practical approach for larger systems is molecular dynamics. In this method, the properties of bonds are determined through a combination of quantum-mechanical simulation and physical experiments, and stored in a database called a (semi-empirical) force field. Then a classical (non-quantum) simulation is done where bonds are modeled as spring-like interactions. Molecular dynamics simulations are appropriate for studies involving the properties of molecules as physical structures and shapes (including "docking" and the catalytic properties of biomolecules in solution, and the structural properties and energy dissipation mechanisms of nanomachine parts in operation).

EARLY COMMERCIAL NANOTECHNOLOGY

Commercial nanotechnology in the 1990s was limited by the lack of a general synthetic capability. It characterized by a plethora of techniques for building nanostructures, including a number of methods based on gas-phase nucleation (e.g., laser pyrolysis, sputtering), methods from synthetic chemistry, including dendrimers and fullerenes, and methods from molecular biology, including DNA synthesis and the use of DNA as a structural material, and protein engineering. In general, the limitations on DNA/protein methods are in the ability to design and predict structures, where the limitations on the other methods are on their physical synthetic capabilities.

Methods capable of detailed atomic manipulation were confined in the 1990s to the laboratory, since they were generally incapable of producing commercially useful amounts of product. There are a few exceptions to this, in applications where a few carefully constructed molecules can be useful, such as chemical sensors, laboratory equipment, and so forth.

THE FUTURE OF NANOTECHNOLOGY

The field of nanotechnology is advancing rapidly, so it is not practical or useful to make short-term predictions about its specific form and capabilities. At most it can be noted that one of the most rapidly advancing subfields is nanoelectronics (sometimes referred to as "molectronics," for molecular electronics).

For longer-term projections, a common practice is to compare the field with others that are advancing as rapidly. Computers, both hardware and software, and biotechnology are cases in point. Both fields bear on nanotechnology and contribute to its advance. In computers, the watershed capability was the stored-program von Neumann architecture. In nanotechnology, it is expected to be another von Neumann design, the self-reproducing system. In practical terms, this simply means that nanomachines need to be capable of making parts for nanomachines, just as a conventional machine shop is capable of making parts for its own machines. Once this is achieved, it will be possible to build commercial quantities of product with atomic precison, and in particular, to have commercial quantities of product that consists of working nanomachines.

J. Storrs Hall

BIBLIOGRAPHY

Crandall, B. C., and Lewis, J., eds. (1992). *Lewis Nanotechnology: Research and Perspectives.* Cambridge, MA: MIT Press.

Drexler, K. E. (1992). *Nanosystems: Molecular Machinery, Manufacturing, and Computation.* New York: Wiley.

ten Wolde, A., ed. (1998). *Nanotechnology: Towards a Molecular Construction Kit.* The Hague: Netherlands Study Centre for Technology Trends.

NATIONAL ENERGY LABORATORIES

The Department of Energy operates seventeen major national laboratories and thirteen minor facilities in the United States that carry out energy research and development, basic science, and defense weapons work. The combined budgets for these laboratories exceed $6 billion annually, with a scientific and technical staff of more than thirty-thousand.

Each of these laboratories is a government-owned/contractor-operated facility selected from industry, academia, and university consortia. As of 1999, the most prominent civilian contractor was Lockheed Martin, the operator of Oak Ridge and Sandia, and the major academic institution was the University of California, administrator of Los Alamos, Lawrence Berkeley, and Lawrence Livermore.

ATOMIC ENERGY COMMISSION YEARS

The origins of the national laboratory network can be traced back to the late 1940s and the beginning of the atomic age. At the end of World War II, the scientific community, particularly the staffs from the Manhattan Project laboratories, lobbied Congress for civilian control of atomic power. Toward this end, the federal government transferred authority from the Army to the newly established Atomic Energy

Commission in 1946. Atomic research and development, both for defense and peaceful use, was to be solely in the hands of the Atomic Energy Commission. Its mission was to "assure the common defense and security," and to "improve the public welfare, increasing the standard of living, strengthening free competition, in private enterprise, and promoting world peace."

After the Russians achieved the first detonation of a nuclear device in 1949, there was a great fear of falling behind in the nuclear weapons race, setting in motion an all-out effort to develop a thermonuclear device. Prominent scientists like Edward Teller and Ernest O. Lawrence were instrumental in establishing the Lawrence Livermore Laboratory to join Los Alamos as a major weapon design laboratories. This strong nuclear science emphasis made the 1950s and 1960s an era of great advances in civilian nuclear energy as well as nuclear weapons and nuclear propulsion for the military. Nuclear weapons not only became more powerful, but more tactical and versatile, including a variety of designs for short, intermediate, and long-range missiles.

The laboratories at Argonne and Oak Ridge were busy designing nuclear reactors in the 1960s. Once the electric utility companies saw that nuclear power could compete with fossil fuel plants on purely economic grounds, more than one hundred reactors were being designed or under construction by the mid-1960s. But confronted by safety and environmental concerns, the nuclear juggernaut started to run into trouble shortly before the energy crisis of 1973—ironically at a time when the need for alternatives to oil and natural gas was greatest.

With so much of the work at the laboratories involving the design of nuclear reactors, the future of many of the laboratories, particularly Argonne and Oak Ridge, looked bleak. But the energy crisis brought plenty of new energy research and development funding to all the laboratories. It also gave many of the laboratories an opportunity to diversify beyond nuclear energy.

DEPARTMENT OF ENERGY YEARS

When the Department of Energy (DOE) was established in 1977, along with its many missions, it was granted authority over a coveted prize: oversight of all the laboratories and research facilities from the predecessor agencies. These included the Bureau of Mines research laboratories at Bartlesville, Morgantown, Pittsburgh, and Laramie from the Federal Energy Administration, and the National Laboratories that were managed by the Atomic Energy Commission (1947–1974) and the Energy Research and Development Administration (1974–1977).

By the late 1990s the number and the size of the laboratories had grown tremendously. From 1977–1997, the federal government's investment in the laboratories was in excess of $100 billion, in seventeen major laboratories and thirteen minor laboratories.

Since the end of the Cold War, the focus of all the laboratories has moved beyond weapons, accelerators, and energy-related research to encompass almost every imaginable field of basic and applied science. Most of the laboratories are increasingly being pushed by Congress to create partnerships with industrial firms to commercialize laboratory-developed technology in the hope that it will improve the overall competitiveness of the U.S. economy.

AMES LABORATORY

To accompany the physics program of the Manhattan Project, Frank H. Spedding, an expert in the chemistry of rare Earth metals, agreed to set up a chemical research and development program in 1942. The Ames Project went on to develop new and far less expensive methods for both melting and casting uranium metal. Based on this successful work, in 1947 the Atomic Energy Commission established Ames Laboratory to produce high-purity uranium metal in large quantities for atomic energy. Although still heavily involved in material research, the research scope expanded considerably through the years to include engineering, environmental, mathematical, and physical sciences.

ARGONNE NATIONAL LABORATORY

The Atomic Energy Commission designated Argonne National Laboratory as the first national laboratory on July 1, 1946. Argonne was the lead laboratory for nuclear reactors, instrumental in designing and building the first nuclear powered submarine, the *USS Nautilus* in 1954, and the first nuclear reactor that completely powered the town of Arco, Idaho, in 1955.

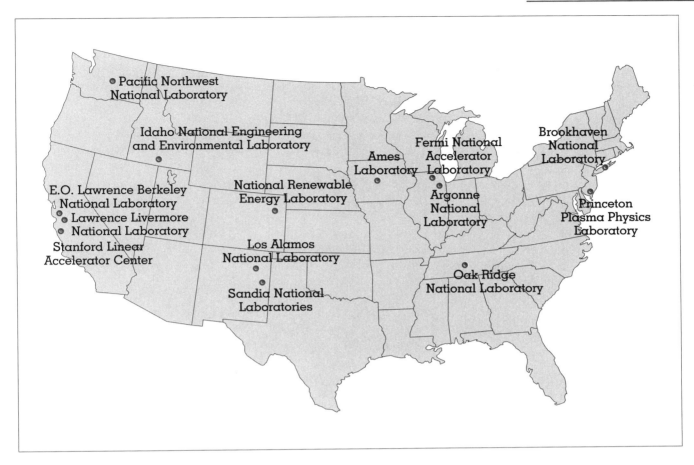

Figure 1.

Locations of the major U.S. National Energy Laboratories.

SOURCE: U.S. Department of Energy website <http://www.doe.gov/people/labsmap.htm>.

Through the 1960s and 1970s, Argonne scientists continued to design and build nuclear reactors, but as nuclear power became engulfed in controversy, federal support dwindled. In response, Argonne began to emphasize other areas such as batteries, magnetohydrodynamics, solar, and fusion in 1977. From the early 1980s through the late 1990s, Argonne continued its diversification, encompassing fields such as materials, medical research, and transportation fuels.

BROOKHAVEN NATIONAL LABORATORY (BNL)

BNL was also founded in 1947 by the Atomic Energy Agency. The management contract was awarded to the nonprofit educational consortium called Associated Universities Inc. with the goal being to create a regional laboratory with the advanced facilities too costly for Universities to build and maintain. Initially BNL concentrated on developing small nuclear reactors for peaceful scientific exploration, and also was one of the pioneers in nuclear medicine. The scope of the laboratory has evolved in the late 1990s beyond the basic sciences to include areas such as environmental research and energy technology.

FERMI NATIONAL ACCELERATOR LABORATORY (FERMILAB)

Fermilab, where groundbreaking for its first linear accelerator began in December 1968, is the premier high energy physics facility in the world. Its mission is to advance the understanding of the fundamental nature of matter and energy. Universities Research Association Inc., a consortium of eighty-six research

Aerial view of the world's largest particle accelerator, located at the Fermi National Accelerator Laboratory in Batavia, Illinois. (Corbis-Bettmann)

universities, operates the laboratory for the DOE. Fermilab is home to the Tevatron, the most advanced accelerator in the world, that can accelerate particles at 0.950 TeV. Huge amounts of electricity, typically running anywhere from $10 to $20 million a year, are needed to power the laboratory and conduct experiments. The future role of Fermilab was uncertain when the Superconducting Super Collider was approved by President Reagan in January 1987. But because federal budgetary problems led to the termination of SSC funding in 1993, it guaranteed the role of Fermilab as the foremost experimental particle research facility in the United States for many years to come.

IDAHO NATIONAL ENGINEERING LABORATORY (INEL)

INEL's origins were as one of the satellite facilities for the Manhattan Project in the 1940s. Since the early 1950s, INEL has developed and operated fifty-two nuclear reactors for the commercial power and national security sectors. INEL has diversified through the years, but the areas of concentration continue to be energy, environmental management (methods for characterizing, treating, and storing radioactive and hazardous waste), and national security. As with the other major laboratories, INEL is committed to the transfer of technology to the private sector.

LAWRENCE BERKELEY LABORATORY (LBNL)

The roots of the LBNL can be traced back to the 1920s, and the pursuit of the secrets of the nucleus. Ernest O. Lawrence, built the first large cyclotron (a particle accelerator) on the Berkeley campus of the University of California in 1931. Unlike most the other labs, LBNL's beginnings depended on the support of philanthropists who saw the promise in Lawrence's work. Seeking private sector support, an onerous chore in good times, was all the more difficult for Lawrence because of the Depression. Nevertheless, by the late 1930s Lawrence had raised enough money to plan his 184-inch diameter cyclotron. It was too big to build on campus, so he moved his Radiation Laboratory to the hill above the campus. After World War II Lawrence and Brobeck built the 7 billion electron volt proton accelerator (called the Bevatron), specifically designed to produce antiprotons, which it did in 1955. But with no more flat land in the Berkeley hills, the lead for more powerful accelerators passed to SLAC (Stanford Linear Accelerator Center), the Fermilab, and to CERN.

As the lead in high-energy accelerators passed to Fermi Laboratory in the 1950s, LBNL moved beyond high energy physics to become a collection of interdisciplinary groups working in many diverse fields. By 1977, because of the energy crisis, the focus continued to change with the addition of two new divisions, one for Energy and Environment and another for Earth Sciences, which included geothermal energy and the disposal of nuclear wastes. By the late 1990s these two divisions together had grown to account for about one quarter of the laboratory's budget.

LAWRENCE LIVERMORE NATIONAL LABORATORY (LLNL)

In an attempt to develop the hydrogen bomb before the Russians, a second weapons laboratory, Lawrence Livermore, was established in July 1952 to handle the additional work that would be necessary to stay ahead of the Russian nuclear weapons program. The administrator chosen was the University of California. For the next forty-five years, this LLNL was a formidable competitor to Los Alamos in the development of nuclear weapons. But much like most of the other major national laboratories, its focus also shifted away from nuclear weapons to basic science to fields like magnetic and laser fusion energy, non-nuclear energy, biomedicine, and environmental science. By the late 1990s, half of the laboratory's budget was nondefense related as the shift away from nuclear weapons continued.

LOS ALAMOS NATIONAL LABORATORY (LANL)

In 1943, under the direction of J. Robert Oppenheimer, the Los Alamos National Laboratory, referred to as

Project Y at the time, was the site chosen to develop the technology and build the bombs that were instrumental in ending World War II. Managed since its beginning by the University of California, the laboratory's original mission to design, develop, and test nuclear weapons has broadened and evolved through the years. During the Cold War, from the 1950s to the 1980s, the work primarily involved nuclear weapons and the Strategic Defense Initiative. But as the Cold War ended, programs in energy, nuclear safeguards, biomedical science, environmental protection and cleanup, material science, and other basic science programs were added and enhanced. National security remains at the core of LANLs mission, yet the laboratory continues to try to diversify toward more basic science research, and the development and commercialization of emerging technologies.

NATIONAL RENEWABLE ENERGY LABORATORY (NREL)

In the belief that solar energy was a key to energy independence, the Solar Energy Research Institute (SERI) was established in 1977. Meanwhile Rockwell International, the operator of the nearby Rocky Flats Plant, was awarded a federal grant to build and test many different small wind turbines in various configurations. SERI and Rocky Flats wind energy programs were merged in 1984 and transferred to SERI. Both programs prospered in the late 1970s because of very generous federal appropriations, but after federal funding for solar energy research declined in the 1980s, SERI, renamed the National Renewable Energy Laboratory in 1991, was forced to expand its mission well beyond photovoltaics, solar industrial technologies, solar thermal electric, and wind energy to a greater emphasis on biomass electricity, biofuels, and geothermal power. In a renewed effort to transfer more technology for commercialization, Battelle Memorial Institute and Bechtel Corp. began managing the laboratory in collaboration with Midwest Research Institute, the original administrator, in November 1998.

OAK RIDGE NATIONAL LABORATORY (ORNL)

Ground breaking for the "Clinton Laboratories," what Oak Ridge was originally called, occurred in February 1943. Whereas the Manhattan Project at Los Alamos served as the center of weapons design,

ORNL devised the techniques to produce and purify the large quantities of fissionable uranium and plutonium. ORNL was a natural choice due to the abundant hydroelectric power that the Tennessee Valley Authority could provide. During the war, the lab was managed by DuPont and the University of Chicago, but in December 1947 the Atomic Energy Commission chose Union Carbide, whose tenure lasted until 1984. Argonne was designated the lead laboratory for reactor design, so ORNL concentrated on isotope production and radiochemical separations, biological research, and specialized reactors for the Navy and Air Force. The promise of electricity "too cheap to meter" in the 1950s and 1960s led to a renewed nuclear reactor emphasis. But as the energy crisis erupted and the promise of nuclear power faded in the 1970s, ORNL diversified into other areas such as energy conservation, synthetic fuels, and solar power. The diversification continued into the 1980s and 1990s with an expansion of environmental research, and an international outreach as ORNL scientists began advising developing countries on ways to secure the energy needed for economic growth.

PACIFIC NORTHWEST LABORATORY (PNL)

Battelle Memorial Institute took over management of what was then called Hanford Laboratories in 1965, and the research facility was separated from Hanford site operations (the Hanford site is where much of the weapons grade plutonium was manufactured from the 1940s to the late 1980s) and renamed the Pacific Northwest Laboratory. PNL is a unique multiprogram laboratory in that part of the facilities are Battelle Memorial Institute (established by the estate of Gordon Battelle, a wealthy industrialist, in 1929) facilities, and part are DOE facilities. When Battelle took over management, they invested $50 million in private research facilities adjacent to the government laboratory.

At first PNL focused on nuclear technology and the environmental and health effects of radiation, but through the years expanded its mission to cover nearly every field of basic science to solve problems in the areas of environment, energy, and national security. Environmental issues and cleanup still encompass two-thirds of PNL work in the 1990s, but PNL has strengthened its role in regional electric power issues as well.

PRINCETON PLASMA PHYSICS LABORATORY (PPPL)

In 1951 Professor Lyman Spitzer, Jr., conceived of an idea of plasma being confined in a figure-eight-shaped tube by an externally generated magnetic field. Called the "stellarator," the Atomic Energy Commission decided to fund it, which established Princton University's controlled fusion effort. Magnetic fusion research was declassified in 1958 so that all nations could collaborate. The PPPL has managed to secure the lion's share of federal plasma and fusion research funding from the 1960s through the 1990s, and with it developed the Tokamak Fusion Test Reactor, that operated at PPPL from 1982 to 1997. Early in the next millenium, an advanced fusion device—the National Spherical Torus Experiment—will begin operation. Although much of the promise of fusion has been unfulfilled to date, some of the knowledge gained in fusion research has found applications in other areas such as materials science, chemistry, and manufacturing.

SANDIA NATIONAL LABORATORIES (SNL)

Sandia National Laboratories began as an ordinance design, testing, and assembly facility in 1945 as part of what is now Los Alamos National Laboratory. AT&T began managing the laboratory on November 1, 1949, and continued to do so until 1993 when the Martin Marietta Corp., now Lockheed Martin, took over. The primary mission of the laboratory was to provide engineering design for all non-nuclear components of the nation's nuclear weapons, but by the late 1990s the laboratory's mission had expanded to include all defense systems, energy security, environmental challenges, and national technological challenges facing industry.

STANFORD LINEAR ACCELERATOR CENTER (SLAC)

The Stanford Linear Accelerator Center, administered by Stanford University, was founded in 1962 as a center for experimental particle physics, but it took until 1966 for its first linear accelerator to be completed. The Stanford Synchrotron Radiation Laboratory, built a decade later, became part of SLAC in 1992. Unlike many of other national laboratories that greatly expanded their mission through the years, SLAC always remained a national basic energy research laboratory.

THE FUTURE OF THE LABORATORIES

The most valuable attribute of the laboratories is the vast human and physical resources that can be called upon to solve national problems of great complexity and scope. However, without a national problem or crisis on the horizon, as was the case in the 1990s, the vast funding needed to maintain this national resource comes into question. And since the laboratories are all funded almost exclusively by federal dollars, the ultimate customer is the U.S. taxpayer, a taxpayer who asks: What have I received for my investment? What can I hope for in the future?

In considering the future course of the laboratories, and the involvement of the federal government, there are two important realities to consider. (1) The number of laboratories and their funding has always increased. (2) No national laboratory has ever been closed. There is a good reason for both realities. The laboratories are located in so many different congressional districts that, much like military bases, finding the political will to close any of the laboratories would prove extremely difficult. Moreover, all the major laboratories have also diversified into many different areas of scientific research to further ensure greater and more stable funding. When there was talk about closing one of the defense laboratories as a "peace dividend" following the breakup of the Soviet Union in the early 1990s, Congress approved more funding, not less.

Critics contend that the laboratories are bureaucratically bloated and inefficient, citing numerous independent reports and audits over the last few decades warning that DOE ownership and operation does not work well. Bureaucracy has made it difficult for the laboratories to generate and carry out a viable mission. And following the Chinese spy scandal of 1999, the DOE has also been widely faulted for a failure to ensure the proper security at its weapons labs.

Some critics recommend that the weapons laboratories (Los Alamos, Sandia, Lawrence Livermore) come under the control of the Department of Defense, and that ownership of the other laboratories be sold to the highest bidder, or turned over to the administrator now running the laboratory. The new owner can then contract with public and private entities in the free market, or shut down the laboratories. They contend that this is the best way for the laboratories to create a vision with value and effectively carry out a mission.

Of course, much of the vision and mission problems of the laboratories can be traced back to funding. Federal funding strongly influences activities of the laboratories. Although much of the $5.0 billion allocated each year for basic science and energy research and development is steered toward the laboratories, the projects supported are decided more for political reasons rather than sound economic and scientific ones. This results in the majority of research dollars flowing toward marginal ideas, which is not the fault of the laboratories, yet the laboratories are held accountable for the dismal record of picking winners and losers that were, in reality, ill-conceived political and bureaucratic decisions. Take electricity generation. The winner—the fuel of choice—turned out to be the fuel that the DOE did not feature in its research and development portfolio: natural gas. Of the $60 billion (in 1996 dollars) spent from 1978 through 1996, only 1 percent ($787 million) went to natural gas, while 99 percent was spent on conservation ($13.3 billion), civilian nuclear ($20.1 billion), coal ($13.3 billion), solar ($5.1 billion), geothermal ($1.8 billion), wind ($900 million), other renewables ($2.8 billion), oil ($1.4 billion), and hydroelectric ($193 million).

Proponents of the laboratories counter that, despite these shortcomings, the laboratories serve a vital mission of undertaking the high risk and expensive investments that the private sector would never agree to invest in. Although natural gas research and development was minimal, DOE support accelerated technological advances on natural gas-fired turbines. Much of the research and development at the laboratories has provided a net social benefit to the nation and economy, work such as safe nuclear reactors and the development of sophisticated defense weapons.

The laboratories undertook the high risk and expensive investments that helped improve the efficiency of current technologies on both the consumption and production end. To lessen dangerous emissions at U.S. coal power plants, the laboratories helped develop better burning systems (e.g., pressurized fluidized bed system), and better scrubber systems. They have also participated in partnerships that have been responsible for a number of fuel cell technology breakthroughs. On the consumption side, the laboratories have helped to develop energy use standards for appliances that have significantly helped lower the nation's overall energy consumption.

Another area of success has been in applied materials research. Because of the integral nature of materials to advances in energy production and consumption, the laboratories have developed a number of toughened ceramics. When used as a replacement for steel, they will improve the energy performance characteristics of high-temperature applications for components of combined-cycle power plants and vehicle engines.

Proponents also point to the fact that thirty-one scientists associated with the laboratories have won Nobel prizes, and that the laboratories have received more "R&D 100" awards (award given annually to technology innovations that hold a strong prospect for commercial success) than any other institution.

John Zumerchik

See also: Government Agencies.

BIBLIOGRAPHY

Brown, E. (1996). *Frontiers: Research Highlights, 1946–1996*. Argonne, IL: Argonne National Laboratory.

Crease, R. P. (1999). *Making Physics: A Biography of Brookhaven National Laboratory*. Chicago: University of Chicago Press.

Crow, M. M., and Bozeman, B. (1998). *Limited by Design: R&D Laboratories in the U.S. National Innovation System*. New York: Columbia University Press.

Hoddeson, L., et al. (1993). *Critical Assembly: A Technical History of Los Alamos During the Oppenheimer Years, 1943–1945*. New York: Cambridge University Press.

Holl, J. M., and Fehner, T. R. (1994). *The Department of Energy, 1977–1994. Energy History Series*. Washington, DC: U.S. Department of Energy.

Johnson, L.; Schaffer, D.; and Schaffer, D. (1994). *Oak Ridge National Laboratory: The First Fifty Years*. Knoxville, TN: University of Tennessee Press.

Plastino, B. J., and Graves, D. P. (1998). *Coming of Age: Idaho Falls and the Idaho National Engineering Laboratory, 1949–1990*. Chelsea, MI: BookCrafters.

Shroyer, J. A. (1997). *The Secret Mesa: Inside Los Alamos National Laboratory*. New York: John Wiley & Sons.

Stelzer, I. (1996). "The Department of Energy: An Agency That Cannot be Reinvented." Washington, DC: American Enterprise Institute.

U.S. House Committee on Science, Subcommittee on Basic Research. (1995). *Alternative Futures for the Department of Energy National Laboratories "The Galvin Report" and National laboratories Need Clearer Missions and Better Management, A GAO Report to the Secretary of Energy*. Washington, DC: U.S. Government Printing Office.

NATIONAL OCEANIC AND ATMOSPHERIC ADMINISTRATION

See: Government Agencies

NATIONAL RENEWABLE ENERGY LABORATORY

See: National Energy Laboratories

NATIONAL SCIENCE FOUNDATION

See: Government Agencies

NATURAL GAS, CONSUMPTION OF

Natural gas is a mixture of naturally-occurring methane (CH_4) with other hydrocarbons and inert gases. The 2.3 trillion cubic meters (Tcm) or 81 trillion cubic feet (Tcf) of gas marketed and consumed globally in 1997 accounted for about 24 percent of the world's primary energy, ranking third among fuels after petroleum liquids (40%) and coal (25%).

The modern natural-gas industry has its origins in the nineteenth century as urban "gas works" that distributed synthesis gas (a mixture of carbon monoxide, hydrogen and carbon dioxide made by the incomplete combustion of coal, oil, or organic wastes in the presence of steam). Gas works illuminated London streets even before 1800, and subsequently

Gas lit lamps in use around 1911. (Library of Congress)

provided lighting, cooking, water- and space-heating for homes, businesses and public buildings. By the late nineteenth century, gas light was common in the central districts of cities and larger towns throughout North America and Western Europe, and even in such places as Buenos Aires, Cairo, St. Petersburg, Shanghai, and Sydney.

Between the World Wars, consumption in North America switched rapidly from synthesis gas to natural gas which, lacking carbon monoxide, was nontoxic and contained three to four times as much energy as synthetic gas per unit of volume. This shift resulted from the advent of thin-walled, seamless welded steel pipe and leak-proof pipe couplings, which permitted highly compressed vapors to be transmitted safely and efficiently over long distances, together with the discovery and production of large volumes of methane as an initially unwelcome byproduct of crude oil. By 1940, almost every major American gas had become a local distribution company engaged in

the resale to retail customers of natural gas purchased mostly from oil companies. In western and central Europe, a similar transition from synthesis gas to natural gas waited until the last third of the twentieth century, when pipelines were laid from natural-gas fields in the North Sea, North Africa, and Western Siberia.

In North America, the local distribution companies that distribute and sell gas to retail customers tend to be distinct from both gas producers and the operators of long-line gas-transmission pipelines, although common ownership of businesses in two or three sectors is not uncommon. Gas distributors are generally treated as public utilities, whose retail prices and other terms of service are regulated by state or provincial authorities. Prior to the 1980s, distribution utilities held a legal monopoly on both the physical delivery and the sales of gas within their local service areas. Recently, however, there has been a trend among state/provincial regulators to unbundle gas

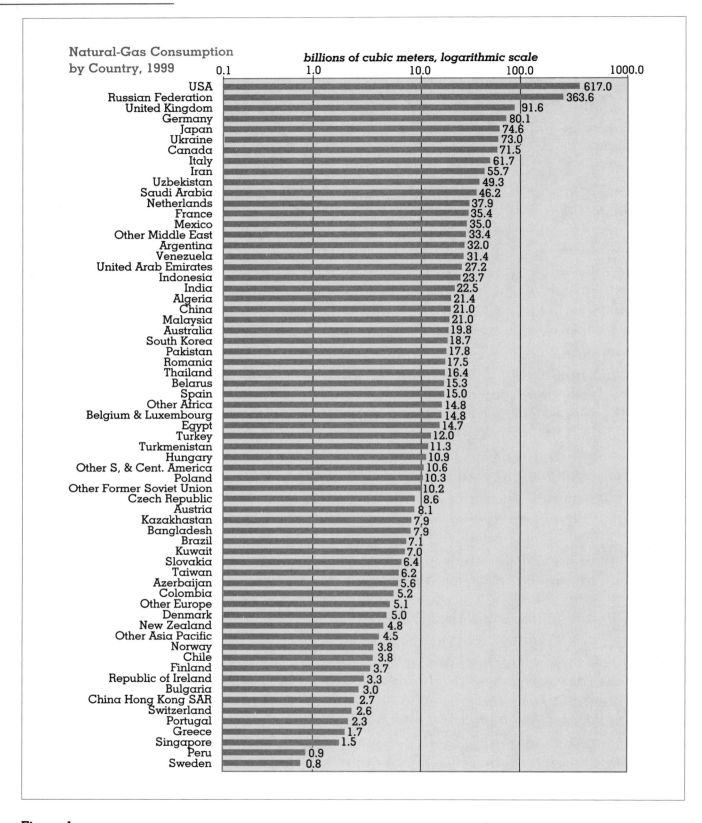

Figure 1.

sales from supposedly naturally monopolistic transport functions of distributors, and to permit competition among gas marketers for retail sales, particularly to industrial and commercial customers.

GAS USE BY COUNTRY

Reliance on natural gas as a primary fuel varies widely among different national economies from the 1999 global average of 25 percent. In absolute quantity of gas consumed (see Figure 1), the United States is the largest single consumer, with natural gas accounting for 25 percent of its total energy consumption. The share of natural gas in total primary energy consumption in the European Union was similar to that of the United States, 23 percent, but varied from a high of 41 percent in the Netherlands, through 34 percent in Italy, 22 percent in Germany, 13 percent in France, to less than 2 percent in Sweden.

The number-two gas-consuming country, Russia, used about the same volume of gas per capita as the United States in 1999 but depended on gas for more than half (54%) of its total energy. It is worth noting that although the Russian Federation was the world's biggest gas producer, and international exporter, the entire eastern third of the country—Eastern Siberia and the Russian Far East, roughly from Krasnoyarsk to Vladivostok and Magadan—consumed less than five percent of its primary energy in the form of natural gas.

Figure 2 shows no systematic correlation between a country's dependence on natural gas and its degree of industrialization or per capita; GDP. Variation in relative dependence on gas was even wider among the less-developed and emerging economies outside of Europe and North America, from a high of 82 percent in Uzbekistan, 70 percent in Bangladesh and Algeria, and 50 percent in Argentina, to only 8 percent in India, 3 percent in China, and none at all in several African and Latin American countries.

NATURAL-GAS CONSUMPTION PATTERNS

In addition to widely differing degrees of dependence on natural gas, different regional economies exhibit dramatically contrasting consumption patterns for the gas that they do consume. Strong distinctions are evident, for example, between the diversified and finely-tuned natural-gas markets of high-income, high-latitude countries in Europe and America; the high-volume but relatively undifferentiated gas industry of Russia and other former Soviet republics; and the specialized liquified natural gas-based gas-consuming sectors of Japan, South Korea, and Taiwan.

North America and the European Union.

Figures 3 and 4 contrast the gas-consuming patterns of the world's two largest economies, the United States and Japan. The uses of gas tend to be the most diverse in high-latitude areas, where seasons are distinct and winters are cold, and in regions such as North America and the European Union, where economies are generally sophisticated. In these areas, a dense network of transmission and distribution pipelines makes gas directly available to a numerically large and finely differentiated population of potential residential, commercial, and industrial customers. There, the sales for residential, commercial, and institutional space heating peak in the winter months when the market value of gas is highest. "In contrast, peak demand for gas to generate electricity in combustion turbines and combined-cycle plants occurs in the summer, driven by power requirements for air-conditioning. Together, these climate-sensitive, seasonal components of gas demand constitute its most valuable portions, generating more than 60 percent of total sales revenue."

More than 60 percent of natural gas physically consumed in the course of a year is nevertheless attributable to purchases at lower, interruptible prices by industrial boiler-fuel users and electrical generators that are capable of substituting natural gas in off-peak months, when gas is available at prices competitive with those of "black fuels" (coal and heavy fuel oil). In addition to these relatively low-value, price-sensitive industrial gas uses is a wide range of intermediate-value demand categories for natural gas, such as in process and feedstock use.

In such diverse and sophisticated gas-using economies, security of supply and the efficient employment of producing assets depend upon an extensive network of pipelines that interconnect regions with diverse climates and diverse consumption patterns: winter-peaking and summer-peaking; demand that is climate-sensitive, business cycle-sensitive, and price-sensitive; customers who place a high premium on continuity of supply, and those who are relatively insensitive to risk of interruption. These parties depend to a different degree, and place

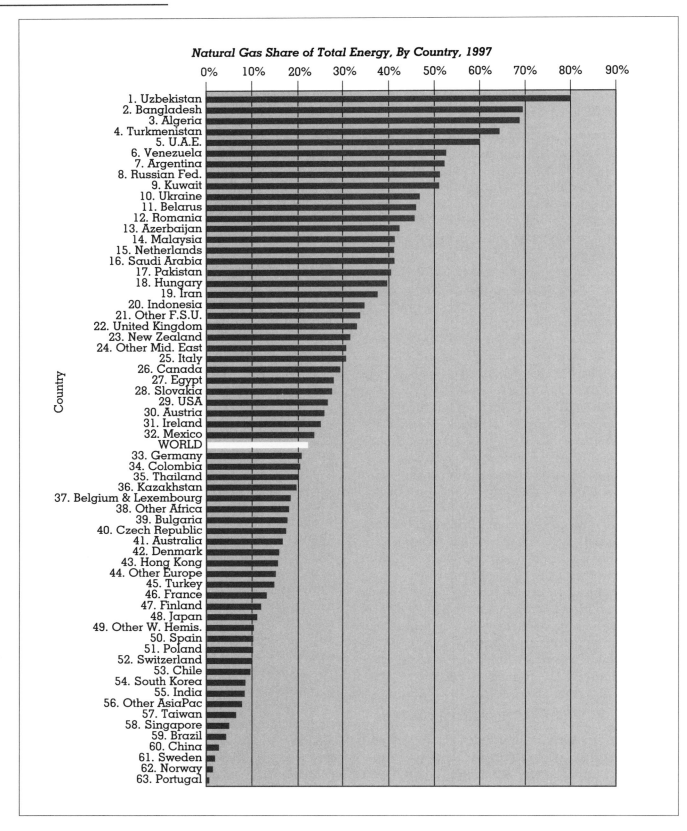

Figure 2.

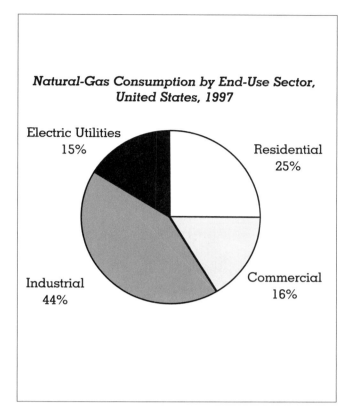

Figure 3.

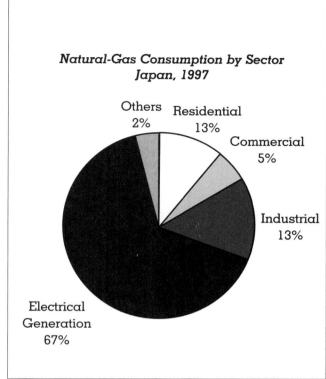

Figure 4.

different values on the access these lines give them to a large and widely distributed system of underground and other gas-storage facilities. The dispatch and allocation of producing, transmission, and storage capacity within the large and diverse population of producing assets and end-users is coordinated by a highly differentiated system of first-sale and wholesale, cash and forward prices for the gas commodity, and for auxiliary services.

Former Soviet Union

Russia, Ukraine, and other entities of the former Soviet Union [FSU] depend even more heavily on natural gas for their primary energy than do the high-income market economies of North America and the European Union. Moreover, their structure of gas demand differs considerably from that of Europe and America, along with the physical and institutional infrastructure of their energy sectors. Because of severe winters, space-heating and domestic water heating doubtless account for an even greater share of total gas consumption than in the West. Rigorous comparison is not possible,

because space- and water-heating by means of gas is seldom served or metered separately for individual homes or flats, offices, or small enterprises. Most dwellings and workplaces do use gas or another primary fuel such as coal for space- and water-heating, but only indirectly through steam or hot water cogenerated at steam-electric stations or generated in stand-alone gas-fired district heat plants. District heat produced in this manner is typically distributed to apartment blocks and office towers through elevated and insulated ducts.

Gas for process and feedstock use, and industrial heat for metallurgy, smelting, and materials drying, and other manufacturing applications in the FSU is often delivered directly to the gas-using enterprise, but neither price nor any general system of end-use priorities is systematically used to dispatch or allocate transmission capacity seasonally or among competing domestic shippers, or gas deliverability among competing domestic users. Thus far, the prevailing strategy for accommodating supply and demand in these areas has been to strive to provide sufficient field deliverability and transmission capacity to accommodate the unconstrained aggregate demand of all connected customers, without causing pipeline pressure to collapse.

Japan, South Korea, and Taiwan

The natural-gas industries of Japan, Korea and Taiwan are based almost entirely on gas imported by tanker as liquified natural gas (LNG). These insular and peninsular economies produce almost no domestic natural gas, and until recent efforts to develop production on the Russian island of Sakhalin immediately north of Japan, import of natural gas by pipeline from nearby has not seemed a realistic alternative. In 1999, LNG imports provided about 12 percent of Japan's primary energy, 9 percent of South Korea's, and 6 percent of Taiwan's. In turn, these three countries together accounted for 79 percent of the world's international movements of LNG.

LNG imports to East Asia from Alaska, Southeast Asia, and the Middle East have generally been targeted to specific base-load electrical generating stations built adjacent to the receiving terminals. To minimize delivered costs per unit of fuel, the entire chain of physical facilities from the producing field, through the liquefaction plant at tidewater in the exporting country, the tankers and import terminals, to the receiving customers are tightly coordinated as to capacity and scheduling. The corresponding chain of commercial transactions is composed almost entirely of long-term "take-or-pay" contracts—which oblige the purchaser to pay for the full contracted volume, even if that volume is not or *can* not be taken—for fixed rates of delivery from specific physical sources to specific end-use facilities.

Japan is the world's second largest economy and the fifth greatest consumer of natural gas, behind the United Kingdom and Germany and ahead of Ukraine and Canada. It nevertheless lacks a network of transmission and distribution pipelines capable of delivering gas to diverse and dispersed customers, or reallocating it among them in response to shifting seasonal or business demand. As a result, delivered gas prices to households and industry (other than for base-load power generation) are nearly the highest in the world. Not surprisingly, consumption of natural gas in the residential, commercial, and industrial sectors, and for peak-load electric generation, is exceptionally small.

In the 1990s, Japan's smaller insular and peninsular neighbors, Taiwan and South Korea, began to create integrated national gas-distribution systems that joined previously separate LNG terminals and the potential markets between them. This strategy, particularly in Korea, is consciously directed at creating a more competitive bulk market for gas and expanding the use of gas for space-heating, electricity peaking supply, and industrial fuel.

COMMERCIALIZATION STRATEGIES FOR "STRANDED GAS"

Natural gas is frequently found in association with crude oil, in the search for crude oil, or fortuitously at great distances from developed gas markets or from existing gas-transmission infrastructure-providing access to such markets. In the first half of the twentieth century, carbon black—high-quality soot used as colorant in printing inks and as an additive to rubber in tires—was a leading "scavenger industry" for stranded gas in North America. Later in the century, manufacture of fertilizer, particularly ammonia and urea, created a major part of the early demand for gas along the U.S. Gulf Coast, in Alaska's Cook Inlet basin and in China. The oil industry has recently devoted great effort to promoting gas-to-liquids (GTM) conversion systems to make stranded gas into motor gasoline or diesel fuel. Several GTM technologies are firmly proved, but thus far appropriate market conditions for their commercial application have been hard to find. All of these initiatives are seeking opportunities to convert abundant, low-cost gas into a higher-valued commodity that is liquid or solid at ambient temperatures, and thus can be moved in "normal" tankers, barges or railcars, rather than requiring costly transcontinental pipelines or cryogenic (super-cooled) transport systems. Other notable applications for stranded gas are the local generation of electricity for local consumption and, particularly in the Middle East and North Africa, desalinization of seawater.

INTERFUEL SUBSTITUTION AND COMPETITION

The broad variance in the amount of energy consumed as natural gas, and the diverse mixes of consumption patterns in different countries, illustrate important characteristics of energy supply and demand generally:

- Neither gas nor any other primary fuel or energy source is technically or economically indispensable to modern civilization. Society's energy "needs" are for heat, light, motive power, information media, and small hydrocarbon building blocks for construction of larger organic-chemical molecules. Primary energy in one form of another is the world's most abundant resource and, in the aggregate and at any human scale, inexhaustible. Economical means are already well established, rapidly proliferating, and improving to transform liquid, solid, or gaseous fuels (and "non-fossil" energy forms) one into another, or into electricity.

- Commercially exploitable natural gas is distributed unevenly in the earth's crust, but is everywhere in relentless competition with other fuels and energy forms over practically all of its actual and potential uses. A substantial share of existing fuel-consuming equipment, mostly in industry or used to generate electricity, has installed dual or multi-fuel capacity. Considering (1) those additional facilities that could economically be retrofitted to use another fuel or energy source, (2) facilities that are nearing the end of their economic lives and subject to replacement by alternatively powered installations, and (3) imminent investment decisions regarding new producer and consumer durables, at least half of the world's energy use is attended by active, near-term interfuel competition.

With appropriate changes in intake, burner, and exhaust hardware, natural gas is readily substitutable for liquid petroleum, coal, and other fuels in almost every stationary (i.e., non-transport) application. Common stationary uses include space heating and cooling, electrical generation, metallurgy, pulp and paper manufacture, petroleum refining, materials drying, and food processing. Natural gas in compressed form or as LNG does indeed serve as transport fuel in motor vehicles, ships, and railway locomotives, and can be adapted even for aircraft, but such mobil employments are less common. Methane and the havier hydrocarbon components of natural gas- ethane and propane-also compete with naphtha, gas oil, and synthesis gas from coal as a feedstock for making the fertilizers ammonia and urea, and "primary petrochemicals" such as ethylene, propylene, methanol, vinyl chloride and acetonitrile. These primary petrochemicals serve as building blocks for further processing into plastics, solvents, pharmaceuticals and other intermediate chemical products.

TRENDS IN GAS CONSUMPTION

Because of the greater difficulty and expense of storing or transporting fuel in gaseous form, markets have historically tended to treat natural gas as less valuable than liquid petroleum products per unit of heating value. However, at the turn of the twenty-first century, world consumption of natural gas is increasing at nearly twice the rate of increase for total primary energy for two reasons: emissions and efficiency.

Natural gas will continue to be substituted for oil and coal as primary energy source in order to reduce emissions of noxious combustion products: particulates (soot), unburned hydrocarbons, dioxins, sulfur and nitrogen oxides (sources of acid rain and snow), and toxic carbon monoxide, as well as carbon dioxide, which is believed to be the chief "greenhouse gas" responsible for global warming. Policy implemented to curtail carbon emissions based on the perceived threat could dramatically accelerate the switch to natural gas.

Natural gas also has an efficiency advantage in electricity generation. The economic and operational superiority of gas-fired combustion turbines and combined-cycle machines (and prospectively, the superiority of gas-powered fuel cells) relative to coal- and nuclear-powered steam turbines made the combination of natural gas and natural gas turbines the supply favorite of most electric utilities in the 1990s.

Arlon R. Tussing

See also: Natural Gas, Processing and Conversion of; Turbines, Gas.

BIBLIOGRAPHY

BP Amoco. (1998). *Statistical Review of World Energy*. London: Author.

Campbell, N., ed. (1995). *Fundamentals of the Natural Gas Industry*. London: The Petroleum Economist.

Energy Information Administration. (1979–2000). *Natural Gas Annual*. Washington, DC: U.S. Department of Energy.

Institute of Energy Economic. (2000). *Handbook of Energy and Economic Statistics in Japan*. Tokyo: The Energy Conservation Center.

Institute of Gas Technology. (1999). *Natural Gas in Nontechnical Language*. Tulsa, OK: PennWell Books.

Tussing, A. R., and Tippee, B (1995). *Natural Gas: Evolution, Structure, and Economics*. Tulsa, OK: PennWell Books.

Tussing, A. R., and Van Vactor, S. A. (1998). "South Korea's Thirst for Gas." *Financial Times Energy Economist*, March.

U.S. Bureau of the Census. (1975). *Historical Statistics of the United States: Colonial Times to 1970*. Washington, DC: U.S. Government Printing Office.

NATURAL GAS, PROCESSING AND CONVERSION OF

Natural gas is an important energy source consisting mainly of methane gas. It is usually found commingled with deposits of crude oil and also in stand-alone deposits where the gas has migrated to, leaving the associated petroleum in some other location. Methane is also produced by decaying vegetation (swamp gas) in some coal mines and in land fills, but these sources generally are not suitable for commercial use.

When natural gas comes out of the ground (see Figure 1), it typically consists of 75 to 95 percent methane, with small quantities of ethane, propane, and butane. It may also contain water vapor, carbon dioxide, nitrogen, oxygen, and sulfurous gases such as hydrogen sulfide. Unlike petroleum, which needs to be separated and refined into a variety of fuels and petrochemical products, the nature and general purity of natural gas makes processing far less complex. Most natural gas from the wellhead needs little processing. Nonhydrocarbons are removed from the contaminated alkanes (hydrocarbons containing only single carbon-carbon bonds) by absorption. The heavier hydrocarbons tend to liquefy at the high operating pressures needed for natural gas pipelining. If the natural gas is to be liquefied (LNG), the hydrocarbons are separated by absorption or low temperature distillation.

Before entering the pipeline, the gas is also adjusted in the field to achieve a uniform heating value of 1,000 Btu (British thermal units) per cubic foot. And for safety, because natural gas is odorless and colorless, an odorant is added to provide a distinctive and disagreeable smell that is easy to recognize.

CONVERSIONS

The clean-burning nature of natural gas has for many years made it the fuel of choice for heating and cooking. If its energy content per cubic meter were comparable to liquid fuels like such as diesel and gasoline, it would be ideal as a transportation fuel as well. However, the void is wide. Whereas gasoline and diesel deliver 110,000 to 120,000 Btu per gallon,

an equivalent volume of natural gas delivers only about 134 Btu per gallon. Thus, there is great interest in finding ways to efficiently compress or liquefy natural gas so that the same low-emission benefits found in the residential, industrial, and electricity generation markets serviced by pipelines can also be enjoyed by areas without pipeline service and by the transportation sector.

Once water vapor, sulfur, and heavy hydrocarbons are removed, natural gas can be compressed or liquefied. As a transportation fuel, the high methane content gives natural gas its high octane rating (120–130) and clean burning characteristics, resulting in the dual benefit of high engine performance and low pollution. There are no sulfur or particulate (smoke) emissions. Currently, there are two types of natural gas vehicles: exclusively natural gas and bifuel. The latter operate on natural gas or either diesel or gasoline, and the fuel can usually be changed with the flip of a switch.

Compression or conversion for greater use in the transportation market is promising for two reasons: First, natural gas is usually cheaper than liquid fuel and, second, there exist large quantities of stranded gas—remotely located natural gas sources that are not economical to use because tanker or pipeline transportation costs can be over four times as much as for crude oil. Often this gas is recompressed and injected back into the oil-producing zones to help maintain reservoir pressure and optimal crude oil flow to the wellhead. In some cases this gas is wasted by being flared, but this practice is increasingly frowned upon. The demand for cleaner-burning transportation fuels, and the advances in gas turbines that have dramatically improved the efficiency of natural gas powered electricity generation have renewed interest in developing ways to compress or liquefy this gas to lower its shipping cost. Liquefaction can mean cooling the natural gas until it condenses at −187°C (at atmospheric pressure) or converting it chemically to a suitable liquid fuel. Both of these schemes entail considerable energy costs.

Some environmentalists have also touted natural gas as a way station on the road to a hydrogen fuel (carbon dioxide-free) economy. As seen in Table 1, per unit of energy released, natural gas generates about 23 percent less carbon dioxide than gasoline and about 30 percent less than heavy fuel oil. This is helpful in reducing greenhouse emissions, but the other excellent properties of natural gas are even

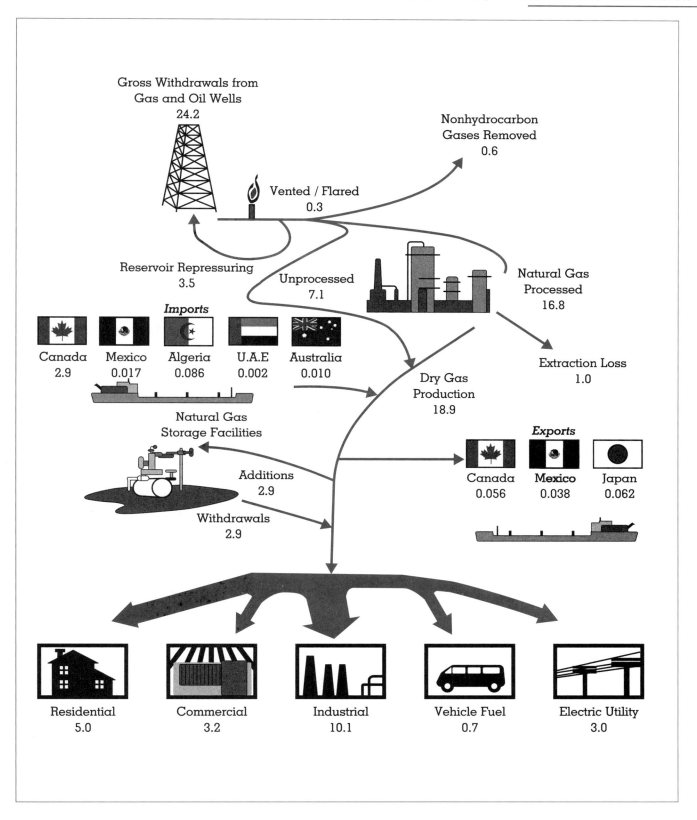

Figure 1.
Natural gas supply and disposition in the United States in trillions of cubic feet.

	Hydrogen	Natural Gas (as Methane)	LPG (as Propane)	Gasoline (as Octane)	Methanol
Molecular Weight	2	16	44	114	32
Heat of Combustion kilocalories / gram	34.16	13.30	12.06	11.46	5.42
Grams of carbon per gram of fuel	0.00	0.75	0.82	0.84	0.38
Grams CO_2 evolved per gram of fuel	0.00	2.75	3.01	3.09	1.38
Grams CO_2 evolved per kilocalorie	0.00	0.207	0.250	0.270	0.254

Table 1.
Fuel Combustion Properties

more important from the standpoint of economic efficiency. However, if the natural gas has to be converted chemically to methanol, the high-octane, clean-burning advantage is maintained, but the carbon dioxide advantage is lost.

As seen in Table 1, on a weight basis, hydrogen has the highest heat of combustion of all fuels. Hydrocarbons are less than half as energetic, the lighter, more hydrogen-rich molecules having somewhat higher heating values than the heavier, more carbonaceous fuels. However, on a volumetric basis, the heavier fuels win out, as their higher density (specific gravity) more than off-sets the difference in heat of combustion per unit weight.

COMPRESSED NATURAL GAS

Table 2 compares the heating value of 20-gal tanks of natural gas at different pressures. Pipelines deliver natural gas at a relatively low pressure of 60 pounds per square inch (psi). A 20-gallon vehicular tank filled at this pressure would provide only about 11,000 Btu versus about 2.4 million Btu for a diesel fuel tank of the same size. Thus, the gas must be compressed and stored in a welded bottle-like tank at 3,000 to 3,600 psi to provide any reasonable range. A 20-gallon 3,000 psi tank will provide about 20 percent of the diesel heating value, so to begin to approach the range of gasoline or diesel vehicles, some of the newer vehicles offer advanced tanks capable of holding gas compressed to 5,000 psi.

Although the heating value of a 20-gal 3,000 psi tank of natural gas is only 20 percent of the same vol-

ume of diesel fuel, the mileage range comparison will likely be better than the volumetric ratios because the natural gas engines can achieve higher performance.

There are two ways of refueling compressed natural gas (CNG): time-fill or fast-fill. For a time-fill compressor, it is necessary to develop a pressure only slightly greater than the vehicle storage pressure—the gas flows from greater pressure to less pressure. Time-fill compressor stations can require a couple of hours to refuel, which is a major inconvenience for most motorists. However, it is the best choice for many fleets that can be refueled overnight with one fill-post for each vehicle to be refueled.

Public stations are of the fast-fill type, typically to satisfy the desire of customers to refuel quickly. The biggest problem for fast-fill operations is the lack of space and high cost. Large capacity high pressure storage is needed to fast-fill vehicles because the larger and greater the pressure, the faster the fill-up.

All CNG ground storage, vehicle storage and refueling equipment must meet stringent industry and government safety standards for both normal operation and crashes. The controls include monitors of critical pressures and temperatures from the pipeline to the storage tank, and the flow of gas from ground storage to vehicle storage. Once the compressor reaches discharge pressure (fill-up complete), a control then automatically turns off the compressor. CNG is then delivered to the engines as low pressure vapor (ounces to 300 psi). Since natural gas cylinders are much thicker and stronger than gasoline or diesel tanks, the safety record of natural gas vehicles is equal or better than conventionally fueled vehicles.

Almost all the major car, bus, and truck manufacturers have developed compressed natural gas engines and vehicles. These manufacturers have been able to offer better performance (due to higher octane) and far lower emissions of nitrogen oxides, carbon monoxide, particulate matter, and carbon dioxide to the atmosphere. In 1998, Honda introduced the cleanest internal combustion engine vehicle ever commercially produced: the natural gas Civic GX with emissions at one-tenth the state of California's Ultra Low Emission Vehicle standard. Primarily due to the high octane of natural gas, Honda achieved these results without sacrificing performance.

Despite the environmental benefits of natural gas vehicles, large numbers of compressed natural gas stations need to be built or compressed natural gas will never be more than a niche fuel servicing large fleets of buses, cabs, and delivery trucks that can be fueled at a central location. Nonroad short-range vehicles such as forklifts, backhoes, street sweepers, and airport ground support equipment are also ideally suited for natural gas use.

The high mileage, local routes, and regular returns to a central refueling point make local transit buses an ideal application for CNG vehicles. Another option for local driving and commuting is home refueling. For around $2,500, a system can be plumbed directly into a home's natural gas supply for refueling in the garage or driveway. Though the benefit of home refueling is a tremendous benefit for the majority of drivers, the CNG disadvantages of shorter range, slower refueling, and few refueling stations far outweigh its advantages. American car buyers want the mobility to go anywhere at any time. Whereas most gasoline cars and trucks can go 250 miles or more on a tank of fuel, natural gas vehicles typically can go about half that distance. Consumers are reluctant to switch to a CNG vehicle that requires refueling twice as often with so few refueling options.

The industry has developed higher compression tanks to expand the range, and more fast-fill stations are becoming available, yet the prospects of the majority of service stations adding compressed natural gas refueling anytime in the near future are bleak. The oil companies, which control most of the service stations and over 60 percent of America's natural gas reserves, are not eager to make the massive infrastructure investment to cannibalize the billions of dollars they have tied up in refineries, pipelines, and service sta-

Pressure	Million BTUs
Ambient (14.7 psi)	0.00267
60 psi	0.01009
3000 psi	0.546
5000 psi	0.910
diesel oil	2.4

Table 2.
Energy in a 20-Gallon Tank of Natural Gas (in million BTUs)

tions designed to deliver gasoline and diesel fuel. However, though not willing to lead, they are certain to follow. If consumers purchase the vehicles, the oil companies will naturally invest in the infrastructure. Supply will follow demand. Moreover, if it turns out that fuel cell vehicles that run on hydrogen are the future, compressed natural gas vehicles could be the logical bridge between petroleum and hydrogen because hydrogen is also a compressed gas. The same infrastructure can deliver both.

More stringent clean-air regulations and enforcement was the primary reason for natural gas vehicle growth in the 1990s, and is highly likely to play a part in future growth. According to the Natural Gas Vehicle Coalition, more than 20 percent of all new orders for transit buses were for natural gas-fueled vehicles in 1998. These new vehicles require a significant capital investment, yet often justify this higher initial investment by reducing air pollutants, lower maintenance costs, and offering fuel savings of about 30 percent compared to gasoline and diesel.

LIQUEFIED NATURAL GAS

Natural gas is liquefied by cooling it to its liquid state (approximately –260°F), at either the wellhead, central facility, or on-site. Since liquefaction reduces its volume by a factor of about 600, it becomes economical to ship by tanker.

To remain a liquid at a reasonably low pressure, liquefied natural gas (LNG) must be maintained at below at least –117°F. Insulated storage tanks alone cannot maintain these very cold temperatures. LNG is stored at its boiling point to take advantage of "autorefrigeration." Just as the temperature of water does not rise above its boiling point (212°F) with increased heat (it is cooled by evaporation), LNG is kept near its boiling point if kept at a constant pressure. As long as LNG vapor boil off is allowed to leave the storage tank, the temperature will remain constant. The pressure and temperature in the tank will rise when the vapor is not drawn off.

During the liquefaction process, usually much of the oxygen, carbon dioxide, sulfur compounds and water are removed so that liquefied natural gas (LNG) is nearly 100 percent methane. LNG takes up one-six-hundredth the volume of natural gas, with a density less than half that of water.

Although LNG is as safe or safer than gasoline and diesel fuel, and emits less harmful emissions when burned, it has three major drawbacks: It is expensive to produce, requires a larger and heavier fuel tank (about 1.5 gallons of LNG per gallon of gasoline and 1.7 gallons per gallon of diesel to achieve the same range), and is not the best fuel for vehicles used rarely or intermittently because of vapor boil-off over time. The best applications for LNG are heavy-duty vehicles (trucks and buses) that are heavily used, and vehicles that can store larger fuel tanks, or are not inconvenienced by need for more frequent refueling.

LNG tanks use low pressure (less than 5 psi), yet need double-wall construction so that insulation between the walls keeps the LNG cool. For the large tanks, a cylindrical design with a domed roof is used, but for smaller quantities (70,000 gallons or less), storage is in horizontal or vertical vacuum-jacketed tanks at pressures any where from less than 5 psi to over 250 psi.

Because much of the world lacks the natural gas resources and transportation pipelines of the United States, remote natural gas must be liquefied and transported by ship. Gas-rich countries want to capture stranded gas by liquefying and shipping it to gas-poor regions as LNG. The gas-poor countries enter into contracts so that a long-term supply is available to warrant the investment in the electricity-generating infrastructure. The overall investment is enormous, not only in the liquefaction plant, but in the refrigerated tankers and the regasification plant at the delivery site.

Sometimes LNG is the only option in regions and countries where political issues constrain pipeline development.

Shipments of LNG began in the early 1960s and continued to expand so that by 1995 there were over 65 ships transporting almost 68 million tons of LNG, with each equipped with a specialized refrigeration system to keep LNG cool enough to stay in its liquefied state. Transportation was estimated to reach 107 million tons by 2,000, with the major exporters being Malaysia, Abu Dhabi, and Qatar, and the major importers being Japan, Korea, and Europe. Since OPEC production quotas limit petroleum production, which by extension limits revenue, LNG has also developed into an attractive export commodity for OPEC countries since current production agreements do not extend to natural gas. Several major projects to expand LNG trade went to contract in the late 1990s. New LNG processing facilities have been built or are under construction in Oman, Qatar, Nigeria, and Trinidad, with Japan, South Korea, Taiwan, and Thailand being the largest customers committing to purchase output from the new facilities.

There has never been an LNG tanker accident; yet, with growing shipments, there is growing concern about a tanker accident since an explosion and fire occurring in a crowded harbor could be disastrous. However, while an LNG tanker may contain the energy equivalent of several Hiroshima atomic bombs, the damage would hardly be comparable because the LNG energy cannot be released quickly. For a detonation, LNG must first be mixed with air in the correct flammability ratio, and near a tank rupture the mix would probably be too rich to explode. Further, the liberation rate of LNG as a gas would be determined by the heat transfer rate to the boiling liquid. Thus, any accident would likely be a large deflagration, not a horrific explosion.

CHEMICAL CONVERSIONS

As an alternate to LNG, natural gas can be chemically converted to methanol, chemical feedstocks (such as ethylene), gasoline, or diesel fuel. Most processes start with the conversion of methane to synthesis gas, a mixture of carbon monoxide and hydrogen. This can be done partial oxidation, an exothermic reaction:

$$CH_4 + 0.5\,O_2 \rightarrow CO + 2\,H_2$$

or by steam reforming, an endothermic reaction:

$$CH_4 + H_2O \rightarrow CO + 3\,H_2$$

Shortly after World War I, Badische Anilin patented the catalytic conversion of synthesis gas to methanol, and Fischer and Tropsch (F-T) announced a rival process in which an iron catalyst converted synthesis gas into a mixture of oxygenated hydrocarbons. Later,

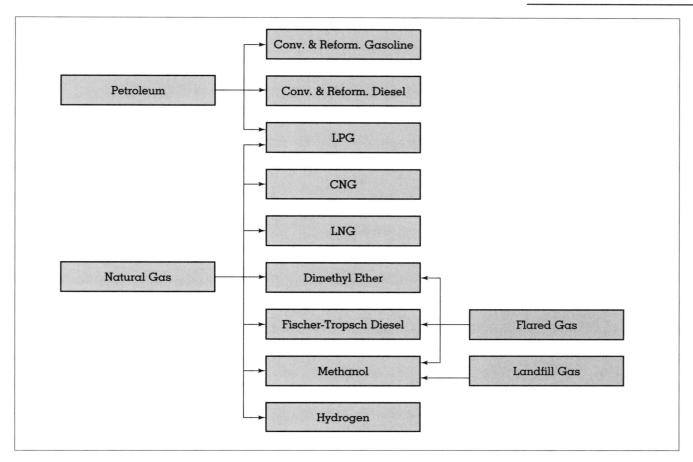

Figure 2.
Alternative fuels derived from natural gas.

improved F-T catalysts produced a liquid resembling a very paraffinic (waxy) crude oil.

Most of the early commercialization used coal as the synthesis gas feed-stock. Use of stranded natural gas feed to a F-T refinery was finally innovated by Shell-Mitsubishi in a small, 10,000 bbl per day refinery in Sarawak, Malaysia in 1993. F-T liquids are refined in the usual manner to produce gasoline and diesel fuel/kerosene of very high quality.

Similarly, a natural gas-to-methanol-to-gasoline process was finally developed by Mobil as a result of the 1973 oil crisis. Methanol is transformed into gasoline range hydrocarbons using proprietary Mobil synthetic zeolite catalysts. This process was commercialized at a small New Zealand refinery in the 1980s. Methanol is also the feedstock of choice in the production of the oxygenated additives needed to produce today's cleaner-burning gasoline blends.

While several other processes have been developed to convert natural gas to liquid fuels (GTL), these technologies are generally uneconomical composed to using the crude oil feedstocks. About one-third of the energy in natural gas is lost in converting it into liquid fuels, so highly distressed gas prices or government subsidies are needed for GTL to be competitive.

Fischer-Tropsch Diesel

One of the most promising GTL fuels is Fischer-Tropsch diesel. Fischer-Tropsch diesel offers lower emissions without compromising fuel efficiency, creating distribution problems (new infrastructure), or requiring a greater investment in equipment for fuel storage and refueling. Depending on the price premium for GTL, the Fischer-Tropsch diesel can be used as a fuel or blended with traditional diesel that is not compliant with Federal or California standards. Used alone, GTL diesel reduces hydrocarbons by over 20 percent, carbon monoxide over 35 percent, nitrous oxide about 5 percent, and particulates around 30 percent. Although GTL diesel is more expensive than traditional diesel fuel, it seems to be a promising short-term solution for the fuel industry to meet the California heavy-duty diesel engine standard that goes into effect in 2004. ARCO, Exxon, Chevron, and Texaco are all in the process of developing pilot plants.

The major advantage of Fischer Tropsch diesel, compared to natural gas, lies in its liquid nature. It does not need special infrastructure and compression like CNG does, and unlike LNG, once converted, it is a liquid fuel that can be treated like any other liquid fuel. However, because the GTL process is more complex than traditional refining, it requires low-cost natural gas priced at less than $1 per million BTUs to remain cost-competitive. Without stranded gas, sources sold at a large discount compared to crude oil, GTL diesel would be considerably more expensive than traditionally refined diesel fuel.

Hydrogen

Many transportation experts feel that hydrogen is the fuel of the future. It has a high energy content and many environmental advantages. However, before hydrogen becomes an economical alternative fuel, ways to produce hydrogen on a large scale will need to be developed. Conversion from natural gas is widely viewed as a promising option for two reasons: Hydrogen-rich natural gas can be converted more cleanly than coal, and natural gas requires less energy input than a conversion from water.

NATURAL GAS AS A FUEL ADDITIVE

Many GTL-derived fuels are being considered for blending with gasoline and diesel to achieve emission reductions of particulate matter (PM), carbon monoxide (CO), nitrogen compounds (NOx) and nonmethane hydrocarbons (NMHC). The most promising fuels converted from natural gas are methanol and ethers such as dimethyl ether (DME) and methyl-t-butyl ether (MTBE).

Like LNG, the natural gas-to-methanol fuel market relies on stranded gas as feedstock. The advantages of conversion to methanol is that it requires far less specialized infrastructure than LNG since the final product is a 110-octane liquid that ships in regular tanks, and does not need regasification. And because of a plentiful natural gas supply in the United States, methanol derived from natural gas as a fuel additive is a promising future market. Methanol has neither the environmental problems of methyl-t-butyl ether (MTBE), nor the evaporating qualities of ethanol.

John Zumerchik
Herman Bieber

BIBLIOGRAPHY

American Petroleum Institute. (1987). *Liquid Fuels from Natural Gas.* Petrol Information, API 34-5250. Washington, DC: Author.

Chen, N. Y.; Garwood, W. E.; and Dwyer, F. G. (1989). *Shape Selective Catalysis in Industrial Applications.* New York: Marcel Dekker.

Dry, M. E. (1990). Fischer-Tropsch Synthesis over Iron Catalysts, Spring 1990 A.I.Ch.E. Meeting, Orlando, Florida. March 18–22, 1990.

Sofranko, J. A. (1988). *Gas to Gasoline: The Arco GTG Process.* Bicentennial Catalyst Meeting, Sydney, Australia.

Tussing, A. R., and Tippee, B. (1995). *The Natural Gas Industry: Evolution, Structure, and Economics,* 2nd ed. Tulsa, OK: PennWell Publishing.

U.S. Energy Information Administration. (1998). *Annual Energy Outlook.* Washington DC: United States Department of Energy.

U.S. Energy Information Administration. (1998). *International Energy Outlook 1998, With Projections Through 2020.* Washington, DC: Author.

Yerchak S., and Wong, S. S. (1992). *Mobil Methanol Conversion Technology Process.* IGT Asian Natural Gas Seminar, Singapore, pp. 593–618.

NATURAL GAS, TRANSPORTATION, DISTRIBUTION, AND STORAGE OF

Transportation of natural gas across state lines from production to consuming areas is a function of interstate pipeline companies. The modern U.S. natural gas industry also includes natural gas exploration and production companies, intrastate pipelines, local distribution companies (LDCs), end-users and, the most recent addition to the industry, marketers.

HISTORICAL BACKGROUND

Transportation of natural gas through pipelines began in the United States in the early part of the nineteenth century. One of the first known uses occurred in 1821 with the building of a system of metallic lead pipes to transport natural gas from a nearby shallow well to commercial establishments in Fredonia, New York. Gas lights—burning gas made from coal—illuminated the streets of Baltimore beginning in 1816.

By 1900, natural gas had been discovered in seventeen states, mostly as a byproduct of oil exploration. Lacking a viable long-distance transportation system to move it to market, however, natural gas, until the 1920s, was used mainly for lighting city streets or was vented into the air when found with oil.

Today, the natural gas industry—responsible for locating, producing, transporting and distributing gas to end-users—is a major contributor to the U.S. economy, employing more than 170,000 workers nationwide. And the fuel, in addition to being used in a variety of commercial and industrial applications—including one of the fastest growing markets, the generation of electricity—is the primary source of energy for space heating in more than fifty million American homes and provides some 25 percent of all the energy consumed in the United States.

This is largely due to the discovery and development of major natural gas fields in the U.S. Southwest, mid-continent, on- and offshore areas of the Gulf of Mexico and Canada—and the development of safe and efficient interstate natural gas transmission pipelines to transport natural gas to markets across the country. Some 77 percent of the natural gas consumers use is produced domestically.

NATURAL GAS PRODUCTION

The first step in the movement of natural gas from production areas to consumers (see Figure 1) begins at the wellhead. Most natural gas wells require pumping to bring the natural gas to the surface. From there, small-diameter gathering lines, connected to clusters or a series of natural gas wells, carry the gas to pipelines or to facilities where it is processed to remove valuable hydrocarbon liquids such as ethane, propane, butane, iso-butane or natural gasoline. Prior to entering the pipeline transmission system, and during actual transmission, the natural gas stream also may undergo processing to remove water vapor, solids and other elements that may interfere with efficient transportation of the fuel.

NATURAL GAS TRANSPORTATION

Natural gas pipelines are generally defined as intrastate, those transporting gas to markets within state boundaries, or interstate, those crossing state lines. Natural gas from Canada, a growing source of natural gas consumed in the United States, also enters the U.S. through interconnections with interstate pipelines.

Today, a network of more than 300,000 miles of interstate natural gas pipelines serves markets across the U.S. Construction of this network began in the 1920s, but large-scale expansion was limited by the technology of the day, the Great Depression and, finally, World War II.

Following War II, advancing technology in steel-making and the manufacture of pipe, along with increased energy demand, led to rapid growth of the nation's natural gas transportation network. Today, large-diameter, high-quality steel pipe, normally ranging from twenty to forty-two inches in diameter, is constructed at pipe mills, generally in twenty and forty-foot "joints" (sections). Transported by truck or train to construction sites, the joints are welded together and buried in trenches to form the long-distance pipelines necessary to transport natural gas from production basins to distribution and other customers across the nation.

Compressor stations, generally located fifty to sixty miles apart along interstate pipelines, house reciprocating or gas turbine engines that drive compressors to propel the gas through the pipeline. The compressor engines, usually powered by a small portion of the natural gas flowing through the pipelines, compress the gas to an average of 700 to 950 pounds per square inch. In the very early years of the natural gas transmission industry, sections of pipe made of iron or steel were bolted together, often resulting in the loss of major amounts of gas in transit. The advent of modern technology, welded pipe seams, and pipeline protection and monitoring programs have reduced the amount of lost and unaccounted for gas during interstate transportation to negligble quantities, generally a fraction of one percent of all the gas transported.

835

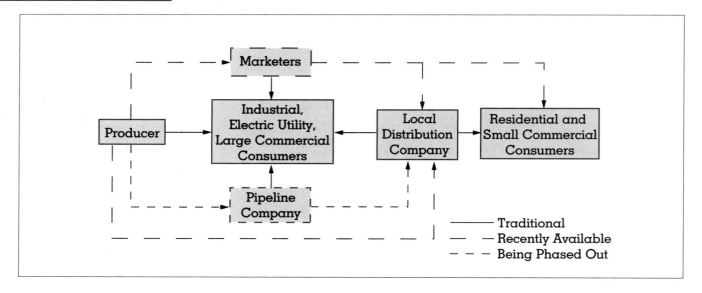

Figure 1.
Transaction paths for natural gas purchases.

BALANCING SUPPLY AND DEMAND

Because of their role as the link between producing and market areas, interstate natural gas pipelines play a crucial role in balancing supply and demand. To do so, the interstate pipelines continually monitor pipeline performance. These companies rely heavily on computers to gather, analyze and retain information on the performance of their pipelines, and to control the flow of gas in remote sections of the line.

Interstate pipelines also use computer simulation programs to calculate pipeline capacity, pressures, horsepower, fuel and other physical characteristics and properties of their systems. Using this information and incorporating variables such as ambient temperatures, facility outages, and changes in market patterns, transmission companies can run daily studies to determine how much natural gas their systems will deliver under expected operating conditions.

Another tool the interstates use to maintain their pipelines is a device known as an intelligent "pig." Propelled through the pipeline with the gas stream, these devices, taking thousands of measurements with electronic sensors that can be analyzed later by computers, can inspect pipeline interior walls for corrosion or other defects and remove accumulated debris from a section of pipeline. Pipelines also use state-of-the-art coating and cathodic protection to battle corrosion.

Advancing technology has had an impact on the interstate pipeline industry in other ways. Interstate pipeline shippers now have access to electronic bulletin boards (EBBs) which are developed and maintained by the pipelines and can be accessed by users to purchase transportation services, check on billing, determine status of pipeline utilization, arrange for the storage of gas transported through a pipeline and gain other information that allows the pipeline's capacity to be used efficiently.

NATURAL GAS STORAGE

Underground storage fields play a key role in smoothing the peaks and valleys of natural gas supply and demand. Natural gas transported by pipeline, for instance, can be injected into storage fields near market areas during periods of low demand in the summer and withdrawn during periods of high demand in the winter, thus reducing the need to build additional pipeline capacity to meet high demand periods and allowing pipelines designed to operate at high load factors to run at efficient levels all year. Storage can also be used to prevent flaring of gas and other waste when production exceeds market demand.

Recent uses for natural gas storage include: utilizing storage to provide transportation balancing services, to take advantage of variations in prices, to maintain wellhead production and to meet new sales commitments by marketers.

A crane moves soil next to sections of 42-inch pipes to be used for a gas pipeline. (Corbis-Bettmann)

The three basic types of underground storage facilities for natural gas are depleted oil and gas formations, salt caverns, and aquifer formations. Depleted oil and gas formations, the most commonly used storage sites, are generally classified as seasonal supply reservoirs, where gas is injected in summer and withdrawn in winter. Salt caverns, the large majority of which are located in the Gulf Coast region, are well-suited to storing gas and provide very high withdrawal and injection rates compared with their working gas capacity (the amount of natural gas inventory that can be withdrawn to meet customer needs). Aquifer formations, water-only reservoirs conditioned to hold natural gas, are generally classified as seasonal supply reservoirs and are generally more expensive to develop and maintain than depleted oil or gas reservoirs.

NATURAL GAS DISTRIBUTION

The final link in the physical transportation of natural gas is distribution, the delivery of natural gas by local distribution customers (LDCs) to local consumers. LDCs deliver gas to consumers from storage or from wellheads accessed by pipelines.

Metering gas as it enters their systems, LDCs lower it to pressures suitable for their customers and add the familiar odor that identifies natural gas, otherwise odorless, to the consumer.

For most of the twentieth century, LDCs, either investor owned or municipally owned, have had exclusive rights or franchises to distribute gas in specified geographic areas. Regardless of ownership, LDCs are regulated, either by state public utility/service commissions or local government agencies, to assure adequate gas supply, dependable service and reasonable prices for consumers.

Today, distribution of natural gas involves some 1,200 LDCs that deliver natural gas to more than 160 million American consumers in all fifty states.

RESTRUCTURING OF THE NATURAL GAS INDUSTRY

Government, recognizing the need to provide dependable service at reasonable prices, has been involved with the gas industry since the mid-1800s. Local regulation gave way to state regulation as the issues surrounding exploration, production, transportation, and distribution became increasingly complex. As improved technology allowed pipeline's to carry large volumes of natural gas over great distances, the federal government increasingly became involved in regulation of this segment of the industry.

The first major federal regulatory event affecting natural gas pipelines was the Natural Gas Act of 1938, which gave the Federal Power Commission jurisdiction to regulate three areas: (1) pipeline sales of gas purchased from producers and resold to local distribution companies in interstate commerce; (2) transportation in interstate commerce; and (3) the facilities used for such sales and such transportation.

The second major federal regulatory event was a Supreme Court decision in 1954 which resulted in the imposition of wellhead price regulation of gas sold in interstate commerce.

Until the 1970s, however, the structure of the industry remained relatively constant. Gas producers explored for and produced natural gas and sold it to pipeline companies. The pipeline companies transported the gas across the country and, for the most part, sold it to local distribution companies, who then sold the gas to end users. The price producers sought for their gas from interstate pipelines was regulated by the federal government, as was the price interstate pipelines imposed on gas sold to LDCs. State or local government agencies in turn regulated the price charged by LDCs to end users.

In 1978, the structure of the industry began a period of dramatic change. Following a severe winter and an energy crisis that closed schools, businesses, and industries—and threw more than a million people out of work, Congress passed the National Energy Act (NGPA) and the Public Utility Regulatory Act. This legislation, assuming the nation was running out of natural gas, said that gas could no longer be used for "low-priority" uses and

that national policy should promote conservation and shift demand to alternative fuels. The NGPA, to provide producers with incentives for exploration and development of new supplies of gas, called for phased and gradual deregulation of the price of newly discovered gas.

In the early to mid-1980s, retail gas prices began to rise as the more expensive "new" gas constituted an increasing percentage of the pipelines' average cost of gas. This drove consumer prices above the level that would exist in a competitive market, and demand for natural gas was subsequently reduced as large industrial customers switched to other fuels. Also reducing demand were the Fuel Use Act, which prohibited the use of natural gas as a boiler fuel, increased conservation by residential and commercial customers, warmer-than-normal winters, and an economic recession.

In 1985, the Federal Energy Regulatory Commission (FERC) began a series of regulatory actions designed to improve the competitiveness of the natural gas market and give the customers of interstate pipeline companies more service options and thus allow ultimate consumers to benefit from deregulation of wellhead prices.

By 1993, the structure of the interstate pipeline industry had undergone dramatic changes. The interstates, which once had acted as both transporters and merchants of natural gas, "bundling" the sale and transmission of gas into one service, were required to separate these services and give pipeline customers the opportunity to contract for only those services they needed.

With "unbundling," interstates no longer own the gas transported on their pipeline systems. They simply transport it for third parties who range from LDCs to natural gas marketers. (Emerging in the 1980s in response to the deregulation of prices, marketers serve as intermediaries between gas buyers and all other segments of the industry.) And buyers of natural gas can now negotiate price and contract terms with different gas suppliers while contracting separately with the interstates for transportation, storage and other services.

Restructuring has also affected LDCs. Many LDCs, for example, given increasing flexibility by regulators, now allow large volume customers to select the most cost-effective and efficient mix of supply, transportation, storage, and other services.

PRICING OF NATURAL GAS

Several factors determine the ultimate price consumers pay for natural gas. They include

Pipeline Transportation Pricing: The Federal Energy Regulatory Commission (FERC) regulates the rates, terms and conditions of service provided by interstate natural gas pipelines. The FERC, with the issuance of its "restructuring" rule, Order 636, in April 1992, set a policy goal to separate the transportation of gas from the sale of gas and to put all natural gas suppliers and gas purchasers on equal footing. Before restructuring, as described earlier, the interstates bought gas from producers and sold it to LDCs and other end users. The rates they charged were "bundled" and included charges for a variety of services including supply, storage, and transportation. Now the interstates must provide all transportation services on an equal basis, whether the gas being transported is purchased from the pipeline, a pipeline marketer or another gas supplier.

Wellhead Pricing: With passage of the Natural Gas Wellhead Decontrol Act of 1989, all remaining wellhead price controls on natural gas were lifted. As a result, federally mandated natural gas wellhead prices no longer exist. With this, the competitive marketplace has become the most important factor affecting natural gas wellhead prices, along with weather, demand, pricing of competitive fuels and competition for supplies.

The Spot Market: This segment of the natural gas market consists of transactions of a short duration that can be rearranged quickly if gas prices change. The spot market developed in the early and mid-1980s, when there was a substantial supply of gas and long-term supply contracts fell into disfavor. Spot market prices fluctuate depending on availability of supply. Most LDCs do not rely exclusively on spot-market transactions, depending on time of year. In the event of unexpected demand surges, however, LDCs and other customers may be forced to purchase spot market gas.

Futures Prices: Natural gas futures contracts began trading on April 3, 1990, on the New York Mercantile Exchange (NYMEX). Sabine Pipe Line Company's Henry Hub near Erath, Louisiana, is the delivery site for these contracts. Henry Hub consists of an interconnection of seven interstate pipelines, two interstates and one gathering system. These interconnections allow natural gas to move from major production areas to major consumption areas. The transportation of natural gas to and from Henry Hub is contracted separately by the seller and buyer, respectively.

Price Indexing: Some gas prices, by contractual agreement, reflect changes in the spot, futures or other markets, such as heating oil. Price indexing is often used to reflect that the parties to the contract believe the price to be paid is related in some way to the particular index used.

Residential Pricing: Residential natural gas rates are for natural gas service and for the gas commodity. The price of the gas commodity comprises about one-third of the total price a residential customer pays, on average. The remainder of the bill includes amounts for transmission and distribution of gas, system maintenance, safety and inspection programs, customer service, metering, billing and other costs.

THE FUTURE OF THE NATURAL GAS INDUSTRY

Additional Regulatory Change

The natural gas industry, undergoing fundamental changes in recent years, continues to face additional change. Current issues include discussion about lifting the federally-set caps on pipeline pricing for certain services and letting the market determine pricing on these services, as well as a number of new pipelines proposed to import significant new volumes of Canadian gas into the United States.

On the state level, several pilot programs now allow customers to choose gas suppliers other than their LDCs.

Supply Outlook

Debate continues over just how much natural gas remains in North America. Part of this debate centers on the definition of gas reserves—the amount of gas in a given area that is recoverable and gas resource, the total amount of gas in the ground. Gas production to date, as might be expected, has most commonly been from easy-to-produce conventional sources, such as the mid-continent or the highly permeable sands of the Gulf Coast.

A considerable amount of gas is located in areas of lower permeability or where the geology is poorly understood, making it more difficult to earn a profit using current technology and production methods. New technology may open these sources to future development and production.

Additionally, new supplies of natural gas from Canada and Alaska could significantly increase available supply, along with imports of Mexican gas and the potential for liquefied natural gas from overseas.

Demand Outlook

Demand for natural gas, in all markets—residential, commercial, and industrial—is projected to grow into the foreseeable future, particularly in the electric power generation market and the industrial sector. Total natural gas use in the United States is projected to grow from 20.1 quadrillion British thermal units in 1992 to 26.1 by 2010, an average growth rate of 1.6 percent per year.

Robert V. McCormick

See also: Natural Gas, Consumption of; Natural Gas, Processing and Conversion of; Oil and Gas, Drilling for; Oil and Gas, Exploration for; Oil and Gas, Production of.

BIBLIOGRAPHY

Capozza, F. C. *Energy Futures Trading in the '90s,* 2nd ed. Red Bank, NJ: Waterman Associates, Inc.

Commoner, B. (1983). "A Nearly Perfect Fuel." *New Yorker Magazine,* May 2.

Energy Policy Act Transportation Study. (October 1995). Washington, DC: Energy Information Administration.

Natural Gas: The Fuel, The Future, The Regulation. Detroit, MI: ANR Pipeline Company.

Wright, J. C., ed. (1994). *The Universal Almanac, 1994.* Kansas City, MO: Andrews and McMeel.

NATURAL GAS REGULATION AND RATES

See: Regulation and Rates for Electricity

NERNST, WALTHER HERMANN (1864–1941)

Walther Hermann Nernst was born in Briesen on June 25, 1864, into a prominent Prussian family. The line of his ancestors can be traced back to the era of the first Prussian king in the seventeenth century. His most famous ancestor is Hermann Nernst, who received an award for conveying the news to the king's family about the Germans victory against Napoleon in the Battle of Waterloo. Nernst lost his mother, Ottilie Nerger, very early. His father, Gustav Nernst, was a district judge. Nernst was a very impulsive person and known for his impatience. He died of a heart attack in his country house in Zibelle on November 18, 1941.

When Nernst attended secondary school, his chemistry teacher awakened in him a passion for chemistry. This left a deep impression on the young pupil, who started early to do his own chemistry experiments at home, in the cellar. Nernst often changed universities during his study of physics, as he wanted to take part in selected lectures by the most influential comtemporary scientists. First he visited the physics lectures of Heinrich Friedrich Weber in Zurich in 1883 and then he went to Berlin for a semester to take thermodynamics lessons from Hermann von Helmholtz, the most famous German physicist at that time. After another semester in Zurich, Nernst studied for the fourth semester in Graz, where Ludwig Boltzmann founded the statistical interpretation of thermodynamics. Nerst, finished his education there and started his Ph.D. program in collaboration with Albert von Ettinghausen, one of Boltzmann's students. Nerst, finished his Ph.D. requirements in Wurzburg in 1887 under the supervision of Friedrich Whilhelm Georg Kohlrausch, who entered a completely new physical field with his research on the ionic conductance of liquids. Nernst adopted Boltzmann's atomistic view (or mechanical picture) of natural phenomena and combined it with the energetic view that he became familiar with during his collaboration with Wilhelm Ostwald in

Walther Hermann Nernst in his laboratory. (Corbis-Bettmann)

Leipzig. It was due to Svante Arrhenius that Nernst became highly interested in the new topic of ionic theory that was established by Ostwald. Fascinated by this topic, Nernst habilitated in Leipzig in 1889 on the electromotive activity of ions. He incorporated into this work Arrhenius' theory of dissociation and Jacobus Henricus van't Hoff's osmotic theory of solutions. Two years later Nernst became an assistant professor in physical chemistry at the University of Gottingen. In 1905 he was offered an appointment as the successor of Hans Landolt and started his work at the University of Berlin. In the same year he discovered the third law of thermodynamics during the time he was lecturing on the thermodynamical treatment of chemical processes. Nernst worked in many fields of physical chemistry—for example the osmotic theory of galvanic elements, the law of distribution, equilibrium at high temperatures and high pressures, specific heats at high and low temperatures, calorimetry, IR radiation, chain reactions, electrochemistry, and photochemistry.

Of fundamental importance to Nernst's discovery of the Third Law of Thermodynamics is Nicholas Leonard Sadi Carnot's work in the middle of the nineteenth century establishing thermodynamics by combining the laws of motion with the concept of action. At the same time Helmholtz found that the principle of energy conservation is also valid for thermal energy and established together with Robert Julius Mayer and James Joule the first law of thermodynamics. A short time later Rudolf Clausius and Carnot introduced the concept of entropy and established the second law of thermodynamics. The chemical affinity—the extent to which a compound is reactive with a given reagent—was known of since attempts of alchemists to produce gold in the Middle Ages. The importance of the concept of chemical affinity was first recognized by van't Hoff. He showed that it was possible to measure the chemical affinity via free energy. Helmholtz and Josiah Gibbs independently established an exact mathematical relationship between the total energy (the energy content of a system) and the free energy (the capacity of a system to perform work). According to Marcellin Pierre Berthelot the total and the free energy should be the same during the electrochemical processes in the galvanic cells Nernst was investigating. Nernst's experiments showed that this statement was not exactly valid for moderate temperatures, and the deviation became larger as the temperature was increased. Therefore he assumed that both energies should be equal at zero value of temperature. Finally, he got the idea that the difference between these two energies asymptotically approaches zero as the absolute temperature approaches zero. This is the first formulation of the new law. Nernst was led to this law, which he and other scientists spent a long time investigating, while he was searching for mathematical criteria for the description of chemical equilibrium and the spontaneity of chemical reactions. That why some chemical reactions are spontaneous while others are not was already known a century before Nernst's own discovery. Since 1900 it was known that a thermodynamical calculation of the chemical equilibrium could not be performed using only the thermal data of the current thermodynamics because the integrated form of the Gibbs-Helmholtz Equation—which allowed calculation of the maximum yield of work during a thermodynamical process—contained an undetermined integration constant.

Nernst's contribution to the solution of this problem was to give this undetermined constant a new interpretation. This new interpretation, however, required an additional assumption for the description of the free energy exchange in the vicinity of the absolute zero value of temperature. He recognized that the work function for the state transition of a system could not be calculated by means of energy differences. Rather, derivatives of the energy and the free energy with respect to temperature were necessary. Furthermore, Nernst assumed that the entropy approaches a constant value provided the absolute temperature approaches zero.

After Nernst's publication of his *Heat Theorem* all resources of the Institute for Physical Chemistry at the University of Berlin were dedicated to its experiments. No experimental result was found to contradict Nernst's heat theorem. Its general validity, which was established by experiments in many subfields of physical chemistry, justified its inclusion among the laws of thermodynamics. This law was regarded by Arnold Sommerfeld as "the most ingenious extension the classical thermodynamics ever experienced in the twentieth century." Expressing Nernst's heat theorem in terms of entropy, this law means that the entropy difference between different states of a system tends to zero as the temperature reaches absolute zero. It can also be expressed as the law of unattainability of absolute zero temperature. Franz Eugen Simon provided Nernst's heat theorem a more elegant theoretical foundation and reformulated it by stating that the entropy differences disappear between all those states of a system that are in internal thermodynamic equilibrium. Very soon after he set forth the third law of thermodynamics, Nernst showed the importance of his discovery by calculating the chemical equilibrium using thermal data only. Nernst was awarded the Nobel Prize in chemistry in 1920 as recognition of his work in thermochemistry.

Tyno Abdul-Redah

See also: Carnot, Nicholas Leonard Sadi; Clausius, Rudolf Julius Emmanuel; Gibbs, Josiah Willard; Heat Transfer; Helmoltz, Herman von; Joule, James Prescott; Ostwald, Wilhelm; Thermodynamics.

BIBLIOGRAPHY

Gillispie, C. C., ed. (1978). *Dictionary of Scientific Biography,* Vol. 15, p. 432. New York: Scribner.

Einstein, A. (1942). "The Work and Personality of Walther Nernst." *Scientific Monthly,* February, p. 195.

Mendelsohn, K. (1973). *The World of Walther Nernst: The Rise and Fall of German Science, 1864–1941,* Pittsburgh: University of Pittsburgh Press.

Nernst, W. H. (1893). *Theoretische Chemie vom Standpunkte der Avogadroschen Regel und der Thermodynamik.* Stuttgart, Germany: F. Enke.

Nernst, W. H. (1906). "Ueber die Berechnung chemischer Gleichgewichte aus thermischen Messungen." *Nachrichten von der Gesellschaft der Wissenschaften zu Göttingen,* pp. 1–40.

Nernst, W. H. (1913). *Experimental and Theoretical Applications of Thermodynamics to Chemistry.* New Haven, CT: Yale University Press.

Partington, J. R. (1953). "The Nernst Memorial Lecture." *Journal of the Chemical Society,* p. 2853.

NEWCOMEN, THOMAS (1663–1729)

Fable or fame? Thomas Newcomen, like many inventors who preceded him in the steam revolution, has been clearly overshadowed in historical circles by the far more famous Scotsman, James Watt, who remains—incorrectly to some—known as the inventor of the steam engine. Watts engines arrived more than fifty years after Newcomen's successful mechanical works, and were considered improved versions of the Englishman's concepts. But this was precisely the basis of many inventors' successes, building upon their predecessors' efforts in the normal course of technological advancement. What is irrefutable is that both men, as well as others, can lay claim as pioneering "fathers" of the Industrial Revolution.

Newcomen came from the ranks of practical tradesmen, unlike many industrial inventors who tended to be noblemen, philosophers and royal protégés. The Newcomen family had had an impressive lineage and had held its manor from the twelfth century until misfortune dropped them into obscurity four centuries later. Yet, a work ethic was instilled by Newcomen's grandfather, who became a merchant venturer (owning several ships), a freeholder of Dartmouth, treasurer for his town, and a staunch Parliamentarian. Elias Newcomen, the father of Thomas, was also a freeholder and a merchant of Dartmouth, trading to distant areas with a ship that he had inherited.

It is reasonably certain that Thomas Newcomen was born in late January or early February 1663 in the family house in Dartmouth in Devon, England. He was schooled at home by the well-known nonconformist scholar, John Flavell, who played a key role in Newcomen's educational thinking. Although throughout his life Newcomen was proud of his common status as an ironmonger, some rivals attempted to credit his success to no more than good luck and chance. Many of his contemporaries doubted that he could be the sole author of so momentous an invention.

But going solo neither accurately summed up the style of Newcomen's inventiveness, nor his ability to learn from others and to work with them. In his teens, he served an apprenticeship before joining his partner, John Calley, in their shop at Dartmouth. By the time Newcomen began solving technical tin mining challenges at the beginning of the eighteenth century, he was already well established as an iron-monger, dealing in annual quantities as high as twenty-five tons per year. He apparently became acquainted with Thomas Savery, or at least his works, observing firsthand the pumping problems in the mines. He may also have read about earlier research on atmospheric pressure done by a Huguenot, Denis Papin, late in the seventeenth century.

Although Newcomen's predecessors may have had superior backgrounds and academic intelligence, his doubters possibly overlooked the practical advantages he and his partner brought to the research table. Through astute knowledge of every variety of metalwork in iron, brass, copper, tin, and lead, he and Calley possessed the right combination of commercial ability and highly practical and versatile craftsmanship. Historical data suggests that these categories may have been deficient among his forerunners in the matter of applying the principles of steam technology.

Yet, Newcomen's success came with great difficulty and frustration, taking ten years before achieving initial success at a South Staffordshire colliery (mine works) at Dudley Castle in 1712. He perfected a variation of Savery's work by creating a dramatic advance of an internal water injection system to cool the heated steam, creating a far more rapid and effective condensation and ensuing vacuum. There is some historical evidence that the arrival of this particular breakthrough came by fortunate accident, whereby an unintended leak may have led him to design his internal water-cooling system.

In another significant change, he connected one end of a large overhead rocking beam, or "great lever," to this piston. Realizing the risks of high pressure steam, he thus wisely relied on two basic forces to drive his engine—atmospheric pressure to plunge the piston down, and simple gravity to lift it back it up, through the weight of the counterbalancing rocker beam.

By modern standards, it was barely an engine at all, but it met the most important criteria of its time by dramatically meeting the challenge of pumping water from the mines, doing the equivalent work of twenty horses and forty men. The use of his excellent pumping engine spread throughout Europe, with over one hundred engines constructed within the life of the patent, which expired four years after his death on August 5, 1729, at the age of sixty-six. Well over fifteen hundred units were eventually built during the eighteenth century, despite their significant cost (over one thousand pounds), high consumption of fuel, and the requirement of an operating license.

The usefulness of Newcomen engines was both startling and unprecedented, with immediate impact benefiting the mining industry for the next century or more. The record for longevity for a Newcomen-type engine was probably set at the South Liberty colliery of the Ashton Vale Iron Company. It was built around 1750, and was still pumping from a depth of seven hundred feet it was until dismantled in 1900. Another remarkable example was at the Cannel Mine at Bardsley, where a Newcomen engine worked from 1760 to 1830, after which it rested in a derelict state for a full century. In 1930 it was acquired by Henry Ford and shipped to his museum in Dearborn, Michigan where it was restored and re-built.

While not nearly as sophisticated as later engines, the quantity of Newcomen engines built vastly exceeded the hundreds of Watt engines, with the more primitive types often used as substitutes for the newer, far more expensive versions.

Dennis R. Diehl

BIBLIOGRAPHY

Rolt, L. T. C., and Allen, J. S. (1997). *The Steam Engine of Thomas Newcomen.* Ashbourne, UK: Landmark Publishing.

NEWTON, ISAAC
(1642–1727)

Isaac Newton was born at Woolsthorpe, near Grantham in Lincolnshire. He entered Cambridge University as a student in 1661. Although much is known of Newton's professional life, little is known about Newton's student life. He studied under Isaac Barrow, the Lucasian professor of mathematics. He was forced by the plague of 1665–1666 to return to Lincolnshire where, during the miraculous year of 1666, he forged the foundations for his considerable achievements in mathematics, optics, and dynamics.

After returning to Cambridge in 1667, Newton was elected Fellow of Trinity College. Two years later he succeeded Barrow as Lucasian Professor. In 1696 Newton moved to London. He served first as Warden and from 1699 to his death in 1727 as Master of the Royal Mint. He was elected a Fellow of the Royal Society of London in 1671, and the President of this society in 1703, a position he retained for the rest of his life. He also served two undistinguished terms as a Member of Parliament for the University of Cambridge (1689–1690 and 1701–1702). He was knighted in Cambridge in 1705.

In the period following the War of the Spanish Succession of 1714, Newton enjoyed a reputation as the most important natural philosopher of his day. His scientific output during the last twenty years of his life was restricted to the revision of his major works and in defending himself against his many critics. In his private life, he was modest, generous, and given to simple tastes; however was buried with great fanfare in Westminster Abbey.

In his scientific life, his behavior and demeanor were far different. Newton was hostile to any criticism and he was capable of ruthless behavior. The priority dispute with the German mathematician Gottfried Wilhelm Leibniz over the invention of the calculus is a case in point. In his capacity as President of the Royal Society, Newton appointed a committee of loyal Newtonians to investigate the matter. He then authored the committee's report, the infamous *Commercium Epistolicum* and submitted it as though it were an utterly impartial report in his own favor.

Newton has been regarded as the very exemplar of the modern scientist, employing a style of reasoning from "the phenomena of motions" that other scien-

Isaac Newton. (Archive Photos, Inc.)

tific disciplines have attempted to emulate with limited success. However, his interests ranged far from mathematical physics. From 1669 to 1696 he pursued alchemy and chemistry with equal or greater passion. His knowledge of the Greek classics was profound. In his *The Chronology of Ancient Kingdoms Amended* (1728), he attempted to reconcile Jewish and pagan dates and to fix them absolutely from an astronomical argument about the earliest constellation figures devised by the Greeks. In *Observations upon the Prophecies of Daniel and the Apocalypse of St John* (1733), Newton also wrote on Judeo-Christian prophecy, sustained by the conviction that its decipherment was essential to the understanding of God.

As to his scientific output, Newton contributed to all extant branches of mathematics, but is renowned for his solutions to the problems in analytical geometry of drawing tangents to curves (differentiation) and defining areas bounded by curves (integration). Newton discovered that these problems were inverse to each other, and designed general methods of resolving problems of curvature—his "method of fluxions" (from the Latin meaning "flow") and "inverse method of fluxions." Fluxions were expressed algebraically, but

Newton made extensive use of analogous geometrical arguments. Late in life, Newton expressed a preference for the geometric style of the Classical Greeks, which he regarded as more rigorous.

During 1669, Newton worked out the details of his discovery of the decomposition of a ray of white light into rays of different colors by means of a prism. Newton's explanation of the theory of the rainbow followed from this discovery. These discoveries formed the subject matter of a series of lectures that he delivered as Lucasian Professor in the years 1669, 1670 and 1671. The results were communicated to the Royal Society in February 1672 and subsequently published in the *Philosophical Transactions*. Reactions to Newton's ideas were negative. The inability of the French physicist Edmé Mariotte to replicate Newton's prism experiments in 1681 entrenched the rejection of Newton's optical theory for a generation. Newton delayed the publication of his *Opticks* (published in 1704, revised in 1706), which was largely composed in 1692, until his critics were dead. The manuscript of his original lectures was printed in 1729 under the title *Lectiones Opticae*.

Through a curious set of circumstances, Newton failed to solve the problem of chromatic aberration, and so, he abandoned the attempt to construct a refracting telescope which should be achromatic, and instead designed a reflecting telescope, probably on the model of a small one that he had constructed in 1668. The form he used is known by his name today.

Newton's dynamics is presented in *Philosophiae Naturalis Principia Mathematica* (published in Latin in 1687, revised in 1713 and 1726, and translated into English in 1729). According to the famous story, on seeing an apple fall in his orchard sometime during 1665 or 1666, Newton conceived that the same force governed the motion of both the Moon and an apple. Newton calculated the force needed to hold the Moon in its orbit, as compared with the force pulling an object to the ground. He also calculated the centripetal force needed to hold a stone in a sling, and the relation between the length of a pendulum and the time of its swing.

These thoughts were put away until correspondence with Robert Hooke (1679–1680) redirected Newton to the problem of the path of a body subjected to a centrally directed force that varies as the inverse square of the distance. Newton calculated this path to be an ellipse, and so informed the astronomer Edmond Halley in August 1684. Halley's

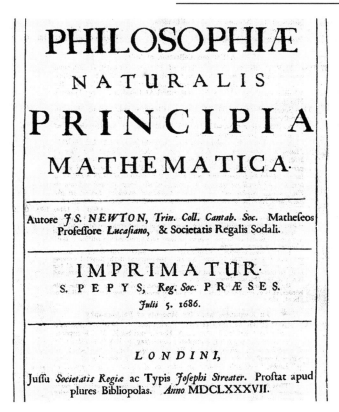

Facsimile of Newton's *Philosophiae Naturalis Principia Mathematica*. (Corbis Corporation)

urgings prompted Newton to return to the problem raised by Hooke.

The result of Newton's labor was *Principia*, which introduced a radically new style of reasoning in science, one that treated scientific problems as though they were exercises in pure mathematics. Newton's work enabled scientists to study forces of different sorts without any inhibiting considerations as to whether such forces can actually (or do actually) exist in nature.

Book I of *Principia* is organized around fundamental laws and axioms, erecting on this foundation the mathematics of orbital motion around centers of force. Gravitation is identified as the fundamental force controlling the motions of the celestial bodies. Newton did not advance an explanation for gravity itself, much to the disappointment of his critics. Its manner of action and its cause were matters that Newton worked hard to distance from the domain of mathematical physics. Further to this, the first half of this book is not entirely original, since it rests upon the achievements of René Descartes, Galileo, and Christian Huygens. Newton's genius shines, however, in the second half of the book where the problem of two bodies is reduced to an equivalent problem of one body attracted to a fixed center. It is here that we find the masterpiece of this book—Newton's treatment of the three-body problem. It is true that his formulation of the general laws of mechanics ruled out a solution to this problem, but Newton's approach inspired future generations of physicists.

Book II investigates the dynamical conditions of fluid motion. Book III displays the law of gravitation at work in the solar system. It is demonstrated from the revolutions of the six known planets, including Earth, and their satellites, though Newton could never quite perfect the difficult theory of the Moon's motion. It is also demonstrated from the motions of comets. The gravitational forces of the heavenly bodies are used to calculate their relative masses. The tidal ebb and flow and the precession of the equinoxes is explained in terms of the forces exerted by the Sun and Moon. These demonstrations are carried out with precise calculations.

Newton's work in dynamics was accepted at once in Britain, though it was strongly resisted on the Continent until 1740 or so when the last of his critics conceded that, while gravitation itself was inconceivable, Newton's arguments were incontestable. During the eighteenth century, Newton's dynamics was extended and perfected by others, but its basic character was unchanged. It was only in the late nineteenth century that Newton's dynamics began to reveal its limitations. Where Newton regarded the law of gravitation as a significant dynamical result, Albert Einstein argued that this law is a geometric result, on a par with Galileo's law of inertia. The movement of a planet around the sun, for example, is to be seen, not in terms of the action of gravity, but in terms of the curvature of space.

Brian S. Baigrie

BIBLIOGRAPHY

Cohen, I. B. (1980). *The Newtonian Revolution.* Cambridge: Cambridge University Press.

Dobbs, B. J. T. (1975). *The Foundations of Newton's Alchemy or "The Hunting of the Greene Lyon."* Cambridge: Cambridge University Press.

Halley, E. (1687). "Review of Newton's Principia." *Philosophical Transactions* 186:291–297.

Herivel, J. W. (1965). *The Background to Newton's Principia. A Study of Newton's Dynamical Researches in the Years 1664–84.* Oxford: Clarendon Press.

Koyré, A. (1965). *Newtonian Studies.* Chicago: University of Chicago Press.

Maclaurin, C. (1748). *An Account of Sir Isaac Newton's Philosophical Discoveries, in Four Books.* London: A. Millar and J. Nourse.

Newton, I. (1934). *Sir Isaac Newton's Mathematical Principles of Natural Philosophy and His System of the World,* tr. A. Motte, rev. F. Cajori. Berkeley: University of California Press.

Newton, I. (1952). *Opticks, or A Treatise of the Reflections, Refractions, Inflections & Colours of Light.* New York: Dover Publications.

Newton, I. (1958). *Isaac Newton's Papers and Letters on Natural Philosophy and Related Documents,* eds. I. B. Cohen and R. E. Schofield. Cambridge: Harvard University Press.

Newton, I. (1959–1977). *The Correspondence of Isaac Newton,* eds. H. W. Turnbull, J. F. Scott, A. R. Hall. Cambridge: Cambridge University Press.

Newton, I. (1962). *The Unpublished Scientific Papers of Isaac Newton: A Selection from the Portsmouth Collection in the University Library, Cambridge,* ed. A. R. Hall and M. B. Hall. Cambridge: Cambridge University Press.

Newton, I. (1967). *The Mathematical Papers of Isaac Newton,* ed. D. T. Whiteside. Cambridge: Cambridge University Press.

Newton, I. (1975). *Isaac Newton's 'Theory of the Moon's Motion' (1702).* London: Dawson.

Pemberton, H. (1728). *A View of Sir Isaac Newton's Philosophy.* London: S. Palmer.

Stukeley, W. (1936). *Memoirs of Sir Isaac Newton's Life,* ed. A. H. White. London: Taylor and Francis.

Westfall, R. S. (1971). *Force in Newton's Physics: The Science of Dynamics in the Seventeenth Century.* London: Macdonald.

NITROGEN CYCLE

All life is dependent on nitrogen; nitrogen is a critical component of amino acids, protein, DNA, and RNA. While the atmosphere is 78 percent nitrogen, animals and plants cannot convert nitrogen directly from the atmosphere into a utilizable form. The nitrogen cycle is the process by which nitrogen from the atmosphere is converted to biological nitrogen compounds in plants and animals and then is returned to the atmosphere. The major steps in the nitrogen cycle are nitrogen fixation, nitrification, nitrogen assimilation, denitrification, and ammonification.

Nitrogen fixation is the conversion of nitrogen (N_2) in the atmosphere to nitrates (NO_3^-) and ammonia (NH_3). In nature, this is done primarily by microorganisms (bacteria and algae). The most important bacteria in this process are symbiotic bacteria that live on nodules on the roots of leguminous plants such as soybeans, peas, and alfalfa. The electricity in lightning also can cause a small amount of nitrogen to be fixed by directly combining nitrogen and oxygen. Nitrification is the process wherein ammonia is further converted to nitrates. Plants use nitrates to make proteins by nitrogen assimilation. Animals consume these plants for subsequent protein production. The cycle is completed by bacteria converting decaying plants, animals, and excretory products back into either atmospheric nitrogen (denitrification) or ammonia compounds (ammonification).

The natural supply of nitrogen available to plants from the nitrogen cycle is limited. To meet the growing demands for agriculture crops, nitrogen is added to soil in the form of fertilizers. An estimated one-third of the human population is fed as a result of the use of synthetic fertilizers.

To "fix" nitrogen in synthetic fertilizers requires hydrogen that is obtained by dissociation from water or from fossil-fuel hydrocarbons. Either way, this process is energy-intrusive. The situation is ominous to some scientists, since the supply of a fundamental world energy source—food—is dependent on nonrenewable energy sources that are being depleted. Although the addition of fertilizers is the largest human impact on the global nitrogen cycle, the burning of fossil fuels also accounts for an increase in the amount of fixed nitrogen in the atmosphere. Anthropomorphic activity is responsible for as much nitrogen fixation as that from natural sources.

There are environmental consequences to human influence on the nitrogen cycle. While the addition of nitrogen to the soil is required to feed the world population, it is estimated that 50 percent of the nitrogen added as fertilizers is washed off the soil and ends up in agricultural runoff. This can cause contamination of drinking water in agricultural areas. Excess fertilizers in waterways also can cause eutrophication, where excess nitrogen leads to events such as algae bloom, which robs other plants and animals of oxygen. Agricultural runoff has altered areas ranging from San Francisco Bay to the Baltic Sea to the Great Barrier Reef. In addition, excess nitrates in the soil can acidify the soil, causing loss of other soil nutrients.

Excess fertilizer and combustion processes also can increase nitrous oxide (N_2O) and nitrogen oxides (NO_x) in the atmosphere. Nitrous oxide is a powerful greenhouse gas, and nitrogen oxides lead to smog and acid rain. The production of fertilizers requires a great deal of energy. The use of fossil fuels to supply the thermal requirements for fertilizer production further increases emission of nitrogen compounds to the atmosphere.

Deborah L. Mowery

See also: Acid Rain; Agriculture; Biological Energy Use, Cellular Processes of; Biological Energy Use, Ecosystem Functioning of; Green Energy.

BIBLIOGRAPHY

Galloway, J. N.; Schlesinger, W. H.; Levy, Hiram, II; Michaels, A.; and Schnoor, J. L. (1995). "Nitrogen Fixation: Anthropogenic Enhancement-Environmental Response." *Global Biogeochemical Cycles* 9:235–252.

Kinzig, A. P., and Socolow, R. H. (1994). "Human Impacts on the Nitrogen Cycle." *Physics Today* 47(11):24–31.

Smil, V. (1997). "Global Population and the Nitrogen Cycle." *Scientific American* 277(1):76–81.

Vitousek, P. M.; Aber, J.; Howarth, R. W.; Likens, G. E.; Matson, P. A.; Schindler, D. W.; Schlesinger, W. H.; and Tilman, G. D. (1997). "Human Alteration of the Global Nitrogen Cycle: Causes and Consequences." *Ecological Applications* 7:737–750.

NORTH AMERICAN ELECTRIC RELIABILITY COUNCIL

See: Government Agencies

NUCLEAR ENERGY, BASIC PROCESSES OF

Nuclear energy, sometimes referred to as atomic energy, originates in the atomic nucleus, which is the extremely dense core at the heart of an atom. A large

Nuclear Power Plant reactor building and cooling tower (left). (Field Mark Publications)

amount of energy can be released by nuclei in two different ways: fission, in which a very large nucleus is induced to break apart into two smaller ones, and fusion, in which two very small nuclei combine.

A great deal of energy is released when a large nucleus undergoes fission, but for most nuclei the fission process is not easy to initiate. There are very few nuclei—uranium-235 and plutonium-239, in particular—that are relatively easy to fission. At present, commercial nuclear reactors use fission of uranium-235 as the energy source. A uranium-235 nucleus can be induced to undergo fission through interactions with a slowly-moving neutron. The uranium nucleus absorbs the neutron, thus becoming a uranium-236 nucleus, which then breaks apart into two smaller nuclei called fission fragments. In addition, neutrons are also released, and these neutrons can then induce fission of other uranium nuclei in a chain reaction. Neutron-absorbing materials are located in the reactor core to ensure that the chain

reaction proceeds at the proper pace. If too few neutrons are available for further fissions, then the reaction slows down or stops. If too many neutrons are available, then the reactor core can overheat.

The energy from nuclear fission is released mainly as kinetic energy of the new, smaller nuclei and neutrons that are produced. This kinetic energy is essentially heat, which is used to boil water to generate steam that turns turbines to drive electrical generators. In a nuclear power plant, the electrical generation area is essentially the same as in a plant that burns fossil fuels to boil the water.

Nuclear fission is also involved in nuclear weapons. To create a bomb, the concentration of the isotope uranium-235 must be increased to at least 85 percent from its natural concentration of only 0.7 percent. This increase of concentration is difficult and expensive. In a typical nuclear reactor the uranium-235 concentration in the fuel is only 3 to 4 percent, and hence a nuclear reactor cannot explode like a bomb. In a nuclear bomb the chain reaction is uncontrolled, and a large number of uranium nuclei undergo fission in a very short period of time, producing a nuclear explosion.

Generation of electricity by nuclear fission power reactors has many advantages. A primary advantage is that very little uranium fuel is required—only about two-hundred tons annually for a typical reactor. To generate the same amount of electricity, a coal-fired plant needs 3 million tons or 15,000 times as much coal. Another advantage is that nuclear power produces no air pollution such as nitrogen oxides and sulphur dioxide, both of which contribute to acid rain, and no greenhouse gases such as carbon dioxide and methane that contribute to global warming. Nuclear power also produces no ozone and no particulate matter in the air, and hence, in terms of air pollutants, a nuclear plant is much "cleaner" than a fossil-fuel electrical plant.

Nuclear reactors, however, do generate highly radioactive waste. This waste, which consists primarily of the fission fragments and their radioactive-decay products, must be stored for many years before its radioactivity decays to a reasonable level, and the safe long-term storage of this waste is a matter of great concern and debate. Fortunately, the volume of waste that is created is only about 20 cubic meters annually from a reactor, compared with 200,000 cubic meters of waste ash from a coal-fired plant. When nuclear weapons were tested in the atmosphere, the radioactive products from the nuclear explosions were released into the air and fell to Earth as radioactive fallout.

Another concern about nuclear power plants is their decommissioning. Nuclear reactors have useful lifetimes of about thirty years, after which they need to be shut down permanently. The remaining fuel, which is very radioactive, will have to be removed and stored for many decades. The entire reactor building structure has also been made somewhat radioactive, as well as all the pipes, valves, etc., in the plant. There are a number of options for dealing with this radioactive material, ranging from immediate dismantling of the plant (with some of the work probably done by robots) to using the normal reactor containment structure as a long-term storage facility. In the United States, the Nuclear Regulatory Commission has required that plant owners set aside sufficient funds for dismantling.

The greatest concern that most members of the general public have about nuclear energy is the possibility of a catastrophic accident such as occurred in 1986 at Chernobyl in the Ukraine. If reactors have been properly designed and the staff trained with safety in mind, then the chance of a major accident is very slight. The Chernobyl disaster was caused by a combination of poor reactor design and insufficient training of the reactor operators, who violated many of the operating procedures related to safety.

Nuclear *fusion* is the energy-producing process that occurs in the sun and other stars. Small nuclei such as those in hydrogen and helium fuse together to produce larger nuclei, releasing energy. Nuclear fusion is not yet commercially viable as an energy source, since extremely high temperatures are required to initiate fusion, and containment of the fusing nuclei at these temperatures is difficult. However, if nuclear fusion ever becomes a usable energy source, a typical fusion power reactor would require less than a ton of fuel annually.

It is often stated that nuclear fusion will produce no radioactive hazard, but this is not correct. The most likely fuels for a fusion reactor would be deuterium and radioactive tritium, which are isotopes of hydrogen. Tritium is a gas, and in the event of a leak it could easily be released into the surrounding environment. The fusion of deuterium and tritium produces neutrons, which would also make the reactor building itself somewhat radioactive. However, the radioactivity produced in a fusion reactor would be much shorter-lived than that from a fission reactor. Although the thermonuclear weapons (that use nuclear fusion), first developed in the 1950s provided the impetus for tremendous worldwide research into nuclear fusion, the science and technology required to control a fusion reaction and develop a commercial fusion reactor are probably still decades away.

Ernie McFarland

BIBLIOGRAPHY

Ahearne, J. F. (1993). "The Future of Nuclear Power." *American Scientist* 81:24–35.

Atomic Energy of Canada Limited (AECL). (1999). *Nuclear Sector Focus: A Summary of Energy, Electricity and Nuclear Data.* Mississauga, Ontario, Canada: AECL

Atomic Energy of Canada Limited (AECL). (2000). <http://www.aecl.ca/english/energy/energy_f.html>.

Cordey, J. G.; Goldston, R. J.; and Parker, R. R. (1992). "Progress Toward a Tokamak Fusion Reactor." *Physics Today* 45(1):22–30.

McFarland, E. L.; Hunt, J. L.; and Campbell, J. L. (1997). *Energy, Physics, and the Environment*, 2nd ed. University of Guelph, Ontario: Department of Physics.

NUCLEAR ENERGY, HISTORICAL EVOLUTION OF THE USE OF

The history of nuclear energy is a story of technical prowess, global politics, unfulfilled visions, and cultural anxiety. The technology's evolution in the second half of the twentieth century progressed through several stages: theoretical development by physicists; military application as atomic weapons in World War II; commercialization by the electrical industry in several industrialized nations; proliferation (for military and non military uses) among less developed nations; crises spawned by power plant accidents, cost overruns, and public protests; and retrenchment and slowdown in the last few decades of the twentieth century. By far the most potent form of energy to be harnessed by humankind, nuclear power has not become the dominant form of energy because of the great economic costs and social risks associated with its use.

MILITARY ORIGINS

The concept of "atoms" dates back to the ancient Greeks, who speculated that the material world was comprised of tiny elemental particles, and for centuries thereafter alchemists attempted to unlock the secrets of the elements. But modern atomic science did not emerge until the turn of the twentieth century. In 1896 Henri Becquerel of France discovered radioactivity, and Albert Einstein calculated the mass-energy relationship ($E = mc^2$) in 1905. By the 1930s, scientists in several countries were making progress toward understanding nuclear reactions, including Ernest Rutherford and James Chadwick in Great Britain; Enrico Fermi in Italy; Niels Bohr in Denmark; and Ernest O. Lawrence in the United States. The key breakthrough came in December 1938, when German physicists Otto Hahn and Fritz Strassmann achieved the first controlled atomic fission, splitting atoms of uranium into lighter elements by bombarding them with neutrons, and releasing enormous amounts of energy in the process.

The news spread quickly, and became charged with implications as Hitler's Nazis began their march through Europe. Two days before the outbreak of World War II in Europe in 1939, Bohr and John

The first page of a letter dated August 2, 1939, from Albert Einstein to President Franklin Delano Roosevelt. In the letter Einstein advises Roosevelt of the possibilities of nuclear research. (Corbis-Bettmann)

Wheeler of Princeton published an academic paper on fission. Several leading physists fled Germany and Stalin's Soviet Union for the United States, including Hungarian refugee Leo Szilard. Fearful that the Nazis might build a powerful atomic bomb, Szilard and fellow Hungarian émigré Eugene Wigner convinced Einstein (then at Princeton University) to write President Roosevelt to warn of the possibilities of atomic weaponry and to suggest U.S. action.

Einstein's letter (dated August 2, 1939) had little impact, however. What stirred the U.S. government into action were reports out of Great Britain in the early 1940s: one from German refugee physists Rudolf Peierls and Otto Frisch in 1940, which discussed the possibility of making a "super-bomb" from a uranium fission chain reaction; and a study from the top-secret British "MAUD committee" in 1941, which deemed a uranium bomb "practicable and likely to lead to decisive results in the war," and which urged the United States to make development of an atomic bomb its "highest priority."

At the time, the scarcity of fissionable material—whether natural uranium or man-made plutonium—seemed the greatest barrier to atomic bomb production. Now determined to win the atomic bomb race, President Franklin Roosevelt approached Colonel (soon-to-be Brigidier General) Leslie Groves of the U.S. Army Corps of Engineers to head what was code-named the "Manhattan Engineer District" (later popularly known as the Manhattan Project), America's atomic bomb project. It was a massive, sprawling effort that ultimately encompassed several leading university laboratories, three giant manufacturing sites, tens of thousands of constructions workers, the world's best scientific talent, several major corporations, and some $2.2 billion of federal funds.

The project's two key scientific advisers were Vannevar Bush, an electrical engineer at head of the Office of Scientific Research and Development, and a former dean of the Massachusetts Institute of Technology; and James Conant, a chemist, chair of the National Defense Research Committee, and president of Harvard University. Groves also recruited Nobel laureate Arthur Compton to the project, who in turn recruited Wigner and Szilard as well as Nobel laureates Fermi and James Franck to his burgeoning research laboratory at the University of Chicago (code-named the "Metallurgical Laboratory"). Meanwhile, the army brought in Boston-based Stone & Webster as principal engineering contractor, and the giant Du Pont chemical firm, which had no experience in plutonium production but took on the work for costs plus $1. After Fermi—working in a racquetball court under the stands at the University of Chicago's Stagg Field—achieved the first self-sustaining nuclear reaction in December 1942, the Manhattan Project settled on water as a coolant and forged ahead with construction plans.

At Hanford, Washington, along the Columbia River, the project built three production piles (reactors) and four separation plants for separating plutonium from other elements. At Oak Ridge, Tennessee (near Knoxville), it built uranium-235 facilities utilizing three processes—thermal diffusion, gaseous diffusion, and electromagnetic separation; the latter, under the direction of Lawrence, proved to be the most successful. And at Los Alamos, New Mexico (near Santa Fe), the Manhattan Project in 1945 began building a facility for making both plutonium and uranium bombs under the direction of the brilliant Harvard- and Gottingen-educated physicist J. Robert Oppenheimer. Like the Chicago-based scientists before him, Oppenheimer and his researchers often clashed with Groves and the project engineers, who preferred to compartmentalize and control information about the project rather than exchange it freely among the scientists. At Los Alamos, Oppenheimer's approach prevailed.

Confident that the uranium bomb would detonate with a gun-type neutron device but less sure of the plutonium bomb's "implosion" detonation device (invented by physicist Seth Neddermeyer), the scientists tested a plutonium bomb—dubbed "Fat Man" in honor of Winston Churchill—at Alamogordo, New Mexico, on July 16, 1945. The awesome power of "Project Trinity" immediately inspired doubts and fears about a nuclear future, even among several Manhattan Project scientists. Szilard, Franck, and others urged the U.S. military to demonstrate the atomic bomb to the Japanese in an uninhabited area; but the momentum of the project, and political pressure for a rapid end to the war, were too great. On August 6, the U.S. B-29 bomber Enola Gay dropped a uranium atomic bomb ("Little Boy") that detonated over Hiroshima, Japan, at 8:15 A.M., killing an estimated 75,000 to 100,000 people instantly, and another 200,000 from radiation over the next five years. Three days later, the United States destroyed Nagasaki with a plutonium bomb, and the Japanese surrendered shortly thereafter. By building the largest government-business-university collaboration in its history, the United States had harnessed atomic energy and brought World War II to a rapid end.

Over the next several years, the atomic bomb helped usher in a new kind of geo political conflict: the Cold War. Rather than clashing with conventional weapons, the two postwar superpowers—the democratic United States and the Communist Soviet Union—relied increasingly on their growing arsenals of nuclear weapons to protect their spheres of influence and to deter encroachment with the threat of mutual destruction. Tensions in the early Cold War were heightened when the United States refused to share atomic bomb technology with its erstwhile ally the Soviet Union. When the USSR successfully tested its own atomic bomb in 1949, the "arms race" between the superpowers to achieve nuclear superiority was on. In October 1952 the United States tested its first hydrogen bomb, at Eniwetok Atoll in the Pacific. By fusing light forms of hydrogen to create helium—the reaction responsible for the Sun's energy—the "H-

Two atomic bombs named "Fat Man" and "Little Boy." The latter was used to destroy Hiroshima, Japan, during World War II. (Library of Congress)

bomb" exploded with 1,000 times the power (the equivalent of some 10 million tons of TNT) of its fission predecessor. The USSR had the H-bomb by 1953. The U.S. then began developing new, solid-fuel intercontinental ballistic missles (ICBMs) to deliver bomb warheads from the U.S. western plains deep into the Soviet Union.

The post–World War II generation was the first to live under the shadow of total and instantaneous annihilation, a reality that—coming on the heels of the holocaust and the massive military horrors of World War II—gave rise to a new nihilism among some philosophers and social thinkers. Psychologists began to probe the possible mental health consequences of life in the atomic age, although many concluded that nuclear apocalyse was too vast and horrible to comprehend. And, not surprisingly, the atomic bomb began to find its way into everyday life and popular culture, sometimes in bizarre and even lighthearted ways.

Communities began to refashion their civil defense procedures to accommodate the bomb. In the United States, this meant designating bomb shelters—usually in the basements of schools and other public buildings, but also involving construction of individual shelters beneath the yards of single-family homes in the nation's burgeoning postwar suburbs. Stocked with canned foods and other provisions, these underground chambers typically were small and spartan, although some realtors seized the opportunity to offer luxury models with modern conveniences. While never ubiquitous, individual bomb shelters nevertheless became normalized, as reflected in a 1959 *Life* magazine story about a newlywed couple who spent two weeks of "unbroken togetherness" honeymooning in their bomb shelter.

The atomic bomb also became a key subject in 1950s American film, particularly in the science fiction genre. While some films spun out scenarios about how the nuclear powers might accidentally bring on Armageddon—as in *Fail-Safe* and Stanley Kubrick's classic tragicomedy *Dr. Strangelove*—most focused on the insidious and little-understood effects of radiation on human and animal life. However implausible the premises of *The H-Man, Attack of the Crab Monsters, The Incredible Shrinking Man, Attack of the Fifty-Foot Woman,* and their ilk, such fantasies resonated with widespread anxieties about genetic damage from atomic fallout.

Although government officials attempted to educate the public and military personnel about atomic civil defense, in retrospect these efforts seem hopelessly naive if not intentionally misleading. Army training films advised soldiers to keep their mouths closed while observing atomic test blasts in order to not inhale radioactive flying dirt. Civil defense films used a friendly animated turtle to teach schoolchildren to "duck and cover" during a nuclear attack—that is, duck under their desks and cover their heads. Such measures, of course, would have offered pitiful protection to those in the blast zone.

Opinion polls showed that American anxiety about the atomic bomb ebbed and flowed in response to geopolitical events. Concerns ran high in the late 1940s in the wake of the atomic bombings of Japan

(many wondered whether the weapon would be used in all military conflicts), and as the Soviet Union undertook a crash program in rocketry and atomic-bomb development. These fears cooled temporarily in the early 1950s, particularly after the death of Soviet dictator Joseph Stalin in 1953 raised hopes for U.S.-Soviet rapprochement. But ICBM development and a wave of H-bomb testing in the Pacific in 1954 stirred up renewed public fears about fallout, especially after milk in heartland cities such as Chicago was found to contain elevated levels of isotopes. The national heart rate spiked during the tense days of the Cuban Missile Crisis in 1962, then slowed after the United States and the Soviet Union signed the Partial Test-Ban Treaty in 1963. Whereas 64 percent of Americans identified the threat of nuclear war as their leading concern in 1959, only 16 percent put the same concern first in 1964.

"ATOMS FOR PEACE": THE ORIGINS OF THE NUCLEAR POWER INDUSTRY

The Atomic Energy Act of 1946 represented the interests of American scientists who wished to see nuclear energy developed for nonmilitary purposes. It called for the establishment of a five-member civilian Atomic Energy Commission (AEC), which could deliver weapons to the military only on presidential order. But the military tensions of the early Cold War delayed civilian nuclear power development until 1948, at which time 80 percent of the AEC's budget went to military ends. In 1951, U.S. civilian nuclear power development consisted of only a small experimental government (liquid metal) reactor in Idaho.

Through the efforts of Captain Hyman G. Rickover, a naval engineering officer, the U.S. Navy made rapid strides in nuclear ship and submarine development. Garnering support from Edward Teller and other key figures outside the navy, Rickover brought in Westinghouse, General Electric, and the Electric Boat Company to construct the *Nautilus,* the world's first nuclear submarine. First tested in 1955, the *Nautilus* ran faster and ten times farther without surfacing than conventional submarines. Nuclear-powered aircraft carriers and other surface ships followed, and by the 1960s U.S. nuclear submarines were equipped with solid-fuel nuclear missiles. In contrast, the U.S. merchant marine did not emphasize use of nuclear power. Its demonstration vessel, the *Savannah,* operated by the U.S. Marine

Corporation beginning in 1959, used a pressurized-water reactor instead of a boiling-water reactor, and required a heavy government operating subsidy.

In 1953 the AEC began planning a full-scale nuclear plant at Shippingport, Pennsylvania. The plant was to be owned and operated by Duquesne Light Company, managed by Rickover, and equipped with a version of the navy's pressurized-water reactor. But rapid U.S. nuclear power development came only in the wake of the new Atomic Energy Act of 1954, which permitted private power companies to own reactors and to patent nuclear innovations; and the Price-Anderson Act of 1957, which limited liability for individual companies to $560 million and gave government subsidies for liability insurance. The first American plants went on line in the early 1960s.

The USSR operated the first nuclear power plant supplying a national grid at Obninsk, south of Moscow in 1954. This modest (5,000 kW) plant was the opening wedge in an aggressive Soviet drive for nuclear energy as reflected in the Bolshevik slogan "Communism equals Soviet power plus electrification of the entire country." Two years later, the United Kingdom operated a full-scale commercial nuclear plant at Calder Hall, a facility designed to produce plutonium for defense with electricity as a by-product. France ran an experimental reactor, at Marcoule, that year, but began its commercial program with the 'Electricite' de France Chinon reactor in 1962. By 1965 the United Kingdom led in nuclear power output with 3.4 million kWh, followed by the United States (1.2 million kWh), the USSR (895,700 kWh); Italy (622,000 kWh); and five other nations producing at least 500,000 kwh each. Within a few years the United States took the lead in technology and total output, its pressurized-water and boiling-water reactors supplied by enriched fuel from Manhattan Project plants.

These were heady times for the nuclear industry. Well-informed experts predicted that electricity soon would become "too cheap to meter." At the 1964–1965 New York World's Fair, General Electric's "Progressland" exhibit featured "the wonders of atomic energy"; and the company's Medallion City claimed to produce fusion energy (using a 0.000006 second, 100 million flash). Fusion energy, said General Eelectric, was going to supply a billion years of electrical energy.

Nuclear power developed unevenly across the globe. In 1987 the United States operated 110 of the world's 418 nuclear plants, the USSR 57, France 49, the United Kingdom 38, Japan 37, Canada 19, and Sweden 12, while some regions—the Arab states, Africa (except South Africa), and most of Central America—had few or none. Although the United States produces a third of the world's electricity, it derives only 15 percent of it from nuclear energy.

In each country, the pattern of nuclear development reflected national technical and economic resources, politics, and culture. In France, for example, national economic planners ensured the rapid growth of nuclear energy by limiting citizen participation and by standardizing reactor design. The Federal Republic of Germany followed a style closer to the United States, and thus sustained greater political challenges by its environmentally sensitive Green Party.

In balancing development with social safety, the Soviet Union was a grim outlier. Recent research has revealed a series of Soviet nuclear accidents, many of them concealed from the outside world, beginning with a dramatic waste dump explosion in 1957 near Kyshtym in the Urals that spread more than 2 million Ci over 20,000 sq miles. The explosion of Unit 4 at Chernobyl, north of Kiev, on April 26, 1986, was the worst recorded nuclear accident in history. It happened when operators were testing a voltage-regulating scheme on turbogenerator 8 when coolant pumps were slowed. Vigorous boiling led to excess steam, slow absorption of neutrons, and increased heat in a "positive feedback" cycle that resulted in meltdown and explosion. By early May, airborne contamination completely covered Europe. Thirty-one people were killed, 200 suffered radiation sickness, hundreds of thousands of people were confined indoors or evacuated, foods that were suspected of being contaminated were banned, and scientists projected some 1,000 extra cancer deaths over the next fifty years. Perhaps most troubling, follow-up investigation faulted Soviet plant design as much as or more than operator error. Following the accident, public opinion shifted sharply against nuclear energy in Europe, but in the United States, the tide already had turned.

NUCLEAR POWER UNDER SIEGE

In the 1950s, widespread anxieties about the atomic bomb in the 1950s spurred few into action. One notable exception in the United States was the Greater St. Louis Committee for Nuclear Information, founded by biologist Barry Commoner and other scientists at Washington University in 1958. This watchdog group published *Nuclear Information* (renamed *Science and the Citizen*) in 1964. For the most part, however, citizens of the two superpowers and their satellites or allies saw nuclear weapons as a necessary tool in the global contest between democracy and communism. Indeed, for many, atomic supremacy was the most accurate measure of national prowess.

This changed dramatically in the 1960s and 1970s as antinuclear opinion switched from weapons to nuclear power and spread from activist groups into a large segment of the middle class. Antinuclear activism was strongest in the two nations with large nuclear power programs and the most open political systems—the United States and West Germany—although there were notable institutional and social differences between the two nations. And while antinuclear activism surely slowed the progress of nuclear power development in many countries, it was one of a cluster of economic, technological, political, and social forces constraining the industry.

Nuclear energy opponents initially targeted thermal pollution—the effect of nuclear power plants on local aquatic life by raising water temperature by several degrees. Attention soon shifted to the question of radioactive contamination of cooling water. When two scientists at Lawrence Livermore Laboratory in Berkeley, California—John Gofman and Arthur Tamplin—argued that the AEC acceptable level of 170 millirems per year would cause some 16,000 additional deaths (they later revised the figure to 74,000), an intense debate ensued that eventually led to a U.S. Senate committee and to a new standard of 25 millirems. Meanwhile, news of nuclear plant accidents in foreign countries reached the United States, which endured an accident of its own at Idaho Falls in 1961. New guidelines for remote nuclear plant citing followed, which resulted in scuttled plans for a Consolidated Edison plant in Queens, New York, and another in Los Angeles. Anti-nuclear activism was strongest in southern California, where protesters managed to stop the construction of a Pacific Gas & Electric plant at Bodega Bay in 1963 because it was near an earthquake fault.

The 1970s were hard times for the nuclear industry. The decade opened with the first Earth Day (April 22), which featured thousands of teaching events, many of them aimed at halting further nuclear power development, and ended with the accident at the Three-Mile Island nuclear plant in Pennsylvania. In

between, the nuclear industry sustained increasingly sophisticated attacks from increasingly better organized opponents, such as the Environmental Defense Fund (1967); Ralph Nader's Critical Mass Project in Washington, D.C.; and the Union of Concerned Scientists, a consortium of nuclear scientists and engineers founded at MIT in 1971. In 1976, California passed a ballot initiative that halted nuclear plant construction until the federal government found a satisfactory way to dispose of radioactive wastes; while in New Hampshire the Clamshell Alliance was formed to oppose the construction of a nuclear plant at Seabrook, an effort that dragged on for years, involved tens of thousands of demonstrators, and helped force the plant's owner (Public Service Company of New Hampshire) to cancel plans for one of its two reactors and to declare bankruptcy in 1988—the first public utility to do so in American history.

In that case, protests caused delays that contributed to large cost overruns. But Seabrook was an exception; most nuclear utilities got into financial trouble with little help from protesters. Although oil prices rose dramatically in the 1970s—a spur to nuclear development—the "stagflation" of the times drove down demand for electricity from 7 percent to 2 percent per annum and drove up interest rates into the double digits. Between 1971 and 1978, nuclear capital costs rose 142 percent, making them more expensive to build per kilowatt-hour of capacity than new fossil fuel plants.

No case better illustrates the travails suffered by the nuclear power industry and its customers than the saga of the Washington Public Power Supply System (WPPSS). Thanks to abundant Columbia River hydropower, the Pacific Northwest enjoyed the nation's lowest electricity rates in the 1950s. By then the prime hydropower sites has been exploited, and demand for electricity continued to rise. In 1968 some one hundred utilities in the region, working through WPPSS, financed a ten-year, $7 billion Bonneville Power Administration plan to improve the region's hydropower and transmission and distribution assets as well as to build seven new thermal power plants, most of them nuclear. Construction costs skyrocketed, however; the second power plant (at Hanford, Washington), projected to cost $352 million in 1977, had consumed nearly $2 billion by 1980. Struggling to avoid bankruptcy, Bonneville boosted its wholesale rates 700 percent between 1979 and 1984. Irate ratepayers dubbed the project "WHOOPS."

Atomic electricity generating station in Shippingport, Louisiana. (Library of Congress)

Viewed in this context, the Three-Mile Island (TMI) accident was the *coup de grace* for an already foundering industry. In spite of the fact that the hydrogen gas bubble that accumulated in Reactor 2 did not explode, although some contaminated gas escaped; and that the commissions who investigated the accident faulted human error rather than equipment failure, TMI caused (as the *New York Times* put it) "a credibility meltdown" for the nuclear industry. (The release of a major motion picture with an eerily similar scenario, *The China Syndrome*, a few weeks before the real accident amplified the public relations crisis.) The TMI cleanup cost roughly $1 billion, less than a third covered by insurance. The Nuclear Regulatory Commission, a federal agency, began to enforce an informal moratorium on new plant licenses pending further investigation of TMI, and soon demanded expensive retrofitting for similar plants. But even without TMI, no new reactors have been ordered in the United States since 1978.

NUCLEAR ENERGY AND NUCLEAR DEFENSE FROM THE LATE 1970S TO THE PRESENT

As they did at the dawn of the atomic age, nuclear weapons have overshadowed nuclear energy for since the late 1970s. The shift began then, when the Carter administration switched its Soviet stance from détente to rearmament and ordered 200 new MX missiles. By then, the nonprofit organizations Union of Concerned Scientists—perhaps seeing civilian nuclear energy as moribund—shifted its attention to weapons. The Cambridge, Massachusetts–based Physicians for Social Responsibility, led by Helen Caldicott, began to publicize the horrors of nuclear war, garnered a large grant from the Rockefeller Foundation, and saw its membership surge from 500 in 1978 to 16,000 in 1980. At the same time, a movement favoring a moratorium on nuclear weapons began to take shape, with 1980 presidential hopefuls Governor Jerry Brown of California and Senator Edward Kennedy of Massachusetts joining the cause.

The election of Ronald Reagan to the White House in 1980 heightened nuclear tensions. Reagan dubbed the Soviet Union an "evil empire" and spoke of first-strike capabilities and strategies of limited nuclear war. Supporters of the 1981 Nuclear Weapons Freeze Campaign, in contrast, called for a verifiable treaty with the Soviet Union to halt nuclear proliferation. In 1983 Regan announced the extravagant Strategic Defense Initiative to shield the United States from nuclear attack using space-based missile interceptors and other yet-to-be-developed technologies. Endorsed by atomic bomb pioneer Edward Teller, the project was so expensive and technically speculative that it became popularly known as "Star Wars," after a Hollywood space fantasy. Although the president called it a purely defensive measure, many foreign-policy experts saw it as a form of destabilizing escalation. Others considered it a ploy to force the USSR into heavy military spending (to achieve parity with Star Wars) at a time when the Soviet empire was undergoing political and economic stress under the reformist regime of Mikhail Gorbachev.

The question of how much Star Wars may have contributed to the 1991 collapse of the Soviet Union continues to generate debate. But the breakup of the postwar nuclear superpower clearly ushered in a new era of nuclear quiescence. To be sure, the fate of the former Soviet Union's tens of thousands of warheads remains a vital concern, especially in light of the region's growing economic pressures. For example will unpaid military officers be tempted to sell nuclear weapons to well-funded terrorist groups? In the post-Soviet world, the threat of nuclear warfare no longer is seen as a matter of global annihilation bur rather as a risk in political hot-spots such as the Middle East, Korea, or—if Iraq's Saddam Hussein should acquire nuclear technology—the United States as a possible target. And dreams of a Star Wars-like defense live on; late in his final term, President Bill Clinton announced plans to consider the development of a nuclear defense system reminiscent of, though less elaborate than, the Reagan plan.

Although reactor-building continues steadily in France and Japan, and may take hold in parts of the less developed world, nuclear power will need a major technological breakthrough (or a fossil fuel energy crisis) to make a comeback in the United States and Germany. One possibility is the fusion reactor, which produces nuclear energy by combining two light elements (such as deuterium) into a single heavier one that weighs less than the sum of its parts, with the difference released in the form of energy. Fusion produces less radioactive waste than fission and faces no fuel constraints (deuterium is found in water). The problem is that scientists have not been able to combine light atoms through collusion—they bounce off each other—but rather do so by accelerating them with heat so intense it either consumes as much energy as the reaction produces, or (at much higher levels) it cannot be contained safely.

Since the earliest days of the atomic age, physicists and engineers have predicted the coming of practicable nuclear fusion within "ten years" or "a generation." History therefore offers many reasons to be skeptical about the promise of nuclear energy. At the same time, this unparalleled form of energy is not going to return to the Pandora's box pried open by the Manhattan Project more than a half century ago.

David B. Sicilia

See also: Einstein, Albert; Emission Control, Power Plant; Energy Management Control Systems; Environmental Problems and Energy Use; Ethical and Moral Aspects of Energy Use; Explosives and Propellants; Historical Perspectives and Social Consequences; Matter and Energy; Military Energy Use, Historical Aspects of; Nuclear Energy; Nuclear Fission; Nuclear Fission Fuel; Nuclear Fusion; Nuclear Waste.

BIBLIOGRAPHY

Boyer, P. S. (1985). *By the Bomb's Early Light*. New York: Pantheon.

Campbell, J. L. (1988). *Collapse of an Industry*. Ithaca NY: Cornell University Press.

Cantelon, P. L.; Hewlett, R. G.; and Williams, R. C., eds. (1991). *The American Atom*. Philadelphia: University of Pennsylvania Press.

Hughes, T. P. (1989). *American Genesis*. New York: Viking.

Joppke, C. (1993). *Mobilizing Against Nuclear Energy*. Berkeley: University of California Press.

Melosi, M. V. (1985). *Coping with Abundance*. Philadelphia: Temple University Press.

Mounfield, P. R. (1991). *World Nuclear Power*. London: Routledge.

Rhodes, R. (1986). *The Making of the Atomic Bomb*. New York: Simon & Schuster.

NUCLEAR FISSION

Nuclear fission is a process in which a heavy nucleus—usually one with a nucleon number of two hundred or more—separates into two nuclei. Usually the division liberates neutrons and electromagnetic radiation and releases a substantial amount of energy. The discovery of nuclear fission is credited to Otto Hahn and Fritz Strassman. In the process of bombarding uranium with neutrons in the late 1930s, they detected several nuclear products of significantly smaller mass than uranium, one of which was identified as ^{137}Ba. The theorectical underpinnings that exist to this day for nuclear fission were proposed by Lise Meitner and Otto Frisch. Shortly after Hahn and Strassman's discovery.

Some heavy nuclei will fission spontaneously. Others can be induced to fission through interaction with a neutron. In both spontaneous nuclear fission and induced nuclear fission the pool of neutrons and protons is conserved. For example, the nucleus ^{252}Cf (Californium) fissions spontaneously. The 98 protons and 154 neutrons in the nucleus of ^{252}Cf are reconfigured into other nuclei. Usually a few neu-

trons are released in the process. Pictorially, a typical spontaneous fission of ^{252}Cf producing two nuclei and three neutrons is shown in Figure 1.

The pictorial depiction of the spontaneous fission of ^{252}Cf can be summarized as an equation:

$$^{252}\text{Cf} \rightarrow {}^{150}\text{Ba} + {}^{99}\text{Mo} + 3\text{n} + \text{energy}$$

The nucleus ^{235}U (uranium) does not fission spontaneously, but it can be induced to fission through interaction with a neutron. Pictorially, a typical neutron-induced fission of ^{235}U producing two nuclei and three neutrons is depicted in Figure 2.

The pictorial depiction of the neutron-induced fission of ^{235}U can be summarized as an equation:

$$^{235}\text{U} + \text{n} \rightarrow {}^{144}\text{Ba} + {}^{89}\text{Kr} + 3\text{n} + \text{energy}$$

Visually, nuclear fission is simply a rearrangement of neutrons and protons, and would seem to be possible for any nucleus. However, energy conservation principles limit significantly the number of possibilities for either spontaneous nuclear fission or induced nuclear fission. Total energy is conserved in a nuclear fission reaction as it is in all nuclear reactions. This means that the total energy before the fission equals the total energy of the fission products. For example, in the induced nuclear fission reaction discussed earlier, the combined energy of the neutron and ^{235}U nucleus equals the combined energy of the ^{144}Ba and ^{85}Kr nuclei and the three neutrons. The energy accounting includes kinetic energy as well as any potential energy associated with the nuclei and the neutrons. The energy associated with mass through the Einstein relation $E = mc^2$ is particularly relevant. The interaction of the neutron and the ^{235}U nucleus takes place when both are essentially at rest and, therefore, have negligible kinetic energy. Yet the reaction products—^{144}Ba, ^{85}Kr, and three neutrons—have significant kinetic energy. The kinetic energy of the products has its origin in the conversion of potential energy associated with the ^{235}U nucleus and the interacting neutron. The gain in kinetic energy is accompanied by a loss of mass in the interacting nuclei (neutron and ^{235}U nucleus). The kinetic energy of the reaction products (^{144}Ba, ^{85}Kr and 3 neutrons) is determined from $E = mc^2$, where m is the mass difference between the interacting nuclei and reaction products. The following example illustrates this principle. In atomic mass units the masses of the

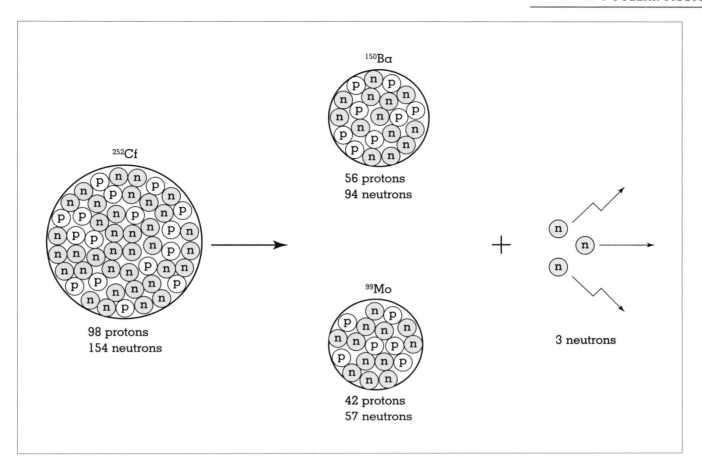

Figure 1.
Spontaneous fission process.

nuclei involved in the illustrative nuclear fission reaction are:

^{235}U	235.0439231 u	^{144}Ba	143.9229405 u
n	1.0086649 u	^{89}Kr	88.9176325 u
		3 n	3.0259947 u

Adding the left and the right columns yields 236.052588 u and 235.8665677 u, respectively. The reaction products have 0.1860 u less mass than the interacting nuclei. One atomic mass unit has an energy equivalent of 931.50 million electron volts (MeV). Therefore, 0.1860 u has an energy equivalent of 173.3 MeV. The energy acquired by the reaction products totals 173.3 MeV. This nuclear fission energy release is huge compared to that in a chemical reaction.

A typical chemical reaction such as that in burning a fossil fuel releases about 10 eV of energy. A release of 170 MeV in one nuclear fission reaction liberates the energy equivalent of 17 million chemical reactions in the burning of a fossil fuel.

A nuclear fission reaction will not occur unless the following occur: (1) the total mass of the reaction products is less than the total mass of the interacting nuclei, and (2) the sum of the neutrons and the sum of the protons in the interacting particles equals the sum of the neutrons and the sum of the protons in the products of the fission.

Any combination of reaction products consistent with these conservation principles is possible. For example, in the neutron-induced nuclear fission of ^{235}U it is possible to produce ^{140}Xe, ^{94}Sr, two neutrons, and 185 MeV of energy. The most likely reaction products are close in atomic number to xenon (Xe) and strontium (Sr), but the possibilities number in the hundreds.

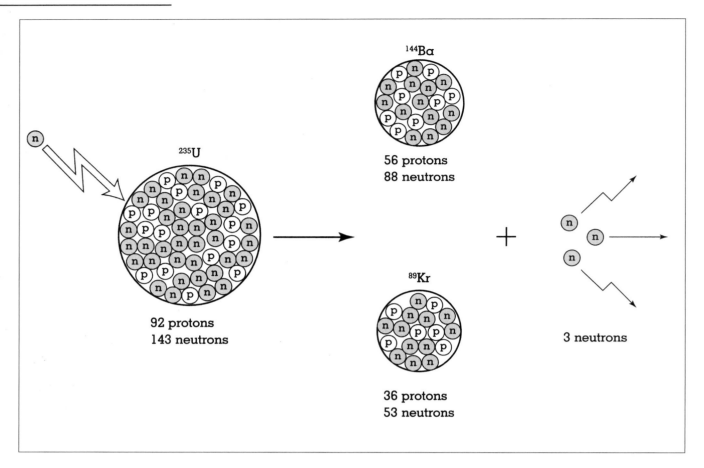

Figure 2.
Neutron-induced fission process.

Describing a neutron-induced nuclear fission reaction such as

$$^{235}\text{U} + \text{n} \rightarrow {}^{144}\text{Ba} + {}^{89}\text{Kr} + 3\text{n} + \text{energy}$$

does not convey the likelihood of it actually happening. When a neutron encounters the ^{235}U nucleus, several nuclear events other than fission may occur. For example, the neutron may bounce off, changing the course of its path much like a marble bouncing off a bowling ball. Or the ^{235}U nucleus may capture the neutron and disintegrate in a way that does not involve nuclear fission. Whatever the event, there is only some chance that the event occurs when the neutron and the nucleus interact, and this probability depends strongly on the energy of the neutron. The fissioning of ^{235}U by a neutron is likely only for low-energy neutrons. Low in this sense means that the energy is of the order of the energy of a molecule in the air in a room.

A drop of water is a collection of molecules bound together by electric forces. Perturbing the drop in some way often causes the drop to separate into two smaller drops. A heavy nucleus such as ^{235}U has 92 protons and 143 neutrons and has features of a liquid drop that are useful for visualizing the dynamics of nuclear fission. The strong nuclear force that tends to bind the protons and the neutrons together competes with electric forces between protons that tend to cause the nucleus to break apart. Competition between these forces produces collective oscillations that are analogous to oscillations of a liquid drop. The energy added to the nucleus when it captures a neutron enhances the oscillations, and on some occasions the nucleus separates into two parts. The model is useful for visualizing the nuclear fission mechanism as well as for doing quantitative calculations involving the probability aspects of the process.

NUCLEAR FISSION CHAIN REACTION

A chain reaction results when some event triggers a sequence of identical events. A chain reaction on a crowded highway develops when a car runs into the rear of another car, the struck car collides with a car in front of it, and so on. An expanding chain reaction develops if a car hits two cars, each struck car moves on to strike two cars, and so on. A chain reaction is central to the operation of a nuclear fission reactor and a nuclear fission bomb. It begins with a neutron-inducing fission in a nucleus. Neutrons liberated in the fission in turn induce fission in other nuclei, and so on (see Figure 3). Each fission releases energy, and unless there is some purposeful control, the energy release proceeds uncontrolled. Controlled or uncontrolled, the chain reaction does not propagate unless (1) the fuel is properly arranged and (2) there is sufficient likelihood that a neutron induces fission.

This is tantamount to saying that a chain reaction of cars does not propagate unless (1) the cars are arranged properly and (2) there is a sufficient chance that a car will make other collisions.

NUCLEAR FUEL

Commercially, there are only two nuclear species that will function in a self-sustaining nuclear fission chain reaction; ^{235}U and ^{239}Pu (plutonium). Uranium occurs naturally in two predominant forms, ^{235}U and ^{238}U. The bulk, 99.3 percent, of natural uranium is ^{238}U. Even though ^{238}U differs from ^{235}U by having only three additional neutrons, the likelihood of neutron-induced fission of ^{238}U is so low that it cannot be used in a self-sustaining chain reaction. Thus the great bulk of natural uranium is not of a form that can be used in self-sustaining chain reactions. ^{239}Pu is not found naturally in appreciable amounts. Rather, ^{239}Pu is made through nuclear transmutations. The process begins by having ^{238}U capture a neutron to form ^{239}U. ^{239}U is unstable and undergoes radioactive decay, with the end product being ^{239}Pu.

NUCLEAR FISSION REACTOR PRINCIPLES

The fuel in a nuclear fission reactor is generally ^{235}U atoms arranged appropriately in a reactor vessel. Neutrons instigate fission of nuclei of ^{235}U atoms and liberate energy. The energy output may be controlled either by regulating the fuel and/or adjusting the neu-

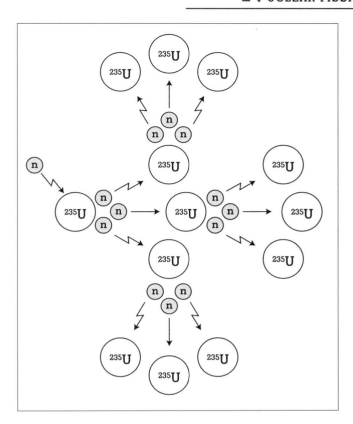

Figure 3.
A nuclear fission chain reaction.

tron supply. Generally, the energy output is controlled by regulating the number of neutrons. Cylindrical control rods made of materials that readily absorb neutrons are moved up or down in strategically arranged spaces in the fuel assembly. Water circulating around the nuclear fuel removes heat produced by nuclear fission reactions and reduces moderates) the speed of neutrons liberated by fission. The main components of a nuclear fission reactor are shown in Figure 4.

Moderator

Neutrons produced in nuclear fission reactions are very energetic and are referred to as fast neutrons. To optimize the chain reaction, the speed of the reaction neutrons must be reduced. Water molecules surrounding the fuel help slow the neutrons through collisions with atoms. A proton is an optimum target for a neutron. Hydrogen atoms in water are a good source of protons because each water molecule has two hydrogen atoms and each atom has a single proton in its nucleus. A neutron produced in a nuclear fission reaction bounces around in a pool of water,

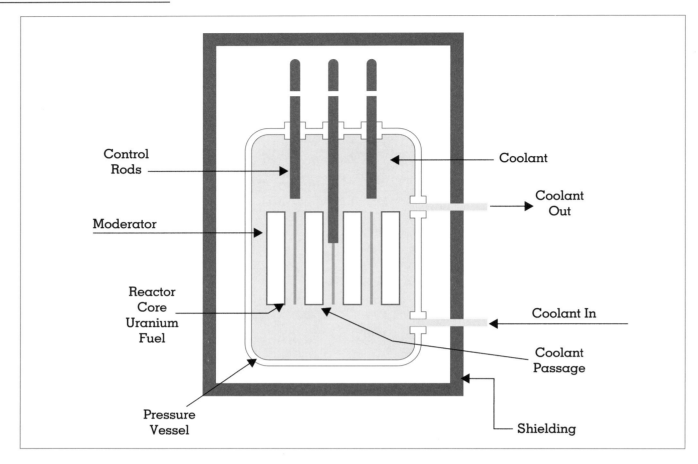

Figure 4.
Main components in a nuclear fission reactor.

losing energy with each collision with a proton. After its energy is reduced sufficiently, it may react with a ^{235}U nucleus to induce fission. Reducing the energy is referred to as moderation and the material used in the moderation is called a moderator. An object colliding head-on with a stationary object loses the greatest amount of kinetic energy when the masses of the two objects are the same. This is apparent when a billiard ball collides head-on with another billiard ball at rest. A hydrogen atom has roughly the same mass as a neutron. Accordingly, a neutron loses virtually all its kinetic energy in a head-on collision with a hydrogen atom. Because a water molecule (H_2O) has two hydrogen atoms, water is an efficient moderator for neutrons produced in nuclear fission reactions. Water is the moderator in nearly all reactors.

Control Rods

A nuclear chain reaction can be controlled by inserting materials within the fuel that deprive the fuel of neutrons. These materials are usually rod-shaped and are referred to as control rods. Control rods are made of materials that are more likely than the fuel to absorb neutrons. Boron and cadmium are favorite materials for control rods. To decrease the energy output, the control rods are inserted closer to the fuel. To increase the energy output, the control rods are pulled away from the fuel. Boric acid (H_3BO_3), a boron-containing liquid that strongly absorbs neutrons, is often included in the water moderator to provide further control of the energy output of the reactor.

Fuel

Uranium is a metal that is found naturally as a constituent of chemical compounds in minerals such as pitchblende. Uranium ore is mined much like coal: Open pits are used to mine shallow deposits, deeper deposits require shaft mining. Commercial ores yield 3 to 5 lb of uranium compounds per ton of ore. A material called yellow-cake is produced that is rich in the uranium compound U_3O_8. About 7 of every 1,000 uranium atoms in the yellow cake are of the variety ^{235}U.

Some nuclear fission reactors are designed to use natural uranium having 0.7 percent ^{235}U and 99.3 percent ^{238}U. CANDU reactors, manufactured in Canada, are examples. These reactors do not use ordinary water for the moderator. Most nuclear fission reactors use ordinary water for a moderator which requires that the fuel be about 3 percent ^{235}U and about 97 percent ^{238}U. Achieving this enrichment requires that the solid uranium compounds in the yellow cake be converted to gaseous uranium hexafluoride (UF_6). Following enrichment, gaseous UF_6 is converted to solid uranium oxide (UO_2) for fabrication of fuel elements for a nuclear reactor.

Uranium oxide fuel is formed into solid pellets about ⅜ in. in diameter and ½ in. long. Fuel pellets are stacked in tubes about 12 ft long. Neutrons must be able to penetrate the tube walls, called cladding, to interact with the uranium, and the walls must withstand the temperature required for heating water for a steam turbine. Stainless steel or a material called zircaloy is used for the cladding. Some 200 tubes are packed into a fuel bundle. About 175 appropriately arranged fuel bundles form the reactor core, where the energy is produced. A cylindrical steel vessel about 20 ft in diameter and about 40 ft high houses the core. Energy and neutrons are liberated from nuclear fission reactions within the fuel pellets. About 83 percent of the energy is in kinetic energy of the fission fragments. Fission fragments convert their kinetic energy to heat in the process of coming to rest in the fuel pellets. Heat is conducted from the pellets by water circulating around the fuel cladding. Water has two very important roles: conducting heat from the reactor core and moderating neutron speeds.

Energy Production

There are about 100 tons of uranium in the fuel bundles. With the control rods in their lowest position, there is no chain reaction, no energy produced, and no neutrons liberated from nuclear fission reactions. Energy production starts by raising the control rods and inserting a source of neutrons. Nuclear fission reactions begin producing energy and neutrons. Neutrons escape from the fuel elements into the water, where they lose speed and change direction with each collision with a proton. After losing sufficient speed, a neutron may reenter a fuel element and

induce a ^{235}U nucleus to fission to help foster the chain reaction. If on the average one neutron from a nuclear fission reaction goes on to produce another nuclear fission reaction, the chain reaction is self-sustaining, and energy is produced at a steady rate. Raising the control rods allows more nuclear fission reactions and more energy production. Higher, but steady, energy production occurs at the new control rod position. Control of the chain reaction depends on

- the arrangement of the fuel elements
- the quality of the moderator
- the quantity of ^{235}U
- the neutron energy required for a high probability for inducing nuclear fission
- the material surrounding the reactor core that is used to minimize the escape of neutrons.

Although protons are very efficient neutron moderators, they also efficiently capture neutrons to form bound proton-neutron pairs called deuterons. Reactors using ordinary water for the moderator compensate for neutron capture by using fuel enriched to about 3 percent ^{235}U.

Usually atoms resulting from nuclear fission are radioactive. There are also radioactive atoms produced from neutron capture by both ^{235}U and ^{238}U. Both types of radioactive atoms remain in the nuclear fuel. It is these radioactive atoms that comprise the nuclear wastes that require disposal in an environmentally acceptable manner.

Boiling-Water Reactors and Pressurized-Water Reactors

Current light-water reactors have two basic designs. One is termed a boiling-water reactor (BWR), the other a pressurized-water reactor (PWR). The principles of the PWR design is illustrated in Figure 5. Water comes to a boil, producing steam inside the containment vessel of a boiling-water reactor. Water is kept under pressure and does not boil inside the containment vessel of a pressurized- water reactor. Steam for the turbine in a nuclear power plant using a pressurized water reactor comes from water brought to a boil in a steam generator garnering heat from water circulating around the reactor core. In a boiling- water reactor, steam is produced in the reactor vessel. In a pressurized-water reactor, steam comes from a system separated from the reactor vessel. Pressure and temperature in a boiling-water reactor are about 1,000 lb per sq in. and 545°F (285°C). These

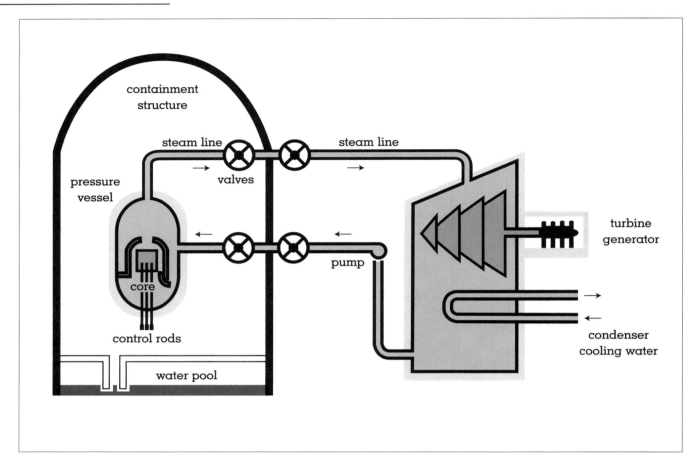

Figure 5.
The water-steam circuit in a pressurized water reactor.

numbers rise to about 2,250 lb per sq in. and 600°F (316°C) in a pressurized-water reactor. Roughly 70 percent of the reactors in the United States are of the PWR type. Essentially, the safety aspects of PWR and BWR reactors are equivalent.

At the beginning of the twenty-first century, 439 nuclear power reactors in 31 countries were generating 352 billion watts of electric power. More than 16 percent of the world's electricity was produced by these electric power plants. On average the power output of one of these plants is roughly 800 megawatts and is comparable to the power output of a large coal-burning electric power plant. Either one of these plants could supply the electricity needs of about a quarter of a million homes. The electric power outputs of solar and wind systems, for example, are usually factors of ten to a hundred times less than that of a large nuclear-fueled or coal-burning electric power plant.

REACTOR CONTROL

The products of nuclear fission reactions are radioactive and disintegrate according to their own time scales. Often disintegration leads to other radioactive products. A few of these secondary products emit neutrons that add to the pool of neutrons produced by nuclear fission. Very importantly, neutrons from nuclear fission occur before those from radioactive decay. The neutrons from nuclear fission are termed prompt. Those from radioactive decay are termed delayed. A nuclear bomb must function on only prompt neutrons and in so doing requires nearly 100 percent pure ^{235}U (or ^{239}Pu) fuel. Although reactor fuel is enriched above its natural 0.7 percent abundance, it is still some thirty times less concentrated than the fuel needed for bombs. A light-water nuclear reactor cannot achieve a self-sustaining condition using only prompt neutrons; delayed neutrons are needed as well. Control rods are slammed into position if a critical excursion from normal operation is sensed. The time involved in delayed neutron emission is crucial for success of the control operation.

A nuclear reactor and its auxiliaries involve a complex assembly of pumps, pipes, and valves for circulating water and steam. Minor malfunctions of the components are inevitable, and routine maintenance is inescapable. Distinct from the routine malfunctions is the possibility of loss of water in the main cooling of the reactor core. This bears the name "loss of cooling accident." Possible causes of such an accident include reactor operator errors, faulty construction, improper maintenance, natural disaster, and sabotage. A loss of cooling terminates the nuclear fission chain reaction because of the need for a moderator. However, radioactivity in the reactor core continues to produce heat at a level of 5 to 7 percent of its rated power. This percentage may seem low, but 5 percent of 3 billion watts of thermal power from normal operation amount to 150 million watts of thermal power that remains after the chain reaction stops. Many of the radioactive products are short-lived, and the power level drops accordingly. Still, there is sufficient heat to begin melting the core in about thirty seconds. Unless checked, this molten radioactive mass could melt its way through the bottom of the containment facilities. Emergency core-cooling systems must operate within this thirty-second time period. Although safety measures make the chance of an accident releasing radioactivity unlikely, it cannot be made zero.

THE NUCLEAR FISSION BREEDER REACTOR

Contemporary nuclear fission reactors are fueled with either ^{235}U or ^{239}Pu. In a nuclear fission reactor fueled with ^{235}U, some ^{238}U in the fuel is converted to ^{239}Pu through nuclear transmutations and becomes part of the fuel inventory. However, the reactor is not designed to optimize the production of ^{239}Pu. A nuclear breeder reactor is designed to both produce energy and convert ^{238}U to ^{239}Pu for fuel. The conversion process is termed breeding.

The fission of ^{235}U or ^{239}Pu liberates, on average, two to three neutrons. One neutron is required to sustain the nuclear fission chain reaction. In a nuclear breeder reactor, the extra neutrons are used to induce nuclear reactions that lead to the production of ^{239}Pu. The sequence begins by arranging for

a neutron to be captured by a ^{238}U nucleus to form ^{239}U. ^{239}U is unstable and disintegrates producing a variety of neptunium, ^{239}Np. ^{239}Np is unstable and decays, forming ^{239}Pu. The breeding principle requires both fissionable fuel (^{235}U or ^{239}Pu) and a breeding material (^{238}U). Both resources are depleted as time passes.

The chance of inducing nuclear fission of ^{235}U is optimum for slow neutrons. Neutron capture by ^{238}U is significant only for energetic (or fast) neutrons. Accordingly, the nuclear fission reactions must be initiated by fast neutrons even though it is more difficult. Water cannot be used as a coolant, as in a light-water nuclear reactor, because of its moderating effect on neutrons. The alternative to water is usually liquid sodium. These reactors are termed liquid metal (for cooling) fast (for energetic neutrons) nuclear breeder reactors.

Nuclear Breeder Reactor Projects

An experimental nuclear breeder reactor has been operating at the Argonne National Laboratory near Idaho Falls, Idaho, since 1963. However, widespread commercial utilization of nuclear breeder reactors has not materialized. A commercial nuclear breeder reactor electric power plant was built in Monroe County, Michigan, in 1963, but was permanently terminated in 1966 after a blockage in a sodium cooling line produced a partial meltdown of its core. Plans were announced in February 1972 for a major nuclear fission breeder electric power plant to be built on the Clinch River in eastern Tennessee and planned for completion in 1980, but the project ultimately was not funded by the U.S. Congress. France and Japan have constructed major nuclear breeder reactor electric power plants, but at the turn of the century all had been closed. Reprocessing nuclear fuel to recover ^{239}Pu is necessity in nuclear fission breeder technology. This is a major stumbling block because of concerns for proliferation of ^{239}Pu for making nuclear bombs.

Joseph Priest

See also: Electric Power, Generation of; Environmental Problems and Energy Use; Explosives and Propellants; Meitner, Lise; Military Energy Use, Historical Aspects of; Molecular Energy; Nuclear Energy; Nuclear Energy, Historical Evolution of the Use of; Nuclear Fission Fuel; Nuclear Fusion; Nuclear Waste.

BIBLIOGRAPHY

Bodansky, D. (1996). *Nuclear Energy: Principles, Practices, and Prospects.* Woodbury, NY: AIP Press.

Clark, R. W. (1980). *The Greatest Power on Earth: The Story of Nuclear Fission.* London: Sidgewick & Jackson.

El Wakil, W. W. (1984). *Powerplant Technology.* New York: McGraw-Hill.

Forsberg, C. W., and Weinberg, A. (1990). "Advanced Reactors, Passive Safety, and Acceptance of Nuclear Energy." *Annual Review of Energy* 15: 133–152.

Golay, M. W., and Todreas, N. E. (1990). "Advanced Light Water Reactors." *Scientific American* 262(4):82–89.

Graetzer, H. G., comp. (1981). *The Discovery of Nuclear Fission.* New York: Arno Press.

Hafele, W. (1990). "Energy from Nuclear Power." *Scientific American* 263(8):136–143.

Krane, K. S. (1987). *Introductory Nuclear Physics.* New York: John Wiley & Sons.

Priest, J. (2000). "Nuclear Energy." In *Energy: Principles, Problems, Alternatives,* 5th ed., ed. J. Priest. Dubuque, IA: Kendall/Hunt.

Ruzic, D. N. (1998). "Light-Water Reactors and Their Advances." Urbana, IL: University of Illinois. <http://starfire.ne.uiuc.edu/ne201/course/topics/light_water_reactors/> August 28, 2000.

NUCLEAR FISSION FUEL

A nuclear power plant generates electricity in a manner similar to a fossil fuel plant. The fundamental difference is the source of heat to create the steam that turns the turbine-generator. A fossil plant relies on the combustion of natural resources (coal, oil) to create steam. A nuclear reactor creates steam with the heat produced from a controlled chain reaction of nuclear fission (the splitting of atoms).

Uranium is used as the primary source of nuclear energy in a nuclear reactor, although one-third to one-half of the power will be produced from plutonium before the power plant is refueled. Plutonium is created during the uranium fission cycle, and after being created will also fission, contributing heat to make steam in the nuclear power plant. These two nuclear fuels are discussed separately in order to explore their similarities and differences. Mixed oxide fuel, a combination of uranium and recovered plutonium, also has limited application in nuclear fuel, and will be briefly discussed.

URANIUM

Uranium (symbol: U; atomic number: 92) is the heaviest element to occur naturally on Earth. The most commonly occurring natural isotope of uranium, U-238, accounts for approximately 99.3 percent of the world's uranium. The isotope U-235, the second most abundant naturally occurring isotope, accounts for another 0.7 percent. A third isotope, U-234, also occurs naturally, but accounts for less than 0.01 percent of the total naturally occurring uranium. The isotope U-234 is actually a product of radioactive decay of U-238.

Nuclear power is now the only substantial use for uranium. But before uranium can be used in a nuclear reactor, it must undergo several processes. After uranium is mined from geological mineral deposits, it is purified and converted into uranium hexafluoride (UF_6). The UF_6 is next enriched, increasing the concentration of U-235 by separating out UF_6 made with U-238 atoms. The enriched UF_6 is then converted into uranium dioxide (UO_2), and pressed into fuel pellets for use in the nuclear reactor.

Uranium Mining and Milling

Uranium is found in most rock, in a concentration of two to four parts per million (ppm). Substantially greater average concentrations can be found in mineral deposits, as high as 10,000 ppm, or 10 percent. Most uranium deposits suitable for mining, however, contain an average of less than 1 percent uranium. Uranium is a metal, and thus its acquisition is not unlike the mining of any other metallic ore. Although uranium is found nearly everywhere on the earth, Canada leads the world in uranium production, mostly due to its heavy financial investment in uranium exploration, and to a few sizable deposits in the Saskatchewan territory. Table 1 depicts the total world uranium production in 1997.

Although the nucleus of the uranium atom is relatively stable, it is radioactive, and will remain that way for many years. The half-life of U-238 is over 4.5 billion years; the half-life of U-235 is over 700 million years. (Half-life refers to the amount of time it takes for one half of the radioactive material to undergo radioactive decay, turning into a more stable atom.) Because of uranium radiation, and to a lesser extent other radioactive elements such as radium and radon, uranium mineral deposits emit a finite quantity of radiation that require precautions to protect workers at the mining site. Gamma radiation is the

Country	Tons of Uranium
Canada	12,029
Australia	5520
Niger	3497
Namibia	2905
USA	2170
Russia	2000
Uzbekistan	1764
South Africa	1100
Kazakhstan	1000
France	748
Czech Republic	590
Gabon	472
China	500
Ukraine	500
others	897
Total world	**35,692**

Table 1.
1997 Uranium Mine Production
SOURCE: Uranium Institute, Core Issues 3/98; and Uranium Institute, 1998, Global Nuclear Fuel Market, Supply & Demand 1998–2020.

most prevalent, followed by beta and alpha radiation. Because airborne particulates can be ingested or inhaled, thus subjecting workers to internal doses of radiation, the most important step taken to protect mine workers is to minimize the amount of airborne particulate matter. In areas where airborne particulate is of great concern, the uranium ore can be wetted to prevent releasing dust to the air. In areas of even greater risk, workers can be outfitted with respiratory equipment. Radon gas is also emitted from uranium, but proper ventilation provides sufficient dilution of the gas so that workers are not affected.

In general, there are three methods available to mine uranium deposits: open-pit, underground, and in-situ leaching. Open-pit mining is used when the uranium deposit is relatively shallow. Underground mining is used when the deposit is relatively deep. But for underground mining to be cost-effective, the deposit must be relatively high-grade and abundant with uranium. For both of these mining methods, the uranium ore is crushed and ground into a fine powder, releasing the uranium from the material surrounding it in the mineral deposit. Next, the crushed uranium is leached from the ore with a solution designed to dissolve only the uranium. After the uranium is dissolved, the solution is separated. The ura-nium-rich liquid is run off into a collection facility, and the solid waste product is left behind. Note that this waste product, referred to as tailings, can be highly radioactive, and can accumulate in substantial quantity. Since the uranium is often found in deposits with other elements having substantially shorter half-lives, these short-lived isotopes are left behind in a concentrated waste mass. The waste water is evaporated from the sludge, and the remaining metals and minerals are stored, often in an open pit at the mill site, for the life of the facility.

A less environmentally invasive method of obtaining the uranium is through a process called in-situ leaching (ISL). Rather than crush the uranium mineral deposit in order to expose the uranium and then dissolve it, the dissolving solution is pumped directly into the uranium deposit. This way, the dissolved uranium solution is obtained without disturbing the minerals surrounding the uranium in the ore, and all of the waste solids are left behind. This assumes the mineral deposit is able to facilitate the leaching process. Uranium deposits must be permeable, like those found in sand and sandstone, to utilize ISL in order for the dissolving solution to pass easily into the deposit to extract the ore. Between 10 and 15 percent of the world uranium supply is mined via in-situ leaching.

Regardless of the method used to obtain the uranium—open pit, underground, or in-situ leaching—the result is a uranium-rich solution. The uranium is recovered from the solution through ion exchange or solvent extraction, and concentrated into a fine precipitate. The uranium-free solution can then be returned to the mine to extract more uranium. The uranium in the precipitate is a blend of uranium dioxide (UO_2) and uranium trioxide (UO_3), combined in a general uranium oxide (U_3O_8). The precipitate is usually dried, forming a yellow cake of uranium, simply called "yellow cake." This concentrated uranium is typically 70 to 90 percent uranium. In order to use the yellow cake as a nuclear fuel, the uranium must go through a process of conversion.

Uranium Conversion

Since the uranium from the milling process is still in an unusable form, the yellow cake is broken down once again. The uranium trioxide is reduced to uranium dioxide at very high temperatures. Refining of the product also takes place. Now the uranium product consists almost entirely of UO_2.

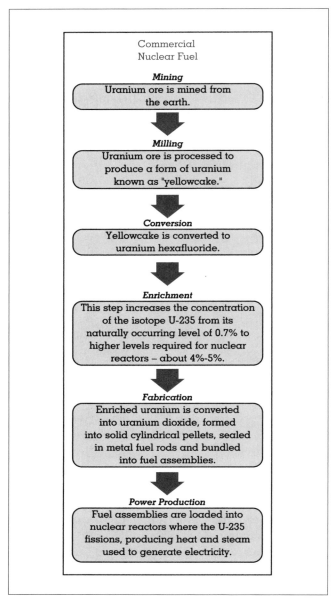

Figure 1.

Processes used in refining uranium for use in reactors

SOURCE: USEC, Inc.

During the conversion process, the object is to create uranium hexafluoride (UF_6), a highly corrosive substance that is gaseous at high temperatures, but is a white crystalline solid at lower temperatures. Uranium hexafluoride is easily transported in its crystalline form to an enrichment facility (the step taken after conversion), but the gaseous form is well suited for the enrichment process, itself. First, the

UO_2 is combined with hydrogen fluoride (HF) to form uranium tetrafluoride (UF_4). This can be accomplished by reacting the UO_2 with either gaseous or liquid HF. The UF_4 is combined with gaseous fluorine to produce UF_6, and is shipped to a facility for enrichment.

Uranium Enrichment

Uranium-235 and U-238 behave differently in the presence of a controlled nuclear reaction. Uranium-235 is naturally fissile. A fissile element is one that splits when bombarded by a neutron during a controlled process of nuclear fission (like that which occurs in a nuclear reactor). Uranium-235 is the only naturally fissile isotope of uranium. Uranium-238 is fertile. A fertile element is one that is not itself fissile, but one that can produce a fissile element. When a U-238 atom is struck by a neutron, it likely will absorb the neutron to form U-239. Through spontaneous radioactive decay, the U-239 will turn into plutonium (Pu-239). This new isotope of plutonium is fissile, and if struck by a neutron, will likely split.

Nuclear fuel always contains a mixture of these uranium isotopes. The U-235, being naturally fissile, will sustain a controlled chain reaction in the nuclear reactor early in the fuel cycle. The U-238, being fertile, will eventually turn into plutonium, continuing to sustain the nuclear chain reaction late in the fuel cycle. The amount of U-235 initially blended into the nuclear fuel will dictate certain performance characteristics. Since naturally-occurring uranium is only approximately 0.7 percent U-235, any nuclear fuel containing more than 0.7 percent U-235 is said to be enriched. Light-water reactors, like those in the United States, use a fuel that has been enriched up to 5 percent. Some reactors are designed to use natural (unenriched) uranium, but they must rely on a medium other than light water to moderate the chain reaction. (These types of reactors typically use heavy water or graphite as a moderator for the nuclear reaction.) For these reactors, the U_3O_8 need only be refined and converted directly to UO_2 for use in fuel pellets, and enrichment is not necessary.

Uranium-235 is less massive than U-238, because each atom of U-235 contains three fewer neutrons than U-238. Although this difference appears insignificant, this 1 percent difference is precisely what allows the UF_6 to be divided into those molecules containing U-235 and those containing U-238. By separating the two quantities, enrichment plants can control the level of enrichment of the end product. Two processes for dividing the uranium isotopes are in large-scale commercial use. These are gaseous diffusion and centrifuge. A third enrichment process, laser enrichment, is being developed.

Gaseous Diffusion. Gaseous diffusion is responsible for producing over 90 percent of the world's enriched uranium. Gaseous diffusion has the ability to process large quantities of gaseous UF_6 at a time. Gaseous UF_6 is fed through a series of porous membranes that tend to filter out the heavier UF_6 molecules containing U-238 atoms. Gas that passes through the membrane will have a slightly higher percentage of U-235 molecules in the UF_6 than when it started, since some of the U-238 molecules did not pass through the membrane. The slightly enriched UF_6 is drawn off and forced through another series of membranes. Each iteration produces a slightly higher enrichment of U-235. After more than 1,500 iterations, the UF_6 is finally enriched to a usable form, at a U-235 concentration of 3 to 4 percent.

Centrifuge Enrichment. In a centrifuge, the UF_6 molecules are spun in a rotating cylinder. Because of the difference in mass between the UF_6 molecules containing U-235 and U-238, the lighter molecules remain at the center of the cylinder while the heavier molecules migrate toward the outside. The enriched UF_6 is drawn off the center of the centrifuge, and then centrifuged again. Similar to the diffusion process, each iteration produces a slightly higher enrichment of U-235. The centrifuge process is much more efficient than the diffusion process, producing a suitable enrichment after only ten to twenty iterations, and requiring significantly less energy to operate the enrichment plant than does a comparable diffusion plant. The capacity of a centrifuge facility is substantially less than the capacity of a diffusion plant, however, severely limiting its commercial application.

Laser Enrichment. Laser enrichment attempts to ionize gaseous U-235 atoms. By using a laser of a specific frequency, valence electrons of U-235 atoms can be removed without ionizing U-238 atoms. The positively charged U-235 atoms are attracted to a negatively charged metal plate, and liquid U-235 is collected. This process does not yet work with the UF_6 created from the existing uranium mining process, so its application is currently limited. Future laser enrichment processes will focus on using UF_6 in its enrichment technique.

Fuel Pellet Fabrication

Finally, the enriched uranium of converted back into UO_2. The UO_2 is pressed into small fuel pellets and packaged in a metal tube (made of a zirconium alloy) for use in a nuclear reactor.

Nuclear fuel can also be fabricated from existing stockpiles of fuel sources. Spent fuel can be reprocessed to extract the remaining U-238 for use in the next fuel cycle. Spent fuel also contains Pu-239, a direct replacement for the fissile U-235 in the enrichment process. Even military grade plutonium and highly enriched uranium can be reprocessed and diluted into a form usable in most nuclear reactors. The reprocessed isotopes are generally manufactured into mixed oxide (MOX) fuels and loaded into conventional nuclear reactors as fuel. The cost of increasing the fissile isotopes in MOX fuel represents a great cost savings over uranium enrichment, and thus reprocessing is increasingly becoming economically viable.

PLUTONIUM

Plutonium (symbol: Pu; atomic number: 93) is not a naturally occurring element. Plutonium is formed in a nuclear reaction from a fertile U-238 atom. Since U-238 is not fissile, it has a tendency to absorb a neutron in a reactor, rather than split apart into smaller fragments. By absorbing the extra neutron, U-238 becomes U-239. Uranium-239 is not very stable, and undergoes spontaneous radioactive decay to produce Pu-239.

Plutonium-239 is a fissile element, and will split into fragments when struck by a neutron in the nuclear reactor. This makes Pu-239 similar to U-235, able to produce heat and sustain a controlled nuclear reaction inside the nuclear reactor. Nuclear power plants derive over one-third of their power output from the fission of Pu-239. Most of the uranium inside nuclear fuel is U-238. Only a small fraction is the fissile U-235. Over the life cycle of the nuclear fuel, the U-238 changes into Pu-239, which continues to provide nuclear energy to generate electricity.

Not all of the Pu-239 will fission during the fuel cycle in a nuclear reactor. Some of the plutonium will not experience neutron bombardment sufficient to cause fission. Other plutonium atoms will absorb one or more neutrons and become higher numbered isotopes of plutonium, such as Pu-240, Pu-241, etc. Plutonium comprises just over 1 percent of nuclear reactor spent fuel—the fuel removed from the nuclear reactor after it has produced all the nuclear fission it can economically provide. Of the plutonium remaining in the spent fuel, slightly less than 60 percent will be fissionable Pu-239.

Because of the high concentration of Pu-240, the plutonium in spent fuel is unsuitable for military application (nuclear warheads). Plutonium-240 is a constant source of neutron decay and is highly unstable. Moreover, the Pu-240 cannot sustain a chain reaction. Military-grade plutonium consists of over 90 percent fissionable plutonium. Because it is impractical to separate the two isotopes of plutonium, special reactors are used to create Pu-239 with extremely small quantities of Pu-240. Even where nations are desperate for electric-generating capacity, the fear of rogue nations teaming to build and operate these special facilities to create military-grade plutonium is a main reason the International Atomic Energy Agency closely safeguards the transfer and use of nuclear energy technology.

Now that much of the world has agreed to nuclear disarmament, scientists and world leaders are searching for uses for the surplus weapons-grade plutonium. Instead of pursuing disposal options, one option is the use of mixed-oxide (MOX) nuclear fuel.

MIXED OXIDE FUEL

In order to deplete stockpiles of weapons-grade plutonium, it is possible to "burn" the fuel in a nuclear reactor. This option has two distinct advantages. First, the fissionable plutonium is a direct replacement for fissionable uranium in a nuclear reactor. Therefore, the plutonium will sustain a controlled chain reaction inside the reactor and provide the heat necessary to create steam for a power plant. Second, the plutonium not undergoing fission will become increasingly contaminated with other isotopes of plutonium (e.g., Pu-240) so that it will no longer be suitable for weapons applications. The longer the plutonium remains inside the nuclear reactor, the more Pu-240 is created as Pu-239 absorbs neutrons. The spent fuel from a reactor using MOX fuel will contain diminished amounts of Pu-239 and increased amounts of Pu-240, effectively removing weapons-grade plutonium from circulation.

Mixed oxide fuel is not appropriate for all nuclear reactors. Plutonium requires faster neutrons in order to operate in a sustained chain reaction. Light-water reactors operate in a highly moderated environment,

and thus U-238 and Pu-239 do not make excellent fuel sources. Light-water reactors can handle up to approximately 30 percent MOX fuel. Heavy-water reactors, such as those in Canada, and other fast neutron reactors, can operate with 100 percent MOX fuel.

Other options for eliminating weapons-grade plutonium are to seal it permanently in solid radioactive waste and dispose of it in waste repositories, and to use the plutonium to fuel fast neutron reactors (without reprocessing the plutonium into a MOX fuel).

Plutonium has a much shorter half-life than uranium (24,000 years for Pu-239; 6,500 years for Pu-240). Plutonium is most toxic if it is inhaled. The radioactive decay that plutonium undergoes (alpha decay) is of little external consequence, since the alpha particles are blocked by human skin and travel only a few inches. If inhaled, however, the soft tissue of the lungs will suffer an internal dose of radiation. Particles may also enter the blood stream and irradiate other parts of the body. The safest way to handle plutonium is in its plutonium dioxide (PuO_2) form because PuO_2 is virtually insoluble inside the human body, greatly reducing the risk of internal contamination.

WASTE DISPOSAL

The main drawback to nuclear power is the production of radioactive waste. Spent fuel from a nuclear reactor is considered a high-level radioactive waste, and remains radioactive for a very long time. Spent fuel consists of fission products from the U-235 and Pu-239 fission process, and also from unspent U-238, Pu-240, and other heavy metals produced during the fuel cycle. That is why special programs exist for the handling and disposal of nuclear waste.

Brian F. Thumm

See also: Nuclear Fission; Nuclear Waste.

BIBLIOGRAPHY

Considine, D. M., ed. (1995). *Van Nostrand's Scientific Encyclopedia*, 8th ed. New York: Van Nostrand Reinhold.
Edwards, C. R. (1992). "Uranium Extraction Process Alternatives." *CIM Bulletin* 85(58):112–136.
Hotta, H. (1987). "Recovery of Uranium from Seawater." *Oceanus* (Spring):30.
Hunter, J. (1991). "Highland In-Situ Leach Mine." *Mining* (August):58–63.
Katz, J. J. (1951). *The Chemistry of Uranium*. New York: Dover.
Lide, D. R., ed. (1996). *CRC Handbook of Chemistry and Physics,* 77th ed. Boca Raton, FL: CRC Press.

Mann, C. A. (1979). *Nuclear Fuel*. East Sussex: Wayland.

Patton, F. S. (1963). *Enriched Uranium Processing*. New York: Oxford.

Sax, N. R., and Lewis, R. J., Sr. (1992). *Dangerous Properties of Industrial Materials*, 8th ed. New York: Van Nostrand Reinhold.

Seaborg, G. T. (1945). "The Chemical and Radioactive Properties of the Heavy Elements." *Chemical Engineering News* 23:2190–2193.

Singleton, A. L., Jr. (1968). *Sources of Nuclear Fuel*. Oakridge, TN: US Atomic Energy Commission, Division of Technical Information.

Uranium Information Centre, Ltd. (1997). "Plutonium." Melbourne: Author.

Uranium Information Centre, Ltd. (1997). "Uranium Enrichment." Melbourne: Author.

Uranium Information Centre, Ltd. (1998). "World Uranium Mining." Melbourne: Author.

The Uranium Institute. (1998). "Uranium: From Mine to Mill." London: Author.

NUCLEAR FUSION

The nuclear fusion of light elements powers the sun and other stars. The fusion process in stars and supernovas creates from primordial hydrogen heavier elements, including those needed for chemistry and life. Nuclear fusion is analogous to the chemical reaction of burning, such as the joining (fusing) of hydrogen and oxygen to produce water vapor and release energy. Both processes produce something new while releasing energy.

Our planet contains large quantities of light elements that are theoretically suitable for providing energy from fusion reactions in central power plants. The potential energy resource from fusion is vast—so vast that the world's energy needs could be met for billions of years at current rates of consumption. Laboratory experiments, as well as nuclear explosions, have released fusion energy on the Earth, and it may be feasible to provide useful energy from fusion in the future in a controlled and safe manner. While the fusion process is a nuclear reaction, the results of the reaction do not include many of the problems with fission: First, there are no products like long-lived fission products that cause many of the problems associated with nuclear waste from fission. Second safeguarding uranium and plutonium is not an issue with fusion because those materials are not used. Fusion does not release compounds such as carbon dioxide that contribute to global warming through the greenhouse effect. Third, starting and maintaining a fusion reaction in a controlled manner appropriate for producing useful energy is difficult and always involves a limited amount of fuel at any given moment so that an accidental nuclear runaway reaction in a large quantity of reactants is not possible.

Fusion energy comes from rearranging electrostatic and nuclear forces inside the nucleus, and in this respect fusion is like fission. The rearrangements of both fission and fusion release energy in proportion to the decrease of mass during the reaction according to the famous formula from Albert Einstein's relativity, $E = mc^2$, where E is the energy released, m is the mass lost, and c is the speed of light.

Chemical reactions, such as the burning of carbon in coal with oxygen, also get energy from rearranging electrostatic forces, but those forces are much smaller and consequently chemical energies released per atom are much smaller than nuclear energies released per nucleus.

Nuclei suitable for fusion must come near each other, where "near" means something like the nuclear radius of 10^{-12} cm. For positively charged nuclei to make such a close approach it requires large head-on velocities, and therefore multimillion-degree Celsius temperature. In contrast, fission can occur at normal temperatures, either spontaneously or triggered by a particle, particularly an uncharged neutron, coming near a fissionable nucleus.

Above a temperature of 10,000 degrees or so, matter starts to become ionized, meaning that electrons separate from atoms. Ionized matter, containing ions and electrons, is called "plasma"—a term not to be confused with the "plasma" of physiology and medicine, which means a liquid in which cells are suspended.

Stars are giant spheres of plasma. Stars fuse hydrogen into helium for their primary source of energy. Omitting details of the catalytic role of carbon in stars, four hydrogens combine to yield one helium, two neutrinos, and two positively charged electrons (positrons) with a total kinetic energy of approximately twenty-eight MeV. The unit "MeV," or million electron volts, is convenient for discussions of nuclear energy, as is the "eV" or electron volt (or calorie and Btu) for treatments of chemical reactions. A million

Particle beam fusion accelerator II (PBFA-II) was designed to deliver at least 100 trillion watts of power and was the first machine with the potential to ingnite a controlled laboratory fusion reaction. (U.S. Department of Energy)

electron volts is 1.6×10^{-13} joule, 3.8×10^{-14} calorie, 1.5×10^{-16} Btu, 4.5×10^{-20} kWh (thermal), which makes the MeV sound like a very small unit until one considers that an MeV of released energy is typically associated with approximately one atomic mass unit (AMU), of which there are 6×10^{23} in a gram.

The fusion reaction of four hydrogens to make a helium yields 185,000 kilowatt-hours (thermal) per gram of original hydrogen fuel—an enormous result for such a small mass. Fusion occurs mostly in the core of stars at a temperature near 15 million degrees and the energy released migrates to the much cooler 6,000 degree surface to produce the radiation emitted, some of which is visible light.

While stars shine from the fusion of plain hydrogen, that reaction is unsuitable for use on the earth. In substellar sizes the energy released leaves the reacting region as electromagnetic radiation much too quickly.

Releasing fusion energy on a human scale requires fusion fuels that react with each other more rapidly than the slow-burning basic fuels of stars.

REACTIONS FOR USEFUL ENERGY

The most plausible fusion reaction for producing energy commercially involves two isotopes of hydrogen, deuterium (D) and tritium (T), or $^1H^2$ and $^1H^3$. Deuterium contains one proton and one neutron for an atomic number of two. Tritium contains one proton and two neutrons for an atomic number of three. The reaction is

$$D + T \rightarrow {}^2He^4 + n$$

with the release of 3.5 and 14.1 MeV in the helium and neutron, respectively.

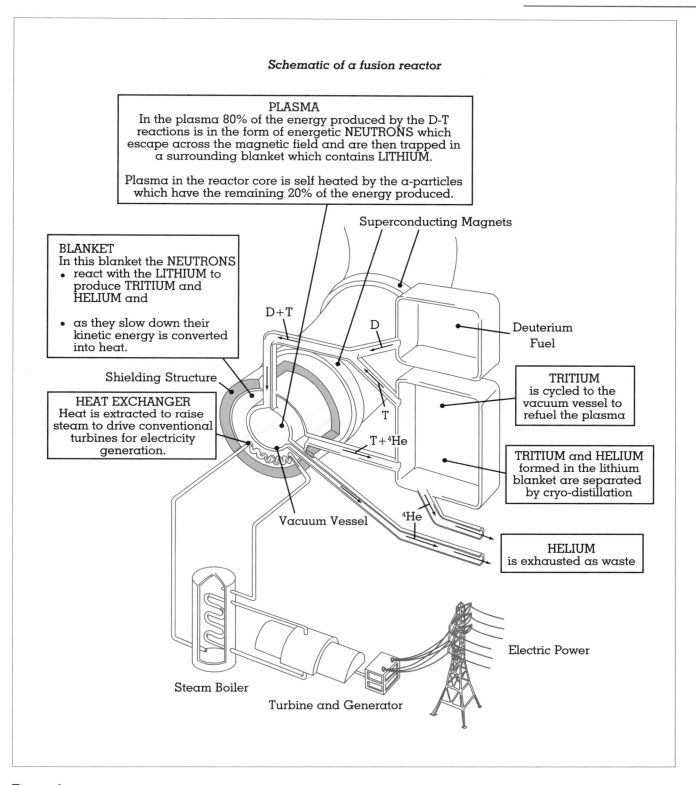

Figure 1.

Schematic of a fusion reactor, assuming a generally toroidal shape of the plasma and magnetic fusion. The principles emphasized are central hot core (red) at 100 million degrees, blanket and heat exchanger, shield, energy conversion, and the handling of D, T, and the "ash" He.

SOURCE: JET Joint Undertaking, Abingdon OX143EA, UK; Graphics Department; JG91.230.

Deuterium occurs naturally, mixed in with plain hydrogen in the tiny proportion of 0.015 percent; in other words, plain hydrogen is the more common isotope by a factor of 6,600. Tritium for fusion energy can be created from another nuclear process involving the interaction of the neutron (in the equation above) with lithium:

$$n + {}^3Li^6 \rightarrow {}^2He^4 + T$$
$$n + {}^3Li^7 \rightarrow {}^2He^4 + T + n$$

with the release of 4.8 MeV in the first reaction and the absorption of 2.5 MeV in the second reaction. The fuel resources for D-T fusion are, therefore, lithium and deuterium. Lithium occurs in brines, mines, and, at the high dilution of only 0.17 parts per million by weight, in sea salts. The resource of lithium in oceans dominates all other sources by a factor of approximately 10,000. And although lithium is plentiful, far fewer lithium atoms than deuterium atoms exist on the Earth making lithium the limiting the resource for D-T fusion.

An essential idea in fusion research is the following: The 3.5 MeV of energy carried by the $^2He^4$, which is also called an α-particle, stays in the fusing region to maintain the high temperature and keep the reaction going. The concept is called "ignition" by analogy to common chemical ignition of fuels in which a small initial spark starts a flame that continues on its own.

Other fusion reactions such as D plus $^2He^3$ and D plus D (not to mention the H plus H of stars) require far more difficult physical conditions than D plus T, but offer potential advantages in reduced neutron production, and even larger reserves of potential energy in the case of D-D.

CONDITIONS FOR USEFUL ENERGY

D-T fusion in substellar sizes on a human scale requires ultrahigh temperatures, approximately 100 million degrees, for the power produced to exceed the fundamental and irreducible cooling effect of power lost to electromagnetic radiation caused by the particles colliding with each other (bremsstrahlung, or braking radiation). Approximations to data on D-T reactivity lead to a simple and often quoted figure of merit for progress on fusion research, ntT, where n is the fuel particle density, t is an average time that energy stays in the fusion region, and T is the temperature.

The best present experiments create ntT approaching the requirements for producing net energy, or 10^{21} m^{-3}-second-keV. (A keV is a kilo electron volt, which is approximately 11 million degrees.) An equivalent statement is that in some present experiments the ratio of fusion energy produced to the energy needed to assemble the hot particles, usually called Q, is near unity. A Q of unity is often called "scientific breakeven," to make clear the contrast with commercial feasibility which requires that fusion energy released be much more than the total energy to operate the machinery in the reactor power station.

The goal of research on nuclear fusion is to produce useful energy in the future as follows: A mixture of D and T is heated to ignition at 100 million degrees, and the fusion energy is released, largely carried by the 14 MeV neutrons produced, streams outwards to a sufficiently thick "blanket" containing lithium and a heat exchange medium, which could in fact be lithium or contain lithium. The neutrons interact with lithium to produce tritium that is retained at the reactor site for later use as fuel. The heat, which comes largely from absorbing the neutron energy in the blanket, is transferred to an external turbine engine to produce electricity with the standard efficiency of 35 to 40 percent. Low temperature heat— constituting of 60 to 65 percent—is not used for electricity. But it can be used for industrial processes, including space heating and refrigeration. The blanket absorbs most neutrons, and a shield outside the blanket absorbs any remaining neutrons and other escaping radiation. The region outside the blanket-shield is then largely free of radioactivity. The blanket, shield, and any material that must remain inside to operate the fusion reactor does become radioactive because of the interaction with the neutrons. The strength and nature of the radioactivity inside the outer surface of the shield has to do with the details of design and the materials used, but remote handling tools will be needed for maintenance. Equipment needed to create and maintain the reacting plasma is outside the shield: for example, electrical power supplies, lasers, ion beams, and superconducting magnets at a few degrees above absolute zero.

At the simplest level there is a remarkable feature of the fusion reactor: a 100 million degree reacting region is near the industrial heat source—the blanket, operating at several hundred degrees. Some kind of insulation must separate the reacting fusion region and the blanket. The two insulation options being explored are

Inertial Confinement Fusion (ICF) concept

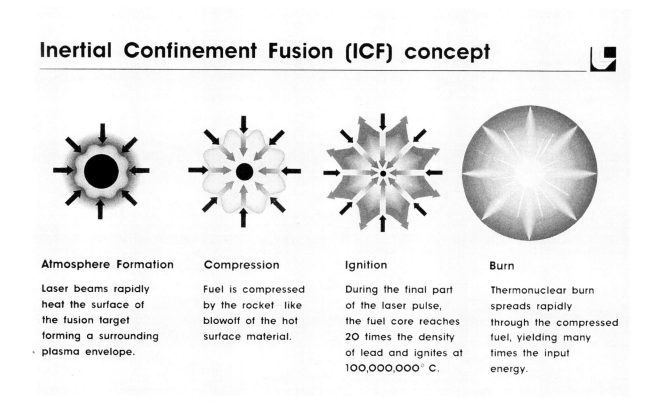

Atmosphere Formation

Laser beams rapidly heat the surface of the fusion target forming a surrounding plasma envelope.

Compression

Fuel is compressed by the rocket like blowoff of the hot surface material.

Ignition

During the final part of the laser pulse, the fuel core reaches 20 times the density of lead and ignites at 100,000,000° C.

Burn

Thermonuclear burn spreads rapidly through the compressed fuel, yielding many times the input energy.

Representation of the inertial confinement fusion (ICF) concept. (U.S. Department of Energy)

magnetic fields for "magnetic fusion" or space for "inertial fusion." The magnetic insulation is much like fiberglass in a walk or ceiling. Magnetic fields slow heat transfer from the motion of plasma particles, much as the fiberglass and trapped air slow the flow of heat through a wall. Space insulation is similar to that obtained in moving back from a blazing fire.

Enhancement of the fusion reaction has been explored in attempts to avoid the extreme requirements for temperature, plus the stringent conditions on density and time. An "atom" of a D or T nucleus and the short lived muon, which is 207 times more massive than the electron, has a much smaller size than a normal D or T atom, so much so that it might be possible to bring the D and T nuclei close enough for fusion at modest conditions. Certain metals can absorb hydrogen so densely that some spontaneous fusioning of D and T might occur. The latter idea is behind the announcement in 1989 of "cold fusion" in an electrolytic cell having platinum and palladium electrodes—an announcement that numerous investigations subsequently proved to be in error. Catalysis, or enhancement of fusion rates has not been pursued as an energy source in recent years.

INERTIAL APPROACH

Inertial confinement fusion has long succeeded in the context of military explosions—the hydrogen bomb. In the military application a fission bomb produces x-rays that drive an implosion of D-T fuel to enormous temperatures and densities such that fusion reactions occur during the short time that inertia keeps the fusing nuclei densely packed and hot.

For civilian purposes the inertial concept is to compress and heat a small sphere of D-T fuel with an external "driver" such as laser light, x-rays produced by lasers, or high energy ion beams. The resulting hot and extremely dense plasma burns the D-T mixture rapid-

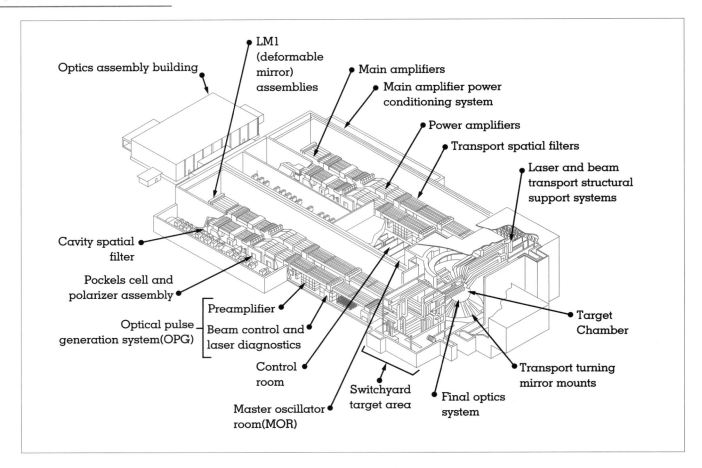

Figure 2.
Layout of the laser and target area building for the National Ignition Facility at the Lawrence Livermore
    National Laboratory in Livermore California.

SOURCE: LLNL ICF Annual Report 1997, p. 95.

ly before the microexplosion expands appreciably. Time scales are nanoseconds (billionth of a second) for each microexplosion, repeated steadily to produce power. Density scales are thousands times solid density. The mission of inertial fusion in the United States is primarily to address the science of high energy densities for the Defense Department program in stewardship of nuclear weapons. Applications to civilian energy needs are a secondary mission.

Much of the driver energy goes into ablation, or blowing-off the surface of the sphere of fuel to force the compression (implosion) by the rocket effect. As a result, the ntT needed for scientific break-even for inertial confinement is around twenty times higher than for magnetic confinement.

The largest inertial fusion program in the United States is the National Ignition Facility (NIF) now under construction at the Lawrence Livermore National Laboratory (LLNL) in Livermore, California (Figure 2). Completion of the stadium-sized construction project is planned for 2003 at an approximate total cost of $1 billion. The goal is to investigate the interaction between spherical targets and ultraviolet laser light producing x-rays in a "hohlraum"—cavity surrounding the target—at a power of 500 tera watts (trillion watts) for several nanoseconds for a total energy of 1.8 MJ (million Joules). Many other significant installations in the United States and around the world are investigating the potential of inertial fusion.

MAGNETIC APPROACH

Magnetic confinement has been pursued for civilian purposes since the 1950s. The idea is that a magnetic field, some of which might be generated by internal heating currents, slows the transfer of energy from the hot plasma core to the surroundings while extra heating power is applied from particle beams or radio frequency transmissions. Time scales are seconds for the energy replacement time, and indefinitely long for an energy-producing burn; density scales are ten thousands below atmospheric gaseous density.

In about 1968 the Soviet Union (now Russia) under the leadership of Academician Lev Artsimovitch took a large step forward in magnetic fusion with a concept called the "tokamak." The term is a Russian acronym for toroidal (inner-tube shaped) magnetic chamber. Since the mid 1970s the tokamak has dominated worldwide magnetic fusion research, culminating in highly successful D-T burning experiments in the United States with the Tokamak Fusion Test Reactor (TFTR) at Princeton University in New Jersey and the Joint European Torus (JET) in the United Kingdom. Numerous other tokamaks in many countries, including Japan (JT-60), France (Tore Supra) , Germany (ASDEX, TEXTOR), Italy (FTU), and the United States (D-III-D, Alcator-C), confirm and extend the progress while not using tritium.

Two essential ideas behind the tokamak are the externally supplied toroidal magnetic field that closes on itself repeatedly, and a slight twist to the field added by currents inside the reacting region. The closed magnetic field lines in the hot plasma do not strike a wall, so that energy is not instantly conducted to a wall. The twist is essential for good and stable confinement of energy. Prior to the emergence of the tokamak idea, the twist needed in closed, toroidal systems was provided often by external coils in the "stellarator" concept developed at Princeton University by astrophysicist Lyman Spitzer.

Many other magnetic fusion concepts have been pursued over the years: for example, the "magnetic mirror" with open field lines that always strike a wall somewhere but with good stability properties; linear and toroidal "pinches" featuring rapid compression and heating. Today the stellarator is enjoying a significant revival with large programs in

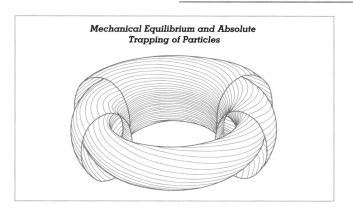

Figure 3.

Schematic representation of the magnetic structure of the Tokamak magnetic confinement device. The lines on the shells represent the direction of the total magnetic field, most of which comes from external coils. The portion that gives the twist, however, comes from current inside the hot plasma itself. The twisting is necessary for stable confinement.

SOURCE: *Transport and Structural Formation in Plasmas* by K. Itoh, S.-I. Itoh, and A. Fukuyama, Institute of Physics Publishing Ltd., 1999. p. 24.

Ring-shaped nuclear fusion research reactor Tokamak 15 at the Kurchatov Institute, Moscow, Russia. (Photo Researchers Inc.)

877

Japan (LHD) and in Germany (W-7-AS), plus smaller experiments elsewhere. At a basic research level, there is interest in a broad array of magnetic approaches.

Still, the tokamak is getting most of the attention, particularly regarding the matter of a large project to follow JET and TFTR and burn D-T so successfully that the plasma ignites—continues to burn on its own—thereby demonstrating in a convincing manner that with further work, the tokamak can be developed into a commercially viable source of energy. The working name for the ignition project is the International Thermonuclear Experimental Reactor (ITER). In 1998, they completed an integrated engineering design, supported by research and development on crucial components, and a reliable budget and schedule. The term "thermonuclear" refers to the multi-million degree temperatures needed to produce energy from nuclear fusion. The four international parties to ITER—Japan, the European Union, the United States, and the Confederation of Independent States (Russia)—had intended to proceed with cooperatively funded construction at a site to be chosen jointly. The cost estimate of approximately 10 billion and political factors have caused ITER to reexamine the design goals, particularly toward a less costly device, but one retaining the most basic and interesting scientific and technical goals. ITER management hopes to go ahead with construction after the year 2000, and operate ignited plasmas in a decade or so.

For both inertial or magnetic fusion energy the long range, but nonspecific, plan is the following: develop concepts to achieve ignition of D-T, whether in ITER, NIF, or other facilities, and proceed to a demonstration reactor (often called DEMO) and then to commercial implementation. DEMO and commercialization are decades away. Even though the ignition DEMO path is being followed by most people working on fusion, researchers generally agree on the need to continue developing the science of fusion and alternative approaches because uncertainties continue about what will eventually prove to be a successful approach.

SUMMARY

Creating useful energy from fusion on the earth has proved difficult: The forty year effort looks like it still has decades to go. The following inherent potential advantages tend to encourage continued effort: abundant fuel in lithium and virtually inexhaustible fuel in deuterium; absence of the greenhouse emissions of coal, oil, natural gas; no issues regarding the control of a nuclear chain reaction; radioactivity induced by fusion neutrons depends on the design of the device, not the fundamentals of nuclear physics.

If and when projects producing energy on an industrial scale begin to appear in the middle of the twenty-first century, the theoretical advantages will be subject to confirmation. Fusion, magnetic or inertial, might then join the mix of energy resources.

D. W. Ignat

BIBLIOGRAPHY

Bromberg, J. L. (1982). *Fusion: Science, Politics and the Invention of a New Energy Source.* Cambridge, MA: MIT Press.

Fowler, T. K. (1997). *The Fusion Quest.* Baltimore: Johns Hopkins University Press.

Häfele, W.; Holdren, J. P.; Kessler, G.; and Kulcinski, G. L. (1977). *Fusion and Fast Breeder Reactors.* Laxenburg, Austria: International Institute for Applied Systems Analysis.

Heppenheimer, T. A. (1984). *The Man Made Sun: The Quest for Fusion Power.* Boston: Little, Brown and Company.

Herman, R. (1990). *Fusion: The Search for Endless Energy.* New York: Cambridge University Press.

Huizenga, J. R. (1992). *Cold Fusion: The Scientific Fiasco of the Century.* Rochester: University of Rochester Press.

Lindl, J. D. (1998). *Inertial Confinement Fusion: The Quest for Ignition and Energy Gain Using Indirect Drive.* New York: Springer-Verlag.

McPhee, J. (1974). *The Curve of Binding Energy.* New York: Farrar, Straus and Giroux.

Niu, K. (1989). *Nuclear Fusion.* New York: Cambridge University Press.

Rhodes, R. (1995). *Dark Sun: The Making of the Hydrogen Bomb.* New York: Simon & Schuster.

Serber, R. (1992). *The Los Alamos Primer: The First Lectures on How to Build an Atomic Bomb.* Berkeley: University of California Press.

Sheffield, J. (1994). "The Physics of Magnetic Fusion Reactors." *Reviews of Modern Physics* 66:1015–1103.

NUCLEAR WASTE

KINDS OF NUCLEAR WASTE

Nuclear waste is radioactive material that has no immediate use. In the United States nuclear waste generally is divided into two main categories: high-level nuclear waste (HLW) and low-level radioactive waste (LLW). Two less common categories are tailings from uranium mining and milling, and a special category derived from particular aspects of nuclear weapons production and defense-related activities. This latter, less commonly discussed, category is called defense wastes and, because of its makeup, is sometimes called transuranic wastes; it makes up about half of the HLW in the United States. This article focuses on HLW and LLW from nuclear power reactors.

HLW comprises most of the radioactivity associated with nuclear waste. Because that designation can cover radioactive waste from more than one source, the term spent nuclear fuel (SNF) will be used to discuss HLW originating from commercial nuclear reactors. LLW comprises nearly 90 percent of the volume of nuclear waste but little of the radioactivity. Nuclear power reactors produce SNF and most of the nation's LLW, although there are approximately 20,000 different sources of LLW. The name SNF is a bit of a misnomer because it implies that there is no useful material left in the fuel, when in fact some fissionable material is left in it.

CONCERN FOR RADIOACTIVE WASTES

Nuclear waste is a concern due to the levels of radioactivity in it, especially in HLW. Relative to the wastes associated with other major methods of power production, the volumes are small. The concern is for the amount of radioactivity and the time of its duration. The time of radioactivity duration is measured in half-lives (the time required for half of the atoms of a given substance to disintegrate, at which time it becomes the new starting amount with which to begin the count again). Generally LLW not only has lower levels of radioactivity, it also has very low concentrations of long-lived radioactive substances. There are short-lived and long-lived substances in nuclear waste, with half-lives varying from seconds to thousands of years.

In the United States the federal government has taken responsibility for HLW, for both SNF and defense wastes, and for mill and mine tailings. Remediation of the effects of such tailings is well under way. Each state or groups of states (compacts) are responsible for the safe isolation of their own LLW. Three departments of the federal government have prime responsibility for matters related to nuclear waste: Department of Energy (DOE) and its predecessor agencies, which created most of the defense wastes; the Nuclear Regulatory Commission (NRC); and the Environmental Protection Agency (EPA). Regulations regarding care of nuclear wastes are also set by individual state agencies, especially for the LLW.

The federal government not only has responsibility for the safe isolation of SNF but also taxes on the generation of electric power to cover the cost of waste isolation at the rate of one mil/kWh ($0.001/kWh). Although the nuclear waste fund has been folded into the general federal budget, it comes from a special levy that can be changed to accommodate the needs of SNF handling. The office that manages this fund and that is responsible for SNF is in the U.S. DOE and is called the Office of Civilian Waste Management (OCRWM).

ORIGINS OF THE RADIOACTIVITY IN SPENT NUCLEAR FUEL

The nuclear waste generated from production arises when uranium atoms in the fuel split. Nuclear power reactors are fueled with assemblies of rods containing pellets, each about the size of a pencil eraser. About 3 percent of the uranium in these pellets is the fissionable isotope, or form, of uranium, U-235. This is the rarer of the two naturally occurring isotopes of uranium. In nature, the U-235 isotope is found in less than 1 percent of uranium ore—usually 0.5 to 0.7 percent of that ore. After uranium ore is mined and processed, the isotopes are separated, in a process called uranium enrichment. This is not a simple task and is mostly done by a gaseous diffusion plant. Usually the process is stopped when the mixture is about 3 to 3.5 percent of the fissionable form in the mix. (The majority uranium, U-238, is radioactive in its own right, emitting a short-range alpha particle, but it is not fissionable. When the fissionable isotope is extracted from uranium, the remaining material is called depleted uranium and has its own other uses,

Radioactive waste piled and tarped in Cleveland. (Greenpeace)

which will be addressed below.) The exact makeup of the reactor fuel varies some what but is mostly in the form of uranium dioxide, with the uranium being about 3 percent fissionable U-235.

There are three main products in the fissioning of uranium-235: tremendous amounts of energy, neutrons that perpetuate the chain reaction, and two new atoms. The splitting is such that eventually every element found on the periodic chart of the elements is made, especially in the flux of speeding neutrons that induced and perpetuated the fission. Every isotope of every element can be found in the waste. Many of the new isotopes produced are highly unstable—that is, they are radioactive. Many decay in seconds, but some exist in their unstable form much longer, eventually decaying into stable isotopes, though often with intermediates that are also radioactive. This results in radioactive decay chains. These chains of radioactive decay mean that SNF will be radioactive for a long time. It has been estimated that it will be approximately 7,000 years before the level of radioactivity in SNF drops to that of natural radioactivity, that of Earth itself. Since some parts of Earth will be radioactive forever, it can be said that SNF will be radioactive forever. In practice, one must decide what levels of radioactivity to call dangerous. That debate influences decisions about nuclear waste isolation. The debate about what is "safe" is not part of this particular topic. The amended version of the Nuclear Waste Policy Act (NWPA) calls for the safe isolation of HLW from the human environment for 10,000 years.

Besides fission products, the various forms of known but newly formed elements in the spent nuclear fuel, there is a small but significant amount of fissionable, or fissile, material in the SNF. This is quite important. There is some unused, unfissioned U-235 that has become too dilute to use. Like natural uranium ores in which chain reactions do not

Solid transuranic interim waste storage at the U.S. Department of Energy's Idaho National Engineering Laboratory in Idaho Falls, Idaho. (U.S. Department of Energy)

occur, the fuel will not sustain a chain reaction. But new fissile material also has been made in the intense neutron irradiation. Elements are made that are heavier than naturally occurring uranium, and because of their location on the Periodic Chart are referred to as transuranics. There are isotopes of plutonium and uranium made that can fission. None is present in concentrations that are great enough for a self-sustained chain reaction, but they do represent new fuel. The concentrations of U-239, Pu-239, Pu-240, Pu-241, and some short-lived intermediates are 1–2 percent of the SNF. Some of these are fissile material. Breeder nuclear reactors can be designed and run to produce significant amounts of these isotopes. "The SNF from a breeder reactor is rich in newly produced fissionable isotopes but it must undergo extensive reprocessing to become new reactor fuel. That reprocessing will adjust the concentration of fission-

able isotopes and eliminate some of the fission products that tend to quench fission process."

HANDLING OF SNF

Workers wearing gloves can handle nuclear fuel that is being freshly installed into a nuclear reactor. This fuel assembly becomes dangerously and highly radioactive SNF after a short time in an operating nuclear reactor. Upon removal from a reactor (after about 18 months), a spent fuel assembly is now HLW and is far too "hot" thermally and radiologically to handle directly. It is removed remotely with a crane and stored in a bay of cooling water beside the reactor.

HLW still looks like a fuel assembly, a collection of long, skinny rods, each filled with fuel pellets, held in a rack that allows water to pass through and pick up thermal energy. The assembly is kept underwater to cool and also to shield the workers from the longer-

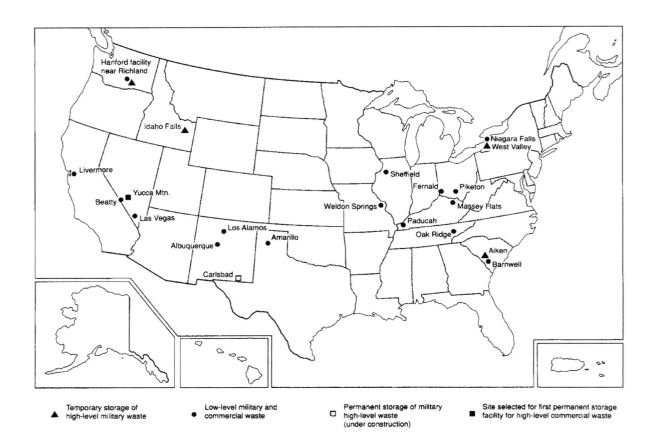

▲ Temporary storage of
high-level military waste

● Low-level military and
commercial waste

☐ Permanent storage of military
high-level waste
(under construction)

■ Site selected for first permanent storage
facility for high-level commercial waste

U.S. radioactive waste disposal sites. (McGraw-Hill, Inc.)

range gamma radiation. The water also contains cor-
rosion-inhibiting chemicals as well as chemicals to
absorb the few neutrons emitted from lingering
atoms of U-235 and other fissile material. Generally
a ten-year cooling-off period allows that much of the
heat to be emitted and the radioisotopes with short
half-lives to have undergone decay. For power plants
with limited pool storage there are two choices when
the cooling pools are filled: rod consolidation and dry
cask storage.

Rod consolidation can permit space saving by
dismantling the fuel assembly racks. The rods are
still maintained underwater. Rod consolidation is
not a routine practice due to issues of heat dissipation and criticality, the possibility of a chain reaction continuing. Dry cask storage is the more
common approach for the oldest fuel assemblies
and calls for removal of the fuel assemblies from the
cooling pool. Designs of the casks vary in detail, but

they usually accommodate several spent fuel assemblies in a sealed steel container that is enclosed in a
concrete box and/or metal canister. Usually these
containers are stored and monitored at the plant
site. As with all major operations at nuclear power
plants in the United States, hearings before the
NRC and the public are conducted on temporary
SNF storage. Almost all dry cask storage occurs at
the power plant site, where monitoring of the containers can be routine.

ULTIMATE ISOLATION OR REPROCESSING OF SNF

The ultimate disposition of HLW or SNF is a matter
of significant importance and is controversial for
some. In the United States all HLW, including SNF,
has always been a federal responsibility, beginning
with the Atoms for Peace Program in the 1950s. A

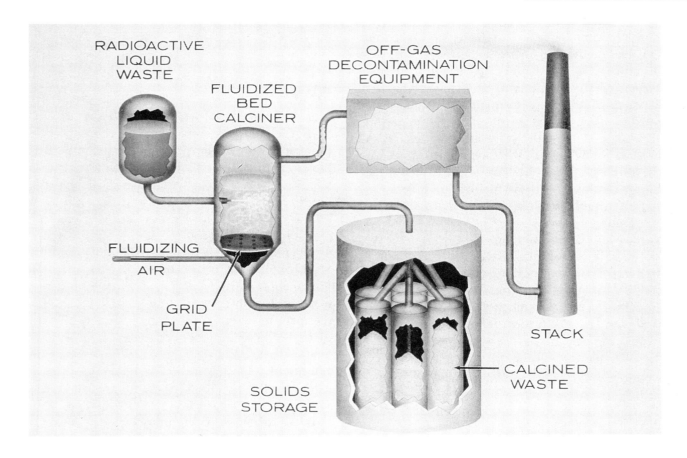

Cutaway view of the new waste calcining facility (NWCF) process. (U.S. Department of Energy)

study done by the National Academy of Sciences in 1957 indicated that nuclear waste disposal would not provide insurmountable technical problems and recommended deep geological isolation. Every country that has nuclear power plants has considered the safe isolation or disposal of HLW and has concluded that deep geologic isolation is the safest method. In the United States eight options, including deep-ocean-trench burial, space launches, Antarctica burial, and transformation into materials with shorter half-lives have all been studied. The media into which the HLW is to be isolated varies widely from country to country and include granite, salt beds, and tuffaceous material, welded volcanic ash, most commonly called tuff. United States attention is focused on Yucca Mountain, at the western edge of the Nevada Test Site, the location of above-and below-ground nuclear weapons testing in the past. It is a mountain of tuff.

YUCCA MOUNTAIN, NEVADA

The amended version of the NWPA has determined that only Yucca Mountain is to be characterized as a possible geologic repository for SNF and possibly some defense waste. It is approximately 100 miles northwest of Las Vegas, Nevada. The planned location of the SNF is approximately 1,000 feet above any groundwater and about the same distance below the surface. Characterization studies of the viability of Yucca Mountain are under way and have been since the early 1990s. The NWPA also prohibits studies of granite as a host rock. (Canada is considering SNF isolation in granite. Most of the eastern half of the United States is on granite.) Both private and governmental leaders—most concerned about tourism—in the state of Nevada are resisting the Yucca Mountain characterization studies. Furthermore, although the majority of scientists studying the mountain find no clear problems, a few find fault with the studies ad the site as a possible repository. The most difficult part of the studies involves predictions about the behavior of the tuff environment for 10,000 years and beyond. The fact that Yucca Mountain is in a region of ancient volcanic activity complicates the assessments. Most attention is devoted to water movement through the tuff and the vulnerability of the underlying earth to undergo any significant "stretching" in the foreseeable future. The Yucca Mountain project has a website listed in the bibliography for this article, with photos and links to scientific data being accumulated as well as links to opposition points of view.

Yucca Mountain, if it becomes the site for the isolation of SNF, will be laced with tunnels, waste in storage casks and monitoring equipment. A waiting period is planned while better isolation alternatives are sought. If Yucca Mountain is not used, it is to be refilled with the tuff material removed earlier. In the United States the SNF that would be isolated in Yucca Mountain would be waste that has not been reprocessed; it would be material that has come out of nuclear reactors and has been cooled at the plant site.

WASTE REPROCESSING

Since the amount of fissile material in the fuel assemblies is only about 3 percent of the uranium present, it is obvious that there cannot be a large amount of radioactive material in the SNF after fission. The neutron flux produces some newly radioactive material in the form of uranium and plutonium isotopes. The amount of this other newly radioactive material is small compared to the volume of the fuel assembly. These facts prompt some to argue that SNF should be chemically processed and the various components separated into nonradioactive material, material that will be radioactive for a long time, and material that could be refabricated into new reactor fuel. Reprocessing the fuel to isolate the plutonium is seen as a reason not to proceed with this technology in the United States.

Congress has decided that reprocessing will not be practiced in this country so that we will not be in the plutonium production business. This seems like a safe thing to do since this action will minimize terrorism threats. Reprocessing generates chemical wastes but greatly reduces the volume of the highly radioactive waste. It also isolates plutonium and unused fuel for possible use as new fuel.

Reprocessing means that the volume of material calling for long-term isolation is reduced by one-third what it would have been before reprocessing.

THE VOLUME OF SNF

If one just concentrates on the radioactive material in SNF, the volume is very small, especially compared to waste from other power production practices. However, one can only discuss the separated radioactive material if it has undergone extensive reprocessing. If SNF is to be isolated, as in a place such as Yucca Mountain, with perhaps 70 miles of tunnels, the volume is that of the interior of this minor mountain. Isolation of up to 100,000 metric tons of SNF in Yucca Mountain means that for the United States, approximately all the SNF made to date and that expected in the operating lifetime of all current reactors can be put there. Approximately 2,000 metric tons of SNF are produced each year in the United States. Waste volume and placement depend on the amount of compaction and consolidation at the sites. The plans for the Yucca Mountain present a realistic and understandable picture of the volume of SNF.

A useful perspective may be seen about the amount of SNF by comparing it to the volume of other waste streams associated with power production. More than half of the electricity made in the United States comes from coal burning, where each large power plant generates bottom and fly ash in volumes measured in acre-feet annually. Each of these plants generates its own small mountain of ash. The gaseous wastes from coal burning and from methane and oil burning result in tons of carbon dioxide daily. This carbon dioxide is an infrared active gas and is thought by many to be contributing to global warming and climate change phenomena. The wastes associated with nuclear power are small in comparison, this is not surprising, considering the tremendous power in the nucleus of an atom.

LOW-LEVEL RADIOACTIVE WASTE

The other major category of nuclear waste, LLW, is generally that which is neither HLW nor the transuranic part of defense waste. Generally, LLW is generated in private and public labs, hospitals, and commercial enterprises and can involve lab clothing, paper, packing material and radioactive isotopes used in medical procedures, both diagnostic and therapeutic. Significant amounts of LLW are associated with defense wastes, but mostly are isolated at U.S. DOE facilities. Many pharmaceutical products require extensive use of radioactive tracers. Academic sources are minor. The major sources of LLW in volume and amount of radioactivity are nuclear power plants. Aside from the decommissioning of nuclear power plants, the volume of LLW is decreasing every year. This is because new technologies have become available to separate the radioactive material from that which is not radioactive. Also, generators of LLW are being more selective in the work that they do so that they generate less LLW each year. These changes have become about partly because the cost of isolating LLW has increased.

The LLW from nuclear power plants contains ion exchange resins as well as clothing, tools, and chemicals. Ion exchange resins, which comprise the majority of this LLW, are used to filter the water circulated in nuclear power plants. The ion exchange resins isolate and trap dissolved materials, much of which can be radioactive. Approximately three-fourths of commercially or privately generated LLW is in the form of the contaminated plastic beads that make up ion exchange resins.

The NRC categorizes LLW into Class A, B, or C. The wastes in Classes A and B contain materials that decay to safe levels in about 100 years. Class C wastes will require about 500 years to reach safe levels and contain mostly ion-exchange resins and filters. (There is a minor amount of LLW referred to as beyond class C material.) The majority of LLW is dry and is stored in cardboard or wooden crates. Some are in metal drums. LLW is most likely given shallow burial, compared to the deep geological isolation required for HLW. A difficult issues in LLW disposed involves a small amount of "mixed waste," where the radiological material is mixed with chemically or biologically hazardous substances. The contaminated chemicals may require special handling, since hazardous chemical wastes do not decay as do radioactive materials. In addition, a small amount of LLW consists of animal carcasses or waste and also needs special handling. Currently there are three commercial sites handling LLW for most of the states and compacts. The largest volume of LLW will come with the decommissioning and dismantling of nuclear power plants. In 1995 nearly 700,000 cubic feet of LLW were handled at commercial sites in the United States.

LLW is placed in sturdy, sealed containers. At the isolation site these containers are first put in concrete and/or metal vaults or bunkers and then buried in shallow trenches before being covered with backfill. Some of these are then paved over to prevent rainwater from entering the waste.

The LLW Forum coordinates information and regulations among and between the various state compacts and states. It maintains a website that is overseen by the Idaho Operations Office, which is part of the National Engineering Lab in Idaho. Although LLW is a responsibility of each compact, or of state for the LLW generated in that state, the regulations governing it are those of the U.S. DOE, the NRC, and the EPA. The regulations governing its isolation allow individual states to set additional requirements for handling the material. After being established, any particular compact can refuse to receive the LLW from other states.

OTHER RADIOACTIVE-WASTE SITES

The Waste Isolation Pilot Plant (WIPP) is in an excavated salt cavern in southern New Mexico, twenty-seven miles from Carlsbad. The WIPP site is 2,000 yards underground, and defense waste is being placed. There are plans to place there about 6 million cubic feet of material there containing fewer than five million curies of radio activity.

The most technically difficult category of HLW is that belonging to the U.S. DOE at its major plutonium production facilities in Hanford, Washington, and at the Savannah River facility near Aiken, South Carolina. Just sampling the material for characterization presents problems since much of it consists of radioactive acidic liquids. The rest of it in sludge form, and some is solid. A plant for the vitrification of that part of the material that can be made into glass bars is in operation at the Savannah River facility. The vitrification process encases the waste in glass "logs" to immobilize it and make it easier to handle. A conservative estimate is that it will take decades to clean up these facilities.

NUCLEAR WASTE IN OTHER COUNTRIES

The United States has the most radioactive nuclear waste and the most complicated array of waste types. Reprocessing of SNF is also practiced in some countries. Although costly, this practice reduces the volume of HLW requiring deep geologic disposal. (In the wide variety of elements in the fission products making up SNF, some are toxic heavy metals.) Reprocessing and the general lack of economy of scale in many other countries help to explain why there is less activity or progress related to HLW isolation elsewhere. Interim monitored retrievable storage is generally the focus of activity abroad. Different philosophies and cultures also mean that a wide variety of approaches to the problem are found. Some feel that the generation deriving the benefits of nuclear power is the one responsible for a solution. Others say that succeeding generations may have better technology or ideas about this problem than current generations: Current generations should not do anything permanent, committing future generations to the solutions seen by people living today. Among the technology that wold drastically reduce the cost of cleanup of waste sites robotics.

All the countries that produce nuclear waste have chosen the same alternative for the ultimate disposition of HLW, deep geological isolation, and they did so independently of one another. The United States has the most radioactive nuclear waste and the most complicated array of waste types of any "nuclear" country. Only in the United States can one find the same economy of scale for waste handling. Thus, it leads the world in most activities aimed at safe isolation. In France, Japan, and Great Britain, however, reprocessing is routinely practiced. Those countries reprocess HLW for many other countries. As mentioned above, reprocessing is not currently allowed in the United States.

France and Germany, local protests have dramatically slowed the choice of a final isolation site. As a result, interim storage is widely practiced in other countries. In Sweden, the waste is mixed in molten copper. The radioactive waste is immobilized in copper logs that are easily handled and stored in large underground caverns until a more permanent isolation is chosen. While most feel that those making the waste have the primary responsibility for its isolation, in Sweden they do not wish to commit future generations to solutions that might be vitrified; the glass "logs" are in concrete bunkers.

In the former U.S.S.R. vast areas of the country are contaminated by poor handling of nuclear waste, especially from that associated with the manufacturer of weapons. Some radioactive waste, especially from nuclear submarines has been isolated at the bottom of the Baltic Sea by the former Russian Navy.

CONCERNS

Transportation of HLW is among the most immediate concerns for the general public. Most experts agree that nuclear wastes, especially SNF, are being kept in places not intended for long-term storage and they acknowledge that the problem must be addressed. There is general agreement that SNF should be kept far from population centers (where the electricity is generated and used) and should be kept in a dry place, since water the most likely medium for its movement into the human environment. Movement of SNF is necessary, most likely by rail or roadway. Groups of U.S. DOE and state regulators are working together, continually revising transportation plans and designing and testing transportation casks. Some wish to see a combined transportation and burial cask, while others want very different qualities in each sort of cask. Planners build on past experience handling and transporting radiological materials. Extensive experience was gained when the first commercial nuclear power plant, at Shippingport, Pennsylvania, was decommissioned in 1989 and the reactor vessel was moved to Hanford, Washington, where it is buried.

Despite the challenges, many see nuclear waste issues as being mostly political and social. There is a growing awareness that technical answers alone will not solve the political and social concerns. Some consider nuclear waste issues small in comparison to the volume and challenges associated with other kinds of wastes, whether generated by power plants or other human activities.

Donald H. Williams

See also: Environmental Problems and Energy Use; Government Agencies; Nuclear Energy; Nuclear Energy, Historical Evolution of the Use of; Nuclear Fission; Nuclear Fission Fuel; Nuclear Fission.

BIBLIOGRAPHY

American Nuclear Society. <http://ans.org>.
American Nuclear Society. (annual). *High Level Radioactive Waste Management: Proceedings for the International Topical Meeting of the American Nuclear Society and the American Society of Civil Engineers.* Chicago: American Nuclear Society.

Murray, R. L. (1994). *Understanding Radioactive Waste*, 4th ed. Columbus, OH: Battelle Press.

Physics Today. (1998). 50(6).

Radwaste Magazine: Progress in Radioactive Waste Management and Facility Remediation. (monthly).

Savage, D., ed. (1995). *The Scientific and Regulatory Basis for the Geological Disposal of Radioactive Waste*. New York: Wiley & Sons.

Study Committee Home Page. <http://www.study committee.org>.

U.S. Department of Energy. (1996) "Integrated Data Base Report 1995: U.S. Spent Nuclear Fuel and Radioactive Waste Inventories, Projections and Characteristics, DOE/RW-0006, Revision 12." Washington, DC: Author.

U.S. Department of Energy: Office of Civilian Radioactive Waste Management. *The Yucca Mountain Project*. <http://www.ymp.gov>.

U.S. Environmental Protection Agency. *Yucca Mountain Home Page*. <http://www.epa.gov.radiation/yucca>.

O

OAK RIDGE NATIONAL LABORATORY

See: National Energy Laboratories

OCEAN DRILLING

See: Oil and Gas, Drilling for

OCEAN ENERGY SYSTEMS

Two-thirds of Earth's surface is covered by oceans. These bodies of water are vast reservoirs of renewable energy. In a four-day period, the planet's oceans absorb an amount of thermal energy from the sun and kinetic energy from the wind equivalent to the world's known oil reserves. Several technologies exist for harnessing these vast reserves of energy for useful purposes. The most promising are ocean thermal energy conversion (OTEC), wave power plants, and tidal power plants. All of these produce electricity from the oceans' reserves of renewable energy. Because the ultimate source of energy from the oceans is solar radiation (or the gravitational force of the sun and the moon in the case of tidal energy), ocean energy systems are renewable, have no fuel costs, and are relatively nonpolluting when compared

to conventional sources of energy, such as fossil fuels. To date, the technologies for harnessing the oceans' energy on a large scale are still in the early stages of development and have high initial costs, making them more expensive than conventional alternatives.

OCEAN THERMAL ENERGY CONVERSION

Each day, tropical oceans absorb the energy equivalent to 250 billion barrels of oil. If less than one-tenth of 1 percent of this energy could be converted into electricity it would supply twenty times the amount of electricity, consumed daily in the United States. Unfortunately, this energy is spread out over 23 million square miles of ocean, providing a large volume of slightly heated water.

Ocean thermal energy conversion (OTEC) power plants generate electricity by exploiting the difference in temperature between warm water at the ocean surface and colder waters found at ocean depths. To effectively capture this solar energy, a temperature difference of 35°F or more between surface waters and water at depths of up to 3,000 feet is required. This situation can be found in most of the tropical and subtropical oceans around the world that are in latitudes between 20 degrees north and 20 degrees south.

The 35 degree temperature difference is necessary because extracting a little bit of heat from a large volume of water is inherently inefficient. In order to harness thermal energy, two reservoirs of heat at different temperatures are required. In such a "heat engine," heat from the high-temperature reservoir (high-grade energy) flows to the low-temperature reservoir (low-grade energy).

The maximum possible efficiency at which a heat engine can work is defined by the Carnot efficiency equation $E = (T2–T1)/T2$, where E is the efficiency of the heat engine, T1 is the temperature of the cold reservoir (Kelvin), and T2 is the temperature of the hot reservoir (Kelvin). Given tropical surface water temperatures of 80°F (300 K) and cold deep ocean temperatures of 40°F (278 K), this yields a maximum theoretical efficiency of about 7 percent. In comparison, the temperature differences found in conventional steam turbines result in maximum theoretical efficiencies of around 60 percent.

An OTEC facility, like any electric power plant, must use energy to run pumps and other electrically driven devices. It requires large amounts of electric energy to pump huge amounts of water from great depths. When the typical operating losses associated with a real-world OTEC power plant are taken into account, it is a challenge for engineers to design a facility that will be a net producer of electricity.

The concept of OTEC was first envisioned by the French physicist Jacques-Arsene d'Arsonval, in 1881. The first working OTEC system was built in Cuba in 1930, by the physicist Georges Claude, who also invented the neon lamp.

OTEC power plants can be located either onshore or at sea. The electricity generated can be transmitted to shore by electrical cables, or used on site for the manufacture of electricity-intensive products or fuels (such as hydrogen). For OTEC plants situated on shore to be economical, the floor of the ocean must drop off to great depths very quickly. This is necessary because a large portion of the electricity generated by an OTEC system is used internally to pump the cold water up from the depths of the ocean. The longer the cold water pipe, the more electricity it takes to pump the cold water to the OTEC facility, and the lower the net electrical output of the power plant.

There are three potential types of OTEC power plants: open-cycle, closed-cycle, and hybrid systems. Open-cycle OTEC systems exploit the fact that water boils at temperatures below its normal boiling point when it is under lower than normal pressures. Open-cycle systems convert warm surface water into steam in a partial vacuum, and then use this steam to drive a large turbine connected to an electrical generator. Cold water piped up from deep below the oceans surface condenses the steam. Unlike the initial ocean water, the condensed steam is desalinated (free of salt) and may be collected and used for drinking or irrigation.

Closed-cycle OTEC systems use warm surface waters passed through a heat exchanger to boil a working fluid that has a low boiling point, such as

ammonia or a chlorofluorocarbon. The vapor given off is passed through a turbine/generator producing electricity. Then cold deep ocean water is used to condense the working fluid and it is returned to the heat exchanger to repeat the cycle. Hybrid OTEC systems combine both technologies and produce both electricity, with a closed-cycle system, and fresh water, with an open-cycle system.

Unlike electrical generation from most other forms of renewable energy that vary with weather and time of day, such as solar and wind energy, OTEC power plants can produce electricity 24 hours per day, 365 days per year. This capability makes OTEC an attractive alternative to conventional baseload electric power plants powered by fossil fuels or nuclear fission. Fresh water production is just one of the potentialy beneficial by-products of OTEC. The cold deep ocean water can also be used for aquaculture (fish farming), as it is pathogen free and nutrient rich. It can also be the source of air conditioning and refrigeration in nearby buildings.

OTEC power plants can have some negative impacts on the natural environment, but overall they are a relatively clean source of electricity compared to conventional options. Cold water released at the oceans surface releases trapped carbon dioxide, a greenhouse gas, but emissions are less than 10 percent of those from a fossil fuel power plant producing a similar amount of power. Also, discharging the cold water at the oceans surface could change local concentrations of nutrients and dissolved gases. This can be minimized by discharging the cold water at depths of greater than 200 feet and/or by using the cold water for air conditioning or refrigeration before it is released.

Despite the fact that OTEC systems have no fuel costs and can produce useful by-products, the initial high cost of building such power plants (up to $5,000 per kilowatt) currently makes OTEC generated electricity up to five times more expensive than conventional alternatives. As such, at the present time OTEC systems are largely restricted to experimental and demonstration units. One of the major facilities for OTEC research is the Natural Energy Laboratory of Hawaii Authority at Keahole Point on the island of Hawaii. An experimental OTEC facility located there has had a maximum *net* power production of 100 kilowatts, at the same time producing 5 gallons per minute of desalinated water.

The most promising markets for OTEC are more experience in building OTEC power plants and stan-

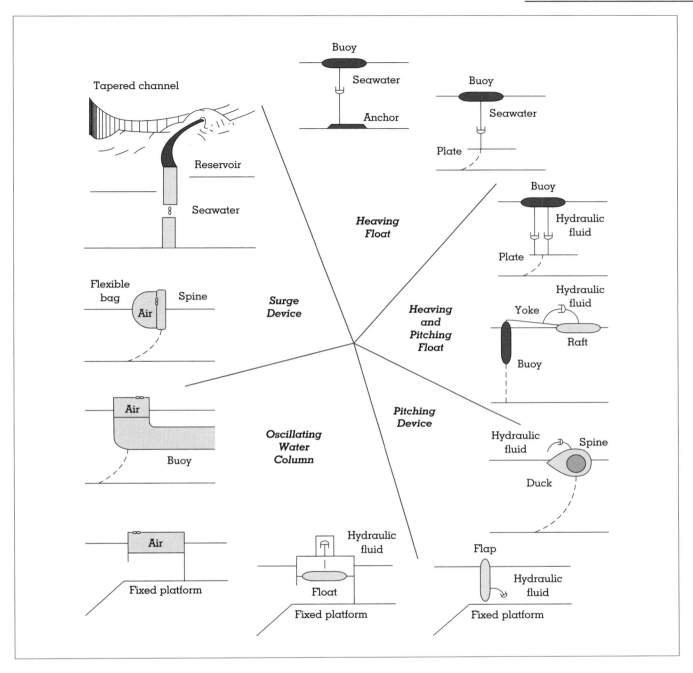

Figure 1.
Symbolic representation of various types of wave energy devices.

dardized plant designs could bring OTEC costs down in the future. However, the initial high costs, coupled with the risk of storm damage to expensive offshore or coastal OTEC facilities, may mean that OTEC electricity generation will never be competitive with conventional sources of electricity

WAVE POWER

Winds blowing across the surface of the world's oceans are converted into waves. The total amount of power released by waves breaking along the world's coastlines has been estimated to be 2 to 3 million megawatts, equivalent to the output of about 3,000 large power plants. Although this vast amount of energy is spread out along thousands of miles of coasts, in favorable locations, the energy density can average 65 megawatts per mile of coastline, an amount that can lead to economical wave-generated electricity. As of 1996, 685 kilowatts of grid-connected wave-generating capacity was operating worldwide.

Wave size is determined by wind speed and "fetch," the distance over the oceans surface which the wind travels. Favorable wind energy sites are generally western coastlines facing the open ocean such as the Pacific Coast of North America and the Atlantic Coast of Northern Europe. Norway, Denmark, Japan, and the United Kingdom are the world leaders in wave energy technologies.

Although a wide variety of devices have been constructed to capture wave energy, commercial electrical generation from wave power is derived from one of three general types of devices: surface-following devises, oscillating water columns, and focusing, or surge devices.

Surface-following devices use a mechanical linkage between two floating objects or between a floating and a fixed object to produce useful mechanical power. This mechanical power can either be connected directly to a generator or transferred to a working fluid, such as water or air, which drives a turbine generator.

One high-efficiency-surface following device that has been tested is the "Salter duck," named after its developer Steven Salter at the University of Edinburgh. It looks like a series of floating ducks that pivot about a stiff shaft; this pivoting motion drives a hydraulic fluid to produce electricity. Tests on the Salter ducks show that they are capable of capturing 80 percent of the energy carried by incoming waves.

Oscillating water columns (OWC) use the force of waves entering a fixed device to perform work or generate electricity. The simplest examples of these are navigational buoys; waves entering the anchored buoy compress air in a vertical pipe. This compressed air can be used to blow a warning whistle, or to drive a turbine generator producing electricity for a light. Japan has installed hundreds of OWC-powered navigational buoys along its coastlines since 1965. Larger OWC power plants include the United Kingdom's experimental OSPREY (Ocean-Swell Powered Renewable Energy), a 2-megawatt offshore facility designed so that waves enter a fixed, submerged chamber, pushing air through a turbine with the rise and fall of each wave.

Focusing, or surge, devices are humanmade barriers that channel and concentrate large waves into a small area, drastically increasing their height. The elevated waves are channeled into an elevated reservoir. This water then passes through hydroelectric turbines on its way back to sea level, thus generating electricity. A 350-kilowatt grid-connected power plant has been operating on the North Sea coast of Norway since 1986. It uses a "tapered channel" design to focus the waves. The tapered channel system uses a narrowing concrete channel to funnel waves up into a reservoir located above sea level. This water then flows through a turbine on its way out, thereby generating electricity.

One of the problems associated with wave power plants is that, during severe storms, the energy unleashed by breaking waves can be over ten times that of the average wave for a coast. The original OSPREY power plant was destroyed by a storm shortly after its construction. The need to design wave power plants to survive such power adds a great deal to the cost of wave energy facilities. Although focusing devices are the most expensive of the three types of wave power plants to construct, their robust nature makes them less susceptible to damage by storms. The Norwegian Tapered Channel design has proven to be the most durable of any wave energy plants. In the late 1990s, this led to the production of two commercial multimegawatt power plants based on this design-one in Java, Indonesia, and one on King Island located between Tasmania and the Australian mainland.

Wave energy power plants consume no fuel during their operation and, as such, do not emit harmful pollutants. However, large-scale wave energy facilities can have an impact on nearby coastal environments. Offshore wave energy facilities such as surface followers or OWCs reduce the height of waves reaching the shore, potentially changing patterns of erosion and sedimentation. Focusing devices result in more erosion where the waves areconcentrated and more sedimentation in adjacent areas. The effect that these changes in erosion and sedimentation could potentially have on local ecosystems is variable and not fully understood by researchers.

TIDAL ENERGY

In coastal areas that have large tides, flowing tidal waters contain large amounts of potential energy.

The principle of harnessing the energy of tides dates back to eleventh-century England when tides were used to turn waterwheels, producing mechanical power. More recently, rising and falling tides have been used to generate electricity, in much the same manner as hydroelectric power plants.

Tides, the daily rise and fall of ocean levels relative to coastlines, are a result of the gravitational force of the moon and the sun as well as the revolution of Earth. The moon and the sun both exert a gravitational force of attraction on Earth. The magnitude of the gravitational attraction of an object is dependent upon the mass of an object and its distance. The moon exerts a larger gravitational force on Earth because, although it is much smaller in mass, it is a great deal closer to Earth than the sun. This force of attraction causes the oceans to bulge along an axis pointing toward the moon. Tides are produced by the rotation of Earth beneath this bulge in its watery coating, resulting in the rhythmic rise and fall of coastal ocean levels.

The gravitational attraction of the sun also affects the tides in a similar manner as the moon, but to a lesser degree. As well as bulging toward the moon, the oceans also bulge slightly toward the sun. When Earth, the moon and the sun are positioned in a straight line (called a full, or new, moon), the gravitational attractions are combined, resulting in very large "spring" tides. At half-moon, the sun and the moon are at right angles, resulting in lower tides called "neap" tides. Coastal areas experience two high and two low tides over a period of slightly longer than twenty-four hours. The friction of the bulging oceans acting on spinning Earth results in a very gradual slowing down of Earth's rotation, a phenomenon that will not have any significant effect on the planet for billions of years.

Certain coastal regions experience higher tides than others, which is a result of the amplification of tides caused by local geographical features such as bays and inlets. In order to produce practical amounts of electric power, a difference of at least 20 feet between high and low tides is required; the higher the tides, the more electricity can be generated from a given site, and the lower the cost of electricity produced. Worldwide, there are about forty sites with this magnitude of tidal range, and approximately 3000 gigawatts of power is continuously available from the action of tides. Due to the constraints of harvesting tidal energy, estimate that only 2 percent, or 60 gigawatts, can potentially be recovered for electricity generation.

The technology required to convert tidal energy into electricity is very similar to the technology used in traditional hydroelectric power plants. The first requirement is a dam or "barrage" across a tidal bay or estuary. Building dams is an expensive process, therefore the best tidal sites are those that exist where a bay has a narrow opening, thus reducing the length of dam which is required. At certain points along the dam, gates and turbines are installed. When there is an adequate difference in the elevation of the water on the different sides of the barrage, the gates are opened. This "hydrostatic head" that is created causes water to flow through the turbines, turning an electric generator to produce electricity.

Electricity can be generated by water flowing both into and out of a bay. Because there are two high and two low tides each day, electrical generation from tidal power plants is characterized by periods of maximum generation every twelve hours, with no electricity generation at the six-hour mark in between. Alternatively, the turbines can be used as pumps to pump extra water into the basin behind the barrage during periods of low electricity demand. This water can then be released when demand on the system its greatest, thus allowing the tidal plant to function with some of the characteristics of a "pumped storage" hydroelectric facility.

The demand for electricity on an electrical grid varies with the time of day, with demand highest during the day and lowest at night. Without pumped storage, the supply of electricity from a tidal power plant, which is dependent upon the slowly shifting times of the tides, can never match the demand on a system. But tidal power, although variable, is reliable and predictable and can make a valuable contribution to an electrical system that has a variety of sources. When tidal electricity is used to displace electricity that would otherwise be generated by fossil fuels or nuclear fission, it results in reduced emissions of greenhouse and acid gases, the risks associated with nuclear power.

Although the technology required to harness tidal energy is well established, because tidal power is expensive there is only one major tidal generating station in operation in the world today. This is a 240-megawatt power plant at the mouth of the La Rance River estuary on the northern coast of France, generating power roughly equal to the annual consumption of the nearby town of Rennes, which has a population of 200,000.

The Barrage de la Rance was the first hydroelectric power plant to generate energy using tidal power. (Corbis-Bettmann)

In operation since 1967, the La Rance generating station has been a very reliable source of electricity, producing electricity at a cost of 3.7 cents per kilowatt hour. Initially, La Rance was designed to be the first of many tidal power plants in France, but the country's nuclear program was greatly expanded in the late 1960s, thereby prohibiting the development of multiple tidal power plants. Elsewhere there is a 20-megawatt experimental facility at Annapolis Royal in Nova Scotia, and a 400-kilowatt tidal power plant near Murmansk in Russia.

Studies have been undertaken to examine the potential of several other tidal power sites worldwide. Scientists estimate that a barrage across the Severn River in western England could supply as much as 10 percent of the country's electricity needs (12 gigawatts). Similarly, several sites in the Bay of Fundy, in Cook Inlet, Alaska, and the White Sea in Russia have been found to have the potential to generate large amounts of electricity.

Like wave and OTEC power plants, one of the main barriers to the increased use of tidal energy is the initial cost of building tidal-generating stations. It has been estimated that the construction of the proposed facility on the Severn River in England would have a construction cost of $15 billion.

The major factors in determining the cost effectiveness of a tidal power site are the size (length and height) of the barrage required, and the difference in height between high and low tide. These factors can be expressed in what is called a site's "Gibrat" ratio. The Gibrat ratio is the ratio of the length of the barrage in meters to the annual energy production in kilowatt hours. The smaller the Gibrat site ratio, the more desirable the site. Examples of Gibrat ratios are La Rance at 0.36, Severn at 0.87, and Passamaquoddy in the Bay of Fundy at 0.92.

Tidal energy is a renewable source of electricity that does not result in the emission of gases responsible for global warming or acid rain, which are associated with fossil fuel-generated electricity. Use of tidal energy could also decrease the need for nuclear power. Changing tidal flows by damming a bay or an estuary could, however, result in negative impacts on aquatic and shoreline ecosystems, as well as on navigation and recreation.

OTHER OCEAN ENERGY TECHNOLOGIES

Theoretical concepts for generating electricity from ocean currents such, as the Gulf Stream, and salinity gradients (differences in salt content) are being investigated. More research and development is required before these concepts reach the stage of demonstration power plants.

Stuart E. Baird

BIBLIOGRAPHY

Avery, W. H., and Berl, W. G. (1997). "Solar Energy from the Tropical Oceans." *Issues in Science and Technology* 14(2):41–42.

Bequette, F. (1998). "Harnessing Ocean Energy." *UNESCO Courier* 51(7/8):33–35.

Bernshtein, L. B. (1995). "Tidal Power Development—a Realistic, Justifiable and Topical Problem of Today." *IEEE Transactions on Energy Conversion* 10:591–599.

Bryden, I. G.; Naik, S.; and Fraenkel, P. (1998)."Matching Tidal Current Plants to Local Flow Conditions." *Energy* 23:699–709.

Carless, J. (1993). *Renewable Energy: A Concise Guide to Green Alternatives.* New York: Walker and Company.

Cavanagh, J. E.; Clarke, J. H.; and Price, R. (1993). "Ocean Energy Systems." In *Renewable Energy,* ed. L. Burnham. Washington, DC: Island Press.

Curran R.; Stewart, T. P.; and Whittaker T. J. T. (1997). "Design Synthesis of Oscillating Water Column Wave Energy Converters: Performance Matching." *Journal of Power and Energy* 211:489–505.

D'Monte, D. (1997). "Construction Begins on India's First Tidal-Energy Power-Plant Prototype." *Solar Letter* 7(3):42–43.

DiChristina, M. (1995). "Sea Power." *Popular Science* 246(5):70–74.

Edwards, R. (1998). "The Big Break." *New Scientist* 160(2154):30–34.

Fraenkel, P. (1998). "Marine Current Power—An Emerging Energy Resource for the Millenium." *Renewable Energy World* 1(2):64–69.

Golob, R., and Brus, E. (1993). *The Almanac of Renewable Energy.* New York: Henry Holt.

Grant, A. D. (1997). "Simplified Tidal Barrage for Small-Scale Applications." *Journal of Energy Engineering* 123:11–19.

Pierce, F. (1998). "Catching the Tide." *New Scientist* 158(2139):38–41.

Patterson, W. C. (1990). *The Energy Alternative.* London: Boxtree Ltd.

Sanders, M. M. (1991). "Energy from the oceans." In *The Energy Sourcebook,* eds. Ruth Howes and Anthony Fainberg. New York: American Institute of Physics.

Sing, M. (1998). *The Timeless Energy of the Sun.* New York: Sierra Club Books/UNESCO.

OCTANE ADDITIVES

See: Gasoline and Additives

OCTANE NUMBER

See: Gasoline and Additives

OERSTED, HANS CHRISTIAN (1777–1851)

Hans Christian Oersted, the son of an impoverished pharmacist, made the great discovery that electricity and magnetism are related. Oersted was born on the small Danish island of Langeland, about halfway between Copenhagen and Hamburg. There was no school in Langeland, so he and his younger brother, Anders Sandoe, went to the homes of neighbors who taught the boys to read and write. Later the town surveyor taught them mathematics, and the mayor taught them English and French. When he was twelve, Hans began to help his father in the pharmacy, and the work stimulated his interest in science.

In 1794 Oersted and his brother Anders matriculated at the University of Copenhagen. Hans studied the sciences, and Anders, who eventually became a leading jurist and a minister of state, studied law. The

Hans Christian Oersted. (Library of Congress)

brothers were recipients of a small state scholarship, but largely supported themselves at the university by tutoring. They lived together, shared costs, and devoted themselves wholeheartedly to their studies. In 1797 Hans Christian Oersted was awarded a degree in pharmacy and, in 1799, received a Doctor of Philosophy degree.

After graduation, Oersted secured a position as a part-time lecturer at the university; he also managed a Copenhagen pharmacy. Word of Alessandro Volta's discovery of a way of producing a continuous electric current reached Copenhagen in 1800, and Oersted began experimenting with acids and alkalis using a voltaic pile. The following year, he left Copenhagen and visited a number of famous scientists during the traditional year of travel taken by European students following graduation. If all the scientists he met, Oersted was most influenced by Johann Wilhelm Ritter, an eccentric German physicist whom he visited for several weeks in Jena. Ritter had also begun his career as a pharmacist and had already discovered ultraviolet light, thermoelectric currents, and the process of electroplating. Oersted returned to

Copenhagen in 1803 and applied to the university for a position as professor of physics, then called "natural philosophy," but was refused. He continued lecturing at the university in the schools of medicine and pharmaceuticals, and at the same time managed the pharmacy, carried on electrochemical experiments, and published his results. In 1806 he was finally made a professor of physics at the University, although he not become a full professor until 1817.

From 1803 to 1820, Oersted's life centered around the cultural and academic life of what was then the small city of Copenhagen. He took part in political and academic debates, participated in a royal geological expedition in Denmark, and became a popular public lecturer. He was knighted and achieved the position of secretary of the Royal Society of Copenhagen. In 1813 he again visited Germany and France and published a book about electrochemical forces. In it Oersted commented on magnetic forces and clearly stated that the connection between electricity and magnetism should be investigated. Before the publication of Oersted's book, it had indeed been suspected by many scientists that magnetism and electricity were somehow related. It was known that iron rods were magnetized by the action of the electric currents passing through them as the result of lightning strikes. However, no scientific verification or understanding of the relation existed.

Oersted's discovery of the relation between magnetism and electricity in 1820 is often described as the result of a lucky accident occurring during the course of a laboratory demonstration. However, Oersted declared that he had actually prepared the experiment before the demonstration and only carried it out during the demonstration to some advanced students because that was the first opportunity. What Oersted observed was that a wire carrying an electric current caused a nearby magnetic compass needle to assume a position perpendicular to the wire, and if the current were reversed, the needle would reverse position.

After this initial discovery, Oersted waited three months, apparently for the construction of a more powerful current source. He then carried out sixty experiments to show that the magnetic field due to the current in a wire is circular around the wire. He showed that the effect is independent of the type of wire, and that it is independent of any intervening common materials. Later, he proved that the effect is proportional to the current in the wire.

On July 21, 1820, Oersted published a four-page Latin monograph, "Experiments on the Effect of a Current of Electricity on the Magnetic Needle." He distributed the monograph to leading scientists throughout Europe and, in the following months, the monograph was reprinted in translation in the most important scientific journals throughout Europe and Britain. A whole new field of investigation and technology was opened. Within a year the laws of electrodynamics were formulated, the electromagnet was invented, and the first primitive electric motor was demonstrated. The first electric telegraph and the first primitive electric generator would soon follow.

Oersted was named a fellow of several learned societies, presented with medals, and awarded cash prizes. At home, Oersted became Denmark's leading citizen. He continued his research, but as an international figure he traveled extensively, became fluent in many languages, and met with the leading scientists of the time. He gave frequent public lectures and became a director of the Royal Polytechnic Institute of Copenhagen. He also had a lifelong interest in literature and, in 1829, he founded a literary journal to which he frequently contributed articles about science. In 1850, the fiftieth anniversary of his appointment at the university was celebrated as a national holiday, and he was given a country home by the government. When he died in 1851, more than 200,000 people joined the funeral procession.

Oersted had a kindly and sympathetic personality. He had a successful marriage and a large family. In 1819 he befriended a poor fourteen-year-old boy who over the years became virtually another member of the Oersted family. The boy, Hans Christian Andersen, was to become the great Danish storyteller. Andersen often referred to himself as "little Hans Christian" and to Oersted as "great Hans Christian."

Leonard S. Taylor

See also: Electricity; Electricity, History of; Electric Motor Systems; Electric Power, Generation of; Magnetism and Magnets.

BIBLIOGRAPHY

Dibner, B. (1962). *Oersted.* New York: Blaisdell Publishing Company.

Meyer, H. W. (1971). *A History of Electricity and Magnetism.* Cambridge, MA: MIT Press.